ESSENTIALS OF
Anatomy & Physiology

Seventh Edition

Jason LaPres

Beth Kersten

ESSENTIALS OF ANATOMY & PHYSIOLOGY, SEVENTH EDITION

Published by McGraw-Hill Education, 2 Penn Plaza, New York, NY 10121. Copyright © 2019 by McGraw-Hill Education. All rights reserved. Printed in the United States of America. Previous editions © 2016, 2013, and 2010. No part of this publication may be reproduced or distributed in any form or by any means, or stored in a database or retrieval system, without the prior written consent of McGraw-Hill Education, including, but not limited to, in any network or other electronic storage or transmission, or broadcast for distance learning.

Some ancillaries, including electronic and print components, may not be available to customers outside the United States.

This book is printed on acid-free paper.

1 2 3 4 5 6 7 8 9 LMN 21 20 19 18

ISBN 978-1-260-31687-2
MHID 1-260-31687-4

Instructor's Edition
ISBN 978-1-260-31698-8
MHID 1-260-31698-X

Senior Portfolio Manager: *Michael Ivanov*
Senior Product Developer: *Elizabeth Sievers*
Executive Marketing Manager: *Jim Connely*
Senior Content Project Managers: *Sherry Kane/Danielle Clement*
Senior Buyer: *Sandy Ludovissy*
Lead Design: *David Hash*
Content Licensing Specialist: *Shawntel Schmitt*
Cover Image: *© DreamPictures/Jensen Walker/Blend Images*
Compositor: *MPS Limited*

All credits appearing on page or at the end of the book are considered to be an extension of the copyright page.

Library of Congress Cataloging-in-Publication Data

Kersten, Beth, author. | LaPres, Jason, author.
  Essentials of anatomy & physiology/Jason LaPres, Beth Kersten.
  Gunstream's anatomy & physiology | Anatomy and physiology essentials
  Seventh edition. | New York, NY : MHE, [2019] | Revised edition of: Gunstream's anatomy & physiology : with integrated study guide/Beth Kersten, Jason LaPres, Yong Tang. Sixth edition. 2016. | Includes index.
  LCCN 2017033647 | ISBN 9781260316872 (alk. paper)
  LCSH: Human physiology–Textbooks. | Human physiology–Study guides. | Human anatomy–Textbooks. | Human anatomy–Study guides.
  LCC QP34.5 .G85 2019 | DDC 612–dc23 LC record available at https://lccn.loc.gov/2017033647

The Internet addresses listed in the text were accurate at the time of publication. The inclusion of a website does not indicate an endorsement by the authors or McGraw-Hill Education, and McGraw-Hill Education does not guarantee the accuracy of the information presented at these sites.

mheducation.com/highered

ABOUT THE AUTHORS

Courtesy of Jason LaPres

Jason LaPres
Lone Star College–CyFair

Jason LaPres received his Master's of Health Science degree with an emphasis in Anatomy and Physiology from Grand Valley State University in Allendale, Michigan.

Over the past 15 years, Jason has had the good fortune to be associated with a number of colleagues who have mentored him, helped increase his skills, and trusted him with the responsibility of teaching students who will be caring for others. Jason began his career in Michigan, where from 2001 to 2003 he taught as an adjunct at Henry Ford Community College, Schoolcraft College, and Wayne County Community College, all in the Detroit area. Additionally, at that time he taught high school chemistry and physics at Detroit Charter High School. Jason is currently Dean of Instruction and Professor of Biology at Lone Star College–CyFair in Houston, Texas. He has been with LSC since 2003. In his capacity with LSC he has served as Faculty Senate President for two of the six LSC campuses. His academic background is diverse and, although his primary teaching load is in the Human Anatomy and Physiology program, he has also taught classes in Pathophysiology and mentored several Honor Projects.

Prior to authoring this textbook, Jason produced dozens of textbook supplements and online resources for many other Anatomy and Physiology textbooks.

Courtesy of Beth Kersten

Beth Kersten
State College of Florida

Beth Ann Kersten is a tenured professor at the State College of Florida (SCF). Though her primary teaching responsibilities are currently focused on Anatomy and Physiology I and II, she has experience teaching comparative anatomy, histology, developmental biology, and nonmajor human biology. She authors a custom A&P I laboratory manual for SCF and sponsors a book scholarship for students enrolled in health science programs. She coordinates a peer tutoring program for A&P and is working to extend SCF's STEM initiative to local elementary schools. Beth employs a learning strength specific approach to guide students in the development of study skills focused on their learning strengths, in addition to improving other student skills such as time management and note taking.

Beth graduated with a PhD from Temple University where her research focused on neurodevelopment in zebrafish. Her postdoctoral research at the Wadsworth Research Center focused on the response of rat nerve tissue to the implantation of neural prosthetic devices. At Saint Vincent College, she supervised senior research projects on subjects such as the effects of retinoic acid on heart development in zebrafish and the ability of vitamin B12 supplements to regulate PMS symptoms in ovariectomized mice. Beth also maintains a membership in the Human Anatomy & Physiology Society.

Beth currently lives in North Port, Florida, with her husband John and daughter Melanie. As former Northerners, they greatly enjoy the ability to swim almost year round both in their pool and in the Gulf of Mexico.

CONTENTS

Preface viii

PART ONE
Organization of the Body 1

■ CHAPTER ONE
Introduction to the Human Body ... 1
Chapter Outline 1
Selected Key Terms 2
1.1 Anatomy and Physiology 2
1.2 Levels of Organization 2
1.3 Directional Terms 6
1.4 Body Regions 6
1.5 Body Planes and Sections 6
1.6 Body Cavities 9
1.7 Abdominopelvic Subdivisions . 13
1.8 Maintenance of Life 13
Chapter Summary 17

■ CHAPTER TWO
Chemicals of Life 24
Chapter Outline 24
Selected Key Terms 25
2.1 Atoms and Elements 25
2.2 Molecules and Compounds ... 28
2.3 Substances Composing the Human Body 33
Chapter Summary 46

■ CHAPTER THREE
Cell 48
Chapter Outline 48
Selected Key Terms 49
3.1 The Human Cell 49
3.2 Cell Structure 49
3.3 Transport Across Plasma Membranes 55
3.4 Cellular Respiration 59
3.5 Protein Synthesis 61
3.6 Cell Division 62
Chapter Summary 65

■ CHAPTER FOUR
Tissues and Membranes 68
Chapter Outline 68
Selected Key Terms 69
4.1 Introduction to Tissues 69
4.2 Epithelial Tissues 69
4.3 Connective Tissues 74
4.4 Muscle Tissues 79
4.5 Nerve Tissue 82
4.6 Body Membranes 83
Chapter Summary 85

PART TWO
Covering, Support, and Movement of the Body 87

■ CHAPTER FIVE
Integumentary System 87
Chapter Outline 87
Selected Key Terms 88
5.1 Functions of the Skin 88
5.2 Structure of the Skin and Subcutaneous Tissue 88
5.3 Skin Color 92
5.4 Accessory Structures 93
5.5 Temperature Regulation 96
5.6 Aging of the Skin 98
5.7 Disorders of the Skin 99
Chapter Summary 100

CHAPTER SIX
Skeletal System — 102
Chapter Outline — *102*
Selected Key Terms — *103*
6.1 Functions of the Skeletal System — 103
6.2 Bone Structure — 103
6.3 Bone Formation — 106
6.4 Divisions of the Skeleton — 108
6.5 Axial Skeleton — 108
6.6 Appendicular Skeleton — 119
6.7 Joints — 124
6.8 Disorders of the Skeletal System — 126
Chapter Summary — *131*

CHAPTER SEVEN
Muscular System — 134
Chapter Outline — *134*
Selected Key Terms — *135*
7.1 Types of Muscle Tissue — 135
7.2 Structure of Skeletal Muscle — 136
7.3 Physiology of Skeletal Muscle Contraction — 140
7.4 Actions of Skeletal Muscles — 146
7.5 Naming of Muscles — 146
7.6 Major Skeletal Muscles — 146
7.7 Disorders of the Muscular System — 156
Chapter Summary — *158*

PART THREE
Integration and Control — 161

CHAPTER EIGHT
Nervous System — 161
Chapter Outline — *161*
Selected Key Terms — *162*
8.1 Introduction to the Nervous System — 162
8.2 Divisions of the Nervous System — 162
8.3 Nerve Tissue — 163
8.4 Neuron Physiology — 167
8.5 Protection for the Central Nervous System — 171
8.6 Brain — 172
8.7 Spinal Cord — 178
8.8 Peripheral Nervous System (PNS) — 180
8.9 Autonomic Division — 184
8.10 Disorders of the Nervous System — 188
Chapter Summary — *190*

CHAPTER NINE
Senses — 192
Chapter Outline — *192*
Selected Key Terms — *193*
9.1 Introduction to the Senses — 193
9.2 Sensations — 193
9.3 General Senses — 194
9.4 Special Senses — 197
9.5 Disorders of the Special Senses — 212
Chapter Summary — *214*

CHAPTER TEN
Endocrine System — 217
Chapter Outline — *217*
Selected Key Terms — *218*
10.1 Introduction to the Endocrine System — 218
10.2 The Chemical Nature of Hormones — 219
10.3 Pituitary Gland — 223
10.4 Thyroid Gland — 226
10.5 Parathyroid Glands — 228
10.6 Adrenal Glands — 230
10.7 Pancreas — 232
10.8 Gonads — 235
10.9 Other Endocrine Glands and Tissues — 236
Chapter Summary — *237*

PART FOUR
Maintenance of the Body — 239

CHAPTER ELEVEN
Blood — 239
Chapter Outline — *239*
Selected Key Terms — *240*
11.1 General Characteristics of Blood — 240
11.2 Red Blood Cells — 241
11.3 White Blood Cells — 243
11.4 Platelets — 247
11.5 Plasma — 247
11.6 Hemostasis — 249

11.7 Human Blood Types	250
11.8 Disorders of the Blood	254
Chapter Summary	255

CHAPTER TWELVE
Cardiovascular System — 257
Chapter Outline	257
Selected Key Terms	258
12.1 Anatomy of the Heart	258
12.2 Cardiac Cycle	265
12.3 Conducting System of the Heart	266
12.4 Regulation of Heart Function	267
12.5 Types of Blood Vessels	269
12.6 Blood Flow	272
12.7 Blood Pressure	273
12.8 Circulation Pathways	275
12.9 Systemic Arteries	275
12.10 Systemic Veins	281
12.11 Disorders of the Heart and Blood Vessels	285
Chapter Summary	286

CHAPTER THIRTEEN
Lymphatic System and Defenses Against Disease — 289
Chapter Outline	289
Selected Key Terms	290
13.1 Lymph and Lymphatic Vessels	290
13.2 Lymphoid Organs	291
13.3 Lymphoid Tissues	294
13.4 Nonspecific Resistance	296
13.5 Immunity	298
13.6 Immune Responses	302
13.7 Rejection of Organ Transplants	303
13.8 Disorders of the Lymphoid System	303
Chapter Summary	305

CHAPTER FOURTEEN
Respiratory System — 307
Chapter Outline	307
Selected Key Terms	308
14.1 Introduction to the Respiratory System	308
14.2 Structures of the Respiratory System	308
14.3 Breathing	314
14.4 Respiratory Volumes and Capacities	316
14.5 Control of Breathing	318
14.6 Factors Influencing Breathing	319
14.7 Gas Exchange	320
14.8 Transport of Respiratory Gases	320
14.9 Disorders of the Respiratory System	322
Chapter Summary	323

CHAPTER FIFTEEN
Digestive System — 326
Chapter Outline	326
Selected Key Terms	327
15.1 Introduction to the Digestive System	327
15.2 Digestion: An Overview	327
15.3 Alimentary Canal: General Characteristics	328
15.4 Mouth	330
15.5 Pharynx and Esophagus	334
15.6 Stomach	334
15.7 Pancreas	337
15.8 Liver	339
15.9 Small Intestine	341
15.10 Large Intestine	344
15.11 Nutrients: Sources and Uses	346
15.12 Disorders of the Digestive System	351
Chapter Summary	353

CHAPTER SIXTEEN
Urinary System — 356
Chapter Outline	356
Selected Key Terms	357
16.1 Overview of the Urinary System	357
16.2 Functions of the Urinary System	357
16.3 Anatomy of the Kidneys	359
16.4 Urine Formation	361
16.5 Excretion of Urine	367
16.6 Maintenance of Blood Plasma Composition	369
16.7 Disorders of the Urinary System	373
Chapter Summary	373

PART FIVE
Reproduction — 376

CHAPTER SEVENTEEN
Reproductive System — 376
Chapter Outline	376
Selected Key Terms	377
17.1 Introduction to the Reproductive System	377

17.2 Male Reproductive System	377
17.3 Male Sexual Response	384
17.4 Hormonal Control of Reproduction in Males	384
17.5 Female Reproductive System	386
17.6 Female Sexual Response	391
17.7 Hormonal Control of Reproduction in Females	391
17.8 Mammary Glands	395
17.9 Birth Control	395
17.10 Disorders of the Reproductive System	398
Chapter Summary	*400*

■ CHAPTER EIGHTEEN
Development, Pregnancy, and Genetics — 402

Chapter Outline	*402*
Selected Key Terms	*403*
18.1 Fertilization and Early Development	403
18.2 Embryonic Development	405
18.3 Fetal Development	408
18.4 Hormonal Control of Pregnancy	409
18.5 Birth	411
18.6 Cardiovascular Adaptations	413
18.7 Lactation	415
18.8 Disorders of Pregnancy, Prenatal Development, and Postnatal Development	416
18.9 Genetics	416
18.10 Inherited Diseases	420
Chapter Summary	*422*

ADDITIONAL RESOURCES

Appendices — 424

A	Keys to Medical Terminology	424
B	Common Medical Abbreviations	429
C	Healthy Values for Common Blood Tests	431
D	Healthy Values for Common Urine Tests	432

Glossary	*433*
Index	*450*

PREFACE

ESSENTIALS OF ANATOMY & PHYSIOLOGY Seventh Edition is designed for students who are enrolled in a one-semester course in human anatomy and physiology. The scope, organization, writing style, depth of presentation, and pedagogical aspects of the text have been tailored to meet the needs of students preparing for a career in one of the allied health professions, or taking the course as a general education requirement.

Acknowledgments

The development and production of this seventh edition has been a team effort. Our dedicated and creative teammates at McGraw-Hill Education have contributed greatly to the finished product. We gratefully acknowledge and applaud their efforts, and it has been a pleasure to work with these gifted professionals at each step of the process: Elizabeth Sievers (Senior Product Developer), Kate Scheinman (Contract Product Developer), Michael Ivanov (Senior Portfolio Manager), Sherry Kane (Senior Content Project Manager), and James Connely (Executive Marketing Manager).

Student-Centric Revision

Students taking a one-semester course in anatomy and physiology have diverse backgrounds, including limited exposure to biology and chemistry, and this presents a formidable challenge to the instructor. To help meet this challenge, this text is written in a clear and concise manner, free from excess jargon, and simplifies the complexities of anatomy and physiology in ways that enhance understanding without diluting the essentials of the subject matter.

In preparation for this seventh edition, we surveyed 50 students (in a variety of majors, including allied health professions) and obtained detailed insight into how they would ideally engage with course materials. Stemming from those results, we adjusted the print and digital delivery of the content to align with student preferences.

Also, we are very pleased to incorporate real student data points and input, derived from thousands of our SmartBook™ users, to help guide our revision. SmartBook™ Heat Maps provided a quick visual snapshot of usage of portions of the text and the relative difficulty students experienced in mastering the content. With this data, we honed not only our text content revision but also the SmartBook™ probes.

Course Guide and Textbook

The previous edition of this title combined two elements: the *Textbook* and the *Study Guide*. For this new seventh edition, the two elements are split into two separate printed products. The *Textbook* content is updated and revised, and the *Study Guide* is expanded and enhanced to serve as a more robust *Course Guide*, ISBN 978-1-259-86463-6.

The intention of the *Course Guide* is to be a 1:1 workbook study partner as students read the *Textbook*. Through the student survey, we uncovered their ideal mix of print and digital course materials. With the strong integration of the Connect™ online assessment tools, including SmartBook™, we worked to create an optimal delivery package of the *Course Guide* and Connect™ with the option to purchase a printed version of the *Textbook* through Connect™ at a discounted rate.

Textbook Themes and Organization

There are two unifying themes in the content presentation: (1) the relationship between structure and function of body parts, and (2) the mechanisms of homeostasis. In addition, the interrelationship of organ systems is noted where appropriate and useful.

The sequence of chapters and content within each chapter progresses from simple to complex. Chapters covering an organ system begin with anatomy to ensure that students are well prepared to understand the physiology that follows. Each organ system chapter concludes with a brief consideration of common disorders that the student may encounter in the clinical setting.

Chapter Opener and Learning Objectives

Each chapter begins with a list of major topics discussed in the chapter, along with an opening vignette and image that introduces and relates the content theme of the chapter. Under each section header within every chapter, the learning objectives are noted. This informs students of the major topics to be covered and their minimal learning responsibilities.

Key Terms

Several features have been incorporated to assist students in learning the necessary technical terms that often are troublesome for beginning students.

1. A list of Selected Key Terms with definitions, and derivations where helpful, is provided at the beginning of the chapter to inform students of some of the key terms to watch for in the chapter.
2. Throughout the text, key terms are in bold or italic type for easy recognition, and they are defined at the time of first usage. A phonetic pronunciation follows for students who need help in pronouncing the term. Experience has shown that students learn only terms that they can pronounce.
3. Keys to Medical Terminology in Appendix A explains how technical terms are structured and provides a list of prefixes, suffixes, and root words to further aid an understanding of medical terminology.

Figures and Tables

Over 350 high quality, full-color illustrations are coordinated with the text to help students visualize anatomical features and physiological concepts. Tables are used throughout to summarize information in a way that is more easily learned by students.

Clinical Insight

Numerous boxes containing related clinical information are strategically placed throughout the text. They serve to provide interesting and useful information related to the topic at hand. The Clinical Insight boxes are identified by a medical cross for easy recognition.

Check My Understanding

Review questions at the end of major sections challenge students to assess their understanding before proceeding.

Chapter Summary

The chapter summary is conveniently linked by section and briefly states the important facts and concepts covered in each chapter.

Changes in the Seventh Edition

The seventh edition has been substantially enhanced and improved.

Global Changes

- Revised all chapter text to focus on healthy conditions rather than "normal" conditions.
- Approximately 70 figures and tables were revised or are completely new.
- Revised descriptive language to improve the overall readability of the text. Terminology and phrasing more commonly used by students outside the classroom have been added where appropriate. By making the text easier to read, students will have an easier time grasping more complex anatomical and physiological content.
- Added more Check My Understanding sections to better assess student learning throughout the chapters.
- The Critical Thinking sections at the end of the chapters have been moved to the *Course Guide* to consolidate all of the assessment content into one resource, except for the Check My Understanding sections; these remain in the text to offer students opportunities to test their understanding before moving on in the chapter.
- *Course Guide* figures were updated to align with the figures within the lecture text. Figure labeling activities were also redesigned to provide the students with a more hands-on labeling experience.
- Revised each chapter's Selected Key Terms definitions to better align with the definitions within the chapter text.
- Updated art to create a more vibrant and consistent style.
- Updated terminology to align with the *Terminologia Anatomica*, *Terminologia Histologica*, and *Terminologia Embryologica*.
- Revised figure legends to include a descriptive title and separate legend.

Chapter 1

- Updated and simplified regional names. For example, the olecranal and antecubital regions were removed and replaced with the cubital region.
- Removed references to the dorsal cavity and ventral cavity. The body cavity section was rewritten to focus on the cranial cavity, vertebral canal, thoracic cavity, and abdominopelvic cavity.
- Revised figure 1.10 to more concisely illustrate homeostatic normal range and set point.
- Replaced military time in figure 1.13 with time using the 12-hour clock, which is more commonly used by students outside of the classroom.

Chapter 2

- Expanded the description of atom reactivity and valence shells. It is now noted that hydrogen atoms react to fill their valence shells with two electrons, while atoms of other elements normally found in the body react to fill their valence shells with eight electrons.
- Added the mass of an electron to aid in student understanding of why atomic mass is calculated based upon only the number of protons and neutrons.

- Revised the terminology used to describe chemical substances. The term *organic compound* was changed to *organic molecule* to emphasize the fact that the atoms are joined by only covalent bonds.
- Revised figure 2.5 to better demonstrate a triple nonpolar covalent bond.
- Simplified the description of acids and bases to improve student learning.
- Added simple sugar and complex carbohydrates to the carbohydrate discussion. These terms are more commonly known owing to their use in nutritional literature. This addition will help students link the nutritional terms with the specific types of carbohydrates introduced in the chapter.

Chapter 3
- Revised the description of *plasma membrane structure* to better explain the relationship between its structure and function.
- Revised the membrane protein discussion and figure 3.2 to specifically include channel proteins, carrier proteins, receptor proteins, and cell identity markers.
- Revised figures 3.1, 3.3, and 3.4 to accurately represent the structure of the rough endoplasmic reticulum and nuclear pores.
- Revised table 3.1 to better reflect the functions of the plasma membrane, cytosol, and organelles discussed within the chapter text.
- Simplified *mitochondrial structure* to promote student learning.
- Revised the discussions of channel and carrier-mediated diffusion and carrier-mediated active transport for better clarity.
- Revised figure 3.11 to match the appearance of sugar molecules with those in figure 2.13.
- Revised the discussion of isotonic solutions, hypertonic solutions, and hypotonic solutions for better accuracy and clarity.
- Updated figure 3.12 extensively to better visually demonstrate the effects of isotonic, hypertonic, and hypotonic solutions on red blood cells.
- Updated figure 3.13 to align the sodium-potassium pump style with the style used in figure 8.7.
- Updated figure 3.15 to match the style in figure 7.7.
- Simplified the discussion of mitosis and meiosis by eliminating the terms *mitotic cell division* and *meiotic cell division*.

Chapter 4
- Updated figure 4.19 by incorporating a higher-quality photomicrograph of the formed elements.
- Added the functions of intercalated discs to the discussion of cardiac muscle tissue.
- Revised figure 4.24. The serous pericardium was added to the human figure showing all of the body membranes. The cutaneous membrane was relabeled to reflect only the types of tissue composing it.
- Revised figures 4.11 and 4.18. New inset images were added to help students identify the locations of reticular tissue, compact bone, and spongy bone.

Chapter 5
- Revised temperature regulation discussion to focus on a person's temperature "set point" rather than on a specific temperature to account for variability in the human population.
- Revised the description of *dermal papillae* and *epidermal ridges* for better accuracy. The labeling within figure 5.4 was also updated to align with the new text description.
- Updated the skin cancer Clinical Insight box to directly relate squamous cell carcinoma and basal cell carcinoma with their specific sites of origination within the epidermis.

Chapter 6
- Added common bone names when appropriate to facilitate the learning of the biological bone names.
- Added examples to the section on bone types.
- Updated the pectoral girdle discussion to include the presence of two pectoral girdles.
- Revised figure 6.8 to show correct labeling for the alveolar process on the maxilla and the alveolar arches on the maxillae and mandible.
- Updated to alveoli on the maxillae and mandible to dental alveoli.
- Added the role of the atlas in the ability to nod head.
- Added the sacral promontory to the text and figure 6.18 because of its importance as an obstetric landmark.
- Revised the description of rib to thoracic vertebrae articulations to improve its accuracy.
- Re-created the sternoclavicular joint in figures 6.20 and 6.21 for anatomical accuracy.
- Revised the labeling of the pelvic brim in figure 6.22 for better accuracy.
- Added a new joint section focusing on the structural types of joints.
- Clarified the difference between patellar ligament and patellar tendon in the knee joint discussion.

Chapter 7
- Added the temperature regulation function for the muscular system.
- Revised figure 7.2 for better histological accuracy. The sarcomere in (c) was revised to create

appropriate overlap between thin myofilaments and thick myofilaments throughout the entire image. The thick myofilament in *(d)* was extended to show appropriate overlap with thin myofilaments.
- Revised figure 7.4 for better text alignment and histological accuracy. The terminal boutons were adjusted so that each terminal bouton makes contact with only one muscle fiber. The structure of the terminal bouton was also adjusted to show no internal branching.
- Revised the labeling within figure 7.5 to correlate better with text presentation of excitation-contraction coupling.
- Updated *lactic acid* to *lactate* and *cellular respiration* to *aerobic respiration* in figure 7.7.
- Updated *epicranius* to *occipitofrontalis* in figures 7.12, 7.13, and 7.14.
- Re-created the left forearm in figures 7.12 and 7.13 for better anatomical accuracy.

Chapter 8

- Updated *nerve impulse* to *action potential*. This update will help students to more easily distinguish between action potentials in axons and impulses in dendrites and cell bodies.
- Revised the discussion of the functional division of the nervous system. The text now reflects that the functional divisions apply to the peripheral nervous system only. The motor division of the peripheral nervous system consists of a somatic division and an autonomic division, which in turn is divided into a sympathetic part and parasympathetic part.
- Updated the term *unipolar neuron* to *pseudounipolar neuron* to align with what is known about the development of these neurons in humans. A better description of the structure of pseudounipolar neurons was also added.
- Revised figure 8.8 with larger, clearer arrows to better illustrate how the sodium-potassium pump is moving ions across the plasma membrane.
- Updated the labels in figure 8.12 to include the subarachnoid space and replace *posterior root ganglion* with *spinal ganglion*.
- Re-created the cross-sectional image of the spinal cord in figures 8.17, 8.20, and 8.21 for better anatomical accuracy. The anterior median fissure was redrawn to be slightly wider than the posterior median sulcus and have a consistent visual appearance in all three figures.
- Color-coded the cranial nerves in figure 8.18 to help students discern what cranial nerves are motor nerves, sensory nerves, and mixed nerves.
- Revised the labels within figure 8.19 to include the lumbar plexus and sacral plexus as components of the lumbosacral plexus.
- Updated figure 8.20 to a more modern style that matches the art within the text. The neurons were also color-coded to make it easier to distinguish the different components of the reflex arc.
- Color-coded the neurons within figure 8.21 to make it easier to distinguish the different components of the neural pathways.
- Re-created the lumbar and sacral splanchnic nerves in figure 8.22 for better anatomical accuracy.
- Updated figure 8.23 to a more modern style that matches the art within the text. The pre- and post-ganglionic connections within the ganglia were also redrawn for better anatomical accuracy.

Chapter 9

- Revised the image labeling within figure 9.1 for better visual flow.
- Revised the highlights for the external ear, middle ear, and internal ear in figure 9.6 to better match the text description of these ear parts.
- Revised the description of tympanic membrane structure for better histological accuracy.
- Added the cranial nerve innervations of extrinsic eye muscles to help students see the correlation with table 8.3 in chapter 8.
- Re-created the anterior portion of the retina and the labeling bracket for the iris in figure 9.15 for better anatomical accuracy.

Chapter 10

- Added the eicosanoid discussion to the paracrine signaling discussion.
- Revised the color of the hormones in figure 10.1 to match the color scheme used in the other chapter figures.
- Added a new figure 10.2 showing the locations of the endocrine glands and the hormones produced by each gland. The hormone names are also color-coded to help students visually distinguish between nonsteroid hormones and steroid hormones.
- Added the effects of thyroid hormones on mitochondria to the discussion of steroid hormones.
- Emphasized the hormonal control, neural control, and humoral control of hormone secretion throughout the chapter.
- Revised figure 10.6 to match the style in figure 7.7.
- Revised the arrows in figure 10.7 to match the style in figure 7.7.
- Revised the ACTH negative-feedback mechanism for better physiological accuracy.
- Added a figure to the Clinical Insight box on stress to illustrate the role of cortisol in stress management.

Chapter 11

- Updated figure 11.1 by incorporating a new photomicrograph of formed elements.
- Added a new paragraph describing what is transported in blood.
- Updated figure 11.5 by incorporating new photomicrographs of white blood cells.
- Added reasons for why females have lower hematocrits than males.
- Updated *hemoglobin percentage* to *hemoglobin concentration* and the healthy values for hemoglobin concentration.
- Created a new introduction for the hemostasis section that included descriptions of procoagulants and anticoagulants.
- Revised the discussion of anti-Rh antibodies for better clarity and accuracy.
- Expanded the discussion on causes of polycythemia.

Chapter 12

- Revised the description of epicardium for simplicity and easier understanding.
- Updated the terms for the components of the conducting system of the heart within the text, as well as in figure 12.9, to align with the *Terminologia Histologica*.
- Revised figure 12.11 to better illustrate the limbic system input, to update the sympathetic chain to the sympathetic trunk, and to emphasize the type of stimulus detected by the carotid bodies and carotid sinuses.
- Updated the location of the carotid bodies.
- Moved the first two paragraphs in the "Exchange of Materials" section to immediately follow the histological description of the capillary wall.
- Added a definition of peripheral resistance and a discussion of how the cross-sectional diameter of a vessel impacts peripheral resistance. The changes in peripheral resistance were then correlated to changes in blood velocity.
- Added definitions of venoconstriction and vasoconstriction.
- Expanded table 12.5 to include all of the lower limb arteries and revised what the arteries supply to match the text.
- Revised table 12.6 to match what the veins drain with the text.
- Expanded table 12.7 to include all of the lower limb and hepatic portal veins and revised what the veins drain to match the text.

Chapter 13

- Identified each lymphoid organ as either a primary or secondary lymphoid organ.
- Moved the description of the role of mucus in nonspecific defense from the chemical barrier section to immediately follow mucous membranes in the mechanical barrier section.
- Added the skin to the chemical barriers, owing to its acidic pH.
- Added a new natural killer cell section to the nonspecific resistance section.
- Revised the description of antigen.
- Expanded the description of what is attacked by T cells to include transplanted cells.
- Added a new Clinical Insight box on cancer immunotherapy for cancer treatment.
- Expanded the description of allergies to include those caused by environmental factors rather than by identified antigens.

Chapter 14

- Revised the description of the glottis for better anatomical accuracy.
- Updated cellular respiration to aerobic respiration as appropriate.
- Refined the locations of the respiratory mucosae.
- Updated *quiet breathing* to *resting breathing*.
- Revised figure 14.4. The radiographic image was vertically flipped to show the correct clinical orientation.
- Updated the color of the smooth muscle on the bronchi in figure 14.5 to the standard red color used in the text for muscle tissue.
- Replaced mm Hg with cm of H_2O when describing the changes in intra-alveolar and intrapleural pressure during breathing for better physiological accuracy. Figure 14.7 was updated to match this change in the text.

Chapter 15

- Revised the description of digestion and absorption in the chapter introductory paragraph.
- Emphasized the role of chemical digestion in breaking large, nonabsorbable nutrients to small, absorbable nutrients.
- Highlighted the boxes in figure 15.1 to visually separate the structures of alimentary canal and accessory organs.
- Revised the text to consistently use italics to highlight the substrates and products associated with each digestive enzyme.
- Adjusted the pharynx outline in figure 15.4 to include the entire laryngopharynx.
- Created a new stomach inset image for figure 15.9 using the stomach image in figure 15.8.
- Divided the description of the liver into separate structure and function sections for better readability.

- Revised figure 15.17 for better clarity. The image was substantially enlarged. Only the names of the lipids and lipid-soluble vitamins were used when illustrating their absorption and incorporation into a chylomicron for better clarity.
- Updated *cellular respiration* to *aerobic respiration* where appropriate for more specificity.
- Expanded the definitions of *anaerobic respiration* and *aerobic respiration*.
- Updated figure 15.20 to match the style in figure 7.7. The different organic molecules were color-coded to visually separate them. Anaerobic respiration and aerobic respiration were also color-coded for easier discernment of the two processes.

Chapter 16

- Revised the description of nephron components to include the collecting duct.
- Expanded the water conservation process to involve both the distal convoluted tubules and the collecting ducts.
- Added the papillary ducts to the microscopic anatomy of the kidney.
- Added a specific explanation for why dilating the afferent glomerular arteriole increases glomerular blood pressure.
- Added a detailed explanation of the role of the juxtaglomerular complex in adjusting the glomerular filtration rate during renal autoregulation.
- Updated *atrial natriuretic peptide* to *natriuretic peptides*.
- Revised targets of aldosterone to include both the distal convoluted tubules and the collecting ducts. Figure 16.9 labeling was updated to match this revision.
- Added factors contributing to the movement of urine from the renal pelvis to the bladder.
- Updated the text and figure 16.10 to reflect the lack of an internal urethral sphincter in females.

Chapter 17

- Revised sperm description to include the function of the neck, principal piece, and end piece.
- Revised figure 17.12 extensively for physiological accuracy. The y-axis was relabeled in 2-day increments allowing the entirety of day 14 to represent ovulation. The changes in follicle-stimulating hormone, luteinizing hormone, estrogens, and progesterone concentrations align with the corresponding text descriptions.
- Revised phrasing within the ovarian cycle text to account for varying cycle lengths between females.
- Updated *mammary gland structure*.
- Revised the image order in figure 17.15 to match the text flow.

Chapter 18

- Revised the definitions of an *embryo, fetus,* and *preembryo* to reflect that they are names given to the developing offspring and not the names of developmental stages.
- Added the length of time of preembryonic development.
- Updated figures 18.4 and 18.5 to a more modern style that matches the art within the text.
- Revised the changes in HCG concentration in figure 18.9 to align with the text description.
- Explained the relationship between weeks of development and weeks of pregnancy.
- Updated figure 18.13 to match the more modern style of figure 18.12. The internal iliac arteries were also added to make the identification of the superior vesical arteries easier.
- Updated the text discussion of the regulation of both prolactin secretion and action on the mammary glands for better physiological accuracy.
- Used attached versus unattached earlobes as the genetic example for introducing dominant and recessive inheritance.

McGraw-Hill Connect® is a highly reliable, easy-to-use homework and learning management solution that utilizes learning science and award-winning adaptive tools to improve student results.

Homework and Adaptive Learning

- Connect's assignments help students contextualize what they've learned through application, so they can better understand the material and think critically.
- Connect will create a personalized study path customized to individual student needs through SmartBook®.
- SmartBook helps students study more efficiently by delivering an interactive reading experience through adaptive highlighting and review.

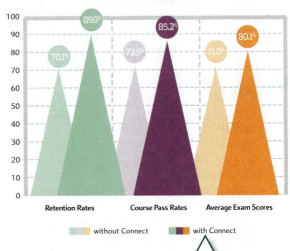

Using **Connect** improves retention rates by **19.8%**, passing rates by **12.7%**, and exam scores by **9.1%**.

Over **7 billion questions** have been answered, making McGraw-Hill Education products more intelligent, reliable, and precise.

73% of instructors who use **Connect** require it; instructor satisfaction **increases** by 28% when **Connect** is required.

Quality Content and Learning Resources

- Connect content is authored by the world's best subject matter experts, and is available to your class through a simple and intuitive interface.
- The Connect eBook makes it easy for students to access their reading material on smartphones and tablets. They can study on the go and don't need internet access to use the eBook as a reference, with full functionality.
- Multimedia content such as videos, simulations, and games drive student engagement and critical thinking skills.

©McGraw-Hill Education

Robust Analytics and Reporting

- Connect Insight® generates easy-to-read reports on individual students, the class as a whole, and on specific assignments.
- The Connect Insight dashboard delivers data on performance, study behavior, and effort. Instructors can quickly identify students who struggle and focus on material that the class has yet to master.
- Connect automatically grades assignments and quizzes, providing easy-to-read reports on individual and class performance.

©Hero Images/Getty Images

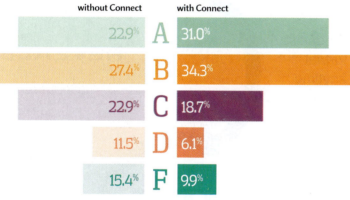

Impact on Final Course Grade Distribution

	without Connect		with Connect
A	22.9%		31.0%
B	27.4%		34.3%
C	22.9%		18.7%
D	11.5%		6.1%
F	15.4%		9.9%

More students earn **As** and **Bs** when they use **Connect**.

Trusted Service and Support

- Connect integrates with your LMS to provide single sign-on and automatic syncing of grades. Integration with Blackboard®, D2L®, and Canvas also provides automatic syncing of the course calendar and assignment-level linking.
- Connect offers comprehensive service, support, and training throughout every phase of your implementation.
- If you're looking for some guidance on how to use Connect, or want to learn tips and tricks from super users, you can find tutorials as you work. Our Digital Faculty Consultants and Student Ambassadors offer insight into how to achieve the results you want with Connect.

www.mheducation.com/connect

50% of the country's students are not ready for A&P

LearnSmart® Prep can help!

Improve preparation for the course and increase student success with the only adaptive Prep tool available for students today. Areas of individual weaknesses are identified in order to help students improve their understanding of core course areas needed to succeed.

LEARNSMART®
Prep for A&P

Students seek lab time that fits their busy schedules. Anatomy & Physiology REVEALED 3.2, our Virtual Dissection tool, allows students to practice anytime, anywhere, and now features enhanced physiology interactives with clinical and 3D animations.

Anatomy & Physiology REVEALED® 3.2
Virtual dissection

Bringing to life complex processes is a challenge. Ph.I.L.S. 4.0 is the perfect way to reinforce key physiology concepts with powerful lab experiments. Tools like physiology interactives, Ph.I.L.S., and world-class animations make it easier than ever.

Ph.I.L.S.
Physiology supplements

The Practice Atlas for Anatomy & Physiology is a new interactive tool that pairs images of common anatomical models with stunning cadaver photography. This atlas allows students to practice naming structures on both models and human bodies, anytime and anywhere.

Practice Atlas for Anatomy & Physiology

Since 2009, our adaptive programs in A&P have hosted 1 million unique users who have answered more than 1 billion questions/probes, providing the only data-driven solutions to help students get from their first college-level course to program readiness.

CHAPTER 1

Introduction to the Human Body

©Randy Faris/Corbis RF

Michael, a freshman in college, overslept and is late for his first anatomy and physiology class. He has been dreading this class but it is necessary for his graduation requirements. Because he does not want to get off to a bad start, he sprints across campus. The combination of the warm day and physical exertion raises his body temperature and, as he throws himself into the nearest seat, sweat is pouring out across his body. Michael begins to feel cooler as he relaxes, and he stops sweating within a few minutes. As his first lecture begins, he is introduced to the concept of homeostasis, which describes the condition of balance within the body, and the feedback cycles responsible for maintaining his internal "normal." He thinks about his morning, the sweat that cooled his body, and realizes just how amazing the human body really is. What a great semester this is going to be!

CHAPTER OUTLINE

1.1 Anatomy and Physiology

1.2 Levels of Organization
- Chemical Level
- Cellular Level
- Tissue Level
- Organ Level
- Organ System Level
- Organismal Level

1.3 Directional Terms

1.4 Body Regions

1.5 Body Planes and Sections

1.6 Body Cavities
- Membranes of Body Cavities

1.7 Abdominopelvic Subdivisions

1.8 Maintenance of Life
- Survival Needs
- Homeostasis

Chapter Summary

Module 1
Body Orientation

SELECTED KEY TERMS

Anatomy (ana = apart; tom = to cut) The study of the structure of living organisms.
Appendicular (append = to hang) Pertaining to the upper and lower limbs.
Axial (ax = axis) Pertaining to the longitudinal axis of the body.
Body region (regio = boundary) A portion of the body with a special identifying name.
Directional term (directio = act of guiding) A term that references how the position of a body part relates to the position of another body part.
Effector (efet = result) A structure that functions by performing an action that is directed by an integrating center.
Homeostasis (homeo = same; sta = make stand or stop) Maintenance of a relatively stable internal environment.
Integrating center (integratus = make whole) A structure that functions to interpret information and coordinate a response.
Metabolism (metabole = change) The sum of the chemical reactions in the body.
Parietal (paries = wall) Pertaining to the wall of a body cavity.
Pericardium (peri = around; cardi = heart) The membrane surrounding the heart.
Peritoneum (ton = to stretch) The membrane lining the abdominal cavity and covering the abdominal organs.
Physiology (physio = nature; logy = study of) The study of the function of living organisms.
Plane (planum = flat surface) Imaginary two-dimensional flat surface that marks the direction of a cut through a structure.
Pleura (pleura = rib) The membrane lining the thoracic cavity and covering the lungs.
Receptor (recipere = receive) A structure that functions to collect information.
Section (sectio = cutting) A flat surface of the body produced by a cut through a plane of the body.
Serous membrane (serum = watery fluid; membrana = thin layer of tissue) A two-layered membrane that lines body cavities and covers the internal organs.
Visceral (viscus = internal organ) Pertaining to organs in a body cavity.

YOU ARE BEGINNING a fascinating and challenging study–the study of the human body. As you progress through this text, you will begin to understand the complex structures and functions of the human organism.

This first chapter provides an overview of the human body to build a foundation of knowledge that is necessary for your continued study. Like the chapters that follow, this chapter introduces a number of new terms for you to learn. It is important that you start to build a vocabulary of technical terms and continue to develop it throughout your study. This vocabulary will help you reach your goal of understanding human anatomy and physiology.

1.1 Anatomy and Physiology

Learning Objective

1. Define anatomy and physiology.

Knowledge of the human body is obtained primarily from two scientific disciplines–anatomy and physiology–and each consists of a number of subdisciplines.

Human **anatomy** (ah-nat'-ō-mē) is the study of the structure and organization of the body and the study of the relationships of body parts to one another. There are two major subdivisions of anatomy. *Gross anatomy* involves the dissection and examination of various parts of the body without magnifying lenses. *Microanatomy*, also known as *histology*, consists of the examination of tissues and cells with various magnification techniques.

Human **physiology** (fiz-ē-ol'-ō-jē) is the study of the function of the body and its parts. Physiology involves observation and experimentation, and usually requires the use of specialized equipment and materials.

In your study of the human body, you will see that there is always a definite relationship between the anatomy and physiology of the body and body parts. Just as the structure of a knife is well suited for cutting, the structure (anatomy) of a body part enables it to perform specific functions (physiology). For example, the arrangement of bones, muscles, and nerves in your hands enables the grasping of large objects with considerable force and also the delicate manipulation of small objects. Correlating the relationship between structure and function will make your study of the human body much easier.

1.2 Levels of Organization

Learning Objectives

2. Describe the levels of organization in the human body.
3. List the major organs and functions for each organ system.

The human body is complex, so it is not surprising that there are several levels of structural organization, as shown in figure 1.1. The levels of organization from simplest to most complex are chemical, cellular, tissue, organ, organ system, and organismal (the body as a whole).

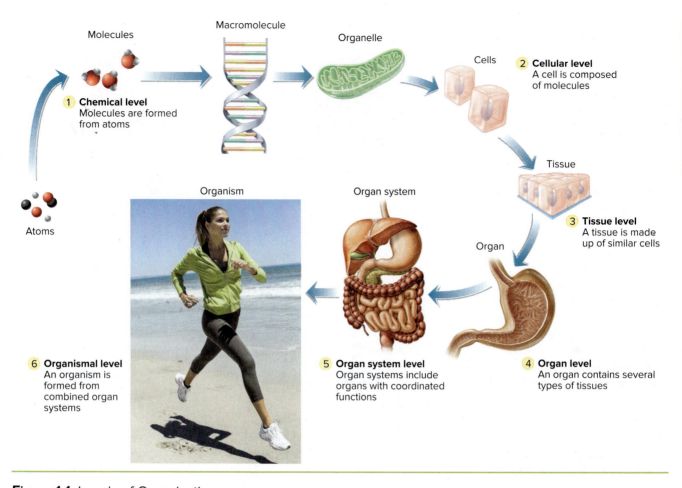

Figure 1.1 Levels of Organization.
Six levels of organization in the human body range from chemical (simplest) to organismal (most complex).
©ONOKY - Fabrice LEROUGE/Getty Images RF

Chemical Level

The chemical level consists of *atoms, molecules,* and *macromolecules.* At the simplest level, the body is composed of chemical substances that are formed of atoms and molecules. Atoms are the fundamental building blocks of chemicals, and atoms combine in specific ways to form molecules. Some molecules are very small, such as water molecules, but others may be very large, such as the macromolecules of proteins. Various small and large molecules are grouped together to form organelles. An **organelle** (or″-ga-nel′) is a complex of macromolecules acting like a "mini-organ" that carries out specific functions within a cell. Nuclei, mitochondria, and ribosomes are examples.

Cellular Level

Cells are the basic structural and functional units of the body because all of the processes of life occur within cells. A cell is the lowest level of organization that is alive. The human body is composed of trillions of cells and many different types of cells, such as muscle cells, blood cells, and nerve cells. Each type of cell has a unique structure that enables it to perform specific functions.

Tissue Level

Similar types of cells are usually grouped together in the body to form a tissue. Each body **tissue** consists of an aggregation of similar cells that perform similar functions. There are four major classes of tissues in the body: epithelial, connective, muscle, and nerve tissues.

Organ Level

Each **organ** of the body is composed of two or more tissues that work together, enabling the organ to perform its specific functions. The body contains numerous organs, and each has a definite structure and function. The stomach, heart, brain, and even bones are examples of organs.

Organ System Level

The organs of the body are arranged in functional groups so that their independent functions are coordinated to perform specific system functions. These coordinated, functional groups are called **organ systems.** The digestive and nervous systems are examples of organ systems. Most organs belong to a single organ system, but a few organs are assigned to more than one organ system.

4 *Chapter 1* Introduction to the Human Body

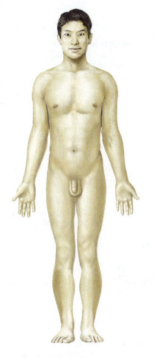

Integumentary system
Organs: skin, hair, nails, and associated glands
Functions: protects underlying tissues and helps regulate body temperature

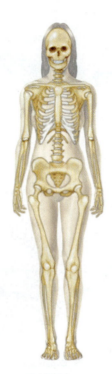

Skeletal system
Organs: bones, ligaments, and associated cartilages
Functions: supports the body, protects vital organs, stores minerals, and is the site of blood cell production

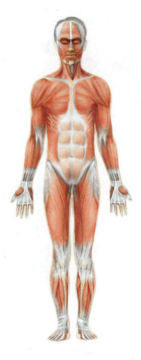

Muscular system
Organs: skeletal muscles and tendons
Functions: moves the body and body parts and produces heat

Figure 1.2 The 11 Organ Systems of the Body. AP|R

For example, the pancreas belongs to both the digestive and endocrine systems.

Figure 1.2 illustrates the 11 organ systems of the human body and lists the major organs and functions for each system. Although each organ system has its own unique functions, all organ systems support one another. For example, all organ systems rely on the cardiovascular system to transport materials to and from their cells. Organ systems work together to enable the functioning of the human body.

Organismal Level

The highest organizational level dealing with an individual is the *organismal level,* the human organism as a whole. It is composed of all of the interacting organ systems. All of the organizational levels from chemicals to organ systems contribute to the functioning of the entire body.

 Check My Understanding

1. What are the organizational levels of the human body?
2. What are the major organs and general functions of each organ system?

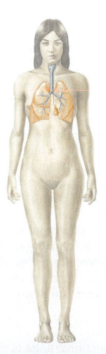

Respiratory system
Organs: nose, pharynx, larynx, trachea, bronchi, and lungs
Functions: exchanges O_2 and CO_2 between air and blood in the lungs, pH regulation, and sound production

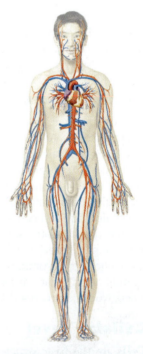

Cardiovascular system
Organs: blood, heart, arteries, veins, and capillaries
Functions: transports heat and materials to and from the body cells

Part 1 Organization of the Body 5

Lymphoid system

Organs: lymph, lymphatic vessels, and lymphoid organs and tissues
Functions: collects and cleanses interstitial fluid, and returns it to the blood; provides immunity

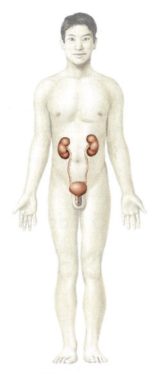

Urinary system

Organs: kidneys, ureters, urinary bladder, and urethra
Functions: regulates volume and composition of blood by forming and excreting urine

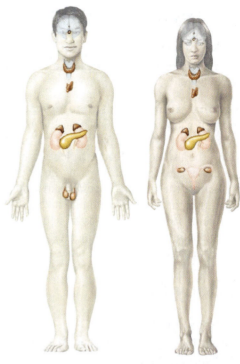

Endocrine system

Organs: hormone-producing glands, such as the pituitary and thyroid glands
Functions: secretes hormones that regulate body functions

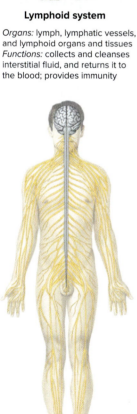

Nervous system

Organs: brain, spinal cord, nerves, and sensory receptors
Functions: rapidly coordinates body functions and enables learning and memory

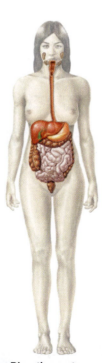

Digestive system

Organs: mouth, pharynx, esophagus, stomach, intestines, liver, pancreas, gallbladder, and associated structures
Functions: digests food and absorbs nutrients

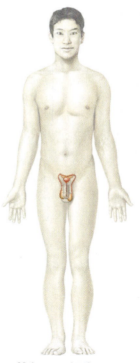

Male reproductive system

Organs: testes, epididymides, vasa deferentia, prostate, bulbo-urethral glands, seminal vesicles, and penis
Functions: produces sperm and transmits them into the female vagina during sexual intercourse

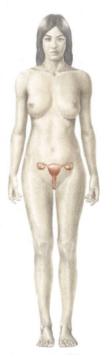

Female reproductive system

Organs: ovaries, uterine tubes, uterus, vagina, and vulva
Functions: produces oocytes, receives sperm, provides intrauterine development of offspring, and enables birth of an infant

1.3 Directional Terms

Learning Objective
4. Use directional terms to describe the locations of body parts.

Directional terms are used to describe the relative position of a body part in relationship to another body part. The use of these terms conveys a precise meaning enabling the listener or reader to locate the body part of interest. It is always assumed that the body is in a standard position, the *anatomical position*, in which the body is standing upright with upper limbs at the sides and palms of the hands facing forward, as in figure 1.3. Directional terms occur in pairs, and the members of each pair have opposite meanings, as noted in table 1.1.

1.4 Body Regions

Learning Objective
5. Locate the major body regions on a diagram or anatomical model.

The human body consists of an **axial** (ak′-sē-al) **portion,** the head, neck, and trunk, and an **appendicular** (ap-pen-dik′-ū-lar) **portion,** the upper and lower limbs and their girdles. Each of these major portions of the body is divided into regions with special names to facilitate communication and to aid in locating body components.

The major **body regions** are listed in tables 1.2 and 1.3 to allow easy correlation with figure 1.4, which shows the locations of the major regions of the body. Take time to learn the names, pronunciations, and locations of the body regions.

1.5 Body Planes and Sections

Learning Objective
6. Describe the three planes used in making sections of the body or body parts.

In studying the body or organs, you often will be observing the flat surface of a **section** that has been produced by a cut through the body or a body part. Such sections are made along specific **planes.** These

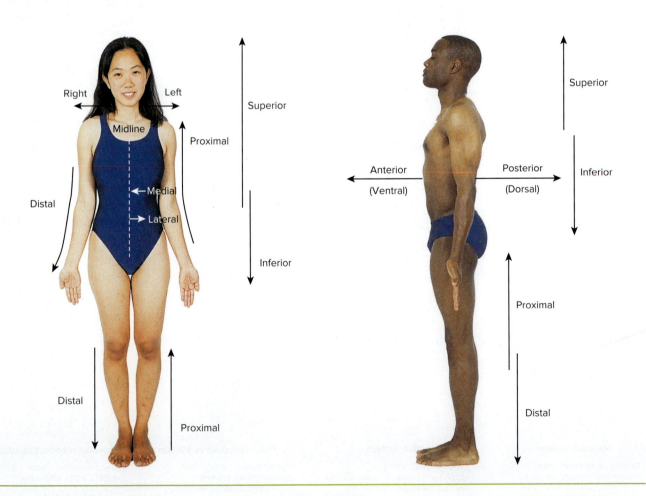

Figure 1.3 Anatomical Position and Directional Terms. AP|R
(a) ©Eric Wise; *(b)* ©Eric Wise/Wise Anatomy Collection

Table 1.1 Directional Terms

Term	Meaning	Example
Anterior (ventral)	Toward the front or abdominal surface of the body	The abdomen is anterior to the back.
Posterior (dorsal)	Toward the back of the body	The spine is posterior to the face.
Superior (cephalic)	Toward the top/head	The nose is superior to the mouth.
Inferior (caudal)	Away from the top/head	The navel is inferior to the nipples.
Medial	Toward the midline of the body	The breastbone is medial to the nipples.
Lateral	Away from the midline of the body	The ears are lateral to the cheeks.
Parietal	Pertaining to the outer boundary of body cavities	The parietal pleura lines the pleural cavity.
Visceral	Pertaining to the internal organs	The visceral pleura covers the lung.
Superficial (external)	Toward or on the body surface	The skin is superficial to the muscles.
Deep (internal)	Away from the body surface	The intestines are deep to the abdominal muscles.
Proximal	Closer to the beginning	The elbow is proximal to the wrist.
Distal	Farther from the beginning	The hand is distal to the wrist.
Central	At or near the center of the body or organ	The central nervous system is in the middle of the body.
Peripheral	External to or away from the center of the body or organ	The peripheral nervous system extends away from the central nervous system.

Table 1.2 Major Regions of the Axial Portion

Region			
Head and Neck	**Anterior Trunk**	**Posterior Trunk**	**Lateral Trunk**
Buccal (bu-kal)	Abdominal (ab-dom′-i-nal)	Dorsum (dor′-sum)	Axillary (ak′-sil-lary)
Cephalic (se-fal′-ik)	Inguinal (ing′-gwi-nal)	Gluteal (glu′-tē-al)	Coxal (kok′-sal)
Cervical (ser′-vi-kal)	Pectoral (pek′-tōr-al)	Lumbar (lum′-bar)	**Inferior Trunk**
Cranial (krā′-nē-al)	Pubic (pyoo-bik)	Sacral (sāk′-ral)	Genital (jen′-i-tal)
Facial (fā′-shal)	Sternal (ster′-nal)	Scapular (skap′-yuh-ler)	Perineal (per-i-nē′-al)
Nasal (nā-zel)	Umbilical (um-bil′-i-kal)	Vertebral (ver-tē′-bral)	
Oral (or-al)			
Orbital (or-bit-al)			
Otic (o-tic)			

Table 1.3 Major Regions of the Appendicular Portion

Region		
Upper Limb	Digital (di′-ji-tal)	Patellar (pa-tel′-lar)
Antebrachial (an-tē-brā′-kē-al)	Palmar (pal′-mar)	Pedal (pe′-dal)
Brachial (brā′-kē-al)	**Lower Limb**	Plantar (plan′-tar)
Carpal (kar′-pal)	Crural (krū′-ral)	Popliteal (pop-li-tē′-al)
Cubital (kū-bi-tal)	Digital (di′-ji-tal)	Sural (sū′-ral)
Deltoid (del-toid)	Femoral (fem′-ōr-al)	Tarsal (tahr-sul)

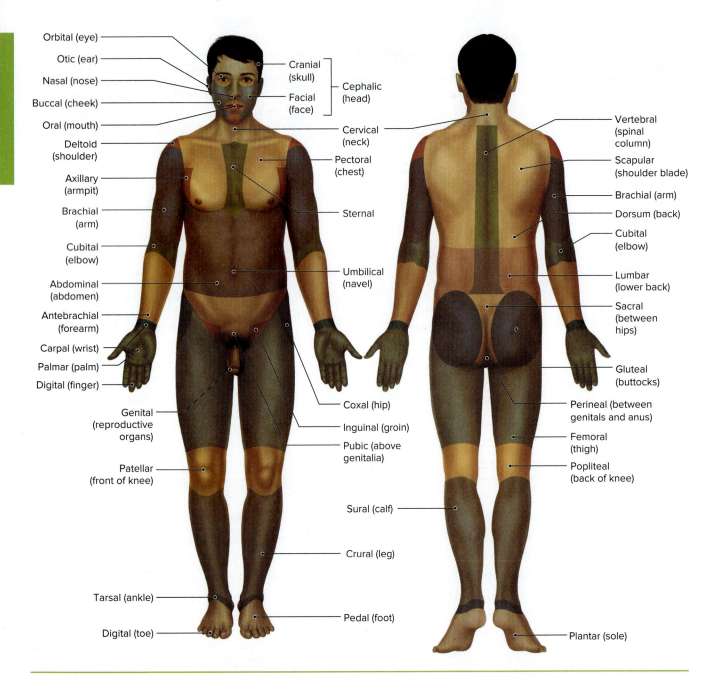

Figure 1.4 Major Regions of the Body. AP|R

well-defined planes–transverse, sagittal, and frontal planes–lie at right angles to each other, as shown in figure 1.5. It is important to understand the nature of the plane along which a section was made in order to understand the three-dimensional structure of an object being observed.

Transverse, or horizontal, **planes** divide the body into superior and inferior portions and are perpendicular to the longitudinal axis of the body.

Sagittal planes divide the body into right and left portions and are parallel to the longitudinal axis of the body. A **median plane** passes through the midline of the body and divides the body into equal left and right halves. A **paramedian plane** does not pass through the midline of the body.

Frontal (coronal) **planes** divide the body into anterior and posterior portions. These planes are perpendicular to sagittal planes and parallel to the longitudinal axis of the body.

Cuts made through sagittal and frontal planes, which are parallel to the longitudinal axis of the body, produce *longitudinal sections*. However, the term longitudinal section also refers to a section made through the longitudinal axis of an individual organ, tissue, or other structure. Similarly,

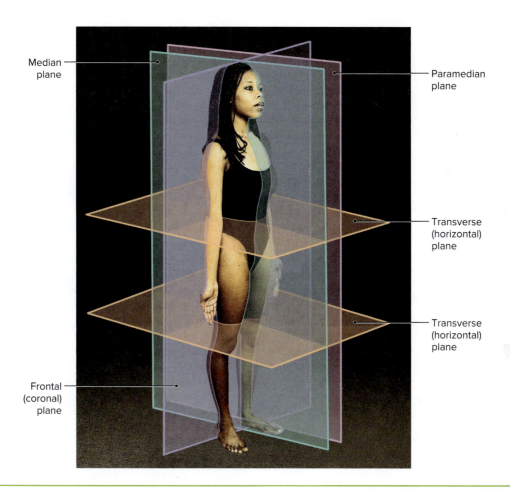

Figure 1.5 Anatomical Planes of Reference. **AP|R**
©McGraw-Hill Education/Joe DeGrandis, photographer

cuts made through the transverse plane produce *cross sections* of the body and can also be produced in organs and tissues when cutting at a 90° angle to the longitudinal axis. *Oblique sections* are created when cuts are made in between the longitudinal and cross-sectional axes.

 Check My Understanding

3. How do sagittal, transverse, and frontal planes differ from one another?

1.6 Body Cavities

Learning Objectives

7. Locate the body cavities and their subdivisions and membranes on a diagram.
8. Name the organs located in each body cavity.

The body cavities protect and cushion the contained organs and permit changes in their size and shape without impacting surrounding tissues. Note the locations and subdivisions of these cavities in figure 1.6.

The **cranial cavity** is enclosed in the bones of the skull and contains the brain. Within the vertebral column is the **vertebral canal,** which contains the spinal cord. Note in figure 1.6 how the cranial bones and the vertebral column form the walls of these cavities and provide protection for these delicate organs.

The two large body cavities in front of the vertebral column are divided by the *diaphragm*, a thin dome-shaped sheet of muscle. Above the diaphragm is the **thoracic cavity** and below it is the **abdominopelvic cavity.** The thoracic cavity is protected by the *thoracic cage* and contains the heart and lungs. The abdominopelvic cavity is subdivided into a superior **abdominal cavity** and an inferior **pelvic cavity,** but there is no structural separation between them. To visualize the separation, imagine a transverse plane passing through the body just above the pelvis. The abdominal cavity contains the stomach, intestines, liver, gallbladder, pancreas, spleen, and kidneys. The pelvic cavity contains the urinary bladder, sigmoid colon, rectum, and internal reproductive organs.

10   Chapter 1   Introduction to the Human Body

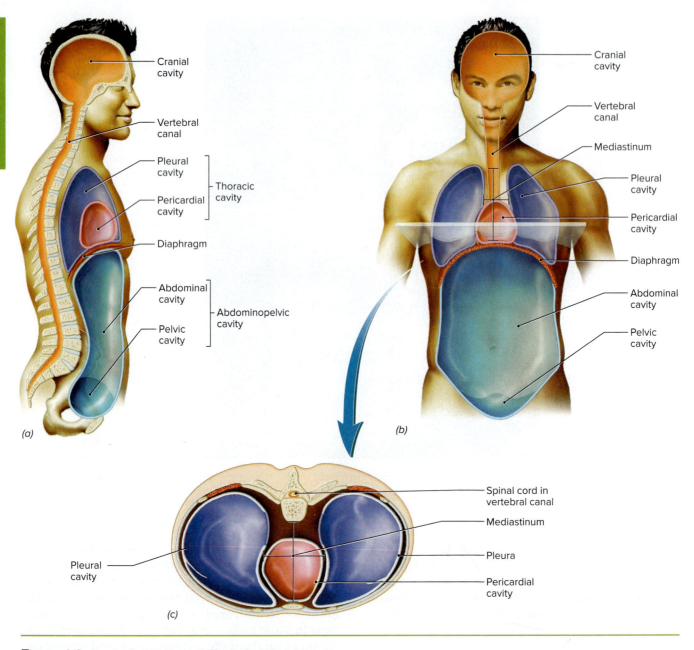

Figure 1.6 Body Cavities and Their Subdivisions.
(a) Sagittal section. (b) Frontal section. (c) Transverse section through the thoracic cavity. AP|R

 Check My Understanding

4. What organs are located in the cranial cavity and vertebral canal?
5. What organs are located in the thoracic and abdominopelvic cavities?

Membranes of Body Cavities

The membranes lining body cavities support and protect the internal organs in the cavities.

Meninges

The cranial cavity and vertebral canal are lined by three layers of protective membranes that are collectively called the **meninges** (me-nin′-jēz; singular, *meninx*). The outer membrane is attached to the wall of the cavity, and the inner membrane tightly envelops the brain and spinal cord. The meninges will be covered in chapter 8.

Serous Membranes

The thoracic and abdominopelvic cavity organs are supported and protected by **serous membranes,** or **serosae**

Clinical Insight

Physicians use certain types of diagnostic imaging systems—for example, *computerized tomography (CT), magnetic resonance imaging (MRI),* and *positron emission tomography (PET)*—to produce images of sections of the body to help them diagnose disorders. In computerized tomography, an X-ray emitter and an X-ray detector rotate around the patient so that the X-ray beam passes through the body from hundreds of different angles. X-rays collected by the detector are then processed by a computer to produce sectional images on a screen for viewing by a radiologist. A good understanding of sectional anatomy is required to interpret CT scans. Transverse sections, such as the image on the left, are always shown in the same way. Convention is to use supine (face up), inferior views as if looking up at the section from the foot of the patient's bed. What structures can you identify in the CT image shown on the right? AP|R

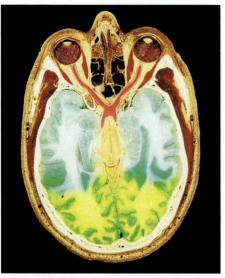
©Scott Camazine

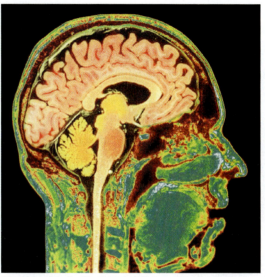
©Du Cane Medical Imaging Ltd./Science Photo Library/Getty Images

(singular, *serosa*). The serous membranes are thin layers of tissue that line the body cavity and cover the internal organs. Serous membranes have an outer *parietal* (pah-rī′-e-tal) *lay*er that lines the cavity and an inner *visceral* (vis′-er-al) *layer* that covers the organ. The parietal and visceral layers secrete a watery lubricating fluid that is generically called *serous fluid* into the cavity formed between the layers. This arrangement is similar to that of a fist pushed into a balloon (figure 1.7). The serous membranes of the body are the pleura, serous pericardium, and peritoneum.

The serous membranes lining the thoracic cavity are called **pleurae** (singular, *pleura*). The walls of the left and right portions of the thoracic cavity are lined by the *parietal pleurae*. The surfaces of the lungs are covered by the *visceral pleurae*. The space between the parietal and visceral pleurae is called the **pleural cavity.** The pleural cavity contains a thin film of serous fluid called *pleural fluid,* which reduces friction as the pleurae rub against each other as the lungs expand and contract during breathing.

The left and right portions of the thoracic cavity are divided by a membranous partition, the *mediastinum* (mē-dē-a-stī′-num). Organs located within the mediastinum include the heart, thymus, esophagus, and trachea.

The heart is enveloped by the **serous pericardium** (per-i-kar′-dē-um), which is formed by membranes of the mediastinum. The thin *visceral pericardium* is tightly adhered to the surface of the heart. The *parietal pericardium* lines the inside surface of a loosely fitting sac around the heart. The space between the visceral and parietal pericardia is the **pericardial cavity,** and it contains serous fluid, called *pericardial fluid,* that reduces friction as the heart contracts and relaxes.

The walls of the abdominal cavity and the surfaces of abdominal organs are covered with the **peritoneum** (per-i-tō-nē′-um). The *parietal peritoneum* lines the walls of the abdominal cavity but not the pelvic cavity. It descends only to cover the top of the urinary bladder. The kidneys, pancreas, and parts of the intestines are located behind the parietal peritoneum in a space known as the *retroperitoneal space*. The *visceral peritoneum*, an extension of the parietal peritoneum, covers the surface of the abdominal organs. Double-layered folds of the visceral peritoneum, the *mesenteries* (mes′-en-ter″-ēs), extend between the abdominal organs and provide support for

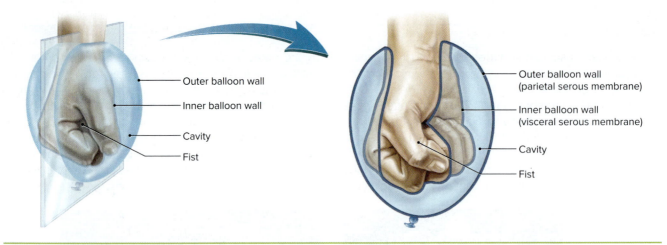

Figure 1.7 Model of a Serous Membrane.
Illustration of a fist pushed into a balloon as an analogy to a serous membrane.

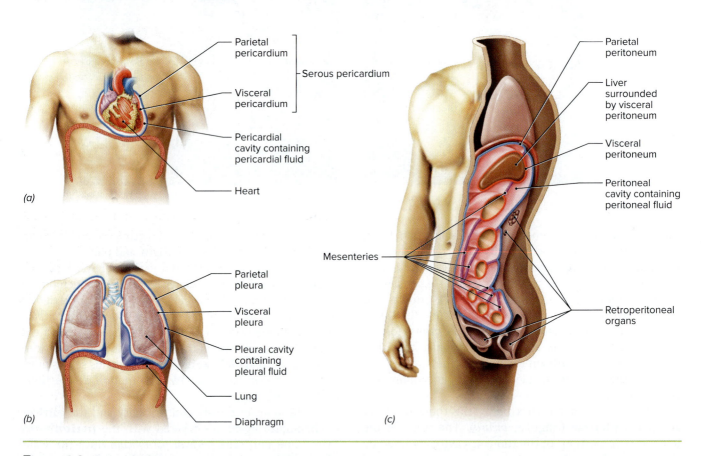

Figure 1.8 Serous Membranes.
(a) Anterior view of serous pericardium. (b) Anterior view of pleurae. (c) Sagittal view of peritoneum.

them (see figure 1.8c). The space between the parietal and visceral peritoneum is called the **peritoneal cavity** and contains a small amount of serous fluid called *peritoneal fluid* (figure 1.8).

> ✓ Check My Understanding
> 6. What membranes line the various body cavities?
> 7. What is the function of serous fluid?

1.7 Abdominopelvic Subdivisions

Learning Objectives
9. Name the abdominopelvic quadrants and regions.
10. Locate the abdominopelvic quadrants and regions on a diagram.

The abdominopelvic cavity is subdivided into either four quadrants or nine regions to aid health-care providers in locating underlying organs. Physicians may feel (palpate) or listen to (auscultate) the abdominopelvic region to examine it. Changes in firmness or sounds may indicate abnormalities in the structures of a quadrant or region.

The four quadrants are formed by two planes that intersect just above the umbilicus (navel), as shown in figure 1.9a. Note the organs within each quadrant.

The nine regions are formed by the intersection of two sagittal and two transverse planes, as shown in figure 1.9c. The sagittal planes extend inferiorly from the midpoints of the collarbones. The superior transverse plane lies just below the borders of the 10th costal cartilages, and the inferior transverse plane lies just below the top of the hip bones.

Study figures 1.8 and 1.9 to increase your understanding of the locations of the internal organs and associated membranes.

Now examine the colorplates that follow this chapter. They show an anterior view of the body in progressive stages of dissection that reveals major muscles, blood vessels, and internal organs. Study these plates to learn the normal locations of the organs. Also, check your understanding of the organs within each abdominopelvic quadrant and region.

Check My Understanding
8. What are the four quadrants and nine regions of the abdominopelvic region?

1.8 Maintenance of Life

Learning Objectives
11. Define metabolism, anabolism, and catabolism.
12. List the five basic needs essential for human life.
13. Define homeostasis.
14. Explain how homeostasis relates to both healthy body functions and disorders.
15. Describe the general mechanisms of negative feedback and positive feedback.

Humans, like all living organisms, exhibit the fundamental processes of life. **Metabolism** (me-tab′-ō-lizm) is the term that collectively refers to the sum of all of the chemical reactions that occur in the body.

There are two phases of metabolism: anabolism and catabolism. **Anabolism** (ah-nab′-ō-lizm) refers to processes that use energy and nutrients to build the complex organic molecules that compose the body. **Catabolism** (kah-tab′-ō-lizm) refers to processes that release energy and break down complex molecules into simpler molecules.

Life is fragile. It depends upon the normal functioning of trillions of body cells, which, in turn, depends upon factors needed for survival and the ability of the body to maintain relatively stable internal conditions.

Survival Needs

There are five basic needs that are essential to human life:

1. **Food** provides chemicals that serve as a source of energy and raw materials to grow and to maintain cells of the body.
2. **Water** provides the environment in which the chemical reactions of life occur.
3. **Oxygen** is required to release the energy in organic nutrients, which powers life processes.
4. **Body temperature** must be maintained close to 36.8°C (98.2°F) to allow the chemical reactions of human metabolism to occur.
5. **Atmospheric pressure** is required for breathing to occur.

Homeostasis

Homeostasis is the maintenance of a relatively stable internal environment by self-regulating physiological processes. Homeostasis keeps body temperature and the composition of blood and interstitial fluids within their normal range. This relatively stable internal environment is maintained in spite of the fact that internal and external factors tend to alter body temperature, and materials are continuously entering and exiting the blood and interstitial fluid.

All of the organ systems work in an interdependent manner to maintain homeostasis. For example, changes in one system tend to affect one or more other organ systems. Therefore, any disruption in one organ system tends to be corrected but may disrupt another organ system. The internal environment is maintained via a *dynamic equilibrium* where there is constant fluctuation taking place in order to maintain homeostasis. Malfunctioning or overcompensation in a homeostatic mechanism can lead to disorders and diseases.

The dynamic equilibrium of homeostasis is primarily maintained by physiologic processes called **negative-feedback mechanisms.** Body fluid composition and other physiological variables fluctuate near a normal value, called a *set point,* and negative-feedback mechanisms are

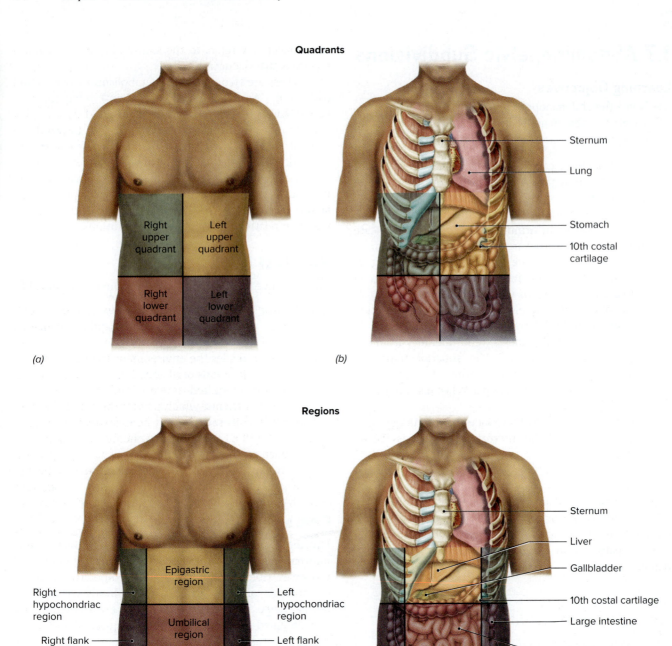

Figure 1.9 Abdominopelvic Subdivisions.
The four quadrants and nine regions of the abdominopelvic cavity. AP|R

used to keep these variables within their normal range (figure 1.10). For a negative-feedback mechanism to work, it needs to be able to monitor and respond to any changes in homeostasis. The structure of the negative-feedback mechanism allows it to function in exactly this manner and is a great example of how anatomical structure complements function. To monitor a physiological variable, a negative-feedback mechanism utilizes a **receptor** to detect deviation from the set point and send a signal notifying the integrating center about the deviation. The **integrating center,** which is the body region that knows the set point for the variable, processes the information from a receptor and determines the course of action that is needed. It then sends a signal that activates an **effector.**

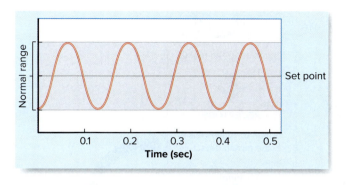

Figure 1.10 Normal Range and Set Point.

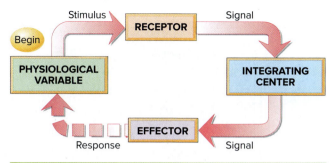

Figure 1.11 A Typical Negative-Feedback Mechanism.

The effector will carry out the necessary response according to the directions of the integrating center and return the variable back toward the set point. In a negative-feedback mechanism, the response of the effector will always be the opposite of the change detected by the receptor (figure 1.11). Once the set point is reached, the negative-feedback mechanism will automatically turn off.

Our body's ability to maintain a relatively constant blood glucose level relies on negative-feedback mechanisms. When the blood glucose level begins to rise, as it does after a meal, there are receptors in the pancreas that can detect this *stimulus* (change). The beta cells of the pancreas act as an integrating center and release the hormone insulin in response to this change. Insulin travels through the blood to several effectors, one of which is the liver. Insulin causes the liver cells to take excess glucose out of the blood and thus decrease the blood glucose level back toward the set point. The pancreas possesses other receptors that can detect decreases in blood glucose, such as occurs between meals. The alpha cells of the pancreas, acting as the integrating center, release the hormone glucagon. Glucagon causes the liver to release glucose into the blood, which will increase blood glucose back toward the set point (figure 1.12).

It is important to note that the response of the integrating center will be stronger if the original stimulus is further from normal. For example, if the blood glucose level rises sharply out of the normal range, causing *hyperglycemia* (blood glucose level above normal), the amount of insulin the beta cells release will be more than the amount released if the blood glucose level is elevated but is still within the normal range. This type of response is called a *graded response* because it can respond on different levels (figure 1.13).

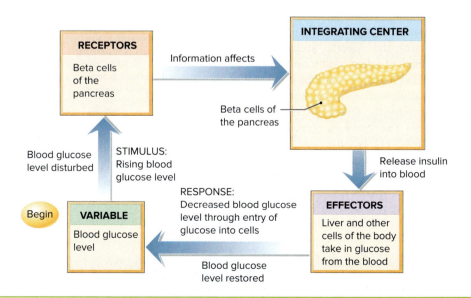

Figure 1.12 An Example of a Negative-Feedback Mechanism.
The negative-feedback mechanism that regulates blood glucose level.

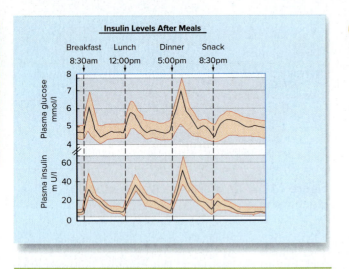

Figure 1.13 The Graded Response.
The graded response of insulin release is based on the amount of blood glucose elevation.
Source: The FASEB Journal C Kendall, vol. 28. no. 1 Supplement 1039.6.

Positive-feedback mechanisms utilize the same basic components as negative-feedback mechanisms. However, the outcome of a positive-feedback mechanism is very different from that of a negative-feedback mechanism. A positive-feedback mechanism is used when the originating stimulus needs to be amplified and continued in order for the desired result to occur. A few examples of positive-feedback mechanisms include fever, activation of the immune response, formation of blood clots, certain aspects of digestion, and uterine contractions of labor. If you think about blood clot formation, blood clots do not form "normally"; when they begin to form, it occurs quickly and completely in order to stop blood loss. This is a necessary mechanism for overall homeostasis. Figure 1.14 illustrates the specific steps of the positive-feedback mechanism of saliva production.

Positive-feedback mechanisms can be harmful because they lack the ability to stop on their own. They will continue to amplify the effect of the original stimulus, which can push the body dangerously out of homeostasis, until the cycle is interrupted by an outside factor. For example, an uncontrolled fever can increase body temperature to a point that is fatal. For this reason, positive-feedback mechanisms are used for rare events within the body, rather than for the daily maintenance of homeostasis.

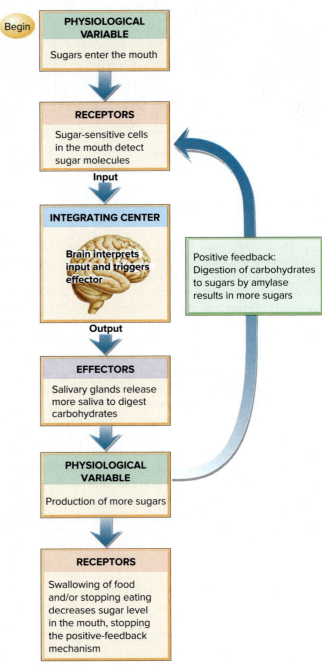

Figure 1.14 A Positive-Feedback Mechanism.
Saliva is produced as needed, and production increases with continued stimulation.

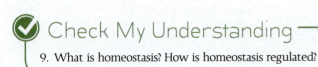

Check My Understanding

9. What is homeostasis? How is homeostasis regulated?

Chapter Summary

1.1 Anatomy and Physiology
- Human anatomy is the study of body structure and organization.
- Human physiology is the study of body functions.

1.2 Levels of Organization
- The body consists of several levels of organization of increasing complexity.
- From simple to complex, the organizational levels are chemical, cellular, tissue, organ, organ system, and organismal.
- The organs of the body are arranged in coordinated groups called organ systems.
- The 11 organ systems of the body are
  - integumentary
  - skeletal
  - muscular
  - nervous
  - endocrine
  - digestive
  - cardiovascular
  - lymphoid
  - respiratory
  - urinary
  - reproductive

1.3 Directional Terms
- Directional terms are used to describe the relative positions of body parts.
- Directional terms occur in pairs, with the members of a pair having opposite meanings.
  - anterior–posterior
  - superior–inferior
  - medial–lateral
  - central–peripheral
  - proximal–distal
  - external–internal
  - parietal–visceral

1.4 Body Regions
- The body is divided into two major portions: the axial portion and the appendicular portion.
- The axial portion is subdivided into the head, neck, and trunk.
- The head and neck contain cervical, cranial, and facial regions. The cranial and facial regions combine to form the cephalic region.
- The facial region consists of orbital, nasal, oral, and buccal regions.
- The trunk consists of anterior, posterior, lateral, and inferior regions.
- Anterior trunk regions include the abdominal, inguinal, pectoral, pubic, sternal, and umbilical regions.
- Posterior trunk regions include the dorsal, gluteal, lumbar, sacral, scapular, and vertebral regions.
- Lateral trunk regions are the axillary and coxal regions.
- Inferior trunk regions are the genital and perineal regions.
- The appendicular portion of the body consists of the upper and lower limbs.
- The upper limb is attached to the trunk at the shoulder. Regions of the upper limb are the antebrachial, brachial, carpal, cubital, deltoid, digital, and palmar regions.
- The lower limb is attached to the trunk at the hip. Regions of the lower limb are the crural, digital, femoral, patellar, pedal, plantar, popliteal, sural, and tarsal regions.

1.5 Body Planes and Sections
- Well-defined planes are used to guide sectioning of the body or organs.
- The common planes are transverse, sagittal, and frontal.
- The common planes produce longitudinal sections and cross sections of the body.

1.6 Body Cavities
- The cranial cavity is located within the skull, and the vertebral canal is located within the vertebral column.
- The thoracic cavity lies above the diaphragm. It consists of two lateral pleural cavities and the mediastinum, which contains the pericardial cavity.
- The abdominopelvic cavity lies below the diaphragm. It consists of a superior abdominal cavity and an inferior pelvic cavity.
- The body cavities are lined with protective and supportive membranes.
- The meninges consist of three membranes that line the cranial cavity and vertebral canal and that enclose the brain and spinal cord.
- The parietal pleurae line the walls of the thoracic cage, while the visceral pleurae cover the surfaces of the lungs.
- The pleural cavity is the space between the parietal and visceral pleurae.
- The parietal pericardium is a saclike membrane in the mediastinum that surrounds the heart. The visceral pericardium is attached to the surface of the heart.
- The pericardial cavity is the space between the parietal and visceral pericardia.
- The parietal peritoneum lines the walls of the abdominal cavity but does not extend into the pelvic cavity. The visceral peritoneum covers the surface of abdominal organs.
- The peritoneal cavity is the space between the parietal and visceral peritoneum.
- The mesenteries are double-layered folds of the visceral peritoneum that support internal organs.
- Kidneys, pancreas, and parts of the intestines are located behind the parietal peritoneum in the retroperitoneal space.

1.7 Abdominopelvic Subdivisions

- The abdominopelvic cavity is subdivided into either four quadrants or nine regions as an aid in locating organs.
- The four quadrants are
  - right upper
  - right lower
  - left upper
  - left lower
- The nine regions are
  - epigastric
  - left hypochondriac
  - right hypochondriac
  - umbilical
  - left flank
  - right flank
  - hypogastric (pubic)
  - left inguinal
  - right inguinal

1.8 Maintenance of Life

- Metabolism is the sum of all of the body's chemical reactions. It consists of anabolism, the synthesis of body chemicals, and catabolism, the breakdown of body chemicals.
- The basic needs of the body are food, water, oxygen, body temperature, and atmospheric pressure.
- Homeostasis is the maintenance of a relatively stable internal environment.
- Homeostasis is regulated by negative-feedback mechanisms.
- Negative-feedback mechanisms consist of three components: receptors, integrating center, and effectors.
- Positive-feedback mechanisms promote an ever-increasing change from the norm.

Improve Your Grade

Connect Interactive Questions Reinforce your knowledge using multiple types of questions: interactive, animation, classification, labeling, sequencing, composition, and traditional multiple choice and true/false.

SmartBook Proven to help students improve grades and study more efficiently, SmartBook contains the same content within the print book but actively tailors that content to the needs of the individual.

Anatomy & Physiology REVEALED® Dive into the human body by peeling back layers of cadaver imaging. Utilize this world-class cadaver dissection tool for a closer look at the body anytime, from anywhere.

COLORPLATES OF THE HUMAN BODY

The five colorplates that follow show the basic structure of the human body. The first plate shows the anterior body surface and the superficial anterior muscles of a female. Succeeding plates show the internal structure as revealed by progressively deeper dissections.

Refer to these plates often as you study this text in order to become familiar with the relative locations of the body organs. ■

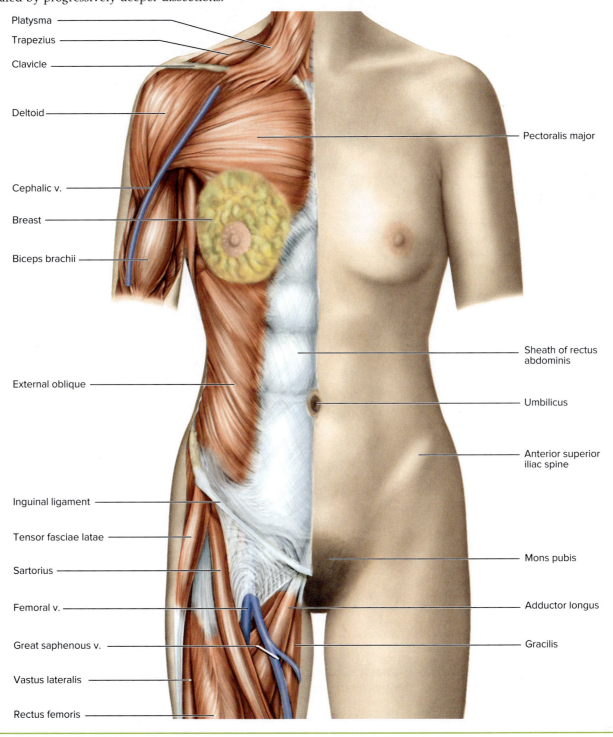

Plate 1 Superficial Anatomy of the Trunk (Female).
Surface anatomy is shown on the anatomical left, and structures immediately deep to the skin on the right (v. = vein).
Source: Kenneth Saladin, *Human Anatomy & Physiology: The Unity of Form and Function.* Copyright © 2015 McGraw-Hill Education.

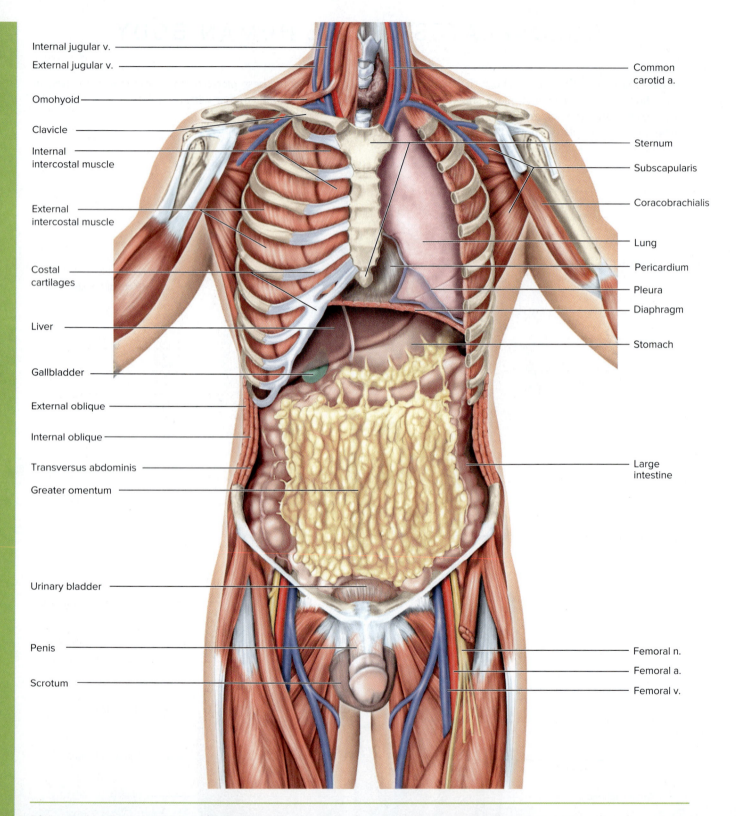

Plate 2 Anatomy at the Level of the Thoracic Cage and Greater Omentum (Male).
The anterior body wall is removed, and the ribs, intercostal muscles, and pleurae are removed from the anatomical left (a. = artery; v. = vein; n. = nerve).
Source: Kenneth Saladin, *Human Anatomy & Physiology: The Unity of Form and Function.* Copyright © 2015 McGraw-Hill Education.

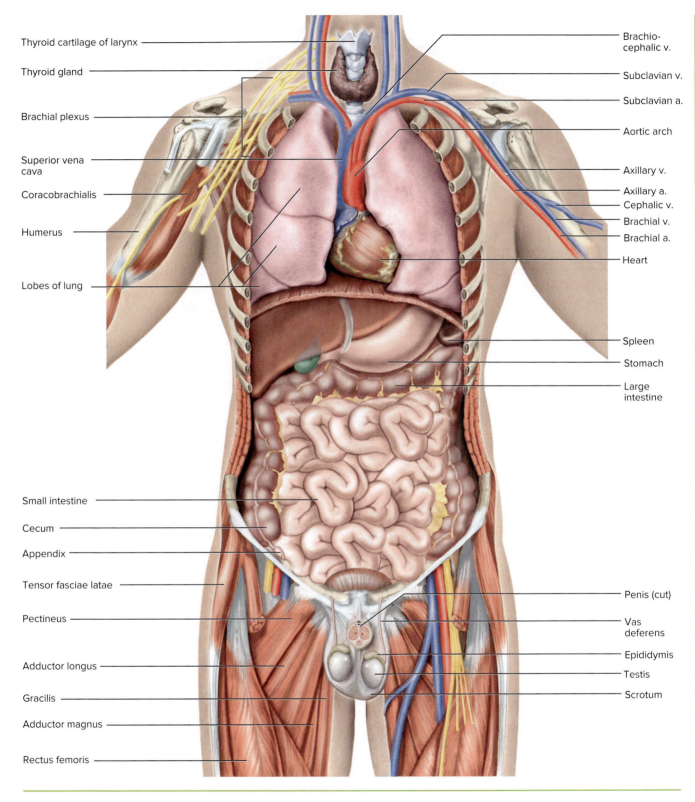

Plate 3 Anatomy at the Level of the Lungs and Intestines (Male).
The sternum, ribs, and greater omentum are removed (a. = artery; v. = vein).
Source: Kenneth Saladin, *Human Anatomy & Physiology: The Unity of Form and Function.* Copyright © 2015 McGraw-Hill Education.

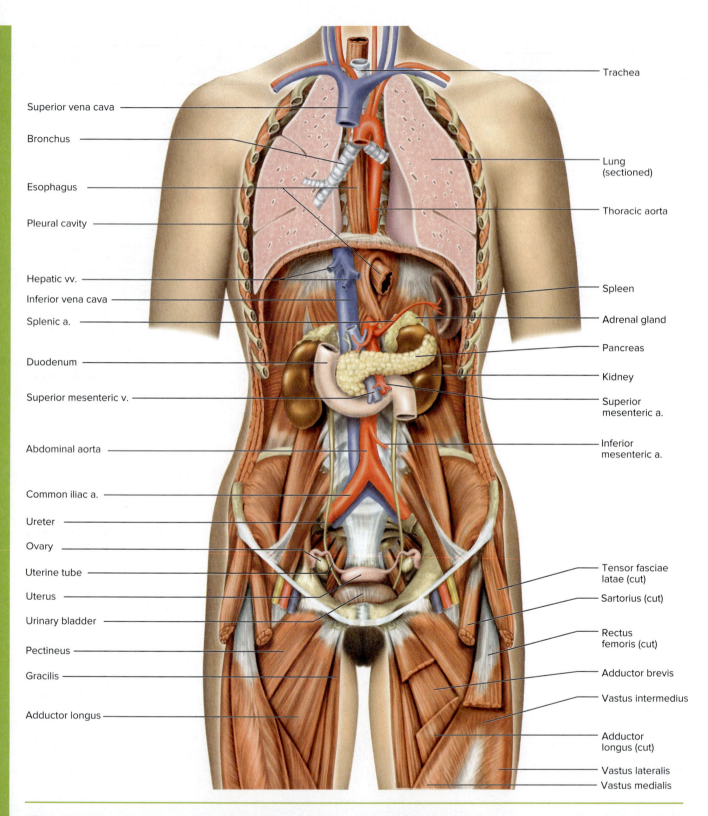

Plate 4 Anatomy at the Level of the Retroperitoneal Viscera (Female).
The heart is removed, the lungs are frontally sectioned, and the viscera of the peritoneal cavity and the peritoneum itself are removed (a. = artery; v. = vein; vv. = veins).

Source: Kenneth Saladin, *Human Anatomy & Physiology: The Unity of Form and Function.* Copyright © 2015 McGraw-Hill Education.

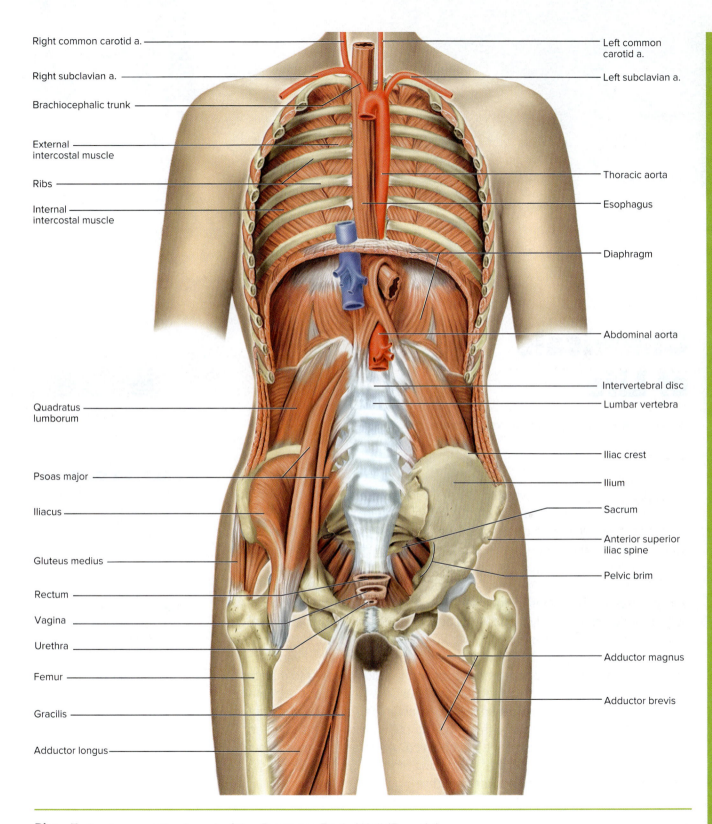

Plate 5 Anatomy at the Level of the Posterior Body Wall (Female).
The lungs and retroperitoneal viscera are removed (a. = artery).
Source: Kenneth Saladin, *Human Anatomy & Physiology: The Unity of Form and Function.* Copyright © 2015 McGraw-Hill Education.

CHAPTER

2

Chemicals of Life

©Purestock/Superstock RF

Have you ever wondered why the USDA (United States Department of Agriculture) recommends a certain number of protein, grain, fat, fruit, vegetable, dairy, and water servings every day? The answer is simple. You are what you eat. For example, protein-rich foods such as meats and nuts provide necessary building units for the production of new proteins within your body. Your body uses glucose, a carbohydrate, as its main energy source. The grains in your diet are rich in carbohydrates and help to replenish the body's glucose supply. Many of the body's chemical reactions require the presence of specific vitamins and minerals, which are obtained through fruits and vegetables, to occur normally. Even the beverages consumed every day help provide the fluids needed to maintain the percentage of the body composed of water. It is clear that the homeostasis of the human body is dependent upon chemicals and the constant supply of these chemicals through the nutrients in our diet. This chapter introduces you to the wonders of this chemical world and creates a foundation for better understanding of everything from cellular functions to organ system physiology in later chapters.

CHAPTER OUTLINE

2.1 Atoms and Elements
- Atomic Structure
- Isotopes

2.2 Molecules and Compounds
- Chemical Formulae
- Chemical Bonds
- Chemical Reactions

2.3 Substances Composing the Human Body
- Major Inorganic Substances
- Major Organic Substances

Chapter Summary

Module 2
Cells & Chemistry

SELECTED KEY TERMS

Atom (atomos = indivisible) The smallest unit of an element.
Carbohydrate (carbo = carbon; hydr = water) An organic molecule composed of carbon, hydrogen, and oxygen, with a 1:2:1 ratio.
Chemical bond (bond from band = fasten) Joining of chemical substances using attractions between electrons.
Chemical formula (formula = draft or small form) Shorthand notation showing the type and number of atoms in a molecule.
Chemical reaction (re = again; actionem = put into motion) Process involving the formation and/or breakage of chemical bonds resulting in new combinations of atoms.

Compound (componere = to place together) A substance formed by atoms from two or more elements combined by ionic or covalent bonds.
Element A substance that cannot be broken down into simpler substances by ordinary chemical means.
Enzyme (en = in; zym = ferment) A protein that catalyzes chemical reactions.
Inorganic substance (in = not) Small, simple substance that usually does not have carbon and hydrogen in the same substance.
Lipid (lip = fat) An organic macromolecule containing mostly carbon and hydrogen, with small amounts of oxygen. Lipids do not mix with water.

Molecule (molecula = little mass) A substance formed by two or more atoms bonded together by covalent bonds.
Nucleic acid (nucle = kernel) A complex organic macromolecule composed of nucleotides.
Organic substance (organon = from living things) Large, complex substances that contain both carbon and hydrogen in the same molecule, usually with oxygen too.
Protein A group of nitrogen-containing organic macromolecules formed of amino acids.

A BASIC UNDERSTANDING OF CHEMISTRY is necessary for health-care professionals because the human body is composed of chemicals and the processes of life are chemical interactions.

2.1 Atoms and Elements

Learning Objectives
1. Describe the basic structure of an atom.
2. Distinguish between atoms, isotopes, and radioisotopes.

Anything that occupies space is **matter.** Chemistry is the scientific study of matter and the interactions of matter. The entire physical universe, both living and nonliving, is composed of matter. All matter is composed of **elements,** substances that cannot be broken down into simpler substances by ordinary chemical means. Carbon, hydrogen, and nitrogen are examples of chemical elements.

New elements are being discovered relatively frequently as technology continues to advance. As of the writing of this textbook, there were 118 elements in the periodic table. Most scientists consider 92 of these elements to be "naturally occurring," which generally means they can be found in samples of soil, air, and water. The remaining elements in the periodic table are man-made. The average person has detectable traces of approximately 60 elements in his or her body, but by most current definitions only 24 are recognized as being involved in maintaining life.

Figure 2.1 highlights the 12 elements of the human body that occur in significant amounts (totaling 99.9%). The four elements isolated in figure 2.1 (oxygen, carbon, hydrogen, and nitrogen) make up approximately 96% of the human body and are found making up the body's major organic molecules, discussed later in the chapter. Other remaining elements occur in very small amounts and are referred to as *trace elements*.

Atomic Structure AP|R

An **atom** (a′-tom) is the smallest single unit of an element. Atoms of a given element are similar to each other, and they are different from atoms of all other elements. Atoms of different elements differ in size, mass, and how they interact with other atoms.

Atoms are composed of three types of subatomic particles: protons, neutrons, and electrons. Each **proton** has a positive electrical charge. Each **neutron** is neutral, meaning it has no electrical charge. Protons and neutrons are located in the **nucleus** at the center of an atom. Each **electron** has a negative electrical charge.

Electrons orbit, or revolve around, the nucleus at high speeds in *electron shells* that are located at various distances from the nucleus. The first shell of electrons, the shell closest to the nucleus, can hold a maximum of two electrons, as is the case for hydrogen. Atoms react to fill their **valence** (outermost) shell. Therefore, hydrogen reacts to gain one electron to fill its valence shell with two electrons. All other atoms typically found in the human body react to fill the valence shell with eight electrons. Atoms always fill the lowest electron shells first. See the diagram of the atomic structures of hydrogen and carbon in figure 2.2.

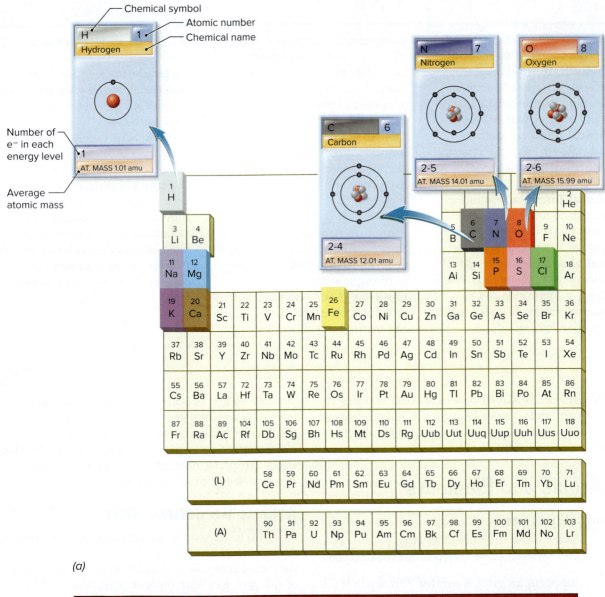

Figure 2.1 The Periodic Table of Elements.
(a) Detailed description of how to read the table. (b) The 12 most abundant elements in the human body.

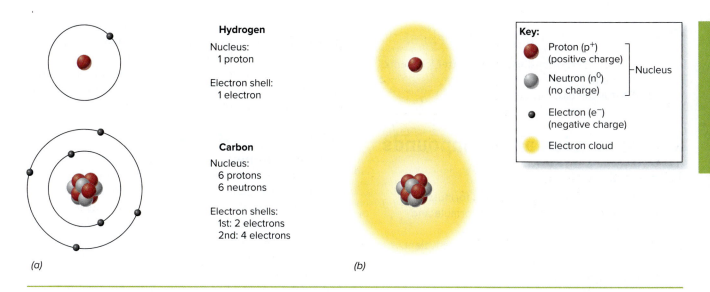

Figure 2.2 Atomic Structures of Hydrogen and Carbon.
(a) Electron shell models. *(b)* Electron cloud models. These models show the most likely locations of the electrons.

An atom is electrically neutral because it has the same number of protons as electrons, although the number of neutrons may vary. Most atoms are not stable in this state and have characteristic ways of losing, gaining, or sharing electrons to achieve stability, which is key to forming chemical bonds.

The atoms of each element are characterized by a specific atomic number, chemical symbol, and atomic mass. These characteristics are used to identify the element. The **atomic number** indicates the number of protons, and because they match, also the number of electrons in each atom. The **chemical symbol** is a shorthand way of referring to an element or to an atom of the element. The mass of either a proton or a neutron is defined as one *atomic mass unit* (amu). Because an electron's mass is only 0.005 amu, the **atomic mass** of an atom is simply the sum of the number of protons plus the number of neutrons in each atom. For example, an atom of carbon has an atomic number of 6, a chemical symbol of C, and an atomic mass of 12. From this information, you know that an atom of carbon has six protons, six electrons, and six neutrons.

Isotopes

As mentioned in the preceding section, all atoms of an element have the same number of protons and electrons. However, some atoms may have different numbers of neutrons. An atom of an element with a different number of neutrons is called an **isotope** (ī′-so-tōp). For example, hydrogen has three isotopes: ^1H, ^2H, and ^3H (figure 2.3). All isotopes of an element have the same chemical properties because they have the same number of protons and electrons.

Certain isotopes of some elements have an unstable nucleus that emits high-energy radiation as it breaks down to form a more stable nucleus. Such isotopes are called *radioisotopes*. Certain radioisotopes are used in the diagnosis of disorders and in the treatment of cancer. See the clinical insight box later in this chapter.

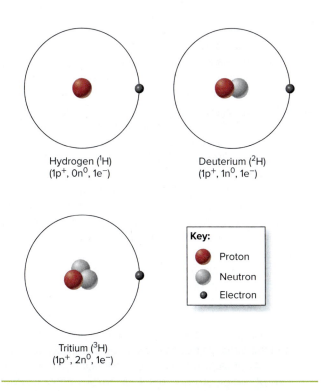

Figure 2.3 The Three Isotopes of Hydrogen.
Notice that only the number of neutrons changes.

Check My Understanding

1. What is the relationship among matter, elements, and atoms?
2. What is the basic structure of an atom?

2.2 Molecules and Compounds

Learning Objectives

3. Explain the meaning of a chemical formula.
4. Compare and contrast molecular formula and structural formula.
5. Compare and contrast ionic, nonpolar covalent, polar covalent, and hydrogen bonds.
6. Compare synthesis, decomposition, exchange, and reversible reactions.

A few elements exist separately in the body, but most are chemically bound to others to form **molecules.** Some molecules are composed of like elements–an oxygen molecule (O_2), for example. Others, such as water (H_2O), are composed of different kinds of elements. **Compounds** are substances composed of atoms from two or more different elements. Thus, the chemical structure of water may be referred to as both a molecule and a compound. Whether a substance is called a molecule or a compound also depends upon the type of chemical bond used to build the substance. Molecules are built by covalent bonds only, while compounds are built by either ionic or covalent bonds. Chemical bonds are discussed in more detail shortly.

Chemical Formulae

A **chemical formula** expresses the chemical composition of a substance. Two major types of chemical formulae exist, the molecular formula and the structural formula. A *molecular formula* expresses the composition of a single molecule. In a molecular formula, chemical symbols indicate the elements involved, while subscripts identify the number of atoms of each element in the molecule. For example, the molecular formula for water is H_2O, which indicates that two atoms of hydrogen combine with one atom of oxygen to form a water molecule. The molecular formula does not describe how the two hydrogen atoms and one oxygen atom in a water molecule are attached to each other. There are many possibilities: H–H–O, H–O–H, or O–H–H, for example. Even if the order in which the atoms are attached is known, the atoms may not be arranged in a straight line as indicated above. A *structural formula* is a diagram that both indicates the composition and number of atoms and illustrates how the atoms

 Clinical Insight

Nuclear medicine is the medical specialty that uses radioisotopes in the diagnosis and treatment of disease. Very small amounts of weak radioisotopes may be used to tag biological molecules in order to trace the movement or metabolism of these molecules in the body. Special instruments can detect the radiation emitted by the radioisotopes and identify the location of the tagged molecules.

In *nuclear imaging,* the emitted radiation creates an image on a special photographic plate or computer screen. In this way, it is possible to obtain an image of various organs or parts of organs where the radioisotopes accumulate. Positron emission tomography (PET) uses certain radioisotopes that emit positrons (positively charged electrons), and it enables precise imaging similar to computerized tomography (CT) scans. PET can be used to measure processes, such as blood flow, rate of metabolism of selected substances, and effects of drugs on body functions. It is a promising technique for both the diagnosis of disease and the study of healthy physiological processes.

Another form of nuclear medicine involves the use of radioisotopes to kill cancerous cells. Certain radioisotopes may be attached to specific biological molecules and injected into the blood. When these molecules accumulate in cancerous tissue, the emitted radiation kills the cancerous cells. A similar effect is obtained by implanting pellets of radioactive isotopes directly in cancerous tissue.

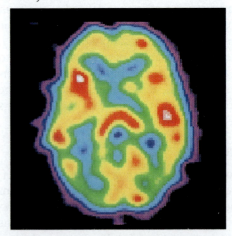

Positron Emission Tomography (PET).
Transverse section through the head. The highest level of brain activity is indicated in red, with successively lower levels represented by yellow, green, and blue.
©McGill University Collection CNRI/Phototake

are linked to one another. Many figures in the text will use structural formulae. Figures 2.5 and 2.7 are good examples.

Up to this point it has been mentioned that molecules are composed of atoms that are "chemically combined." However, no mention has been made as to how this occurs. We will explore this next.

Chemical Bonds APR

Chemicals are combined when electrons interact to form **chemical bonds,** which join atoms together to form a molecule. A chemical bond is a force of attraction between two atoms. An atom combines with another atom in order to fill its valence shell. A full valence shell makes an atom more stable. To do this, atoms either (1) receive or lose electrons, which results in the formation of an ionic bond; or (2) share electrons, which leads to the formation of a covalent bond.

Ionic Bonds

Consider the interaction of sodium and chlorine in the formation of sodium chloride (table salt), as shown in figure 2.4. Sodium has a single electron in its valence shell, while chlorine has seven electrons in its valence shell. Note in step 2 of figure 2.4 that, after transferring an electron from sodium to chlorine, sodium now has 11 protons (+) and 10 electrons (−), while chlorine has 17 protons (+) and 18 electrons (−). Thus, the transfer of an electron from sodium to chlorine causes the sodium atom to have a net electrical charge of +1 and the chlorine atom to have a net electrical charge of −1.

Atoms or groups of atoms with a net electrical charge, either positive or negative, are called **ions.** Thus, the transfer of an electron from sodium to chlorine has (1) resulted in the valence shell of each atom being filled with electrons and (2) produced a sodium ion (Na^+) and a chloride ion (Cl^-). Positively charged ions, such as Na^+, are called *cations*. Negatively charged ions, such as Cl^-, are called *anions*. The force of attraction that holds cations and anions together as an *ionic compound* is an **ionic bond.**

Covalent Bonds

Atoms that form molecules by sharing electrons are joined by **covalent bonds.** The shared electrons orbit around each atom for part of the time so that they may be counted in the valence shell of each atom. Thus, the valence shell of each atom is filled.

A simple example of covalent bonding is found in a molecule of hydrogen gas. A hydrogen atom with a single

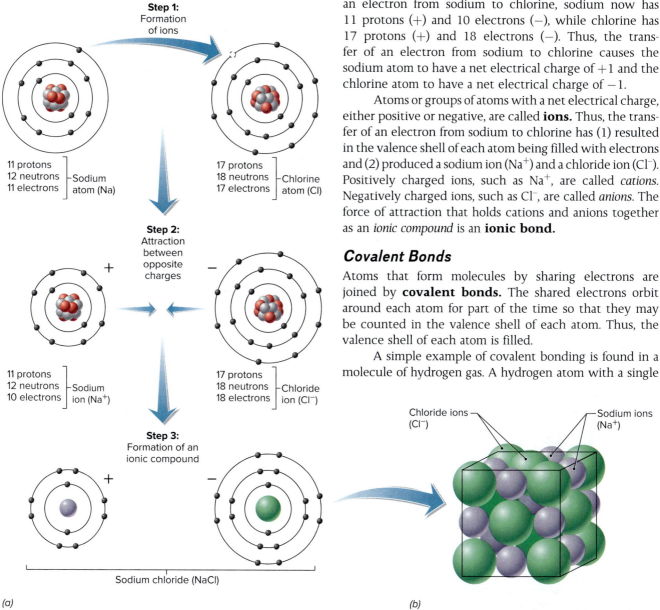

Figure 2.4 The Synthesis of Sodium Chloride by the Formation of an Ionic Bond.
(a) The transfer of an electron from sodium to chlorine converts sodium to a cation and chlorine to an anion.
(b) The attraction between these oppositely charged ions is an ionic bond.

electron requires one more electron to fill its valence shell. Two hydrogen atoms can form a molecule of hydrogen gas (H_2) by sharing their electrons. In this way, the valence shell of both atoms is complete and a single covalent bond is formed. The single covalent bond is shown in a structural formula as a single straight line between chemical symbols for hydrogen (H–H), as is illustrated in figure 2.5a.

Double and triple covalent bonds can also form. A molecule of gaseous oxygen (O_2) is formed when two oxygen atoms share two pairs of electrons. Each oxygen atom requires two electrons to complete its valence shell, so by sharing two pairs of electrons the valence shell of both atoms is complete and a double covalent bond (O=O) is formed (figure 2.5b). Similarly, some molecules contain

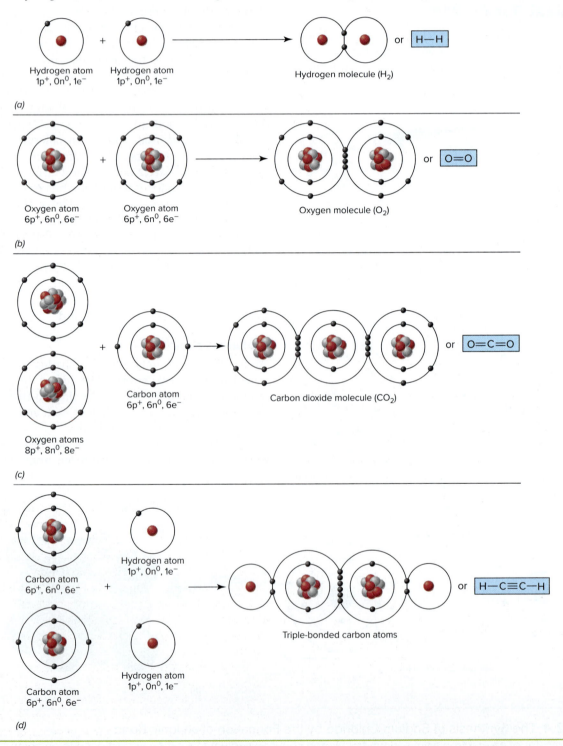

Figure 2.5 Formation of Nonpolar Covalent Bonds.

carbon atoms that are triple bonded. A triple bond is formed when two atoms share three pairs of electrons. Nitrogen gas (N_2) is also formed of triple bonds (N≡N). Figure 2.5d shows a triple bond.

There are two types of covalent bonds: nonpolar covalent and polar covalent. *Nonpolar covalent bonds* are commonly found between atoms of the same type and between C and H. In a nonpolar covalent bond, the shared electrons spend equal time revolving between the two atoms. The equal sharing forms a molecule that is electrically neutral. These **nonpolar molecules** do not mix well with water and are referred to as being **hydrophobic** (hydro = water; phobos = fear). *Polar covalent bonds* involve an unequal sharing of electrons between two atoms. For example, when a hydrogen atom is covalently bonded to an oxygen atom, the shared electrons spend less time near the hydrogen atom and more time near the oxygen atom. This occurs because the oxygen atom has a stronger pull on the electrons, which is referred to as *electronegativity*. In this situation the hydrogen atom becomes slightly positively charged, notated as δ^+, and the oxygen becomes slightly negatively charged, notated as δ^- (figure 2.6). Most molecules formed by polar covalent bonds are called **polar molecules** because different areas of the molecule have a different electrical charge. Polar molecules and ions tend to mix well with water and are thus referred to as **hydrophilic** (philos = loving). A good rule of thumb in determining whether a substance is hydrophobic or hydrophilic is "like mixes with like." For a substance to mix with water, it must be like water, meaning it must also be electrically charged.

It is important to note that a molecule may contain polar covalent bonds and still be a nonpolar molecule. The carbon dioxide molecule in figure 2.5c is a great example; while the C=O bonds are polar covalent bonds, the molecule is linear and symmetrical so that the opposing bonds cancel one another, making the entire structure similar (nonpolar) throughout.

Hydrogen Bonds

A **hydrogen bond** is a weak attractive force between a slightly positive area and a slightly negative area of a polar molecule. These attractions may occur between different sites within the same molecule or between different molecules. They may also occur between polar molecules and ions. Figure 2.7c illustrates how the slightly negative oxygen atom of one water molecule attracts the slightly positive hydrogen atom of a different water molecule to form a hydrogen bond. It is important to note that the bonds between oxygen and hydrogen within one molecule are polar covalent bonds. The hydrogen bonds between water molecules are responsible for many of water's unique characteristics that help support life, which is why space exploration focuses so heavily on identifying other planets with water. For example, based on atomic mass, water should be a gas at room temperature. However, the hydrogen bonds keep water liquid at room temperature. Hydrogen bonds also play an important role in protein and nucleic acid structure, as you will see later in this chapter.

 Check My Understanding

3. How do ionic and covalent bonds join atoms to form molecules or compounds?
4. How are nonpolar covalent bonds and polar covalent bonds different?
5. What are hydrogen bonds?

Chemical Reactions

In **chemical reactions,** bonds between atoms are formed or broken, and the result is a new combination of atoms. There are four basic types of chemical reactions: synthesis, decomposition, exchange, and reversible reactions. Such reactions occur continuously within the body. In general, a chemical reaction begins with one or more substance(s) referred to as the *reactant(s)*. The reaction occurs and the new combination of atoms is (are) called the *product(s)*. Figure 2.8 shows a few examples of chemical reactions.

Synthesis (anabolic) reactions form new chemical bonds. Energy is required to form new bonds. Atoms or simple molecules combine to form a more complex product. The reaction of hydrogen and oxygen (reactants) to form water (product) is an example. Figure 2.8b shows how energy may be used to combine amino acids to form a protein. Synthesis reactions produce complex molecules used

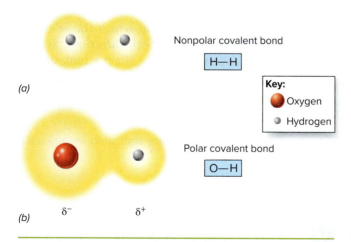

Figure 2.6 Types of Covalent Bonds.
Comparison of electron locations in *(a)* nonpolar versus *(b)* polar covalent bonds.

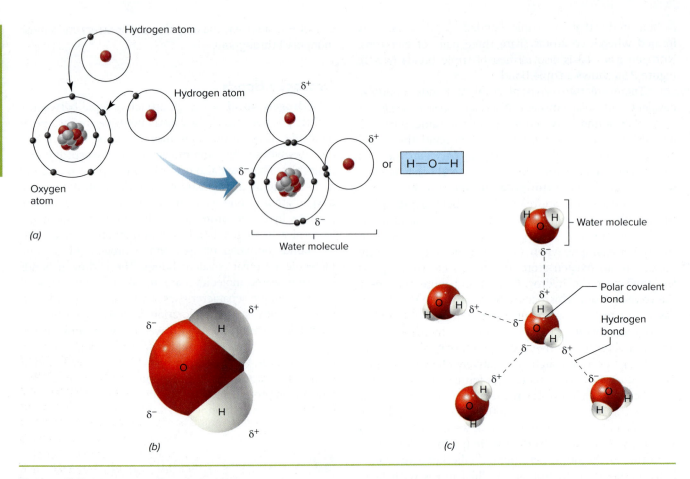

Figure 2.7 Hydrogen Bonds in Water.
(a) The synthesis of a water molecule by the formation of polar covalent bonds. *(b)* Space-filling model of water molecule. *(c)* Hydrogen bonds forming between adjacent water molecules.

in the growth and repair of body parts. Synthesis reactions may be generalized as

$$A + B \rightarrow AB$$

A **decomposition (catabolic) reaction** is the opposite of a synthesis reaction. Chemical bonds of a complex molecule are broken to form two or more simpler molecules, releasing energy in the process. For example, water can decompose to form hydrogen and oxygen. Decomposition reactions are used to break down food molecules to form nutrients usable by body cells. Figure 2.8c shows how glycogen may be decomposed to form glucose, thus releasing energy. Decomposition reactions may be generalized as

$$AB \rightarrow A + B$$

Exchange (rearrangement) reactions occur when different reactants exchange components, resulting in the breakdown of the reactants and the formation of new products. Thus, exchange reactions involve both decomposition of the reactants and synthesis of the new products. The dehydration synthesis and hydrolysis reactions discussed later in this chapter are examples of exchange reactions. Exchange reactions may be generalized as

$$AB + CD \rightarrow AD + CB$$

Reversible reactions exist in which the reactants and products may convert in both directions. Several factors determine the reversibility of a reaction, such as energy available and relative abundance of reactants and products. The reactions of chemical buffers are great examples of exchange reactions that are also reversible reactions. Chemical buffers are discussed later in this chapter. Reversible reactions are indicated by a double arrow:

$$A + B \rightleftharpoons AB$$

 Check My Understanding

6. How do synthesis, decomposition, exchange, and reversible reactions differ?

(a) Chemical reaction

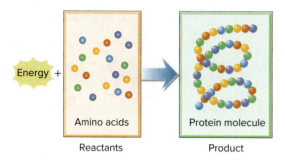

(b) Anabolic (or synthesis) reaction

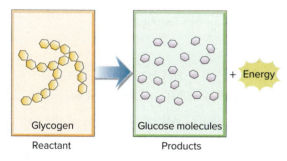

(c) Catabolic (or decomposition) reaction

Figure 2.8 Types of Chemical Reactions. *(a)* General example of a chemical reaction. *(b)* A synthesis reaction uses energy to form bonds, resulting in one or more larger products. *(c)* A decomposition reaction breaks bonds, releasing energy and resulting in numerous smaller products.

2.3 Substances Composing the Human Body

Learning Objectives
7. Distinguish between inorganic and organic substances.
8. Explain the importance of water and its locations in the body.
9. Compare and contrast electrolytes and nonelectrolytes, and acids and bases.
10. Explain the use of the pH scale.
11. Explain the importance of buffers.
12. Distinguish between carbohydrates, lipids, proteins, and nucleic acids and their roles in the body.
13. Explain the role of enzymes.
14. Describe the mechanism of enzymatic action.
15. Describe the structure and function of adenosine triphosphate (ATP).

Substances composing the human body include both inorganic and organic substances. **Inorganic substances** may contain either carbon or hydrogen in the same substance, *but not both.* Bicarbonates, such as sodium bicarbonate ($NaHCO_3$), are an exception to this rule. Molecules of **organic substances** always contain *both* carbon and hydrogen, and they usually also contain oxygen. Recall from chapter 1 that large complex molecules are called *macromolecules.* The carbon atoms form the "backbone" of organic macromolecules.

Major Inorganic Substances

The major inorganic substances in the body are water, most acids and bases, and inorganic salts. Acids, bases, and mineral salts **ionize** to release ions when dissolved in water, as you will see shortly.

Water

Water is by far the most abundant molecule found within cells and in the extracellular fluid. Water composes about two-thirds of the body weight. Water generally occurs within the body as part of a mixture called an **aqueous solution.** A *solution* is a mixture composed of a solvent and one or more solutes. A *solvent* is a liquid used to dissolve or suspend substances. *Solutes* are the substances dissolved or suspended in the solvent. In this aqueous solution, water is acting as a solvent. Recall from earlier in the chapter that water is a solvent for electrically charged substances, such as polar molecules and ions. Chemical reactions that occur in the body take place in this aqueous solution. Water is used to transport many solutes throughout the body, through the plasma membrane of a cell, or from one part of the cell to another. Another function of water is that it is a great lubricant in the body. Water is also important in maintaining a constant cellular temperature, and thus a constant body temperature, because it absorbs and releases heat slowly. Evaporative cooling (sweating) through the skin also involves water. Another function of water is as a reactant in the breakdown (hydrolysis) of organic molecules.

The specific locations where water is found in the body are called *water compartments.* They are

Intracellular fluid (ICF): fluid within cells; about 65% of the total body water.
Extracellular fluid (ECF): all fluid not in cells; about 35% of the total body water.
  Interstitial fluid (tissue fluid): fluid in spaces between cells.
  Plasma: fluid portion of blood.
  Lymph: fluid in lymphatic vessels.
  Transcellular fluids: fluid in more limited locations, such as serous fluid, cerebrospinal fluid (CSF–located within and around the brain and spinal cord), and synovial fluid (in certain joints).

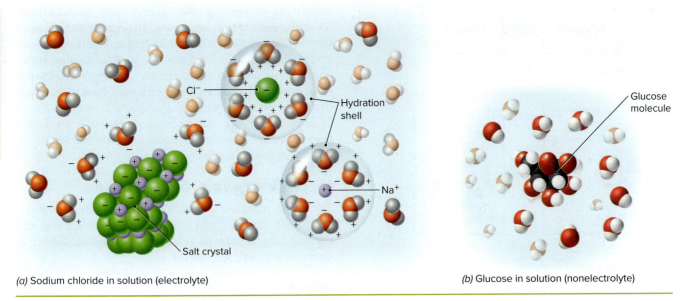

Figure 2.9 Comparison of an Electrolyte and a Nonelectrolyte in an Aqueous Solution.

Electrolytes AP|R

When ionic compounds are dissolved in water they tend to ionize (dissociate), releasing ions. The process of ionization involves the formation of *hydration spheres* by water using its polar nature to separate ions from each other (figure 2.9a). Such compounds are called **electrolytes** (ē-lek′-trō-līz) because when dissolved they can conduct an electrical current. Acids, bases, and salts are types of electrolytes (figure 2.10). Substances that dissolve in water without ionizing and therefore do not conduct electrical current are termed **nonelectrolytes,** such as the glucose molecule in figure 2.9b. Nonelectrolytes are usually organic molecules, which are discussed later in the chapter. The composition and concentration of electrolytes and nonelectrolytes in the body must be kept within narrow limits to maintain homeostasis.

Acids and Bases

An **acid** is a chemical that releases **hydrogen ions (H^+)** into solution.

The stronger the acid, the greater is the degree of ionization, which results in a greater concentration of H^+. Hydrochloric acid (HCl) is a strong acid. It ionizes into hydrogen and chloride ions (H^+ and Cl^-).

$$HCl \rightarrow H^+ + Cl^-$$

A **base** decreases the concentration of H^+ in a solution by combining with the H^+. The stronger the base, the greater is its ability to combine with H^+. Some bases combine directly with H^+. Other bases, such as sodium hydroxide (NaOH), ionize to release hydroxide ions (OH^-), which combine with H^+ to form water.

$$NaOH \rightarrow Na^+ + OH^-$$
$$H^+ + OH^- \rightarrow HOH$$

The symbol **pH** is a measure of the hydrogen ion concentration in a solution.

Measurement of pH Chemists have developed a pH scale that is used to indicate the measure of acidity or alkalinity (basicity) of a solution, meaning the relative concentrations of hydrogen ions (H^+) and hydroxide ions (OH^-) in a solution. The **pH scale** ranges from 0 to 14. These ions are equal in concentration at pH 7, so a solution with a pH of 7 is neither an acid nor a base and is referred to as neutral. For example, pure water has a pH of 7.

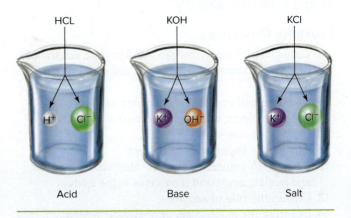

Figure 2.10 Ionization in Water.
Ionization of an acid, a base, and a salt in water.

Acids contain more H^+ than OH^- and have a pH less than 7. Bases contain more OH^- than H^+ and have a pH greater than 7. There is a tenfold difference in the concentrations of H^+ and OH^- when the pH changes by one unit. For example, an acid with a pH of 4 has a concentration of H^+ that is 10 times greater than that of an acid with a pH of 5. Some examples of pH values for body fluids include blood (7.4), stomach acid (1–2), urine (6), and intestinal fluid (8).

Buffers Cells of the body are especially sensitive to pH changes. Even slight changes can be harmful. The hydrogen ion concentration (pH) of blood and other body fluids is maintained within narrow limits by the lungs, kidneys, and buffers in body fluids. A **buffer** is a chemical or a combination of chemicals that either picks up excess H^+ or releases H^+ to keep the pH of a solution rather constant. The carbonic acid–bicarbonate buffer system illustrated in figure 2.11 is the most important buffer system in the body. Notice that carbonic acid and bicarbonate ions react in a reversible reaction whose direction is determined by pH.

Buffers are extremely important in maintaining the healthy pH of body fluids, but they can be overwhelmed by a disruption of homeostasis. The healthy pH of the arterial blood is 7.35 to 7.45. In *acidosis*, the pH is less than 7.35, and a patient could go into a coma. In *alkalosis*, the pH range is greater than 7.45, and a patient may have uncontrolled muscle contractions. Extreme variations, outside of 6.8–8.0, may be fatal.

Salts

Like acids and bases, *salts* are ionic compounds that ionize in an aqueous solution, but they do not produce hydrogen or hydroxide ions. The most important salts in the body are sodium, potassium, and calcium salts. Calcium phosphate is the most abundant salt because it is a main component of bones and teeth. Sodium chloride (NaCl), a common salt in body fluids, ionizes into sodium and chloride ions (Na^+ and Cl^-).

$$NaCl \rightarrow Na^+ + Cl^-$$

Salts provide ions that are essential for healthy body functioning. Physiological processes in which ions play an essential role include blood clotting, muscle and nervous functions, and pH and water balance (table 2.1).

Check My Understanding

7. Where is water located in the body and why is it so important?
8. What is the relationship between acids, bases, and pH?
9. What are salts and why are they important?

Major Organic Substances

The major organic macromolecules of the body are carbohydrates, lipids, proteins, and nucleic acids (table 2.2). Another molecule, adenosine triphosphate (ATP), is also considered here because it plays such a vital role in the transfer of energy within cells.

Before beginning a study of the various organic molecules, it is important to understand one reversible reaction that will help to clarify a great deal about how biochemistry (chemistry in living things) in general works.

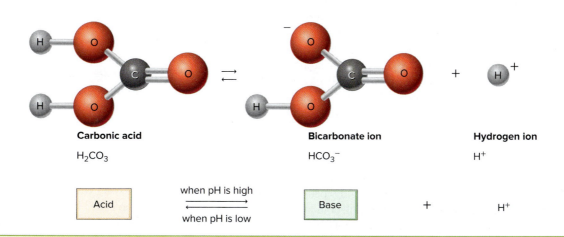

Figure 2.11 Carbonic Acid–Bicarbonate Buffer System.
Carbonic acid ionizes to release H^+ when pH is high. Bicarbonate ion combines with H^+ when pH is low.

Table 2.1   Important Inorganic Ions

Ion	Symbol	Functions
Bicarbonate	HCO_3^-	Helps maintain acid–base balance
Calcium	Ca^{2+}	Major component of bones and teeth; required for muscle contraction and blood clotting
Carbonate	CO_3^{2-}	Major component of bones and teeth; helps maintain acid–base balance
Chloride	Cl^-	Helps maintain water balance
Hydrogen	H^+	Helps maintain acid–base balance
Hydroxide	OH^-	Helps maintain acid–base balance
Phosphate	PO_4^{3-}	Major component of bones and teeth; required for energy transfer; helps maintain acid–base balance
Potassium	K^+	Required for muscle and nervous function
Sodium	Na^+	Required for muscle and nervous function; helps maintain water balance

Table 2.2   Important Organic Substances

Macromolecule	Building Units	Examples	Functions
Carbohydrates	Monosaccharides (simple sugars)	Glucose	Primary energy source for cells
		Starch	Storage form in plants; common in plant foods
		Glycogen	Storage form in animals; stored in liver and muscles
Lipids	Glycerol, fatty acids	Triglycerides (fats)	Energy source and storage
	Diglyceride, phosphate group	Phospholipids	Cell structure
	Cholesterol (4C rings)	Steroids	A variety of functions (e.g., sex hormones promote sexual development)
Proteins	Amino acids	Structural proteins	Cell structure
		Functional proteins	Nonstructural functions (e.g., catalyze chemical reactions and chemical transport)
Nucleic acids	Nucleotides	DNA	Storage of genetic information
		RNA	Processing of genetic information leading to protein synthesis
		ATP	Energy carrier for cellular processes

The reaction involves two processes, dehydration synthesis and hydrolysis. **Dehydration synthesis** literally means "remove water to bond together," while **hydrolysis** means "break with water." Figure 2.12a shows that in dehydration synthesis one molecule gives up a hydrogen atom, which then combines with a *hydroxyl group* (–OH) given up by a second molecule to form water. The two molecules involved will need to re-form bonds where the H and –OH were removed in order to refill their valence shells. They satisfy this need by combining with each other, forming a new bond. In a hydrolysis reaction (figure 2.12b), the opposite occurs. Water "attacks" a chemical bond between two molecules, causing it to break. The resulting molecules have extra electrons that bind to the water molecule and split it, with H bonding to one molecule and –OH bonding to the other. Understanding this mechanism will assist your understanding of how organic macromolecules are synthesized and decomposed.

Carbohydrates

Carbohydrates (kar″-bo-hī′-drātz) are formed of carbon, hydrogen, and oxygen. In each carbohydrate molecule, there are two hydrogen atoms for every carbon and oxygen atom. Therefore, C, H, and O are said to have a 1:2:1 ratio. Carbohydrates are the primary source of nutrient energy for cells of the body. Carbohydrates are classified according to molecular size as monosaccharides, disaccharides, or polysaccharides (figure 2.13).

The simplest carbohydrates are **monosaccharides** (mon-ō-sak′-ah-rī ds). For example, **glucose** ($C_6H_{12}O_6$) is a six-carbon monosaccharide (hexose) that is the major carbohydrate fuel for cells. It is often called blood sugar because it is the form in which carbohydrates are transported to body cells. Fructose and galactose are other hexoses found in foods. Glucose, fructose, and galactose have the same molecular formula ($C_6H_{12}O_6$), but are chemically bonded differently. Molecules

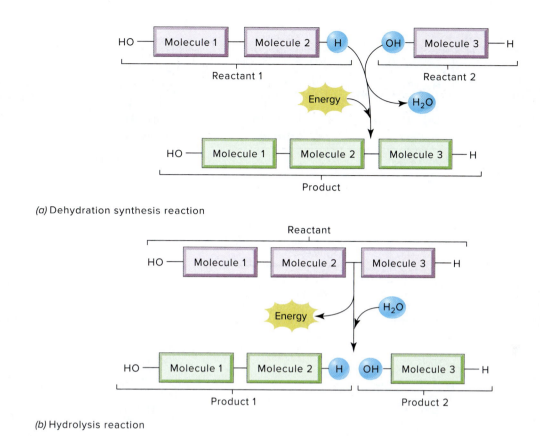

Figure 2.12 Common Reactions Involving Organic Substances. Basic model for *(a)* dehydration synthesis and *(b)* hydrolysis reactions.

with the same molecular formula but different structures are called *isomers.* Monosaccharides are the *building units* that chemically combine to produce larger carbohydrates.

The chemical combination of two monosaccharides forms a **disaccharide** (dī-sak′-ah-rīd), a double sugar. The common disaccharides in foods are *maltose,* or malt sugar (glucose + glucose); *sucrose,* or table sugar (glucose + fructose); and *lactose,* or dairy sugar (glucose + galactose). Collectively, monosaccharides and disaccharides are called *simple sugars* and are used only as a source of energy.

A **polysaccharide** (pol-ē-sak′-ah-rīd) is a macromolecule formed by the chemical combination of many monosaccharide units. Two polysaccharides are important to our study: glycogen and starch. Both are formed of many glucose units. Glycogen and starch are sometimes referred to as *complex carbohydrates.* Complex carbohydrates are a source of energy and also make up the fiber in our diets.

Glycogen is the storage form of carbohydrates in animals, including humans. Some of the excess glucose in blood is converted into glycogen and stored primarily in the liver, but small amounts are stored in muscle cells. Glycogen serves as a reserve energy supply that can be quickly converted into glucose. For example, whenever the level of blood glucose declines, the liver converts glycogen into glucose via catabolic (hydrolysis) reactions to increase the blood glucose level.

Starch is the storage form of carbohydrates in plants, so it is present in many foods derived from plants.

Lipids

Lipids are a large, diverse group of organic macromolecules that consist of carbon, hydrogen, and oxygen atoms. Carbon atoms form the backbone of the molecules, and there are many times more hydrogen atoms than oxygen atoms. The most abundant lipids in the body are triglycerides (fats), phospholipids, and steroids.

Molecules of **triglycerides** (trī-glys′-er-īds), or **fats,** consist of one **glycerol** (glys′-er-ol) molecule and three **fatty acid** molecules joined together (figure 2.14). Triglycerides are the most concentrated energy source found in the body and are the most abundant lipids in our diet. Excess nutrients (energy reserves) are stored as triglycerides in the adipocytes (fat cells) of the body, primarily around internal organs and beneath the skin. Fats are nonpolar molecules that do not mix well with water, meaning they are *hydrophobic.*

Phospholipids (fos′-fō-lip′-idz) are molecules similar to triglycerides. The basic difference is that one of the fatty acids is replaced with a *phosphate group.* Table 2.1 shows a phosphate group, which is a very important component of many organic substances. Unlike fats, phospholipids are partially soluble in water, or *amphiphilic.* The end of the

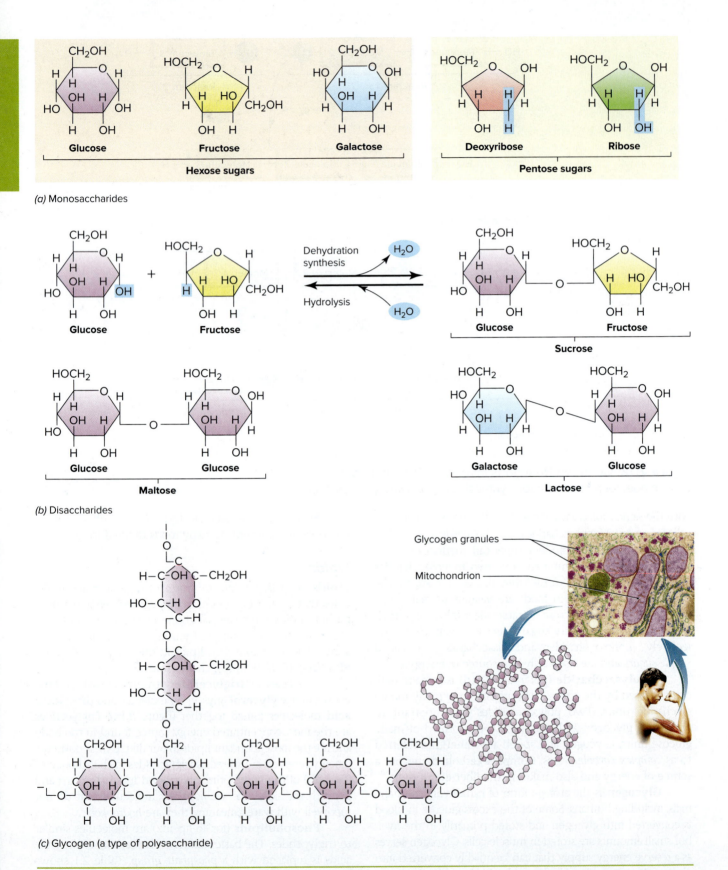

Figure 2.13 Carbohydrates.
(a) The building units of carbohydrates are monosaccharides. *(b)* Two monosaccharides combine to form a disaccharide. *(c)* The combination of many monosaccharide units forms a polysaccharide.

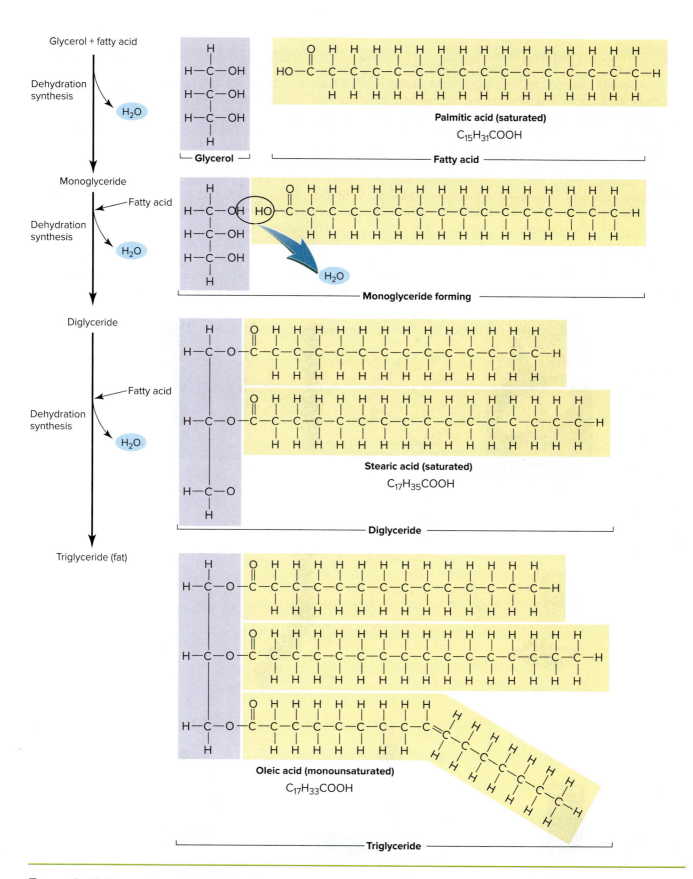

Figure 2.14 Formation of a Triglyceride.
A triglyceride molecule consists of three fatty acids joined to a glycerol molecule by three successive dehydration synthesis reactions.

Clinical Insight

Triglycerides may be classified as either saturated or unsaturated fats. In *saturated fats* (e.g., animal fats), the bonds of the carbon atoms in the fatty acids are saturated (filled) by hydrogen atoms so that the carbon–carbon bonds are all single bonds. Saturated fats, such as butter and lard, are solid at room temperature. Excessive saturated fats in the diet are associated with an increased risk of heart disease (coronary artery disease).

In *unsaturated fats* (plant oils), not all carbon bonds in the fatty acids are filled with hydrogen atoms, and one or more double carbon–carbon bonds are present. Fatty acids of monounsaturated fats (e.g., olive, peanut, and canola oils) have one carbon–carbon double bond; those of polyunsaturated fats (e.g., corn, safflower, and soy oils) have two or more carbon–carbon double bonds. Unsaturated fats occur as oils at room temperature.

Hydrogenation, the process of adding hydrogen atoms to unsaturated fats, converts most carbon–carbon double bonds to carbon–carbon single bonds and changes vegetable oil to a solid (e.g., margarine) at room temperature. This process also changes the bonding pattern of some fatty acids to form *trans fats,* which increase the risk of coronary artery disease even more than do saturated fats. It is better to cook with oils than with lard or margarine because saturated and trans fats are more easily converted into the "bad" cholesterol associated with heart disease.

Saturated fatty acids
©Elena Shashkina/123RF RF

Palmitic acid

Stearic acid

Monounsaturated fatty acids
©David Murray/Getty Images

Oleic acid

Polyunsaturated fatty acids
©Science History Images/
Alamy Stock Photo RF

Linoleic acid

Arachidonic acid

molecule with the phosphate group is polar and therefore soluble in water but not in lipids, while the end with the two fatty acids is nonpolar and therefore lipid-soluble but insoluble in water. Thus, phospholipids can join–or serve as an interface between–a water environment on one side and a lipid environment on the other. They are major components of plasma membranes that surround cells and certain organelles within the cell (see chapter 3). Figure 2.15 shows the basic structure of a phospholipid molecule.

Steroids constitute another group of lipids, and their molecules characteristically contain four carbon rings. Cholesterol (kō-les′-ter-ol), vitamin D, certain adrenal hormones, and sex hormones are examples of steroids. Figure 2.16 shows several examples of steroids. Cholesterol is an essential component of body cells and serves as the raw material for the synthesis of other steroid molecules.

Proteins

Proteins (pro′-tēns) are large, complex macromolecules composed of smaller molecules (building units) called **amino acids.** There are 20 different kinds of amino acids used in building proteins, and each is composed of carbon, hydrogen, oxygen, and nitrogen. Each amino acid consists of a central C atom that is attached to four separate components. Three components are the same in all amino acids. The first is simply a hydrogen atom, and the other two

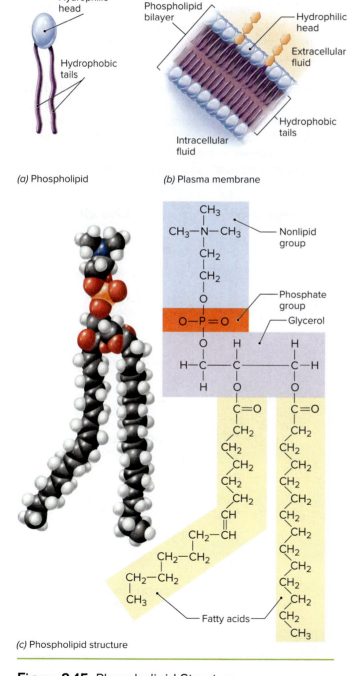

Figure 2.15 Phospholipid Structure.
In a phospholipid, one fatty acid is replaced by a polar phosphate group.

Figure 2.16 Steroid Structure.
Note the four carbon rings, which are characteristic of steroids. Carbon and hydrogen atoms are not all shown in these shorthand structures.

are the components for which an amino acid is named, an *amine group* (–NH$_2$) and an *acid group* (–COOH). However, it is the side chain, called the *R group*, that distinguishes the different kinds of amino acids and their chemical properties. Figure 2.17 depicts the basic structure of an amino acid, as well as examples of the four types of amino acids.

Amino acids are joined by **peptide bonds.** Two amino acids bonded together form a **dipeptide.** A chain of many amino acids forms a **polypeptide.** A chain of more than 50 amino acids forms a protein. As figure 2.18 shows, the sequence of amino acids is the *primary structure* of the protein. The long chain of amino acids forms hydrogen bonds between polar parts of the polypeptide, causing various areas of the molecule to twist (helix) and/or fold (pleated sheet). Helices (plural of helix) and pleated sheets are examples of *secondary structures*. The polypeptide then folds in response to the watery environment of the human body. The folding results in a *tertiary structure* with mostly hydrophobic amino acids at its interior and hydrophilic amino acids exposed to the watery exterior. The result is that each specific polypeptide or protein has a unique three-dimensional shape. In some instances, multiple polypeptides combine to form a protein. In these cases, the protein is said to have a *quaternary structure*.

Proteins may be classified as either structural or functional proteins. *Structural proteins* compose parts of body cells and tissues, where they provide support and strength in binding parts together. Ligaments, tendons, and contractile fibers in muscles are composed of structural proteins. *Functional proteins* perform a variety of different functions in the body. *Antibodies*, which provide immunity, *transport proteins*, which carry substances throughout the body and in and out of cells, and *enzymes*, which speed up chemical reactions, are examples of functional proteins.

Enzymes Without **enzymes,** the body's chemical reactions would occur too slowly to maintain life. Body cells contain thousands of enzymes, and each enzyme *catalyzes* (speeds up) a particular chemical reaction. An enzyme may catalyze synthesis, decomposition, or exchange reactions. A single enzyme may also catalyze a reversible reaction in

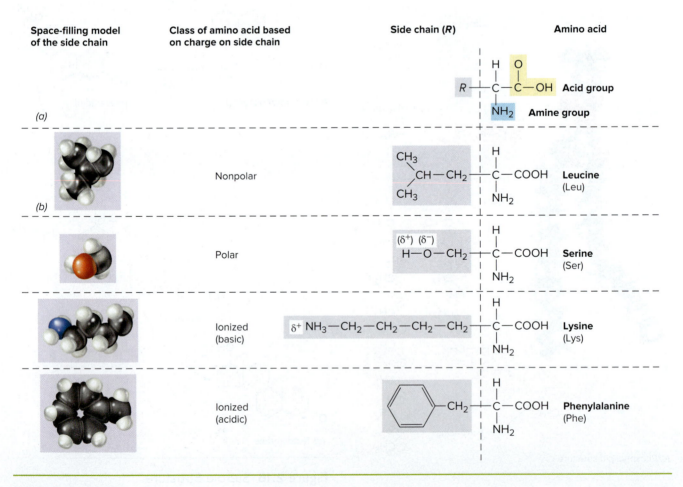

Figure 2.17 Amino Acids.
(a) The basic structure of an amino acid. *(b)* Representative amino acids. Note that the hydrogen atom, the amine group, and the acid group are the same in all amino acids. It is only the R group that is different among different amino acids.

both directions. Figure 2.19a illustrates enzyme action in an exchange reaction.

The enzyme must have just the right shape so the **substrate** (or reactant) molecule(s) can fit onto the **active site** of the enzyme, somewhat like a piece of a puzzle. The active site is where the reaction occurs. The enzyme binds to the substrate to form an *enzyme-substrate complex* where chemical bonds are formed between substrate molecules in synthesis reactions, or chemical bonds of the substrate molecule are broken in decomposition reactions. Once the reaction has occurred, the product separates from the enzyme. The enzyme is not altered in the reaction and may be recycled and used again and again. A single enzyme may catalyze thousands of reactions. An enzymatic reaction may be generalized as

$$E + S \rightarrow ES \rightarrow E + P$$

Like other proteins, the three-dimensional shape of an enzyme is determined by hydrogen bonds. Hydrogen bonds are easily broken by several factors, such as temperature

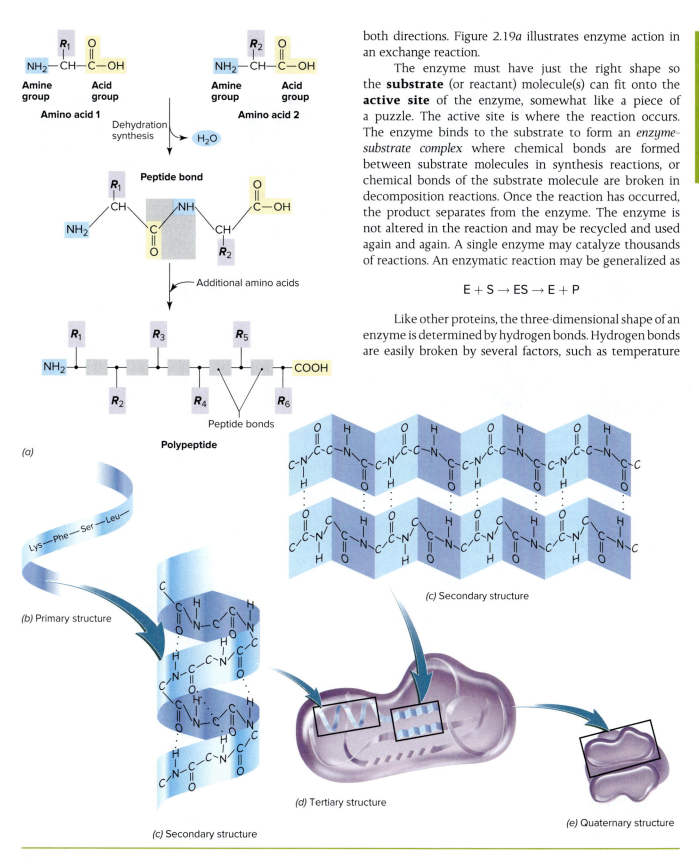

Figure 2.18 Protein Structure.
(a) The formation of a peptide bond. *(b–e)* The primary structure of a polypeptide ultimately determines how it folds into a more complex structure.

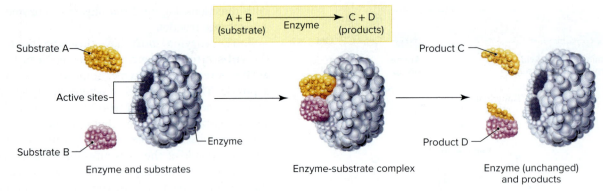

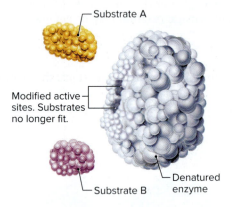

Figure 2.19 A Model of Enzyme Action. (a) An exchange reaction. The substrates bind to the active site of the enzyme where the reaction occurs. Then, the products are released, and the enzyme is recycled. (b) Denaturation inactivates an enzyme by changing the shape of the active sites. APR

and pH changes, poisons, and radiation. If an enzyme's hydrogen bonds are altered, its shape is changed, and the enzyme is denatured (inactivated) (figure 2.19b). Because it cannot bind to the substrate, the reaction it catalyzes will not occur. If the reaction is vital, the result can be fatal.

Nucleic Acids

Nucleic (nū-klā′-ic) **acids** are the human body's largest macromolecules. Two types of nucleic acids occur in cells: DNA and RNA. **Deoxyribonucleic** (dē-ok″-se-rī″-bō-nū-klā′-ic) **acid (DNA)** is located in the cell nucleus. DNA contains the *genetic code*, encoded information that determines hereditary traits and cellular functions. One way the genetic code does this is by determining the structure of proteins. The portions of DNA that encode for specific proteins are called **genes** (jēns).

Ribonucleic acid (RNA) carries the coded instructions from DNA to the cellular machinery involved in protein synthesis.

Both DNA and RNA consist of repeating building units called **nucleotides** (nū′-klē-ō-tīds). Each nucleotide consists of three parts: a five-carbon sugar, a phosphate group, and a nitrogenous base. Figure 2.20 shows the typical structure of a deoxyribonucleotide and a ribonucleotide, as well as the five possible nitrogenous bases used in nucleic acids.

DNA consists of two strands of nucleotides joined by hydrogen bonds that form between the complementary pairing of the nitrogenous bases. It superficially resembles a "twisted ladder" (figure 2.21). The "sides of the ladder" are formed of deoxyribose sugars and phosphate groups. This is commonly referred to as a sugar–phosphate backbone. The "rungs of the ladder" are composed of the paired nitrogenous bases: **adenine (A)** pairs with **thymine (T)**, and **cytosine (C)** pairs with **guanine (G)**. In contrast, RNA consists of a single strand of nucleotides. The backbone is formed of ribose sugar and phosphate, and the nitrogenous bases are adenine, uracil, cytosine, and guanine. Note that **uracil (U)** is present in RNA instead of thymine, which occurs in DNA.

Adenosine Triphosphate (ATP)

Adenosine (ah-den′-ō-sēn) **triphosphate** (trī-fos′-fāt), or **ATP**, is a modified nucleotide that consists of adenosine and three phosphate groups. The last two phosphate groups are joined to the molecule by special bonds called *high-energy phosphate bonds*. In figure 2.22b, these bonds are represented by wavy lines. Energy in these bonds is released to power chemical reactions within a cell. In this way, ATP provides immediate energy to keep cellular processes operating.

Energy extracted from nutrient molecules by cells is temporarily held in ATP and then released to power chemical reactions. When the terminal high-energy phosphate bond of ATP is broken and energy is transferred, ATP is broken down into **adenosine diphosphate (ADP)** and

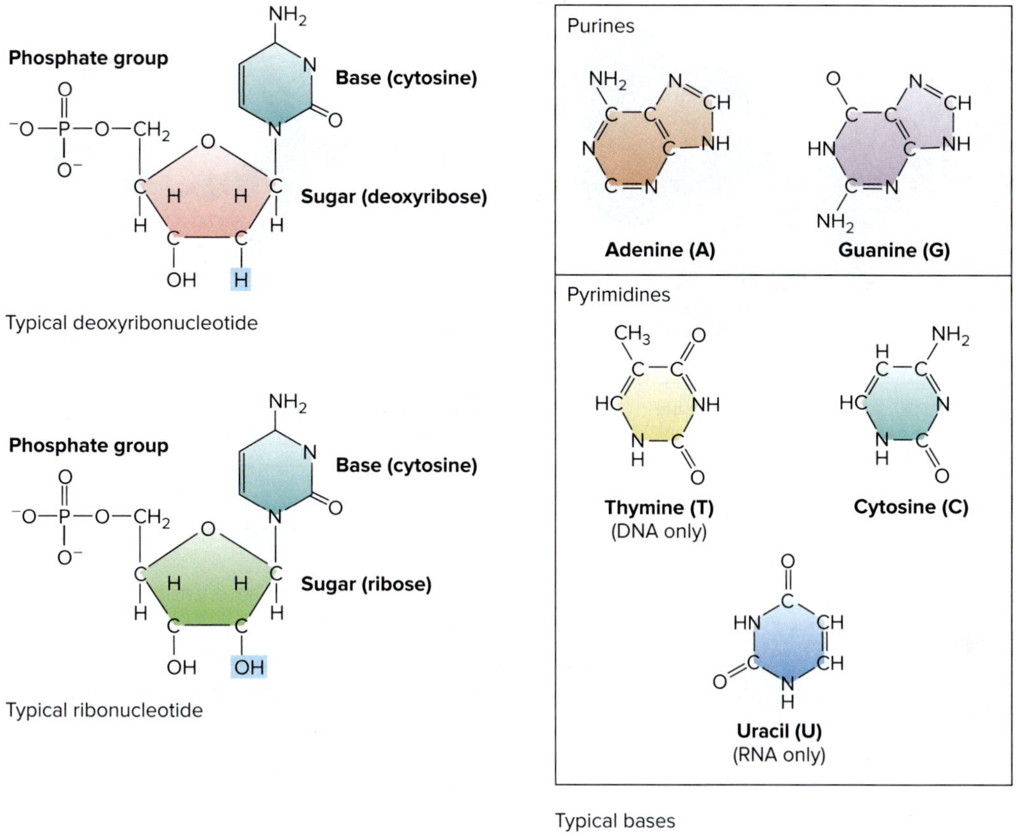

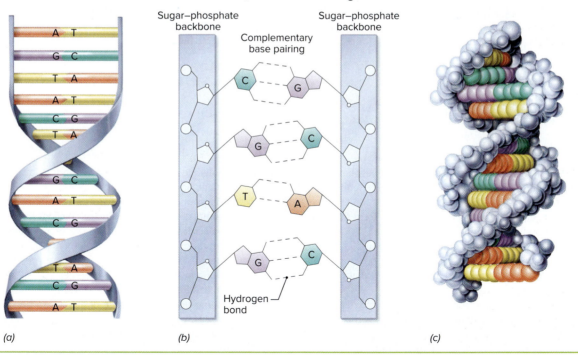

Figure 2.20 Nucleotide Structure.
Chemical structures of nucleotides in DNA and RNA, and the five nitrogenous bases.

Figure 2.21 The Structure of DNA.
(a) A model showing the double helix or "twisted ladder" structure of DNA. (b) A small segment of DNA showing sugar–phosphate molecules forming the backbone of each strand, and the strands joined together by hydrogen bonds formed between the paired nitrogenous bases. (c) A more complex molecular model. AP|R

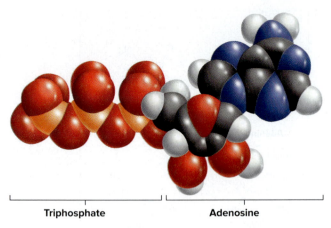

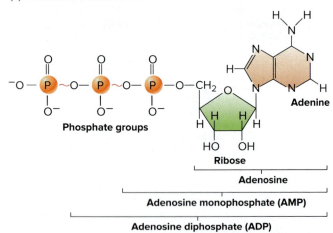

(a) A molecular model of ATP

(b) ATP structure

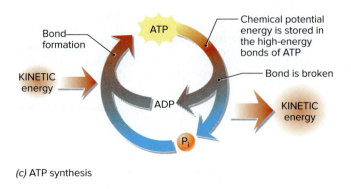

(c) ATP synthesis

Figure 2.22 Adenosine Triphosphate. The breakdown of ATP forms ADP and a phosphate group and releases energy to power cellular reactions. ATP is synthesized from ADP, phosphate, and energy extracted from nutrients.

a low-energy phosphate group (P_i). The addition of a high-energy phosphate group to ADP re-forms ATP. ATP is continuously broken down into ADP to release energy, and it is re-formed as energy is made available from nutrients (figure 2.22c).

 Check My Understanding

10. What distinguishes the chemical structure and functions of carbohydrates, lipids, proteins, and nucleic acids?
11. What is the role of enzymes in body cells?
12. What is ATP, and what is its role in body cells?

Chapter Summary

2.1 Atoms and Elements

- Matter is composed of elements, substances that cannot be broken down into simpler substances by chemical means.
- Oxygen, carbon, hydrogen, and nitrogen form 96% of the human body by weight.
- An atom is the smallest unit of an element.
- An atom consists of a nucleus formed of protons (+1) and neutrons (neutral), and electrons (−1) that orbit the nucleus.
- Electrons fill electron shells from inside to outside.
- The outermost shell containing electrons is the valence shell.
- Elements are characterized by their atomic numbers, chemical symbols, and atomic mass.
- Isotopes of an element have differing numbers of neutrons.
- Radioisotopes emit radiation.

2.2 Molecules and Compounds

- A molecule is formed of two or more atoms joined by covalent bonds.
- A compound is formed of atoms from two or more elements combined by ionic or covalent bonds.
- A molecular formula indicates the types of elements and the number of atoms of each element in a molecule.
- A structural formula adds to a molecular formula by also showing how the atoms fit together.
- Chemical bonds join atoms to form molecules.
- An ionic bond is the force of attraction between two ions with opposite electrical charges. It results from one atom donating one or more electrons to another atom.
- A covalent bond is formed between two atoms by the sharing of electrons in the valence shell.
- Nonpolar covalent bonds share electrons equally; polar covalent bonds share electrons unequally.

- Nonpolar substances are hydrophobic (water fearing); polar substances and ions are hydrophilic (water loving).
- A hydrogen bond is a weak force of attraction between a slightly positive H atom and a slightly negative atom either in the same molecule or in different molecules, or between ions and polar molecules.
- Synthesis reactions combine simpler substances to produce more complex substances.
- Decomposition reactions break down complex substances into simpler substances.
- Exchange reactions involve both decomposition of the reactants and synthesis of new products.
- Reversible reactions may occur in either direction depending on the environment.

2.3 Substances Composing the Human Body

- Inorganic substances do *not* contain both carbon and hydrogen. Organic substances contain *both* carbon and hydrogen.
- Water (H_2O) is the most abundant inorganic substance in the body, and it is the solvent of living systems.
- There are two major water compartments: intracellular fluid (65% of body water) and extracellular fluid (35% of body water).
- Electrolytes ionize (dissociate) in water, producing ions. The resulting solution can conduct electricity.
- Nonelectrolytes are substances that do not produce ions in water, and they do not conduct electricity.
- An acid releases H^+ in an aqueous solution, and a base releases OH^- or absorbs H^+ in an aqueous solution.
- pH is a measure of the relative concentrations of H^+ and OH^- in a solution.
- A buffer keeps the pH of a solution relatively constant by picking up or releasing H^+.
- A salt releases positively and negatively charged ions in an aqueous solution, but they are neither H^+ nor OH^-.
- Organic molecules are synthesized by dehydration synthesis and broken down by hydrolysis.
- Carbohydrates are composed of C, H, and O in a 1:2:1 ratio.
- Monosaccharides are the building units of carbohydrates. Disaccharides consist of two monosaccharides. Polysaccharides are formed of many monosaccharides.
- Lipids are a diverse group of organic macromolecules that include triglycerides, phospholipids, and steroids.
- Triglycerides (fats) consist of three fatty acids bonded to glycerol. Unsaturated fats differ from saturated fats by having one or more double carbon–carbon bonds in their fatty acids. Excess nutrients are stored as fats.
- Phospholipids consist of two fatty acids and a phosphate group bonded to glycerol.
- Steroids are an important group of lipids that includes sex hormones and cholesterol.
- Proteins are macromolecules formed of many amino acids. The 20 different kinds of amino acids are distinguished by their R groups.
- Amino acids are joined by peptide bonds.
- Structural proteins form parts of cells and tissues. Functional proteins include enzymes, transporters, and antibodies.
- Enzymes catalyze chemical reactions.
- Nucleic acids are very large macromolecules formed of many nucleotides.
- A nucleotide consists of a five-carbon sugar (ribose in RNA and deoxyribose in DNA), a phosphate group, and a nitrogenous base.
- There are two types of nucleic acids: deoxyribonucleic acid (DNA) and ribonucleic acid (RNA). DNA determines hereditary traits and controls cellular functions. RNA works with DNA in the synthesis of proteins.
- Adenosine triphosphate (ATP) is a modified nucleotide that temporarily holds energy in high-energy phosphate bonds and releases that energy to power chemical reactions in a cell.

Improve Your Grade

Connect Interactive Questions Reinforce your knowledge using multiple types of questions: interactive, animation, classification, labeling, sequencing, composition, and traditional multiple choice and true/false.

SmartBook Proven to help students improve grades and study more efficiently, SmartBook contains the same content within the print book but actively tailors that content to the needs of the individual.

Anatomy & Physiology REVEALED® Dive into the human body by peeling back layers of cadaver imaging. Utilize this world-class cadaver dissection tool for a closer look at the body anytime, from anywhere.

CHAPTER 3

Cell

(a) ©MedicalRF.com RF; (b) ©MedicalRF.com RF; (c) Source: Dr. Cecil Fox/National Cancer Institute; (d) ©MedicalRF.com RF

As a living structure, a cell possesses numerous organelles that carry out the chemical processes necessary to maintain homeostasis. Learning the relationships between organelle structure and function can be quite challenging for students owing to the organelles' microscopic nature and the complexity of the chemical processes involved. So for a moment, imagine that you are a cell, a small living unit within the body. Every day, numerous challenges must be overcome to survive but, within you, there are structures that do just that. For example, if you need to obtain nutrients or remove wastes, you have an outer boundary that controls the movement of chemicals. If energy is needed to carry out a chemical process, you have structures that make the energy in your nutrients accessible. In order to respond to changes in your surroundings, you have the ability to communicate with other cells by detecting or producing chemical and electrical signals. By making cellular structures and functions relate to you and your real-world experiences, you will create a way to better "visualize" and understand the microscopic world within your body.

CHAPTER OUTLINE

3.1 The Human Cell

3.2 Cell Structure
- The Plasma Membrane
- Cytoplasm
- Organelles

3.3 Transport Across Plasma Membranes
- Passive Transport
- Active Transport

3.4 Cellular Respiration

3.5 Protein Synthesis
- The Role of DNA
- The Role of RNA
- Transcription and Translation

3.6 Cell Division
- Mitosis

Chapter Summary

Module 2
Cells & Chemistry

SELECTED KEY TERMS

Active transport Movement of substances across a plasma membrane, requiring the expenditure of energy by the cell.
Cell (cella = room) The simplest structural and functional living unit of an organism.
Cellular respiration Breakdown of organic nutrients in cells, to release energy and form ATP.
Centrioles (centr = center) Paired cylindrical organelles that form the mitotic spindle during cell division.
Chromosome (chrom = color; soma = body) A threadlike or rodlike structure in the nucleus consisting of DNA and protein.
Cytoplasm (cyt = cell; plasma = molded) The semifluid material located between the nucleus and the plasma membrane.
Cytosol (sol = soluble) The gellike fluid of the cytoplasm.
Diffusion The net movement of a substance from an area of higher concentration to an area of lower concentration.
Endocytosis (end = inside; cyt = cell; sis = condition) The process by which the plasma membrane engulfs, or internalizes, solid particles and droplets of liquid.
Exocytosis (exo = outside) The process by which a cell releases substances by fusion of a secretory vesicle with the plasma membrane.
Mitosis (mit = thread; sis = condition) Process by which a cell divides to form two new daughter cells with the same number and composition of chromosomes as the parent cell.
Nucleus (nucle = kernel) Spherical organelle containing chromosomes that control cellular functions.
Organelle (elle = little) A specialized structure with specific function(s) within a cell.
Osmosis The passive movement of water across a selectively permeable membrane.
Passive transport Movement of substances across a plasma membrane without expenditure of energy by the cell.
Plasma membrane Outer boundary of a cell.
Selectively permeable membrane A membrane that allows only certain substances to enter or exit the cell.

3.1 The Human Cell

Learning Objective

1. Describe human cells.

The human body is composed of about 75 trillion **cells**, the simplest structural and functional living units of an organism. Body cells can be classified into about 300 types, such as neurons, epithelial cells, muscle cells, and red blood cells. Each type of cell has a unique structure for performing specific functions. Although these cells vary in size, shape, and function, they exhibit many structural and functional similarities.

Human cells are very small and are visible only with a microscope. Knowledge of cell structure is based largely on the examination of cells with an electron microscope, a type of microscope that provides magnifications up to 200,000× or more.

3.2 Cell Structure

Learning Objective

2. Describe the structure and function(s) of each part of a generalized cell.

Although human cells are small, they are amazingly complex, with many specialized parts. The composite cell in figure 3.1 illustrates the major structures known to compose human cells. These structures are shown as they appear in electron microscope images. The three common parts found in all cells are the plasma membrane, cytoplasm, and nucleus. The other structures may or may not be present, depending on cell type. As each part of a cell is discussed, note its structure and relationship to other structures in figure 3.1.

The Plasma Membrane

The **plasma membrane** forms the outer boundary of a cell. It maintains the integrity of the cell and separates the intracellular fluid from the extracellular fluid surrounding the cell. The plasma membrane consists of two layers of phospholipids, aligned with their nonpolar (hydrophobic) ends forming the middle of the membrane and their polar (hydrophilic) ends on the surfaces facing the extracellular and intracellular fluids (figure 3.2). This orientation ensures that the plasma membrane has the ability to interact with its watery surroundings. Cholesterol molecules are scattered among the phospholipids, where they serve to increase the stability of the plasma membrane. The nonpolar region of the plasma membrane allows lipid-soluble substances to pass across the membrane but prevents the passage of water-soluble substances. Thus, the plasma membrane serves as a barrier between water-soluble substances in the intracellular and extracellular fluids.

Many different types of protein molecules are embedded in the plasma membrane, and each type has specific functions. **Channel proteins** are tunnel-shaped membrane proteins. They create pores or openings that allow for specific substances, such as water and water-soluble substances, to move across the plasma membrane. **Carrier proteins** physically bind to and transport specific

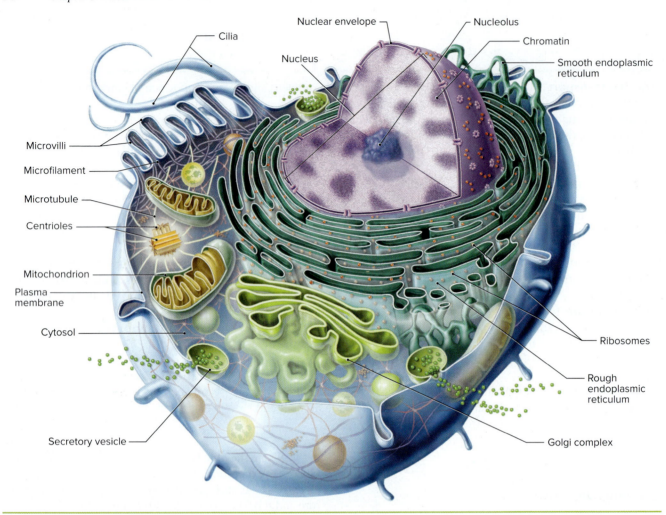

Figure 3.1 Cell Structure.
A composite human cell showing the major organelles. No cell contains all of the organelles shown. AP|R

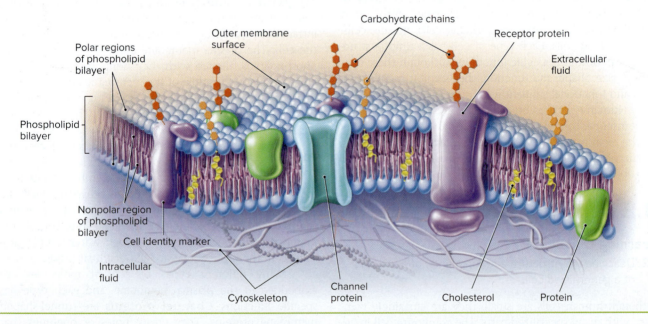

Figure 3.2 Structure of the Plasma Membrane.
The plasma membrane is composed of two layers of phospholipid molecules with scattered embedded protein and cholesterol molecules. The hydrophilic heads of the phospholipids face the extracellular and intracellular fluids, and the hydrophobic tails form the middle of the membrane. AP|R

water-soluble substances across the plasma membrane. **Receptor proteins** bind substances, such as hormones, that influence the function of a cell. Other proteins are enzymes that catalyze metabolic reactions. **Cell identity markers** are proteins, in combination with carbohydrate molecules, serving as identification markers that allow cells to recognize each other. These cell identity markers allow the lymphoid system to recognize "self" cells from "nonself" (foreign) cells, a distinction essential in fighting pathogens.

All materials that enter or exit a cell must pass across the plasma membrane. The plasma membrane is a **selectively permeable membrane** because it allows only certain substances to enter or exit the cell. Whether or not a substance can pass across the membrane is determined by a number of factors that include the substance's size, lipid solubility (electrical charge), and attachment to carrier proteins (discussed later in the chapter).

Cytoplasm

The interior of a cell between the plasma membrane and the nucleus is filled with a semifluid material called **cytoplasm** (sī′-tō-plasm). It is composed of a gellike fluid called **cytosol,** which is 75% to 90% water and contains organic and inorganic substances, and small subcellular structures known as organelles. The cytosol is the site of numerous chemical reactions, including those involved in anaerobic respiration.

Organelles

A variety of **organelles** (or-gah-nel′z), or "mini-organs," are specialized structures with specific functions within a cell. Organelles are distinguished by size, shape, structure, and specific function. Table 3.1 summarizes the structure and functions of the major parts of a cell.

Table 3.1 Summary of Major Parts of a Cell

Component	Structure	Function
Plasma membrane	Phospholipid bilayer with proteins and cholesterol molecules embedded in it	Maintains the integrity of the cell; separates intracellular fluid from extracellular fluid; binds substances like hormones; allows for cell recognition; selectively regulates the movement of materials in and out of the cell.
Cytosol	Gellike fluid surrounding organelles	Site of numerous chemical reactions including those involved in anaerobic respiration
Organelles		
Nucleus	Largest organelle; contains chromosomes and nucleoli	Controls cellular functions
Endoplasmic reticulum (ER)	System of membranes extending through the cytoplasm; RER has ribosomes on the membrane; SER does not	Support for cytoplasm; channels for material transport within cell; site of protein and lipid synthesis
Ribosomes	Tiny granules of rRNA and protein either associated with RER or free in cytoplasm	Sites of protein synthesis
Golgi complex	Series of stacked membranes near the nucleus and ER	Sorts and packages substances in vesicles for export from cell or use within cell; forms lysosomes
Mitochondria	Contain a folded inner membrane within an outer membrane	Sites of aerobic respiration that form ATP from breakdown of nutrients
Lysosomes	Small vesicles containing strong digestive enzymes	Digest foreign substances and worn-out cells and parts of cells
Microfilaments	Thin rods of protein dispersed in cytoplasm	Provide support for cell; play a role in cell movement and cell division
Microtubules	Thin tubules dispersed in cytoplasm	Provide support for cell; move organelles; form cilia and flagella; form spindle fibers during cell division
Microvilli	Numerous, tiny extensions of the plasma membrane on certain cells	Increase the surface area, which aids absorption of substances
Centrioles	Two short cylinders formed of microtubules; located near nucleus	Form mitotic spindle during cell division; form microtubules within cilia and flagella
Cilia	Numerous short, hairlike projections from certain cells	Move materials along the free surface of cells
Flagella	Long, whiplike projections from sperm	Enable movement of sperm
Vesicles	Tiny membranous sacs containing substances	Transport or store substances

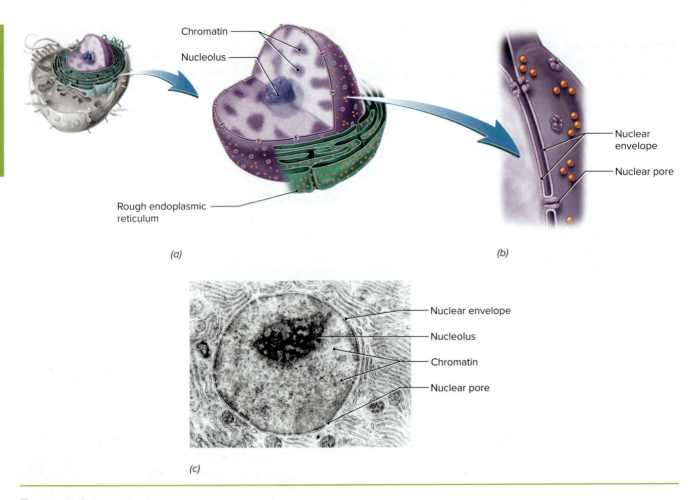

Figure 3.3 The Nucleus.
(a) The nuclear envelope is selectively permeable and allows certain substances to pass. *(b)* Details of the nuclear envelope. *(c)* A transmission electron photomicrograph of a cell nucleus (8,000×). APR
(c) ©Don W. Fawcett/Science Source

Nucleus

The largest organelle is the **nucleus** (nū′klē-us), a spherical or egg-shaped structure that is slightly more dense than the surrounding cytoplasm. It is separated from the cytoplasm by a double-layered **nuclear envelope** containing numerous *nuclear pores* that allow the movement of materials between the nucleus and cytoplasm.

Chromosomes (krō′-mō-sōms), the most important structures within the nucleus, are threadlike or rodlike structures consisting of DNA and proteins. The DNA of chromosomes contains coded instructions, called *genes*, that determine the functions of the cell (see chapter 18 for the details). When a cell is not dividing, chromosomes are extended to form thin threads called *chromatin* (krō′-mah-tin) when viewed microscopically, as in figure 3.3. During cell division, the chromosomes coil, shorten, and become rod-shaped (see figure 3.19). Each human body cell contains 23 pairs of chromosomes, with a total of 46.

A **nucleolus** (nū-klē-ō-lus) is a dense spherical body within the nucleus. It consists of RNA and protein and is the site of ribosome production. A nucleus may possess multiple nucleoli (nū-klē-ō-lē).

Ribosomes

Ribosomes are tiny organelles that appear as granules within the cytoplasm even in electron photomicrographs. They are composed of ribosomal RNA (rRNA) and proteins, which are preformed in a nucleolus before migrating from the nucleus into the cytoplasm. Ribosomes are the sites of protein synthesis in cells. They may occur singly or in small clusters and are located either on the endoplasmic reticulum (figure 3.4) or as free ribosomes in the cytoplasm.

Endoplasmic Reticulum

The numerous membranes that extend from the nucleus throughout the cytoplasm are collectively called the **endoplasmic reticulum** (en″-dō-plas′-mik rē-tik′-ū-lum), or **ER** for short. These membranes provide some support for the cytoplasm and form a network of channels that facilitate the movement of materials within the cell.

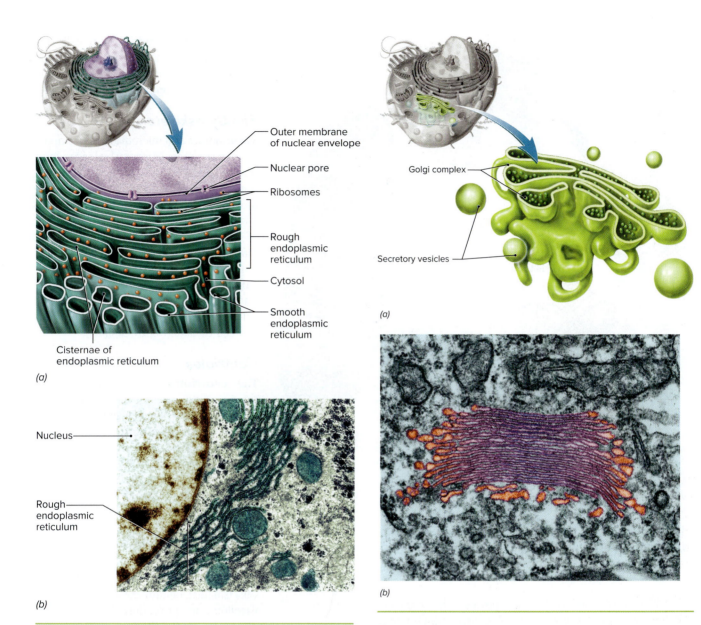

Figure 3.4 The Rough Endoplasmic Reticulum. *(a)* Rough ER is dotted with ribosomes, and smooth ER lacks ribosomes. *(b)* Transmission electron photomicrograph of rough endoplasmic reticulum (100,000×). APR
(b) ©Biomedical Imaging Unit, Southampton General Hospital/SPL/Science Source

There are two types of ER: rough ER and smooth ER. *Rough endoplasmic reticulum (RER)* is characterized by the presence of numerous ribosomes on the outer surface of the membranes, meaning it is a site of protein synthesis. *Smooth endoplasmic reticulum (SER)* lacks ribosomes and serves as a site for the synthesis of lipids (see figure 3.4).

Golgi Complex

This organelle appears as a stack of flattened membranous sacs that are usually located near the nucleus and ER. The **Golgi** (Gol′-jē) **complex** processes and sorts synthesized

Figure 3.5 The Golgi Complex.
(a) The Golgi complex packages substances in vesicles that move within the cell or to the plasma membrane to release the substances outside the cell. *(b)* Transmission electron photomicrograph of Golgi complex (100,000×).
(b) ©Dennis Kunkel Microscopy/Science Source

substances, such as proteins, into vesicles. **Vesicles,** or "little bladders," are tiny membranous sacs that carry substances from place to place within a cell. *Secretory vesicles* transport substances to the plasma membrane and release them outside the cell (figure 3.5).

Mitochondria

The **mitochondria** (mī″-to-kon′-drē-ah, singular, *mitochondrion*) are relatively large organelles composed of outer and inner membranes. The folds of the inner membrane are called *mitochondrial cristae* (singular, crista).

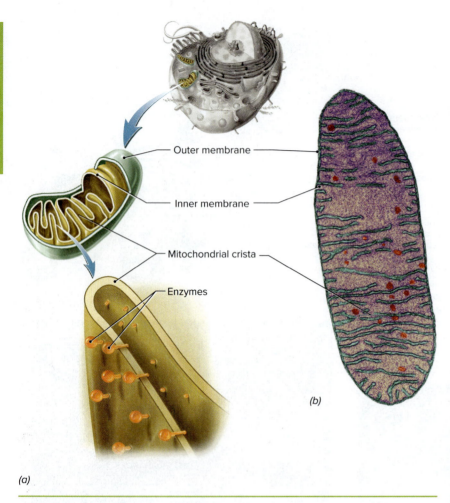

Figure 3.6 The Mitochondrion.
(a) A mitochondrion and its inner membrane. *(b)* Transmission electron photomicrograph of a mitochondrion (40,000×). AP|R
(b) ©Keith R. Porter/Science Source

The mitochondrial cristae possess the enzymes involved in *aerobic respiration*. Aerobic respiration involves the use of oxygen to release energy from nutrients and form ATP. For this reason, mitochondria are sometimes called the "powerhouses" of the cell. Mitochondria can replicate themselves if the need for additional ATP production increases (figure 3.6).

In addition to the nucleus, mitochondria contain a small amount of DNA, known as mitochondrial DNA. The genes carried by this DNA account for less than 0.2% of the total genes in the human body, and are responsible only for the functions of the mitochondria. Mitochondrial DNA cannot be used to establish paternity as with nuclear DNA, because only maternal mitochondrial DNA is passed on to offspring.

Lysosomes AP|R

Lysosomes (lī'-sō-sōms) are formed by the Golgi complex. They are small vesicles that contain powerful digestive enzymes (see figure 3.1). These enzymes are used to digest (1) bacteria that may have entered the cell, (2) cell parts that need replacement, and (3) entire cells that have become damaged or worn out. Thus, they play an important role in cleaning up the cellular environment (figure 3.7a).

The Cytoskeleton

Microtubules and microfilaments compose the cytoskeleton. **Microtubules** are long, thin protein tubules that provide support for the cell and are involved in the movement of organelles. The *spindle fibers* formed during cell division are composed of microtubules. Microtubules are also found within cilia and flagella, which will be discussed shortly. The thinner **microfilaments** are tiny rods of contractile protein that not only support the cell but also play a role in cell movement and cell division (figure 3.7).

Centrioles

The **centrioles** (sen'-trē-olz) are two short cylindrical organelles that are located near the nucleus and are oriented at right angles to each other. Nine triplets of micro-tubules are arranged in a circular pattern to form the wall of each cylinder (see figure 3.1). Centrioles form the *mitotic spindle*, which is composed of spindle fibers, during cell division (see figure 3.19). They are also involved in the formation of microtubules within cilia and flagella.

Cilia, Flagella, and Microvilli

Cilia (singular, cilium) and flagella (singular, flagellum) are projections from cells that are capable of wavelike movement. **Cilia** (sil'-ē-ah) are numerous, short, hairlike projections from cells that, in humans, are used to move substances along the free cell surfaces in areas such as the respiratory and reproductive tracts (figure 3.8). **Flagella** (flah-jel'-ah) are long, whiplike projections from cells. In humans, only sperm possess flagella, and each sperm has a single flagellum that enables movement. Both cilia and flagella contain microtubules that originate from centrioles positioned at the base of these flexible structures.

 Check My Understanding

1. What are the distinguishing features and functions of a mitochondrion, a nucleus, the Golgi complex, and rough endoplasmic reticulum?
2. What organelles enable cell movement or movement of substances along the free surface of the cells?

Part 1 Organization of the Body   55

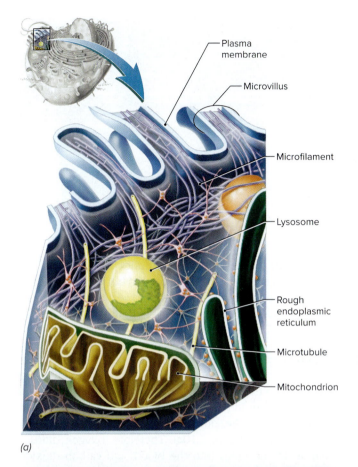

(a)

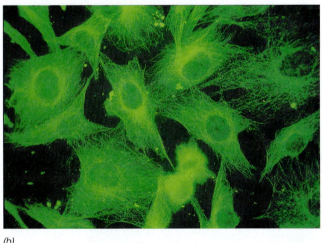

(b)

Figure 3.7 The Cytoskeleton.
(a) Microtubules and microfilaments. (b) A false-color electron photomicrograph (750×) shows the microtubules and microfilaments of the cytoskeleton in green. AP|R
(b) ©Dr. Gopal Murti/Science Source

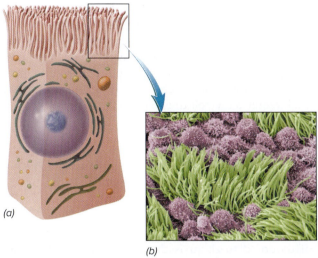

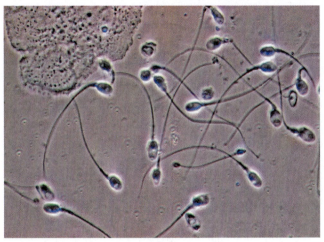

(c)

Figure 3.8 Cilia and the Flagellum.
(a) Cilia are located on the free surface of certain cells. Because these cells are stationary, beating cilia move substances along the free surfaces of the cells. (b) An electron photomicrograph of cilia (10,000×). (c) A light photomicrograph of human sperm (1,000×).
(b) ©Steve Gschmeissner/Science Source RF; (c) ©Kage-Mikrofotografie/AGE Fotostock

Microvilli (singular, *microvillus*) are extensions of the plasma membrane that are smaller and more numerous than cilia. They do not move like cilia or flagella, but they increase the surface area of the plasma membrane and, therefore, aid in the absorption of substances. Microvilli are abundant on the free surface of the cells lining the intestines (see figure 3.7a).

3.3 Transport Across Plasma Membranes

Learning Objectives
3. Compare the mechanisms of passive and active transport of substances across the plasma membrane.
4. Describe osmosis and tonicity, and the effect of tonicity on the cells.

A cell maintains its homeostasis primarily by controlling the movement of substances across the selectively permeable plasma membrane. Some substances pass across the plasma membrane by **passive transport,** which requires no expenditure of ATP by the cell. Other substances move across the plasma membrane by **active transport,** which requires the cell to expend ATP.

Passive Transport

There are three major types of passive transport: diffusion, osmosis, and filtration. Filtration is described in chapters 12 and 16.

Diffusion

Diffusion (di-fū′-zhun) is the net movement of a substance from an area of higher concentration to an area of lower concentration. Thus, the movement of a substance is along a **concentration gradient,** the difference in the concentration of a substance between two areas.

Diffusion occurs in both gases and liquids and results from the constant, random motion of a substance. Diffusion is not a living process; it occurs in both living and nonliving systems. For example, if a pellet of a water-soluble dye is placed in a beaker of water, the dye molecules will slowly diffuse from the pellet (the area of higher concentration) throughout the water (the area of lower concentration) until the dye molecules are equally distributed, meaning they are at equilibrium (figure 3.9). In a similar way, the molecules of cologne, on the skin of a student sitting in the corner of a classroom, will spread throughout the room.

Lipid-soluble molecules, such as lipids, oxygen, carbon dioxide, and lipid-soluble vitamins, are able to diffuse across a plasma membrane because they can dissolve in the nonpolar region of the plasma membrane. This type of

Figure 3.9 An Example of Diffusion.
As a drop of ink gradually dissolves in a beaker of water, the ink molecules diffuse from the region of their higher concentration to a region of their lower concentration. AP|R
©McGraw-Hill Education/Charles D. Winters, photographer

diffusion is called **simple diffusion** (figure 3.10a) because it does not require the help of membrane proteins. For example, the exchange of respiratory gases occurs by simple diffusion. Air in the lungs has a greater concentration of oxygen and a lower concentration of carbon dioxide than the blood does (figure 3.10a). Therefore, oxygen diffuses from air in the lungs into the blood, and carbon dioxide diffuses from the blood into the air in the lungs.

Water-soluble molecules, such as glucose, amino acids, water-soluble vitamins, and ions, cannot be transported by simple diffusion because they cannot dissolve in the nonpolar region of the plasma membrane. Additional methods of transport are required for these substances. In **channel-mediated diffusion,** water-soluble substances

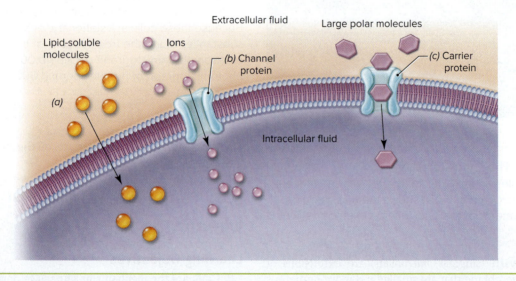

Figure 3.10 Diffusion.
(a) Simple diffusion. *(b)* Channel-mediated diffusion. *(c)* Carrier-mediated diffusion. AP|R

Clinical Insight

Dialysis involves the application of diffusion to remove small solute molecules across a selectively permeable membrane from a solution containing both small and large molecules. Dialysis is the process that is used in artificial kidney machines. As blood is passed through a chamber with a selectively permeable membrane, small waste molecules diffuse from the blood across the membrane into an aqueous solution that has a low concentration of these waste molecules. In this way, waste products in the blood are reduced to healthy levels.

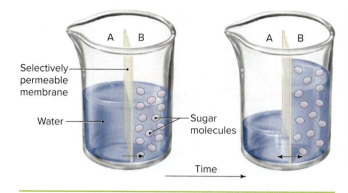

Figure 3.11 Osmosis. AP|R

like ions pass through pores in channel proteins to cross the plasma membrane (figure 3.10b). A channel protein usually allows only substances with a certain size and charge to pass through the pore. In **carrier-mediated diffusion,** carrier proteins bind and move water-soluble substances across the plasma membrane (figure 3.10c). Carrier proteins are "selective," meaning one type of carrier protein binds only one type of substance. Carrier-mediated diffusion is a type of *facilitated transport,* a transport method relying on carrier proteins to facilitate the movement of substances across the plasma membrane. Carrier-mediated active transport, another type of facilitated transport, is discussed later.

Osmosis

The passive movement of water across a selectively permeable membrane is called **osmosis** (os-mo'-sis). Water molecules move across the plasma membrane from an area of higher water concentration (lower solute concentration) into an area of lower water concentration (higher solute concentration), either by crossing the plasma membrane directly or by moving through a channel protein. Osmosis plays a very important role in the functions of the cells and the whole body. Water molecules are the main components of cells and serve as the solvent of the other chemicals. Also, the movement of water molecules into and out of the cells has the ability to significantly affect the volume of cells and the concentration of the chemicals within them.

Figure 3.11 illustrates the process of osmosis. The beaker is divided into two compartments (A and B) by a selectively permeable membrane that allows water molecules but not sugar molecules to pass across it. Because the higher concentration of water is in compartment A, water moves from compartment A into compartment B. Sugar molecules cannot pass across the membrane, so water molecules from compartment A continue to move into compartment B, causing the volume of the solution in compartment B to increase as the volume of water in compartment A decreases.

Like compartment B in figure 3.11, living cells also contain many substances to which the plasma membrane is impermeable. Therefore, any change in the concentration of water across the plasma membrane will result in a net gain or loss of water by the cell and a change in cell volume and shape.

The ability of a solution to affect the tone or shape of living cells by altering the cells' water content is called *tonicity.* A solution that has the same water concentration (same impermeable solute concentration) as the cell is an **isotonic solution.** When surrounded by this solution, a cell exhibits no net gain or loss of water and no change in volume (figure 3.12a). A solution with a higher water concentration (lower impermeable solute concentration) than the cell is called a **hypotonic solution.** A cell placed in this solution will gain water and increase in size, which may eventually lead to rupture of the cell (figure 3.12b). A solution with a lower water concentration (higher impermeable solute concentration) than the cell is known as a **hypertonic solution.** A cell placed in this solution will lose water and shrink, which may lead to cell death (figure 3.12c).

Clinical Insight

Solutions that are administered to patients intravenously usually are isotonic. Sometimes hypertonic solutions are given intravenously to patients with severe edema, or an accumulation of excess fluid in body tissues. The hypertonic solution will help to draw the excess fluid out of the body tissue and into the blood, where it can be removed by the kidneys and excreted in urine. Severely dehydrated patients may be given a hypotonic solution orally or intravenously to increase the water concentration of blood and tissue fluid, by increasing water movement from the digestive tract into the blood and from the blood into body tissues.

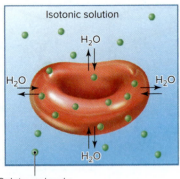
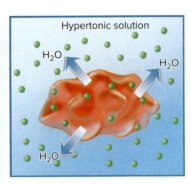

Solute molecule

(a) In an isotonic solution, there is no net gain or loss of water by the cell; the shape of the cell remains unchanged.

(b) In a hypotonic solution, there is a net gain of water by the cell, causing it to swell. Ultimately, the cell may burst.

(c) In a hypertonic solution, there is a net loss of water from the cell, causing it to shrink.

Figure 3.12 The Effect of Tonicity on Human Red Blood Cells. AP|R

Active Transport

Unlike passive transport, active transport requires the cell to expend energy (ATP) to move substances across a plasma membrane. There are three basic active transport mechanisms: carrier-mediated active transport, endocytosis, and exocytosis.

Carrier-Mediated Active Transport

In **carrier-mediated active transport,** carrier proteins use ATP to move substances across the plasma membrane against (opposite to) their concentration gradient, meaning from an area of low concentration to an area of high concentration. Figure 3.13 shows how the *sodium-potassium pump* (Na^+/K^+ pump), a type of carrier protein, establishes and maintains sodium and potassium gradients across the plasma membrane. The Na^+/K^+ pump uses the energy from an ATP molecule to move three sodium ions out of the cell and two potassium ions into the cell. The continuous action of the Na^+/K^+ pumps in the plasma membrane creates a sodium gradient with the highest concentration on the outside of the cell and a potassium gradient with the highest concentration on the inside of the cell. These sodium and potassium gradients are very important to the overall functioning of the human body.

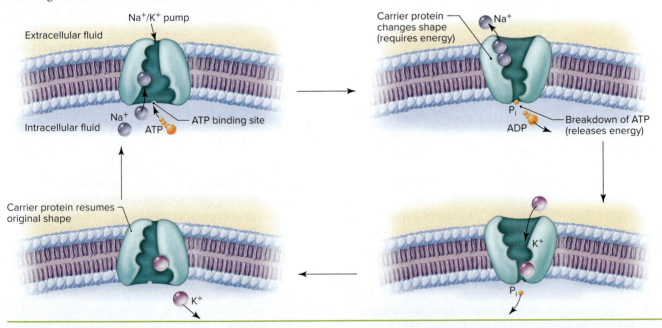

Figure 3.13 Carrier-Mediated Active Transport.
Sodium and potassium ions are moved across the plasma membrane against the concentration gradient by carrier-mediated active transport. AP|R

Endocytosis and Exocytosis AP|R

Materials that are too large to be transported by channel or carrier proteins must enter and exit a cell by totally different mechanisms. **Endocytosis** (en″-dō-sī-tō′-sis) is a process that uses the plasma membrane to engulf, or internalize, solid particles and droplets of liquid.

During endocytosis, the plasma membrane flows around the substance to be engulfed, forms an enveloping vesicle around the substance, and re-forms the plasma membrane so that the vesicle and substance are brought inside the cell (figure 3.14a). There are two types of endocytosis: pinocytosis and phagocytosis. **Pinocytosis** (pī″-nō-sī-tō′-sis) is the engulfment of small droplets of extracellular fluid. **Phagocytosis** (fag″-ō-sī-tō′-sis) is the engulfment of solid particles. Many types of cells use these processes, but phagocytosis is especially important for certain white blood cells that engulf and destroy bacteria as a defense against disease.

Exocytosis is the reverse of endocytosis, in that it is used to remove large substances from cells. A secretory vesicle containing the substance forms within the cell. It then moves to the plasma membrane, fuses with it, and empties its contents outside of the cell (figure 3.14b). The secretion, or release, of enzymes and hormones from cells involves exocytosis. Table 3.2 summarizes the types of transport across the plasma membrane.

Check My Understanding

3. By what means do substances enter and exit living cells?

3.4 Cellular Respiration

Learning Objectives

5. Describe cellular respiration and its importance.
6. Compare aerobic respiration and anaerobic respiration.

Cells require a constant supply of energy to power the chemical reactions of life. This energy is directly supplied by ATP molecules, as noted in chapter 2. Because cells have a limited supply, ATP molecules must constantly be produced by cellular respiration in order to sustain life.

Cellular respiration is the process that breaks down nutrients in the cells to release energy held in their chemical bonds and transfers some of this energy into the high-energy phosphate bonds of ATP. About 40% of the energy in a nutrient molecule is "captured" in this way; the remainder is lost as heat. Glucose, a carbohydrate molecule, is the primary nutrient used in cellular respiration; however, the building units of proteins and lipids are also used (see chapter 15 for the details).

The actual process of cellular respiration of glucose is complex, but may be simplified as the equation below. Note that the breakdown of glucose ($C_6H_{12}O_6$) requires oxygen (O_2) and yields carbon dioxide (CO_2) and water (H_2O). The energy released is used to form ATP from ADP and P_i (phosphate group). Some of the energy is released as heat.

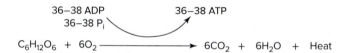

$$C_6H_{12}O_6 + 6O_2 \xrightarrow{\text{36-38 ADP, 36-38 } P_i \rightarrow \text{36-38 ATP}} 6CO_2 + 6H_2O + \text{Heat}$$

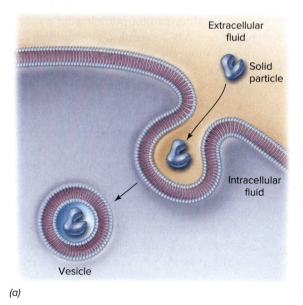
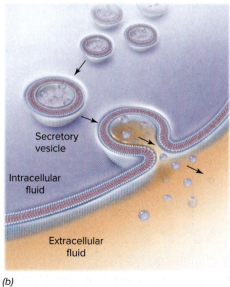

(a)   (b)

Figure 3.14 Endocytosis and Exocytosis.
(a) A particle is engulfed by the plasma membrane and brought into the cell by endocytosis. *(b)* Particles are enclosed in a secretory vesicle and are expelled from the cell by exocytosis.

Table 3.2 Types of Transport Across the Plasma Membrane

Type	Mechanism
Passive Transport	The transport that requires no expenditure of ATP by the cell
Simple diffusion	Transport of lipid-soluble substances across the plasma membrane along their concentration gradient without the help of membrane proteins
Channel-mediated diffusion	Transport of water and water-soluble substances across the plasma membrane along their concentration gradient through channel proteins
Carrier-mediated diffusion	Movement of water-soluble substances across the plasma membrane along their concentration gradient by using carrier proteins that facilitate transport
Osmosis	Movement of water across the plasma membrane toward the area with a lower water concentration (higher solute concentration), either by crossing the plasma membrane directly or by moving through a channel protein
Active Transport	The transport that requires the expenditure of ATP by the cell
Carrier-mediated active transport	Movement of small substances across the plasma membrane against their concentration gradient through the action of carrier proteins (pumps)
Exocytosis	Movement of solid particles out of the cell, by merging the secretory vesicle with the plasma membrane and emptying its contents into extracellular space
Endocytosis	Movement of solid particles and droplets of liquid into the cell, by engulfing the substances with the plasma membrane and forming a vesicle containing the transported substance in the intracellular space
Pinocytosis	The process by which cells engulf droplets of extracellular fluid
Phagocytosis	The process by which cells engulf solid particles

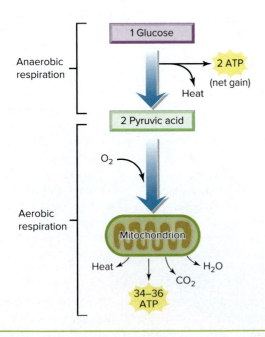

Figure 3.15 Cellular Respiration.
Cellular respiration of a glucose molecule occurs in two major steps. Anaerobic respiration occurs in the cytosol, does not require oxygen, and yields a net of 2 ATP. Aerobic respiration occurs in mitochondria, requires oxygen, and yields a net of 34–36 ATP. About 40% of the energy in the chemical bonds of glucose is captured to form ATP molecules.

Cellular respiration involves two sequential processes: anaerobic respiration and aerobic respiration. Each chemical process involves many steps, with each step requiring a special enzyme. However, the processes can be simplified as shown in figure 3.15. **Anaerobic** (an-a-rō′-bik) **respiration** (1) does not require oxygen and (2) occurs in the cytosol. It breaks down a six-carbon glucose molecule into two three-carbon pyruvic acid molecules to yield a net of two ATP molecules. The low level of ATP production by anaerobic respiration is insufficient to keep a person alive. A person deprived of oxygen or of the ability to use oxygen in aerobic respiration (as in cyanide poisoning) quickly dies because anaerobic respiration does not provide sufficient ATP to sustain life.

Aerobic respiration, the second part of cellular respiration, (1) requires oxygen, (2) occurs only within mitochondria, and (3) is essential for human life. Aerobic respiration releases the energy in the high-energy electrons produced by anaerobic respiration, breaks down the two pyruvic acid molecules produced by anaerobic respiration into carbon dioxide and water, and yields a net of 34–36 ATP molecules. Thus, the respiration of a molecule of glucose yields a net total of 36–38 ATP.

 Check My Understanding

4. What is cellular respiration and why is it important?

3.5 Protein Synthesis

Learning Objectives
7. Describe the process of protein synthesis.
8. Explain the roles of DNA and RNA in protein synthesis.

Proteins play a vital role in the body. Structural proteins compose significant portions of all cells, and functional proteins, such as enzymes and hormones, directly regulate cellular activities. Remember that a protein is a long chain of amino acids joined together by peptide bonds. Protein synthesis involves placing specific amino acids in their correct positions in the amino acid chain.

DNA and RNA are intimately involved in the synthesis of proteins.

The Role of DNA

Recall the structure of DNA described in chapter 2. The two coiled strands of nucleotides are joined by hydrogen bonds between the nitrogenous bases in each strand by *complementary base pairing.* Adenine (A) pairs with thymine (T), and cytosine (C) pairs with guanine (G).

The sequence of bases in a DNA molecule encodes information that determines the sequence of amino acids in a protein. More specifically, a sequence of three nucleotide bases (a triplet) in DNA encodes for a specific amino acid. For example, a sequence of ACA encodes for the amino acid cysteine, while AGG encodes for serine. In this way, inherited information that determines the structure of proteins is encoded in DNA.

The Role of RNA

In contrast to DNA, RNA consists of a single strand of nucleotides. Each nucleotide contains one of four nitrogenous bases: adenine, cytosine, guanine, or *uracil* (U). Note that uracil is present in RNA instead of thymine, which occurs in DNA. RNA is synthesized in a cell's nucleus by using a strand of DNA as a template. Complementary pairing of RNA bases with DNA bases produces a strand of RNA nucleotides whose bases are complementary to those in the DNA molecule. Uracil (U) in RNA pairs with adenine (A) in DNA; adenine (A) in RNA pairs with thymine (T) in DNA.

There are three types of RNA, and each plays a vital role in protein synthesis.

Messenger RNA (mRNA) carries the genetic information from DNA into the cytoplasm to the ribosomes, the sites of protein synthesis. This information is carried by the sequence of bases in mRNA, which is complementary to the sequence of bases in the DNA template.

Ribosomal RNA (rRNA) and protein compose ribosomes, the sites of protein synthesis. Ribosomes contain the enzymes required for protein synthesis.

Transfer RNA (tRNA) carries amino acids to the ribosomes, where the amino acids are joined like a string of beads to form a protein. There is a different tRNA for transporting each of the 20 kinds of amino acids used to build proteins.

Table 3.3 summarizes the characteristics of DNA and RNA.

Table 3.3 Distinguishing Characteristics of DNA and RNA

	DNA	RNA
Strands	Two strands joined by the complementary pairing of their nitrogenous bases	One strand
Sugar	Deoxyribose	Ribose
Bases	Adenine	Adenine
	Thymine	Uracil
	Cytosine	Cytosine
	Guanine	Guanine
Shape	Helix	Straight

Transcription and Translation

The process of protein synthesis involves two successive events: **transcription,** which occurs in the nucleus, and **translation,** which takes place in the cytoplasm.

In *transcription,* the sequence of bases in DNA determines the sequence of bases in mRNA due to complementary base pairing. Thus, transcription transfers the encoded information of DNA into the sequence of bases in mRNA. For example, if a triplet of DNA bases is AGG, which encodes for the amino acid serine, the complementary paired triplet of bases in mRNA is UCC. A triplet of bases in mRNA is known as a **codon,** and there is a codon for each of the 20 amino acids composing proteins. Messenger RNA consists of a chain of codons. Once it is synthesized, mRNA moves out of the nucleus into the cytoplasm where it combines with a ribosome, the site of protein synthesis. AP|R

In *translation,* the encoded information in mRNA is used to produce a specific sequence of amino acids to form the protein. As the ribosome moves along the mRNA strand, tRNA molecules bring amino acids to the ribosome and place them in the correct sequence in the forming polypeptide chain (protein) as specified by the mRNA codons. AP|R

Each tRNA molecule has a triplet of RNA bases called an **anticodon** at one end of the molecule. Because there are 20 different kinds of amino acids composing proteins, there are at least 20 kinds of tRNA whose anticodons can bind with codons of mRNA. A tRNA molecule can only transport the specific amino acid that is encoded by the codon to which its anticodon can bond. For example, a tRNA transporting the amino acid serine has the anticodon AGG that can bond with the mRNA codon UCC to place serine in the correct position in the forming amino acid chain. See figure 3.16.

By transcription and translation, DNA determines the structure of proteins, which, in turn, determines the

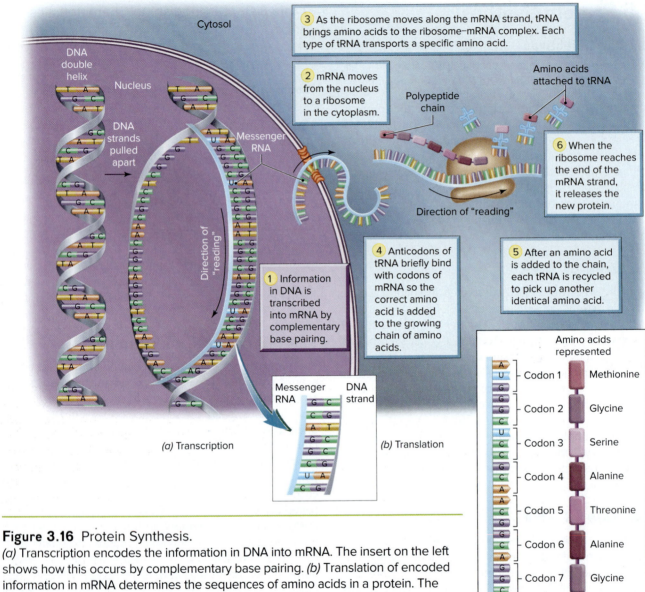

Figure 3.16 Protein Synthesis.
(a) Transcription encodes the information in DNA into mRNA. The insert on the left shows how this occurs by complementary base pairing. *(b)* Translation of encoded information in mRNA determines the sequences of amino acids in a protein. The insert on the right shows a few mRNA codons and the amino acids they encode.

functions of proteins. Transcription and translation may be summarized as follows:

DNA $\xrightarrow{\text{Transcription}}$ mRNA $\xrightarrow{\text{Translation}}$ Protein

Check My Understanding

5. How does chromosomal DNA determine the structure of proteins?

3.6 Cell Division

Learning Objectives
9. Describe the two types of cell division and their roles.
10. Describe each phase of mitosis.

Cells replicate through a process called **cell division.** Two types of cell division occur in the body: mitosis and meiosis. Somatic cells (cells other than sex cells) divide by **mitosis** (mahy-toh´-sis), during which a parent cell divides to form two new daughter cells that have the same number (46) and composition of chromosomes as the parent cell. Mitosis enables growth and the repair of tissues. **Meiosis** (mahy-oh´-sis) occurs only in the production of ova and sperm. In meiosis, a single parent cell divides to form four daughter cells that contain only half the number of chromosomes (23) found in the parent cell. In this chapter, we consider mitosis only. Meiosis is studied in chapter 17.

Mitosis

Starting with the first division of the fertilized egg, mitosis is the process that produces new cells for growth of the new individual and the replacement of worn or damaged cells. Mitosis occurs at different rates in different kinds of cells. For example, epithelial cells undergo almost continuous division, but muscle cells lose the ability to divide as they mature.

In dividing cells, the time period from the separation of daughter cells of one division to the separation of daughter cells of the next division is called the **cell cycle.** Mitosis constitutes only 5% to 10% of the cell cycle. Most of the time, a cell merely is carrying out its average life functions (figure 3.17).

Interphase is the phase that occurs when the cell is not involved in mitosis. When viewed with a microscope, a cell in interphase is identified by its intact nucleus containing chromatin. In cells that are destined to divide, both the centrioles and chromosomes replicate during interphase, while other organelles are synthesized and assembled. There is a growth period before and after replication of the 46 chromosomes.

A chromosome consists of a very long DNA molecule coated with proteins. During interphase, chromosomes are uncoiled and resemble very thin threads within the cell nucleus. Chromosomes replicate during interphase in order to provide one copy of each chromosome for each of the two daughter cells that will be formed by mitosis. Chromosome replication is dependent upon the replication of the DNA molecule in each chromosome. Figure 3.18 illustrates the process of DNA replication.

The two original DNA strands "unzip," and new nucleotides are joined in a complementary manner by their bases to the bases of the separated DNA strands. When the new nucleotides are in place and joined together, each new DNA molecule consists of one "new" strand of nucleotides joined to one "old" strand of nucleotides. In this way, a DNA molecule is precisely replicated so that both new DNA molecules are identical.

Mitotic Phases

Once it begins, mitosis is a continuous process that is arbitrarily divided into four sequential phases: prophase,

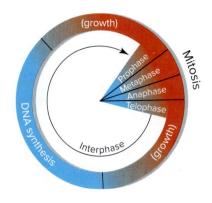

Figure 3.17 The Cell Cycle.
Interphase occupies most of the cell cycle. Only 5% to 10% of the time is used in mitosis. DNA and chromosome replication occur during interphase. AP|R

metaphase, anaphase, and telophase. Each phase is characterized by specific events.

Prophase During **prophase,** the replicated chromosomes coil, appearing first as threadlike structures and finally shortening sufficiently to become rod-shaped. Each replicated chromosome consists of two **chromatids** joined at their **centromeres.** Simultaneously, the nuclear envelope gradually disappears, and each pair of centrioles migrate toward opposite ends of the cell. A **mitotic spindle** is formed between the migrating centrioles. The mitotic spindle consists of spindle fibers that are formed of microtubules (figure 3.19a).

Metaphase During the brief **metaphase,** the replicated chromosomes line up at the equator of the mitotic spindle. The centromeres are attached to spindle fibers (figure 3.19b).

Anaphase During **anaphase,** separation of the centromeres results in the separation of the paired chromatids.

 ## Clinical Insight

Mitosis is usually a controlled process that ceases when it is not necessary to produce additional cells. Occasionally, control is lost and cells undergo continuous division, which leads to the formation of tumors. Tumors may be benign or malignant. *Benign tumors* do not spread to other parts of the body and may be surgically removed if they cause health or cosmetic problems. *Malignant tumors,* or *cancers,* may spread to other parts of the body by a process called *metastasis* (me-tas'-ta-sis). Cells break away from the primary tumor and are often carried by blood or lymph to other areas, where continued cell divisions form secondary tumors.

Treatment of malignant tumors involves surgical removal of the tumor, if possible, and subsequent chemotherapy and/or radiation therapy. Both chemotherapy and radiation therapy tend to kill malignant cells because dividing cells are more sensitive to treatment, and malignant cells are constantly dividing.

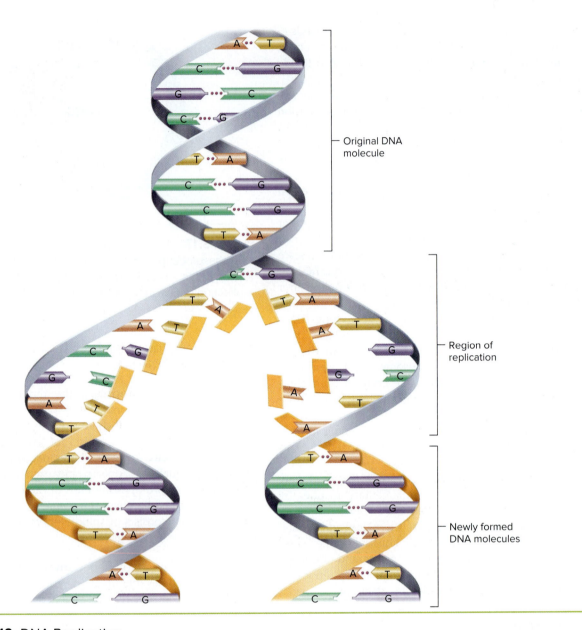

Figure 3.18 DNA Replication.
When a DNA molecule replicates, the two strands unzip. Then, a new complementary strand of nucleotides forms along each "old" strand to produce two new DNA molecules. AP|R

The members of each pair are pulled by spindle fibers toward opposite sides of the cell. The separated chromatids are now called chromosomes, and each new set of chromosomes is identical (figure 3.19c).

Telophase During **telophase,** the spindle fibers disassemble and a new nuclear envelope starts forming around each set of chromosomes as the new nuclei begin to take shape. The chromosomes start to uncoil, and they will ultimately become visible only as chromatin. The new daughter nuclei are completely formed by the end of telophase.

Usually during late anaphase and telophase, the most obvious change is the division of the cytoplasm, which is called **cytokinesis** (si″-toh-ki-nē′-sis). It is characterized by a furrow that forms in the plasma membrane across the equator of the mitotic spindle and deepens until the parent cell is separated into two daughter cells. The formation of two daughter cells, each having identical chromosomes in the nuclei, marks the end of mitosis (figure 3.19d).

 Check My Understanding

6. What are the phases of mitosis, and how is each phase distinguished?

Part 1 Organization of the Body   65

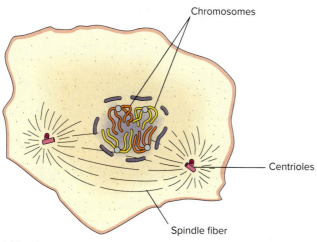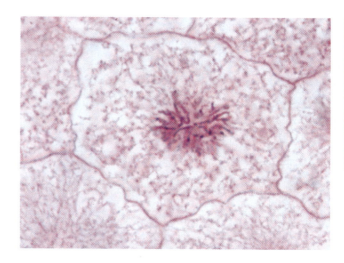

(a) Prophase
Replicated chromosomes coil and shorten; the nuclear envelope disappears; centriole pairs move toward opposite sides of the cell, forming a mitotic spindle between them.

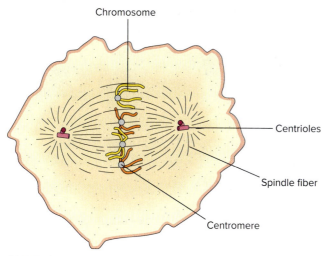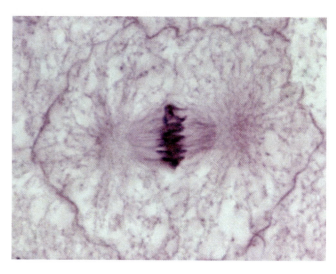

(b) Metaphase
Chromosomes line up at the equator of the mitotic spindle. Each replicated chromosome consists of two chromatids joined at their centromeres.

Figure 3.19 Drawings and Photomicrographs (1,000×) of Mitosis.
(a) ©Ed Reschke; (b) ©Ed Reschke/Getty Images; (c) ©Ed Reschke/Getty Images; (d) ©Ed Reschke

Chapter Summary

3.1 The Human Cell

- The human body is composed of about 300 types of cells.
- Each type of cell has a unique structure for performing specific functions.

3.2 Cell Structure

- The plasma membrane is composed of a double layer of phospholipid molecules along with associated cholesterol and protein molecules. It is selectively permeable and controls the movement of the materials into and out of cells.
- The cytoplasm, which is composed of cytosol and organelles, lies outside the nucleus and is enveloped by the plasma membrane.
- The nucleus is a large, spherical organelle surrounded by the nuclear envelope.
- Chromosomes, composed of DNA and protein, are found in the nucleus. The uncoiled chromosomes appear as chromatin in nondividing cells.
- The nucleolus is the site of ribosome synthesis.

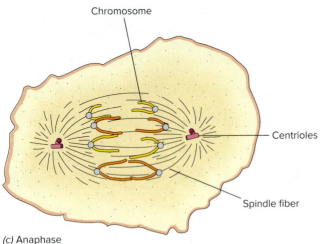
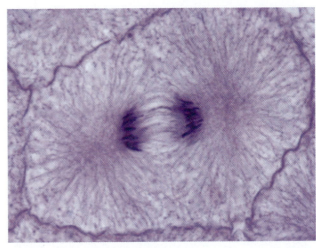

(c) Anaphase
Chromatids separate; members of each chromatid pair are pulled by spindle fibers toward opposite sides of the cell.

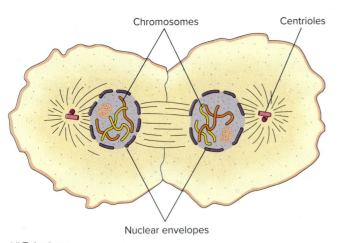
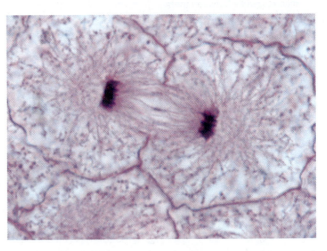

(d) Telophase
A new nuclear envelope forms around each set of chromosomes; spindle fibers disappear; chromosomes uncoil and extend; cytokinesis produces two daughter cells.

- Ribosomes are tiny organelles formed of rRNA and protein. They are sites of protein synthesis.
- The endoplasmic reticulum (ER) consists of membranes that form channels for transport of materials within the cell. RER is studded with ribosomes that synthesize proteins for export from the cell. SER lacks ribosomes and is involved in lipid synthesis.
- The Golgi complex packages materials into vesicles for secretion from the cell or transport within the cell.
- Mitochondria are large, double-membraned organelles within which aerobic respiration occurs.
- Lysosomes are small vesicles that contain digestive enzymes used to digest foreign particles, worn-out parts of a cell, or an entire damaged cell.
- The cytoskeleton is formed by microtubules and microfilaments, and is used in maintaining cell structure and cell movement.
- A pair of centrioles, used in cell division, is present near the cell's nucleus. The wall of each centriole is composed of microtubules arranged in groups of three.
- Cilia are short, hairlike projections on the free surface of certain cells. The beating of cilia moves materials along the cell surface.
- Each sperm swims by the beating of a flagellum, a long, whiplike projection.

3.3 Transport Across Plasma Membranes

- Passive transport does not require the expenditure of energy by the cell.
- Diffusion is the movement of a substance from an area of higher concentration to an area of lower concentration. It is caused by the constant motion of substances in gases and liquids.

- Substances diffuse across the plasma membrane by simple diffusion, channel-mediated diffusion, and carrier-mediated diffusion.
- Osmosis is the passive movement of water across a selectively permeable membrane.
- Hypotonic solutions have a higher water concentration than the cells. Hypertonic solutions have a lower water concentration than the cells. Isotonic solutions have the same water concentration as the cells.
- Cells in hypotonic solutions have a net gain of water. Cells in hypertonic solutions have a net loss of water. Cells in isotonic solutions have no net change of water content.
- Active transport requires the cell to expend energy.
- Active transport mechanisms include carrier-mediated active transport, endocytosis, and exocytosis.

3.4 Cellular Respiration

- Cellular respiration is the breakdown of nutrients in cells to release energy and form ATP molecules, which power cellular processes.
- Cellular respiration of glucose involves anaerobic respiration and aerobic respiration.
- Cellular respiration of a glucose molecule yields a net of 36–38 ATP. A net of 2 ATP is produced during anaerobic respiration, which occurs in the cytosol. A net of 34–36 ATP is produced during aerobic respiration, which occurs in mitochondria.

3.5 Protein Synthesis

- Protein synthesis involves the interaction of DNA, mRNA, rRNA, and tRNA.
- The sequence of bases in DNA determines the sequence of codons in mRNA, which, in turn, determines the sequence of amino acids in a protein.

$$\text{DNA} \xrightarrow{\text{Transcription}} \text{mRNA} \xrightarrow{\text{Translation}} \text{Protein}$$

3.6 Cell Division

- Mitosis produces two daughter cells that have the same number and composition of chromosomes. It enables growth and tissue repair.
- Meiosis results in the production of ova and sperm. Four daughter cells are formed that have half the number of chromosomes as the parent cell.
- Most of a cell cycle is spent in interphase, where cells carry out average life functions. In cells destined to divide, chromosomes and centrioles are replicated in interphase.
- After chromosome replication, mitosis is the orderly process of separating and distributing chromosomes equally to the daughter cells.
- Mitosis consists of four phases: prophase, metaphase, anaphase, and telophase.

Improve Your Grade

Connect Interactive Questions Reinforce your knowledge using multiple types of questions: interactive, animation, classification, labeling, sequencing, composition, and traditional multiple choice and true/false.

SmartBook Proven to help students improve grades and study more efficiently, SmartBook contains the same content within the print book but actively tailors that content to the needs of the individual.

Anatomy & Physiology REVEALED® Dive into the human body by peeling back layers of cadaver imaging. Utilize this world-class cadaver dissection tool for a closer look at the body anytime, from anywhere.

CHAPTER

4

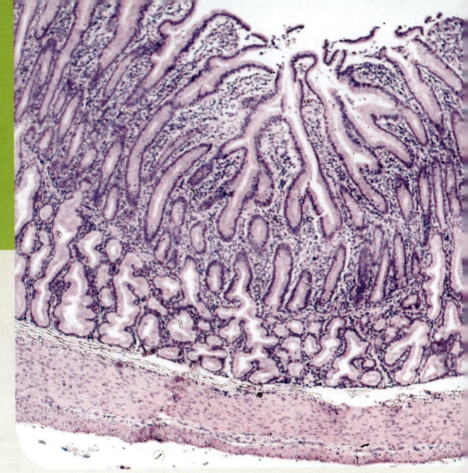

©Victor P. Eroschenko RF

Tissues and Membranes

Your body's ability to maintain homeostasis depends upon on the normal structure and function of body tissues. Consider your ability to move your hand off an environmental hazard, such as a hot surface. The bones of the body are physically hardened due to mineralized bone. Attached to these bones are muscles containing skeletal muscle tissue, which has the ability to contract and create force. When the muscles of the arm contract with force, they pull on the bones in the forearm to create movement at the elbow. As a result, the hand is moved away from the hazard. Nerve tissue detects and processes the pain stimuli from the hand when it contacts the hazard. It then acts to control and coordinate the contraction of the skeletal muscle tissue in response. As you can see, many types of body tissues are involved in all of the body's physiological processes. Gaining an understanding of your body's various tissues and their capabilities will facilitate a better understanding of how your organ systems work to maintain homeostasis within the body in upcoming chapters.

CHAPTER OUTLINE

4.1 Introduction to Tissues

4.2 Epithelial Tissues
- Simple Epithelium
- Stratified Epithelium

4.3 Connective Tissues
- Loose Connective Tissue
- Dense Connective Tissue
- Cartilage Tissue
- Bone Tissue
- Blood

4.4 Muscle Tissues
- Skeletal Muscle Tissue
- Cardiac Muscle Tissue
- Smooth Muscle Tissue

4.5 Nerve Tissue

4.6 Body Membranes
- Epithelial Membranes
- Connective Tissue Membranes

Chapter Summary

Module 3
Tissues

SELECTED KEY TERMS

Adipose tissue (adip = fat) A connective tissue that stores fat.
Bone tissue A hard connective tissue with a rigid matrix of calcium salts and fibers.
Cartilage tissue A connective tissue with a relatively rigid, semisolid matrix.
Connective tissue (connect = to join) A tissue that binds other tissues together.
Epithelial tissue (epi = upon, over; thel = delicate) A thin tissue that covers body and organ surfaces and lines body cavities, and forms secretory portions of glands; epithelium.
Fibroblast (fibro = fiber; blast = germ) A cell that produces fibers and ground substance in connective tissue.
Matrix The extracellular substance in connective tissue.
Mucous membrane Epithelial membrane that lines tubes or cavities of organ systems, with openings to the external environment.
Muscle tissue (mus = mouse) A tissue whose cells are specialized for contraction.
Nerve tissue A tissue that forms the brain, spinal cord, and nerves.
Serous membrane Epithelial membrane that lines the thoracic and abdominopelvic cavities and covers most of the internal organs within these cavities.
Tissue (tissu = woven) A group of similar cells performing similar functions.

4.1 Introduction to Tissues

Learning Objectives
1. Compare embryonic stem cells and adult stem cells.
2. Describe the four basic types of tissue.

The different kinds of cells composing the human body result from the specialization of cells during embryonic development. **Embryonic stem cells** of an early embryo are unspecialized cells containing encoded information in their DNA that enables them to form all types of specialized cells. As these cells divide repeatedly, producing many generations of cells, the daughter cells become partially specialized. Such cells can produce daughter cells for only certain related types of specialized cells. This trend of decreasing potential (increasing specialization) continues through many generations of cells, ultimately producing the highly specialized cells of the human body plus a few partially specialized cells known as **adult stem cells**. Once fully specialized, cells may or may not divide. If they do, they can form only specialized cells like themselves; for example, skin cells divide to produce only skin cells. Because all of a person's cells (except red blood cells, which lack nuclei) contain the same DNA, the transition from unspecialized embryonic stem cells to fully specialized body cells results from cellular mechanisms that turn off specific portions of the encoded information in DNA.

A **tissue** is a group of fully specialized cells that perform similar functions. Most tissues contain a few adult stem cells, which play an important role in tissue repair. Each type of tissue is distinguished by the structure of its cells, its extracellular substance, and the function it performs. The structure of a tissue is a reflection of its function.

The different tissues of the body are classified into four basic types: epithelial, connective, muscle, and nerve tissues.

1. **Epithelial** (ep″-i-thē′-lē-al) **tissue** covers the surface of the body, lines body cavities and covers organs, and forms the secretory portions of glands.
2. **Connective tissue** binds organs together and provides protection and support for organs and the entire body.
3. **Muscle tissue** contracts to provide force for the movement of the whole body and many internal organs.
4. **Nerve tissue** detects changes, processes information, and coordinates body functions via the transmission of nerve impulses.

 Clinical Insight

Adult stem cells from a variety of tissues are used in medical therapies. For example, stem cells in red bone marrow are used to treat leukemia. Medical scientists think that, with more research, stem cells may be used to treat cancer, brain and spinal cord injuries, multiple sclerosis, Parkinson disease, and other injuries and disorders.

4.2 Epithelial Tissues AP|R

Learning Objectives
3. Describe the distinguishing characteristics of epithelial tissues.
4. Identify the common locations and general functions of each type of epithelial tissue.

Epithelial tissues, or epithelia (singular, epithelium), may be composed of one or more layers of cells. The number of cell layers and the shape of the cells provide the basis for

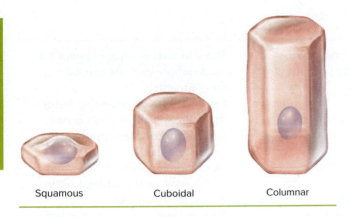

Squamous   Cuboidal   Columnar

Figure 4.1 Shapes of Epithelial Cells.

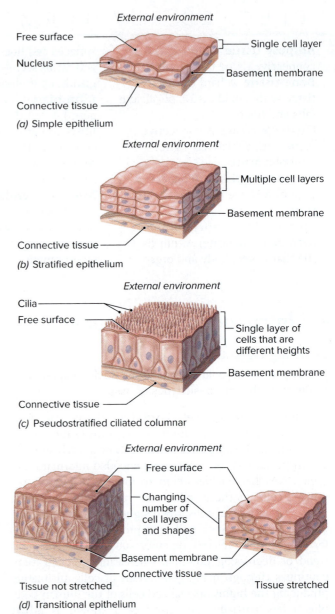

Figure 4.2 Epithelial Classifications Based on the Number of Cell Layers.

classifying epithelial tissues (figures 4.1 and 4.2). Epithelial tissues are distinguished by the following five characteristics:

1. Epithelial cells are packed closely together with very little extracellular material between them.
2. The sheetlike tissue is firmly attached to the underlying connective tissue by a thin layer of proteins and carbohydrates called the **basement membrane.**
3. The surface of the tissue (free surface) opposite the basement membrane is not attached to any other type of tissue and is located on a surface or next to an opening. An opening within a hollow internal organ is called a **lumen.**
4. Blood vessels are absent, so epithelial cells must rely on diffusion to receive nourishment from blood vessels in the underlying connective tissue. Because these tissues are on surfaces, they are prone to damage. The lack of blood vessels prevents unnecessary bleeding.
5. Epithelial cells undergo rapid mitosis to regenerate the tissues. Large numbers of epithelial cells are destroyed and replaced each day.

The functions of epithelial tissues vary with the specific location and type of tissue, but generally they include *protection, diffusion, osmosis, absorption, filtration, secretion,* and *friction reduction.* Certain epithelial cells form *glandular epithelium,* the cells in glands that produce secretions. Two basic types of glands are contained in the body: exocrine and endocrine glands. **Exocrine glands** (exo = outside of; crin = to secrete) have ducts (small tubes) that carry their secretions to specific areas; sweat glands and salivary glands are examples. **Endocrine glands** (endo = within) lack ducts. Their secretions, called hormones, are carried by the blood supply to organs within the body to regulate their function. The thyroid gland and adrenal glands are examples of endocrine glands. Endocrine glands and their hormones will be discussed in more detail in chapter 10.

Simple Epithelium

Simple epithelium consists of a single layer of cells that may be flat (squamous), cubelike (cuboidal), and columnlike (columnar) in shape (see figures 4.1 and 4.2).

These tissues are located where rapid diffusion, secretion, absorption, or filtration occurs in the body.

Simple squamous (skwa⁻′-mus) *epithelium* consists of thin, flat cells that have an irregular outline and a flat, centrally located nucleus. In a surface view, the cells somewhat resemble tiles arranged in a mosaic pattern. Simple squamous epithelium performs a diverse set of functions that include diffusion, osmosis, filtration, secretion, absorption, and friction reduction. Its locations in the body include (1) the air sacs in the lungs, where O_2 and CO_2 diffuse into and out of the blood, respectively; (2) special structures in the kidney where blood is filtered during urine production (see chapter 16); (3) the **mesothelium,** which is part of the serous membranes lining the thoracic and abdominopelvic cavities;

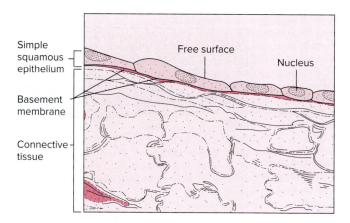

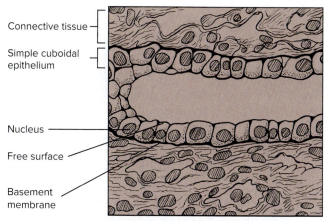

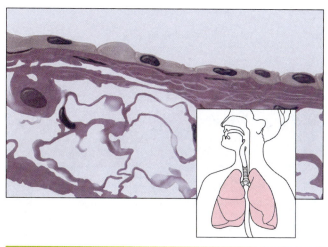

Figure 4.3 Simple Squamous Epithelium (250×). AP|R
Structure: A single layer of squamous cells.
Location: Endothelium, mesothelium, air sacs of the lungs, and filtration units within the kidneys.
Function: Absorption, secretion, filtration, diffusion, osmosis, and friction reduction.
©Ed Reschke

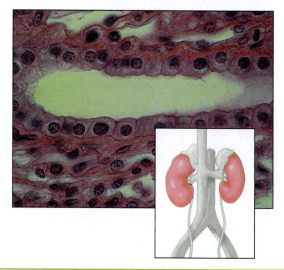

Figure 4.4 Simple Cuboidal Epithelium (250×). AP|R
Structure: A single layer of cuboidal cells.
Location: Forms kidney tubules, secretory portion of some glands, and the outer layer of the ovaries.
Function: Absorption and secretion.
©Ed Reschke

(4) and the **endothelium,** which lines the inner surfaces of the heart, blood vessels, and lymphatic vessels (figure 4.3).

Simple cuboidal epithelium consists of a single layer of cube-shaped cells. The cells have a single, round, centrally located nucleus. Its basic functions are absorption and secretion. Locations for simple cuboidal epithelia include (1) the secretory portion of glands, such as the thyroid and salivary glands; (2) the kidney tubules where secretion and reabsorption of materials occur; and (3) the outer layer of the ovaries (figure 4.4).

Simple columnar epithelium consists of a single layer of elongated, columnar cells with oval nuclei usually located near the basement membrane. Scattered among the columnar cells are **goblet cells,** specialized mucus-secreting cells with a goblet or wine glass shape. Their purpose is to secrete a protective layer of mucus on the free surface of the epithelium. Secretion and absorption are the major functions of this tissue in areas such as the stomach and intestines. The cells lining the intestine possess numerous microvilli on their free surface, often called a "brush border" because of its bristlelike appearance, which greatly increases their absorptive surface area. In areas such as uterine tubes, paranasal sinuses, and ventricles of the brain, this tissue possesses cilia that allow for movement of materials across the tissue surface (figure 4.5).

Pseudostratified ciliated columnar epithelium consists of a single layer of cells. It is said to be **pseudostratified** (pseudo = false) because its structure creates a visual illusion of being multilayered but it really is a simple epithelium. The layered effect results in part from the nuclei being located at various levels within the cells. Also, even though all of the cells are attached to the basement membrane, not all of them reach the free surface (see figure 4.2). Just as in simple columnar epithelium, goblet cells are scattered throughout the tissue. This ciliated epithelium lines the inner surfaces of many of the

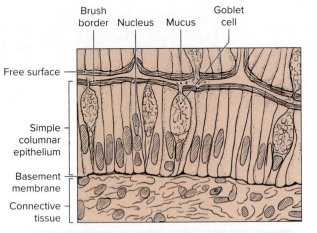
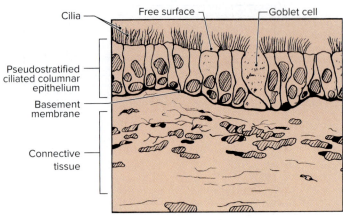
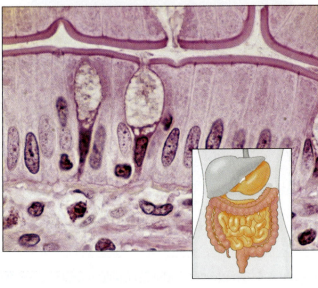
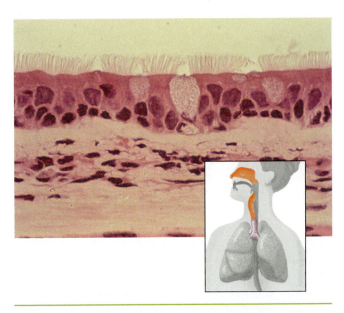

Figure 4.5 Simple Columnar Epithelium (400×). AP|R
Structure: A single layer of columnar cells; contains scattered goblet cells.
Location: Lines the inner surfaces of the stomach and intestines, the ducts of many glands, uterine tubes, paranasal sinuses, and ventricles in the brain.
Function: Absorption, secretion, and protection.
©Ed Reschke/Getty Images

Figure 4.6 Pseudostratified Ciliated Columnar Epithelium (500×). AP|R
Structure: A single layer of ciliated columnar cells that appears to be more than one layer of cells; contains scattered goblet cells.
Location: Lines many respiratory passageways.
Function: Secretion of mucus; beating cilia remove secreted mucus and entrapped particles.
©Ed Reschke/Getty Images

respiratory passageways, where it collects and removes airborne particles. The particles are trapped in the secreted mucus, which is moved by the beating cilia to the throat, where it is either swallowed or expectorated (figure 4.6).

Stratified Epithelium

Stratified epithelium consists of more than one layer of cells, which makes them more durable to abrasion (see figure 4.2). Only the deepest layer of cells produces new cells through mitosis. The cells are pushed toward the free surface of the tissue as more new cells are formed beneath them. Cells in the outermost layer are continuously lost as they die and are rubbed off by abrasion. Protection of underlying tissues is an important function of stratified epithelia. These tissues are named according to the shape of cells on their free surfaces.

Stratified squamous epithelium occurs in two distinct forms: keratinized and nonkeratinized. The keratinized type forms the outermost layer (epidermis) of the skin. Its cells become impregnated with a waterproofing protein, **keratin** (ker′-ah-tin), as they migrate to the free surface of the tissue. This specific type of epithelium is discussed further in chapter 5. The nonkeratinized type lines the mouth, esophagus, vagina, and rectum. Both types provide resistance to abrasion (figure 4.7).

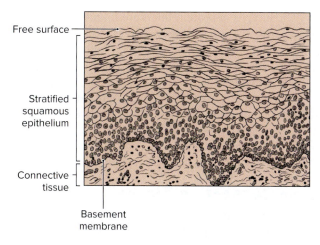

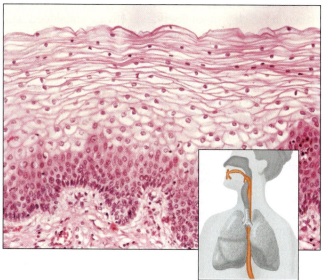

Figure 4.7 Stratified Squamous Epithelium (70×). AP|R
Structure: Several cell layers; cells in the deepest layer are cuboidal in shape but gradually become flattened as they migrate to the surface of the tissue.
Location: The keratinized type forms the epidermis of the skin; the nonkeratinized type lines the mouth, esophagus, vagina, and rectum.
Function: Protection.
©Biophoto Associates/Science Source

Transitional epithelium lines most of the urinary tract and stretches as these structures fill with urine. It consists of multiple layers of cells, with the free surface cells of the unstretched tissue possessing a large and rounded shape. When stretched, the free surface cells become thin, flat cells resembling squamous epithelial cells (figure 4.8, see figure 4.2).

Two relatively rare types of stratified epithelial tissues are not shown. *Stratified cuboidal epithelium* lines larger ducts of certain glands (e.g., mammary and salivary glands). *Stratified columnar epithelium* lines parts of the pharynx and male urethra.

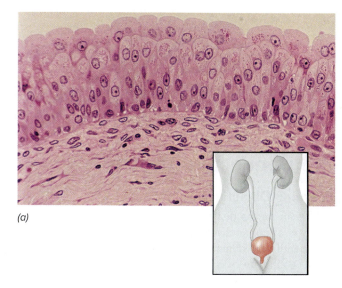

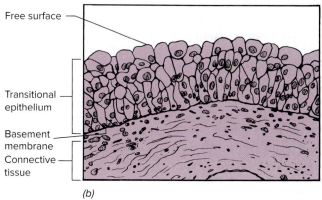

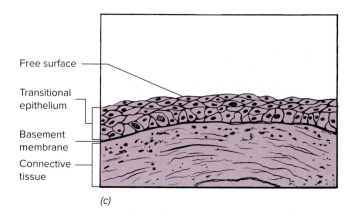

Figure 4.8 Transitional Epithelium. AP|R
(a) A photomicrograph (250×) and *(b)* drawing showing several layers of rounded cells when the urinary bladder wall is contracted. *(c)* When the bladder wall is stretched, the tissue and cells become flattened.
Structure: Several layers of large, rounded cells that become flattened when stretched.
Location: Lines the inner surface of the urinary tract.
Function: Protection; permits stretching of the wall of the urinary tract.
(a) ©Ed Reschke/Getty Images

Check My Understanding

1. What are the general characteristics and functions of epithelial tissues?
2. How are the various epithelial tissues different in terms of structure, location, and function?

4.3 Connective Tissues AP|R

Learning Objectives

5. Describe the distinguishing characteristics of each type of connective tissue.
6. Identify the common locations and general functions of each type of connective tissue.

Connective tissues are the most widely distributed and abundant tissues in the body. As the name implies, connective tissues support and bind together other tissues, so they are never found on exposed surfaces. Like epithelial cells, most connective tissue cells have retained the ability to reproduce by mitosis.

Connective tissues consist of a diverse group of tissues that can be divided into three broad categories: (1) loose connective tissues, (2) dense connective tissues, and (3) connective tissues with specialized functions—cartilage tissue, bone tissue, blood, and lymph. Lymph, a fluid connective tissue, is discussed in chapter 13. Loose and dense connective tissues are sometimes referred to as "connective tissue proper" because they are common tissues that function to bind other tissues and organs together.

All connective tissues consist of relatively few, loosely arranged cells and a large amount of extracellular substance called **matrix** (ma′-triks). Matrix, which is produced by the cells, is used to classify connective tissues. It is composed of ground substance and protein fibers. **Ground substance,** which is composed of water and both inorganic and organic substances, can be fluid, semifluid, gelatinous, or calcified.

Clinical Insight

Because epithelial and connective tissue cells undergo cell division, they are prone to the formation of tumors when normal control of cell division is lost. The most common types of cancer arise from epithelial cells, possibly because these cells have the most direct contact with *carcinogens,* cancer-causing agents in the environment. A cancer derived from epithelial cells is called a *carcinoma.* Malignant tumors that originate in connective tissue are also common types of cancer. A cancer of connective tissue is called a *sarcoma.*

Three types of protein fibers are found in the matrix of connective tissues. **Collagen fibers,** composed of collagen protein, are relatively large fibers resembling cords of a rope. They provide strength and flexibility but not elasticity. **Reticular fibers,** also made of collagen, are very thin and form highly branched, delicate, supportive frameworks for tissues. **Elastic fibers** are made of elastin protein and possess great elasticity, which means they can stretch up to 150% their resting length without damage and then recoil back to their resting length.

Loose Connective Tissue

Loose connective tissues help to bind together other tissues and form the basic supporting framework for organs. Their matrix consists of a semifluid or jellylike ground substance in which protein fibers and cells are embedded. The word "loose" describes how the protein fibers are widely spaced and intertwined between the cells. **Fibroblasts** are the most common cells, and they are responsible for producing the ground substance and protein fibers. There are three types of loose connective tissue: areolar connective tissue, adipose tissue, and reticular tissue.

Areolar Connective Tissue

Areolar (ah-rē′-ō-lar) **connective tissue** is the most abundant connective tissue in the body. Fibroblasts are the most numerous cells, but macrophages are present to help protect against invading pathogens (see chapters 11 and 13). A semifluid ground substance fills the spaces between the cells and protein fibers. Areolar connective tissue (1) attaches the skin to underlying muscles and bones as part of the subcutaneous tissue (see chapter 5); (2) provides a supportive framework for internal organs, nerves, and blood vessels; (3) is a site for many immune reactions; and (4) forms the outermost region of the dermis, which is the deep layer of the skin (figure 4.9).

Adipose Tissue

Large accumulations of fat cells, or **adipocytes,** form **adipose** (ad′-i-pōs) **tissue,** a special type of loose connective tissue. It occurs throughout the body but is more common underneath the skin, within the subcutaneous tissue, and around internal organs. Adipocytes are filled with fat droplets that push the nucleus and cytoplasm to the edge of the cells. In addition to fat storage, adipose tissue serves as a protective cushion for internal organs, especially around the kidneys and behind the eyeballs. It also helps to insulate the body from abrupt temperature changes and, as part of the subcutaneous tissue, to attach skin to underlying bone and muscle (figure 4.10).

Reticular Tissue

Reticular tissue consists of a fine interlacing of reticular fibers and **reticular cells,** the main cell type in this tissue. Reticular tissue forms a supportive network called a *stroma*

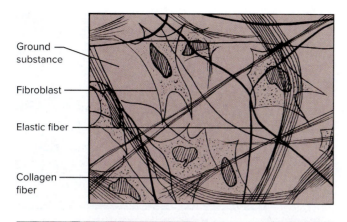
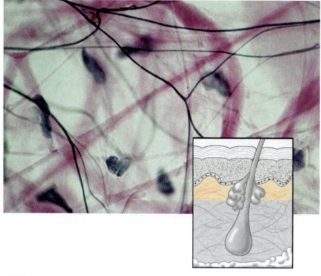

Figure 4.9 Areolar Connective Tissue (250×). AP|R
Structure: Formed of scattered fibroblasts and a loose network of collagen and elastic fibers embedded in a gellike ground substance.
Location & Function: Attaches the skin to underlying muscles and bones as part of the subcutaneous tissue; supports internal organs, blood vessels, and nerves; site for immune reactions; forms the outermost region of the dermis of the skin.
©Ed Reschke/Getty Images

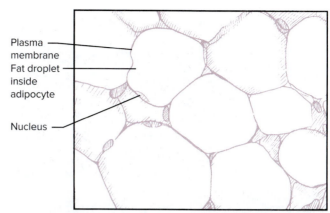
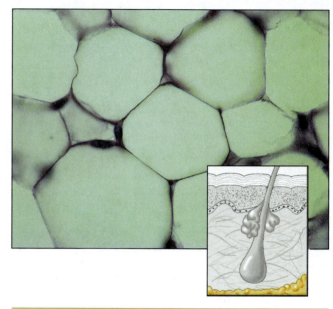

Figure 4.10 Adipose Tissue (250×). AP|R
Structure: Formed of closely packed adipocytes with little matrix. Large fat-containing droplet pushes the cytoplasm and nucleus to the edge of the cell.
Location & Function: Stores excess nutrients as fat; provides insulation and attaches skin to underlying bones and muscles as part of the subcutaneous tissue; provides a protective cushion to bones, muscles, and internal organs.
©Ed Reschke

that assists in maintaining the structure of red bone marrow and organs such as the liver and spleen. Reticular fibers also act as filters in structures like lymph nodes, where they help to remove bacteria from lymph (figure 4.11).

Dense Connective Tissue

Like loose connective tissues, **dense connective tissues** aid in binding tissues together and providing support for organs. However, dense connective tissue has far fewer cells and ground substance and more numerous, thicker, and "denser" protein fibers. These tissues also contain far fewer blood vessels than loose connective tissues. There are three types of dense connective tissue: dense regular connective tissue, dense irregular connective tissue, and elastic connective tissue.

Dense Regular Connective Tissue

Dense regular connective tissue is characterized by an abundance of tightly packed collagen fibers and relatively few cells. The collagen fibers exist in large bundles that are "regularly" arranged, meaning they are generally parallel to each other. Fibroblasts are located in rows between the collagen bundles. This tissue exhibits great strength when stress is applied in the same direction as

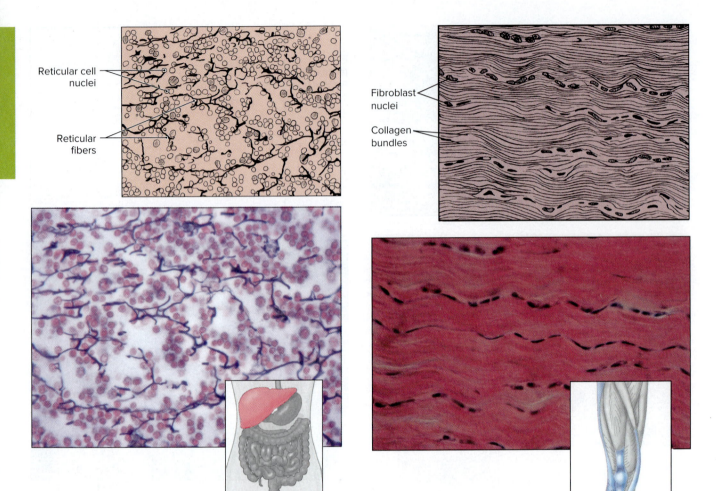

Figure 4.11 Reticular Tissue (400×).
Structure: Formed of reticular cells and a delicate, interwoven network of reticular fibers.
Location & Function: Forms a stroma to maintain the structure of red bone marrow and organs like the liver and spleen; acts as a biological filter in organs like lymph nodes.
©Beth Ann Kersten, Ph.D.

Figure 4.12 Dense Regular Connective Tissue (100×).
Structure: Consists of tightly packed collagen fibers that are separated by scattered rows of fibroblasts.
Location & Function: Strong attachment; forms ligaments attaching bones to bones at joints and tendons attaching muscles to bones, other muscles, dermis, and ligaments.
©Ed Reschke

the collagen bundles, meaning this tissue can withstand damage when stress is applied in one direction but not when stress is applied in multiple directions. Dense regular connective tissue is the main tissue in structures such as (1) ligaments, which attach bones to bones, and (2) tendons, which attach skeletal muscles to bones, other muscles, dermis, and ligaments (figure 4.12).

Dense Irregular Connective Tissue

Dense irregular connective tissue is similar in structure to dense regular connective tissue, except for the organization of the collagen bundles. In this tissue, the collagen bundles are "irregularly" arranged, meaning they are oriented in multiple directions throughout the tissue. The irregular arrangement allows this tissue to resist tearing when stress arrives from multiple directions. Dense irregular connective tissue can be found in (1) the innermost region of the dermis, (2) the joint capsules surrounding freely movable joints, (3) the membranes surrounding bone, cartilage tissue, and the heart, (4) heart valves, and (5) membrane capsules surrounding some internal organs (figure 4.13).

Elastic Connective Tissue

An abundance of elastic fibers in the matrix distinguishes **elastic connective tissue.** Collagen fibers are also present, and fibroblasts are scattered between the fibers. Elastic

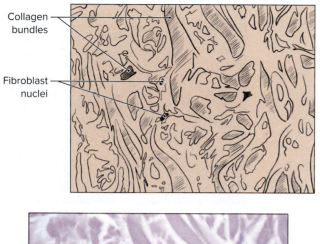

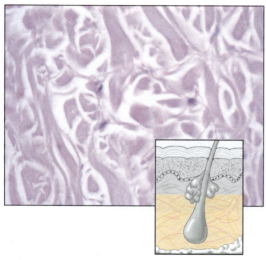

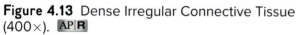

Figure 4.13 Dense Irregular Connective Tissue (400×). AP|R
Structure: Consists of tightly packed, irregularly arranged collagen fibers with scattered fibroblasts between the fibers.
Location & Function: Resists tearing with stress in the innermost region of the dermis; joint capsules of movable joints; membranes surrounding bone, cartilage tissue, heart, and other internal organs; and heart valves.
©Beth Ann Kersten, Ph.D.

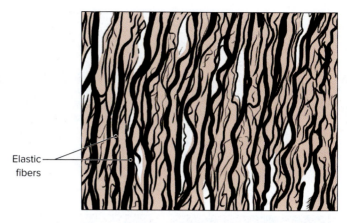

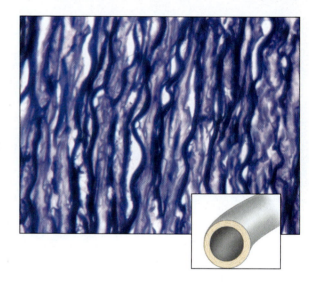

Figure 4.14 Elastic Connective Tissue (400×). AP|R
Structure: Consists of tightly packed, regularly arranged elastic fibers with scattered fibroblasts between the fibers.
Location & Function: Allows for elasticity in structures such as the lungs, air passageways, vocal cords, and arterial walls.
©Beth Ann Kersten, Ph.D.

connective tissue occurs where extensibility and elasticity are advantageous, such as in the lungs, air passages, vocal folds, and arterial walls. For example, elastic connective tissue enables the expansion of the lungs as air is inhaled and the recoil of the lungs as air is exhaled (figure 4.14).

Cartilage Tissue

Cartilage tissue consists of a firm, gelatinous matrix in which cartilage cells, or **chondrocytes** (kon′-drō-sītz), are embedded. The fluid-filled spaces in the matrix that contain the chondrocytes are called **lacunae** (lah-kū′-nē; singular, *lacuna*), which means "little lakes." Cartilage tissue usually lacks blood vessels; this means that these tissues rely on diffusion to obtain needed substances. Because diffusion is slow through cartilage matrix, cellular processes occur at much slower rates. The major functions of cartilage tissue are support and protection. All types of cartilage tissue act as cushions to absorb shock, and their toughness allows them to be deformed by pressure and return to their original shape when the pressure is removed. Three types of cartilage tissue are present in the body: hyaline cartilage, elastic cartilage, and fibrocartilage.

Hyaline Cartilage

Under microscopic examination with standard stains, the matrix of **hyaline** (hī′-a-lin) **cartilage** has a smooth, glassy, bluish white or pinkish white appearance. It contains collagen fibers, but they are not easily visible.

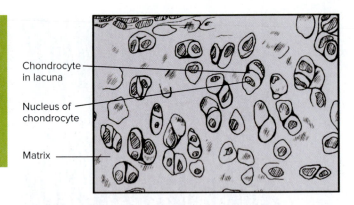

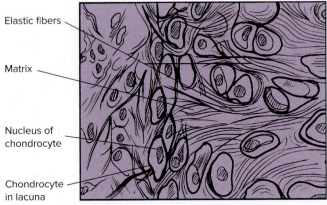

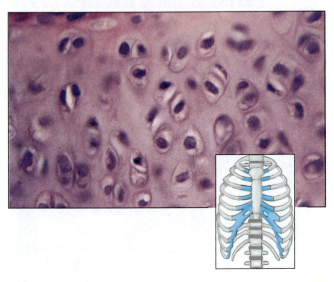

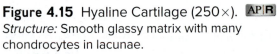

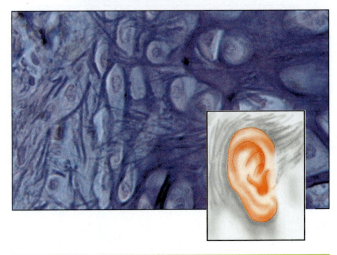

Figure 4.15 Hyaline Cartilage (250×). AP|R
Structure: Smooth glassy matrix with many chondrocytes in lacunae.
Location & Function: Forms protective covering of bones at freely movable joints; forms the larynx and part of the nose; attaches ribs to sternum, and supports walls of air passages.
©Ed Reschke

Figure 4.16 Elastic Cartilage (100×). AP|R
Structure: Consists of numerous chondrocytes occupying lacunae in a gellike matrix containing numerous elastic fibers.
Location & Function: Provides the supporting framework for the external ears; forms the auditory tubes that connect the pharynx to the middle ear; forms the epiglottis, which closes the airway when swallowing.
©Ed Reschke

Numerous chondrocytes in lacunae are present. Hyaline cartilage is the most abundant cartilage tissue in the body and its functions include (1) providing a protective covering on the bone surfaces forming freely movable joints, (2) forming the larynx, or voicebox, and part of the nose, (3) connecting the ribs to the sternum (breastbone), and (4) supporting the walls of air passages. During embryonic development, most bones of the body are initially formed of hyaline cartilage. Subsequently, the hyaline cartilage is gradually remodeled into bone tissue (figure 4.15).

Elastic Cartilage

This tissue is similar to hyaline cartilage, but **elastic cartilage** contains an abundance of elastic fibers that impart greater elasticity and flexibility to the tissue. Elastic cartilage forms (1) the auditory tubes connecting the pharynx (throat) to the middle ear, (2) the epiglottis, a lid that closes the opening into the larynx when swallowing, and (3) the supportive framework for the external ear (figure 4.16).

Fibrocartilage

The matrix of **fibrocartilage** contains many tightly packed collagen fibers that lie between short rows or clumps of chondrocytes. This cartilage tissue forms (1) the intervertebral discs that are located between vertebrae, (2) the cartilaginous pads in the knee joints, and (3) the protective cushion of the pubic symphysis (anterior union of the hip bones). Fibrocartilage is especially tough, and the dense collagen fibers enable it to absorb greater shocks and pressure without permanent damage (figure 4.17).

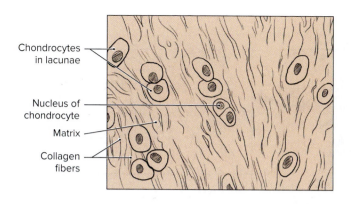
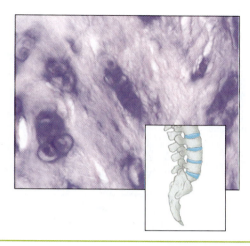

Figure 4.17 Fibrocartilage (250×).
Structure: Consists of rows or clusters of chondrocytes occupying lacunae in a matrix containing tightly packed collagen fibers.
Location & Function: Composes the intervertebral discs between vertebrae, the pubic symphysis, and cartilaginous pads in the knee joint where it serves as a protective shock absorber.
©Beth Ann Kersten, Ph.D.

Bone Tissue

Of all the supportive connective tissues, **bone tissue,** also called *osseous tissue,* is the hardest and most rigid. This results from the minerals, mostly calcium salts, that compose the matrix along with some collagen fibers. Bone tissue provides the rigidity and strength necessary for the skeletal system to support and protect the body. There are two types of bone tissue: **compact bone** and **spongy bone** (figure 4.18).

In compact bone, bone matrix is deposited in concentric rings, called **lamellae** (lah-mel′-e), around microscopic tubes called **central (or osteonic) canals.** These canals contain blood vessels and nerves. A central canal and the lamellae surrounding it form an **osteon,** the structural unit of compact bone. Spongy bone does not possess osteons; rather, the lamellae are organized into thin, interconnected bony plates called **trabeculae.** The spaces between trabeculae are filled with highly vascular red or yellow bone marrow. Bone cells, or **osteocytes,** are located in lacunae that are located between lamellae in both types of bone tissue. The tiny, fluid-filled canals that extend outward from the lacunae are called **canaliculi** (kan″-ah-lik′-u-li; singular, *canaliculus*), and they contain cell processes from osteocytes. Canaliculi serve as passageways for the movement of materials between osteocytes and the blood supply within the central canals and bone marrow. Bone tissue will be discussed further in chapter 6.

Blood

Blood is a specialized type of connective tissue, called a fluid connective tissue. It consists of numerous **formed elements** that are suspended in the plasma, the liquid matrix of the blood. There are three basic types of formed elements: red blood cells, white blood cells, and platelets (figure 4.19).

Blood plays a vital role in carrying materials and gases throughout the body. For example, blood is used to carry nutrients absorbed by the digestive tract to cells throughout the body and wastes produced by body cells to the kidneys for elimination. The formed elements also perform crucial body functions. **Red blood cells** are used primarily to transport O_2 molecules and, to a lesser degree, CO_2 molecules. **White blood cells** carry out various defensive and immune functions throughout the body. **Platelets** play a crucial role in the process of blood clotting. Blood is discussed in more depth in chapter 11.

 Check My Understanding

3. What are the general characteristics and functions of connective tissue?
4. What are the characteristics, locations, and functions of the different connective tissues?

4.4 Muscle Tissues

Learning Objectives

7. Describe the distinguishing characteristics and locations of each type of muscle tissue.
8. Identify the general functions of each type of muscle tissue.

Muscle tissue consists of muscle cells. Muscle cells have lost the ability to divide, so destroyed muscle cells cannot

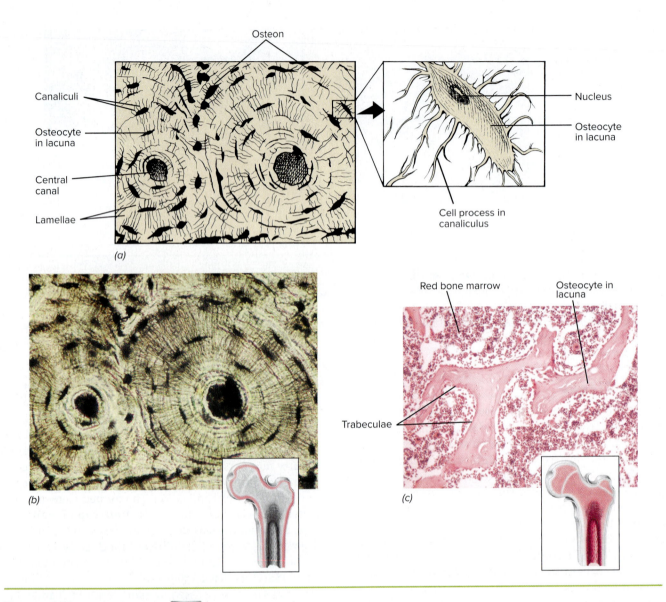

Figure 4.18 Bone Tissue. AP|R
(a) (b) Compact Bone (160×); *(c)* Spongy Bone (100×).
Structure: In compact bone, matrix is arranged in concentric layers around central canals. In spongy bone, the bone layers form thin, bony plates called trabeculae. Osteocytes are found within lacunae located between layers of matrix. Canaliculi, minute channels between lacunae, enable movement of materials between osteocytes in both compact and spongy bone.
Location & Function: Forms bones of the skeleton that provide support for the body and protection for vital organs.
(b) ©Victor B. Eichler, Ph.D.; (c) ©Beth Ann Kersten, Ph.D.

be replaced. In skeletal muscle tissue, muscle cells are called **muscle fibers** owing to their long, cylindrical appearance. The cells in smooth and cardiac muscle tissue are not long and cylindrical, so they are referred to as muscle cells but not muscle fibers. The cells within all three types of muscle tissue are specialized for contraction (shortening). Contraction is enabled by the interaction of specialized protein fibers. The contraction of these tissues enables the movement of the whole body and many internal organs, in addition to producing heat energy.

The three types of muscle tissue are classified according to their (1) location in the body, (2) structural features, and (3) functional characteristics.

Skeletal Muscle Tissue

Named for its location, **skeletal muscle tissue** is usually attached to bones and skin. Its contractions enable movement of the head, trunk, and limbs. The muscle fibers are wide, elongated, and cylindrical. Each skeletal muscle

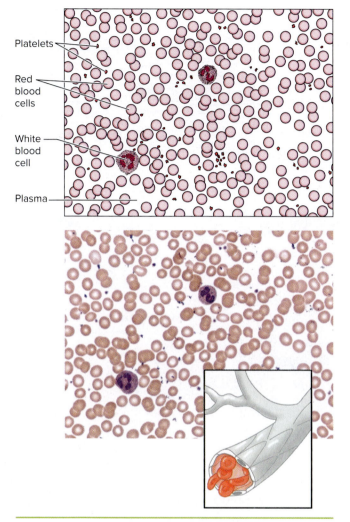

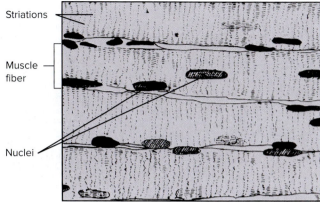

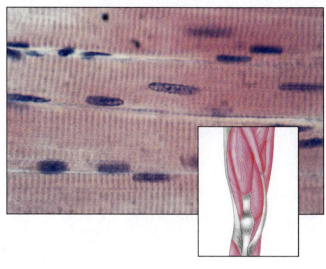

Figure 4.19 Blood (500×).
Structure: Consists of red blood cells, white blood cells, and platelets that are carried in a liquid matrix called plasma.
Location & Function: Located within blood vessels and the heart; transports materials and gases throughout the body, participates in blood clotting process, provides defense against disease.
©McGraw-Hill Education/Al Telser, photographer

Figure 4.20 Skeletal Muscle Tissue (250×).
Structure: Consists of cylindrical muscle fibers that have striations and multiple, peripherally located nuclei.
Location & Function: Composes skeletal muscles that attach to bones and skin; voluntary, rapid contractions.
©Ed Reschke

fiber contains multiple nuclei, which are located along the periphery of the fiber. *Striations,* alternating light and dark bands, extend across the width of the fibers. Functionally, skeletal muscle tissue is considered to be voluntary muscle because its rapid contractions can be consciously controlled (figure 4.20). Skeletal muscle tissue is discussed in greater detail in chapter 7.

Cardiac Muscle Tissue

The muscle tissue located in the walls of the heart is **cardiac** (kar′-dē-ak) **muscle tissue.** It consists of branching cells that interconnect in a netlike arrangement. **Intercalated** (in-ter-kah′-lā-ted) **discs** are present where the cells join together. The discs possess connections that help to prevent cell separation and pass electrical signals between cells. Cardiac muscle cells are striated like skeletal muscle fibers but possess only one centrally located nucleus per cell. The rhythmic contractions of cardiac muscle are involuntary because they cannot be consciously controlled (figure 4.21).

Smooth Muscle Tissue

Smooth muscle tissue derives its name from the absence of striations in its cells. It occurs in the walls of hollow internal organs, such as the stomach, intestines, urinary bladder, and blood vessels. The cells are long and spindle-shaped with a single, centrally located nucleus. The slow contractions of smooth muscle tissue are involuntary (figure 4.22).

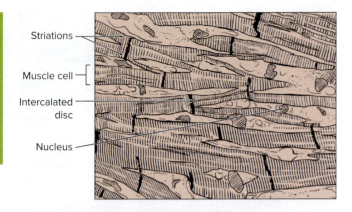

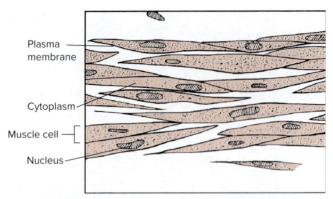

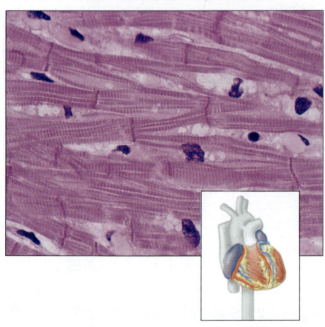

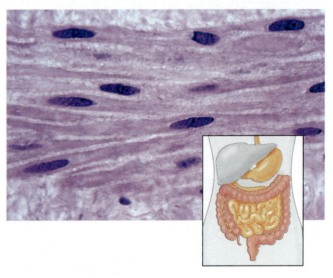

Figure 4.22 Smooth Muscle Tissue (250×). AP|R
Structure: Consists of elongated, tapered cells that lack striations and have a single, centrally located nucleus.
Location & Function: Forms muscle layers in the walls of hollow internal organs; involuntary, slow contractions.
©Ed Reschke/Getty Images

Figure 4.21 Cardiac Muscle Tissue (400×). AP|R
Structure: Consists of striated cells that are arranged in an interwoven network. Intercalated discs are present at the junctions between cells. A single, centrally located nucleus is present in each cell.
Location & Function: Forms the muscular walls of the heart; involuntary, rhythmic contractions.
©Ed Reschke/Getty Images

4.5 Nerve Tissue

Learning Objectives
9. Describe the distinguishing characteristics and general functions of nerve tissue.
10. Identify the common locations of nerve tissue.

The brain, spinal cord, and nerves are composed of **nerve tissue,** which consists of **neurons** (nū′-ronz), or nerve cells, and numerous supporting cells that are collectively called **neuroglia** (nū-rog′-lē-ah). Neurons are the functional units of nerve tissue. They are specialized to detect and respond to environmental changes by generating and transmitting nerve impulses. Neuroglia nourish, insulate, and protect the neurons.

A neuron consists of a **cell body,** the portion of the cell containing the nucleus, and one or more *neuronal processes* extending from the cell body (figure 4.23). There are two types of neuronal processes. **Dendrites** respond to stimuli by generating impulses and transmitting them toward the cell body. An **axon** transmits nerve impulses away from the cell body and dendrites. A neuron may have many dendrites but only one axon. The complex interconnecting network of neurons enables the nervous system to coordinate body functions. Nerve tissue is discussed in more detail in chapter 8.

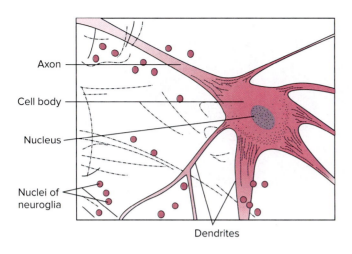
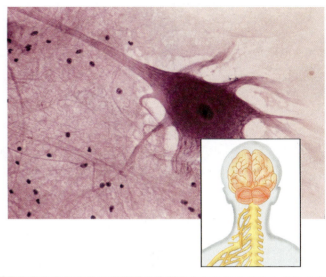

Figure 4.23 Nerve Tissue (50×). AP|R
Structure: Consists of neurons and neuroglia. Each neuron consists of a cell body, which houses the nucleus, and one or more neuronal processes extending from the cell body.
Location & Function: Forms the brain, spinal cord, and nerves; nerve impulse formation and transmission.
©Ed Reschke/Getty Images

Clinical Insight

Following minor injuries, tissues repair themselves by *regeneration*—the division of the remaining intact cells. The capacity to regenerate varies among different tissues. For example, epithelial tissues, loose connective tissues, and bone tissues readily regenerate, but cartilage tissues and skeletal muscle have little capacity for regeneration. Cardiac muscle never regenerates, and neurons in the brain and spinal cord usually do not regenerate.

After severe injuries, repair involves **fibrosis,** the formation of scar tissue. Scar tissue is formed by an excess production of collagen fibers by fibroblasts. Scar tissues that join together tissues or organs abnormally are called *adhesions,* which sometimes form following abdominal surgery.

Check My Understanding

5. What are the distinguishing characteristics, locations, and functions of the three types of muscle tissue?
6. What types of cells form nerve tissue and what are their functions?

4.6 Body Membranes

Learning Objectives

11. Compare epithelial and connective tissue membranes.
12. Describe the locations and functions of each type of epithelial membrane.
13. Identify examples of connective tissue membranes.

Membranes of the body are thin sheets of tissue that line cavities, cover surfaces, or separate tissues or organs. Some are composed of both epithelial and connective tissues; others consist of connective tissue only.

Epithelial Membranes

Sheets of epithelial tissue overlying a thin supporting framework of areolar connective tissue form the epithelial membranes in the body. Blood vessels in the connective tissue serve both connective and epithelial tissues. There are three types of epithelial membranes: serous, mucous, and cutaneous membranes (figure 4.24).

Serous membranes, or *serosae,* line the thoracic and abdominopelvic cavities and cover most of the internal organs within these cavities. They secrete **serous fluid,** a watery fluid, which reduces friction between the membranes. The pleurae, serous pericardium, and serous peritoneum are serous membranes. Recall that the epithelium of a serous membrane is a special tissue called mesothelium.

84 Chapter 4 Tissues and Membranes

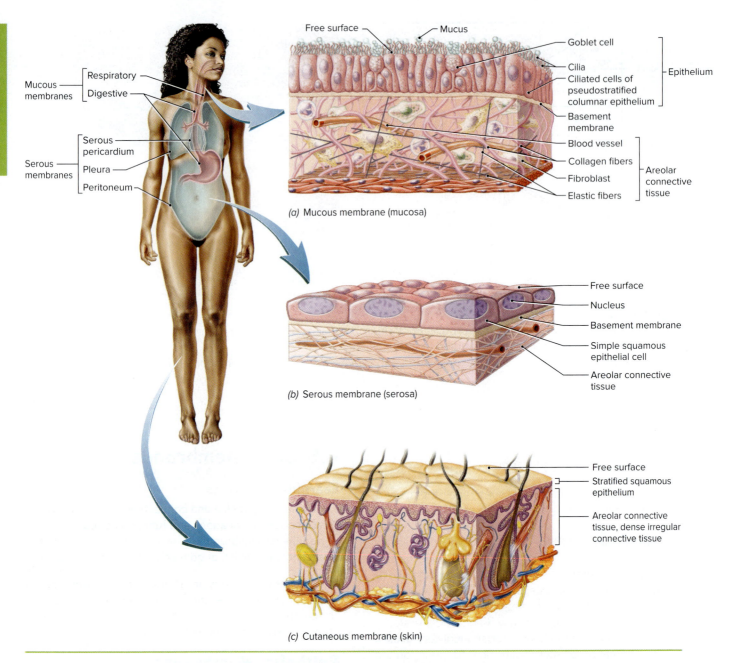

Figure 4.24 Epithelial Membranes.

Mucous membranes, or *mucosae,* line tubes or cavities of organ systems, with openings to the external environment. Their goblet cells secrete **mucus,** which coats the surface of the membranes to keep the cells moist and to lubricate their surfaces. The mucus also helps to trap foreign particles and pathogens, which limits their ability to enter the body. The digestive, respiratory, reproductive, and urinary tracts are lined with mucous membranes.

The **cutaneous membrane** is the skin that covers the body. Unlike other membranes, its free surface is dry, composed of nonliving cells, and exposed to the external environment. The skin is discussed in detail in chapter 5.

Connective Tissue Membranes

Some specialized membranes are formed only of connective tissue, usually dense irregular connective tissue. These are considered in future chapters with their respective organ systems, but here are four examples:

1. The **meninges** are three connective tissue membranes that envelop the brain and spinal cord.
2. The **perichondrium** is a connective tissue membrane covering the surfaces of cartilage tissues. It contains blood vessels, which supply cartilage tissues through diffusion.

3. The **periosteum** is a connective tissue membrane that covers the outer surfaces of bones. It contains blood vessels that enter and supply the bone.
4. **Synovial membranes** line the cavities of freely movable joints, such as the knee joint. They secrete watery synovial fluid, which reduces friction in the joint.

 Check My Understanding

7. What are the two kinds of body membranes?
8. How do the structures, locations, and functions of the epithelial membranes differ?

Chapter Summary

4.1 Introduction to Tissues
- Embryonic stem cells are unspecialized cells that divide to form all types of specialized cells.
- Adult stem cells are partially specialized cells that divide to produce only certain related types of specialized cells.
- A tissue is a group of fully specialized cells that perform similar functions.
- The different tissues of the body are classified into four basic types of tissues: epithelial, connective, muscle, and nerve tissues.

4.2 Epithelial Tissues
- Epithelial tissue covers the surface of the body and the surfaces of organs, and lines the body cavities.
- Epithelial tissue is composed of closely packed cells with little extracellular material.
- Epithelial tissues are attached to underlying connective tissue by a noncellular basement membrane.
- Epithelial tissue lacks blood vessels.
- Epithelial tissues function in absorption, secretion, filtration, diffusion, osmosis, protection, and friction reduction.
- Epithelial tissues are classified according to the number of cell layers and the shape of the free surface cells. The epithelial tissues are

  Simple Epithelium
  - squamous
  - cuboidal
  - columnar
  - pseudostratified ciliated columnar

  Stratified Epithelium
  - stratified squamous
  - transitional

4.3 Connective Tissues
- Connective tissue is composed of relatively few cells located within a large amount of matrix.
- All connective tissues but cartilage tissue are supplied with blood vessels.
- Connective tissue binds other tissues together and provides support and protection for organs and the body.
- Connective tissue is classified according to the nature of the matrix. The connective tissues are

Loose Connective Tissue
- Areolar connective tissue
- Adipose tissue
- Reticular tissue

Dense Connective Tissue
- Dense regular connective tissue
- Dense irregular connective tissue
- Elastic connective tissue

Cartilage Tissue
- Hyaline cartilage
- Elastic cartilage
- Fibrocartilage

Bone Tissue

Blood

4.4 Muscle Tissues
- Muscle tissue is composed of muscle cells that are specialized for contraction.
- Contraction of muscle tissue enables movement of the body and internal organs.
- Muscle tissue is classified according to its location in the body, the characteristics of the muscle cells, and the type of contractions (voluntary or involuntary).
- Three types of muscle tissue are skeletal, cardiac, and smooth muscle tissue.

4.5 Nerve Tissue
- Nerve tissue consists of neurons and neuroglia.
- Neurons consist of a cell body and long, thin neuronal processes, and are adapted to form and conduct nerve impulses.
- Nerve tissue forms the brain, spinal cord, and nerves.

4.6 Body Membranes
- Membranes in the body are either epithelial membranes or connective tissue membranes.
- Epithelial membranes are composed of both epithelial and connective tissues, while connective tissue membranes are composed of connective tissue only.
- There are three types of epithelial membranes: serous, mucous, and cutaneous.
- Examples of connective tissue membranes are meninges, perichondrium, periosteum, and synovial membranes.

Improve Your Grade

Connect Interactive Questions Reinforce your knowledge using multiple types of questions: interactive, animation, classification, labeling, sequencing, composition, and traditional multiple choice and true/false.

SmartBook Proven to help students improve grades and study more efficiently, SmartBook contains the same content within the print book but actively tailors that content to the needs of the individual.

Anatomy & Physiology REVEALED® Dive into the human body by peeling back layers of cadaver imaging. Utilize this world-class cadaver dissection tool for a closer look at the body anytime, from anywhere.

CHAPTER 5

Integumentary System

©Francisco Cruz/Purestock/SuperStock RF

Emily and Chet Roberson and their children have gathered for a picnic on a sunny day in June to celebrate the birthday of their youngest child. Emily, ever aware of the risk of skin cancer, makes sure that all of the children are wearing sunscreen and hats. Even Chet applies sunscreen to avoid getting sunburn. However, Emily left her hat at home and forgets to apply sunscreen to her own skin after taking care of her kids. The entire family enjoys two hours of volleyball before retreating to a blanket in the shade for lunch. Later that evening, Emily assesses the damage to her body that her forgetfulness caused. Luckily, her thick curly hair protected her scalp from the sun's ultraviolet (UV) radiation and limited her scalp sunburn to just the location of the part in her hair. The melanin within Emily's tanned arms and legs provided some protection to UV radiation, although Emily did end up with slight sunburn in these areas. All in all, Emily's skin and its accessory structures did an amazing job of protecting her from the damaging rays of the sun.

CHAPTER OUTLINE

5.1 Functions of the Skin
5.2 Structure of the Skin and Subcutaneous Tissue
- Epidermis
- Dermis
- Subcutaneous Tissue

5.3 Skin Color
5.4 Accessory Structures
- Hair
- Glands
- Nails

5.5 Temperature Regulation
5.6 Aging of the Skin
5.7 Disorders of the Skin
- Infectious Disorders
- Noninfectious Disorders

Chapter Summary

Module 4
Integumentary System

SELECTED KEY TERMS

Apocrine sweat gland (apo = detached; crin = separate off) A sweat gland depositing secretions into a hair follicle.
Cutaneous (cutane = skin) Pertaining to the skin.
Dermal papillae (papilla = nipple) Nipplelike projections of the dermis at the dermis-epidermis boundary.
Dermis (derm = skin) The innermost layer of the skin.
Eccrine sweat gland (ec = out from) A sweat gland depositing secretions onto the skin surface.
Epidermis (epi = upon) The outer layer of the skin.
Hair follicle (folli = bag) An inward, tubular extension of the epidermis containing the hair root.
Integument (integere = to cover) The skin.
Keratin (kerat = horny, hard) Tough, fibrous protein that provides waterproofing and abrasion protection for the epidermis.
Melanin (melan = black) The brown-black pigment formed by melanocytes.
Sebaceous gland (seb = grease, oil) A gland depositing sebum into a hair follicle.
Subcutaneous tissue (sub = below) The loose connective tissue beneath the skin.
Sweat gland A sweat-producing gland.

THE SKIN AND THE STRUCTURES that develop from it—hair, glands, and nails—form the *integumentary* (in-teg-ū-men'-tar-ē) *system*. The skin is a pliable, tough, waterproof, self-repairing barrier that separates underlying tissues and organs from the external environment. Although it often gets little respect, the skin is vital for maintaining homeostasis.

5.1 Functions of the Skin

Learning Objective
1. Explain the functions of the skin.

The skin, or **integument,** is also known as the **cutaneous membrane,** one of the three types of epithelial membranes, as noted in chapter 4. The skin performs six important functions:

1. **Protection.** The skin provides a physical barrier between other body structures and the external environment. It provides protection from abrasion, dehydration, ultraviolet (UV) radiation, chemical exposure, and pathogens.
2. **Excretion.** Sweat, produced by sweat glands, removes small amounts of organic wastes, salts, and water.
3. **Temperature regulation.** When body temperature rises above an individual's set point, blood vessels near the body surface dilate to increase heat loss and cool the body. Sweat production and evaporation also aid in heat loss. When body temperature drops below an individual's set point, blood vessels near the body surface constrict to conserve body heat.
4. **Sensory perception.** The skin contains nerve endings and sensory receptors that detect stimuli associated with touch, pressure, temperature, and pain.
5. **Synthesis of vitamin D.** Exposure to UV radiation stimulates the production of precursor molecules that are needed for the body to form active vitamin D.
6. **Absorption.** The skin is capable of absorbing lipid-soluble vitamins (A, D, E, and K), in addition to lipid-soluble drugs (e.g., topical steroids, nicotine) and toxins (e.g., acetone, lead, mercury).

5.2 Structure of the Skin and Subcutaneous Tissue

Learning Objectives
2. Describe the structures and functions of the two tissue layers forming the skin.
3. Describe the structure and functions of the subcutaneous tissue.

The skin is thickest in areas subjected to wear and tear (abrasion), such as the soles of the feet, where it may be 6 mm in thickness. It is thinnest on the eyelids, eardrums, and external genitalia, where it averages about 0.5 mm in thickness.

The skin consists of two major layers: the epidermis and the dermis. The *epidermis*, the thinner outer layer, is composed of an epithelium. The *dermis*, the thicker inner layer, is composed of connective tissue. The *subcutaneous tissue*, located beneath the dermis, is not part of the skin but is considered here because of its close association with the skin. Figure 5.1 shows the arrangement of the epidermis, dermis, and subcutaneous tissue, as well as accessory organs of the skin. Table 5.1 summarizes these three tissue layers.

Epidermis

The **epidermis** is a keratinized stratified squamous epithelium. Recall from chapter 4 that an epithelium is avascular, meaning it lacks blood vessels. Because the epidermis is prone to injury, the lack of blood vessels prevents unnecessary

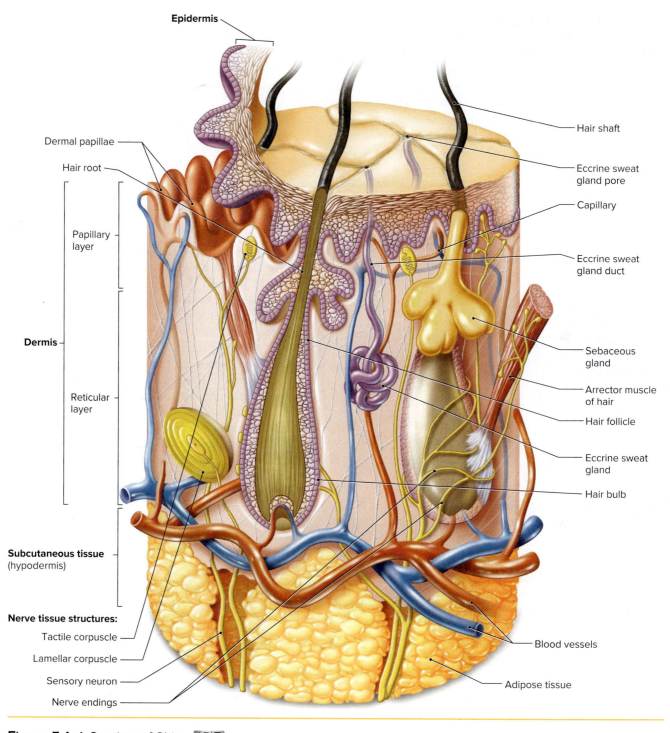

Figure 5.1 A Section of Skin. AP|R

bleeding. It also means that epidermal tissue must obtain nutrients and remove wastes through diffusion with blood vessels in the dermis. However, the epidermis does contain numerous nerve endings, which allows it to easily detect changes on the body's surface. The epidermis is the boundary between the body and the external environment. It protects the body against (1) the entrance of pathogens, (2) UV radiation, (3) evaporative water loss, (4) exposure to environmental chemicals, and (5) abrasion. The epidermis is also involved in the production of the precursor molecules necessary for active vitamin D production.

The epidermis is composed primarily of cells called **keratinocytes,** and these cells are organized into distinct layers (figure 5.2). In the thin skin covering the majority of the body's surface, the epidermis is organized into four layers. Thick skin, which is organized into five layers, is

Table 5.1 The Skin and Subcutaneous Tissue

Layer	Structure	Function
Epidermis	Keratinized stratified squamous epithelium; forms hair and hair follicles, sebaceous glands, and sweat glands that penetrate into the dermis or subcutaneous tissue	Protects against abrasion, evaporative water loss, pathogen invasion, chemical exposure, and UV radiation
Dermis	Areolar and dense irregular connective tissues containing blood vessels, nerves, and sensory receptors	Provides strength and elasticity; sensory receptors enable detection of touch, pressure, pain, cold, and heat; blood vessels supply nutrients to the epidermis
Subcutaneous tissue	Areolar connective tissue and adipose tissue; contains abundant blood vessels and nerves	Provides insulation, protection from impact, and fat storage; attaches skin to underlying organs

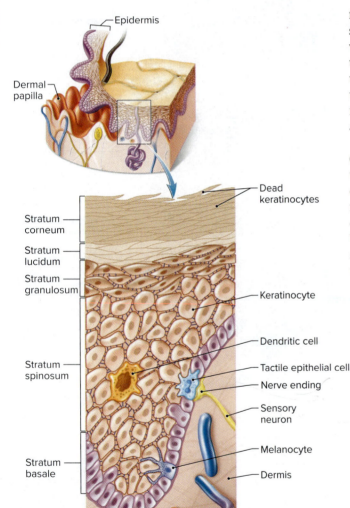

found only in areas subjected to high levels of abrasion, such as the palms and soles (figure 5.3). The keratinocytes within both thin and thick skin are so named because they undergo a process called *keratinization,* which leads to the eventual hardening and flattening of the cells through the production of keratin. **Keratin** (ker′-ah-tin) is a tough, fibrous protein that provides waterproofing and abrasion protection for the epidermis.

The innermost layer of cells, the **stratum basale** (ba-sah′-le), continuously produces new keratinocytes through mitosis. As new cells are produced, the older cells above them are gradually pushed toward the surface. The constant division of cells in the stratum basale enables the epidermis to repair itself when damaged. Upon leaving the stratum basale, keratinocytes enter the **stratum spinosum.** In this layer, keratinocytes begin to produce keratin within their cytoplasm. Neighboring

Figure 5.2 Anatomy of the Epidermis.
(a) The five layers of the epidermis in thick skin and the four types of cells that compose the epidermis. *(b)* Photomicrograph of the epidermis (200×) in thick skin.
(b) ©Lutz Slomianka RF

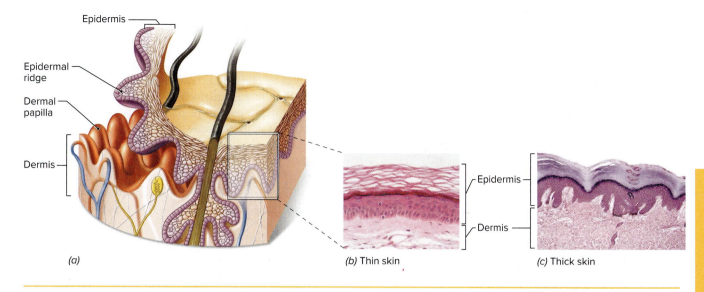

Figure 5.3 Comparison of the Epidermis in Thin and Thick Skin.
(a) A section of the skin. (b) Photomicrograph of thin skin (400×). (c) Photomicrograph of thick skin (100×).
(b) ©Lutz Slomianka RF; (c) ©Victor Eroschenko RF

cells also become physically connected, which creates a "spiny" appearance microscopically in fixed tissue.

As the keratinocytes continue to be pushed toward the epidermal surface, they leave the stratum spinosum and enter the **stratum granulosum.** The cells within this layer possess numerous granules that stain darkly in tissue section and are essential for the formation of keratin inside the cells. The cells within the stratum granulosum also undergo *apoptosis*, which is programmed cell death and involves the destruction of the nucleus and other organelles.

In thick skin, cells move next into the **stratum lucidum** (see figure 5.2). Because the cells in this layer do not have nuclei or organelles, this layer often appears "lucid" or transparent in section. The stratum lucidum is absent in thin skin, meaning cells move directly from the stratum granulosum into the outermost layer of the epidermis, the **stratum corneum** (kor′-nē-um). The stratum corneum is so named because it consists of approximately 20–40 layers of dead, squamous, and keratinized (cornified) cells. The outermost cells of the stratum corneum are continually being sloughed off and replaced by underlying cells moving toward the surface. The journey from stratum basale to stratum corneum usually takes seven to ten days, with cells remaining in the stratum corneum for another two weeks on average.

Three types of specialized cells are also present in the epidermis (see figure 5.2). **Melanocytes** (mel-an′-ō-sītz) are located in the stratum basale. They produce the brown-black pigment that is primarily responsible for skin color. **Dendritic** (*Langerhans*) **cells** are located in the strata spinosum and granulosum of the epidermis and are derived from monocytes, a type of white blood cell (see chapter 11). These cells migrate throughout the epidermis

> ⊕ **Clinical Insight**
>
> Whenever possible, surgeons try to make incisions parallel to the dominant direction of collagen fibers in the dermis because the healing of such incisions forms little scar tissue.

where they use receptor-mediated endocytosis to remove pathogens trying to enter the body and alert the lymphoid system to launch an attack. **Tactile epithelial** (*Merkel*) **cells** in the stratum basale work with nerve endings in the dermis in touch sensation detection (see chapter 9).

Dermis

The **dermis,** the innermost layer of the skin, can be divided into two regions: the outer papillary layer and the inner reticular layer (see figure 5.1).

The **papillary layer** of the dermis is next to the epidermis and is composed of areolar connective tissue. The most notable features of this region are **dermal papillae** (pah-pil′-ē), nipplelike projections of the dermis that extend outward into the epidermis (see figure 5.3). The dermal papillae contain numerous blood vessels that are used to supply nutrients to and remove wastes from the nearby epidermal cells through diffusion. They contain touch receptors called the **tactile** (*Meissner*) **corpuscles** (see figure 5.1 and chapter 9). The *dermal ridges* that produce the fingerprints and toe prints unique to each person are formed by the dermal papillae (figure 5.4). Dermal ridges provide a textured surface that

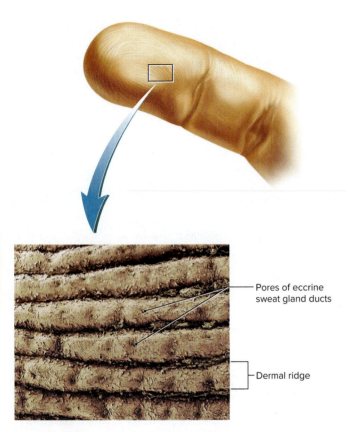

Figure 5.4 Dermal Ridges.
The pattern if dermal ridges are established by the dermal papillae of the papillary layer of the dermis.
©Steve Gschmeissner/Science Source

Labels: Pores of eccrine sweat gland ducts; Dermal ridge

increases traction on these gripping surfaces, in addition to the man-made application of personal identification. *Epidermal ridges* are inward extensions of epidermis between the dermal papillae (see figure 5.3). The dermal papillae and epidermal ridges help to interlock the epidermis and dermis, so that they move as a unit.

The inner **reticular layer** of the dermis is thicker than the papillary layer, making up 70% to 80% of the total thickness of the dermis. The dense irregular connective tissue within this region possesses an abundance of collagen and elastic fibers. The collagen provides the dermis with strength and flexibility, while the elastic fibers provide elasticity (ability to stretch and recoil back to its original shape). Numerous pressure, pain, and temperature receptors are located here. For example, the **lamellar** (*Pacinian*) **corpuscles** that are used to detect pressure are found within the innermost areas of the reticular layer (see figure 5.1 and chapter 9). Free nerve endings responsible for touch, pain, and temperature are located throughout both the dermis and the epidermis (see chapter 9). The blood vessels found within this region play an important role in temperature regulation, which is considered later in this chapter.

Subcutaneous Tissue

The **subcutaneous tissue,** also called the *hypodermis,* attaches the skin to underlying tissues and organs. It consists primarily of areolar connective tissue and adipose tissue. It is the site used for subcutaneous injections and where white blood cells attack pathogens that have penetrated the skin. Subcutaneous adipose tissue absorbs the forces created by impact to the skin, which protects underlying structures, and serves as a storage site for fat. It insulates the body by conserving body heat and limits the penetration of outside heat into the body. Blood vessels and nerves within the subcutaneous tissue give off branches that supply the dermis.

Check My Understanding

1. What are the general functions of the skin?
2. What changes occur in keratinocytes after they are formed?
3. What are the functions of the dermis and subcutaneous tissue?

5.3 Skin Color

Learning Objectives
4. Describe how skin color is determined.
5. Explain how the skin provides protection from ultraviolet radiation.

Skin color results from the interaction of three different pigments: hemoglobin, carotene, and melanin. **Hemoglobin** (he′-mō-glō″-bin) is the red pigmented protein in red blood cells that is used to carry oxygen and carbon dioxide in the blood. **Carotenes** (kair′-ō-tēns) are a group of lipid-soluble plant pigments that range in color from violet, to red-yellow, to orange-yellow. Beta-carotene, which is the most abundant carotene, is found in yellow-orange and green leafy fruits and vegetables. The human body uses carotenes for vitamin A production, which is needed for the maintenance of epithelial tissues and for the production of photopigments in the retina of the eye. Excess carotenes are stored in and add color to the body's fatty areas, such as the subcutaneous tissue, and the stratum corneum. **Melanin** (mel′-ah-nin) is a brown-black pigment that is formed by melanocytes (figure 5.5). Melanocytes insert melanin into adjacent keratinocytes where it forms a protective UV radiation shield over their nuclei.

Most people have the same number and distribution of melanocytes. Generally, melanocytes are equally distributed throughout the epidermis. However, there are areas that have higher numbers of melanocytes and, as a result, greater amounts of melanin and darker coloring. Areas of the skin subjected to more sun exposure, such as the face, neck, and limbs, have greater numbers of melanocytes to hemoglobin is relatively constant, but it is often masked by high concentrations of melanin. Dark-skinned races produce abundant melanin and have a greater protection from UV radiation. Caucasians produce relatively little melanin and are more susceptible to the harmful effects of UV radiation. The reduced amount of melanin in light-skinned Caucasians allows the hemoglobin of blood within dermal blood vessels to show through and give the skin a pinkish hue. Some people of Asian descent have a yellowish-tinged skin color due to the presence of a different form of melanin.

Check My Understanding

4. What relationship exists between skin color and protection against UV radiation?

5.4 Accessory Structures

Learning Objectives

6. Describe the anatomy of each accessory structure formed by the epidermis.
7. Explain the function of each epidermal accessory structure.

The accessory structures of the skin are hair, glands, and nails. These structures originate in either the dermis or the subcutaneous tissue because they develop from inward growths of the epidermis. Table 5.2 summarizes these accessory structures.

Hair

A *hair* is formed of keratinized cells and consists of two parts: a shaft and a root. The *hair shaft* is the portion that projects above the skin surface. The *hair root* lies below the skin surface in a **hair follicle,** an inward, tubular extension of the epidermis. The follicle penetrates into the dermis and usually into the subcutaneous tissue (figure 5.6; see figure 5.1). The region of cell division, where the stratum basale forms new hair cells, is located in the *hair bulb* (base) of the follicle. The hair bulb is enlarged where it fits over a dermal papilla, which contains blood vessels that nourish the dividing epidermal cells. The hair cells become keratinized, die, and become part of the hair root. The continuing production of new cells causes growth of the hair.

Each hair follicle has an associated *arrector muscle of hair* that is attached at one end to the innermost portion of a hair follicle and to the papillary layer of the dermis at the other end (see figure 5.6). Each muscle is a small group of smooth muscle cells. When a person is frightened or very cold, the arrector muscles of hair contract and raise the hairs on end, producing goose bumps or chicken skin. Though of little value in humans, the

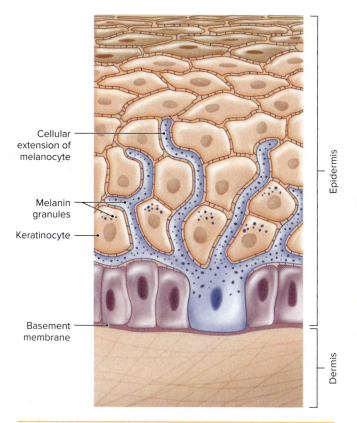

Figure 5.5 A Melanocyte.
A melanocyte with cellular extensions that transfer melanin granules to adjacent keratinocytes.

provide more UV protection to these areas. Greater melanocyte numbers are also found in structures such as the areola surrounding the nipple and the external genitalia, creating darker coloration in body areas that indicate reproductive readiness. Also, melanocytes can occur randomly in small clumps and create *freckles* due to the concentration of melanin in small patches.

The amount of melanin that a person can produce is determined genetically. Some people are genetically predisposed to make melanin at a faster rate, which results in a darker skin color. Other people are predisposed to make melanin at a slower rate, which results in a lighter skin color. A person's rate of melanin production can be influenced by exposure to UV radiation. Exposure to UV radiation will increase melanin production and produce a "tanned" appearance in lighter-skinned individuals. The increased melanin production is a protective, homeostatic mechanism. When exposure to UV radiation decreases, melanin production also decreases and the darker coloration is lost in a few weeks as the "tanned" cells migrate to and are sloughed off the stratum corneum.

The skin colors characteristic of the various human races result primarily from varying amounts of carotene and melanin in the skin and are inherited. The effect of

Table 5.2 Accessory Structures

Structure	Origin	Function
Hair	Fused keratinized epidermal cells formed at base of hair follicles	Scalp hair: protects scalp from excessive heat loss, mechanical injury, and UV radiation
		Eyelashes and eyebrows: protect eyes from sunlight and dust
		Ear and nasal hairs: keep dust and insects out of the external acoustic meatus and nasal cavity
		Across body surface: detect skin contact
Sebaceous glands	Formed from hair follicle outpocketings	Produce sebum to prevent excessive dryness of hair and skin, inhibit microbial growth on skin surface, and reduce evaporative water loss
Sweat glands	Apocrine sweat glands formed from hair follicle outpocketings in axillary, areolar, beard, and genital regions	Produce a sweat that contains human pheromones
	Eccrine sweat glands formed from inward growths from epidermis	Produce sweat that cools the body through evaporation, protects from pathogens through lysozyme, has an acidic pH, provides a flushing action, and makes minor contribution to waste removal
Ceruminous glands	Modified apocrine sweat glands	Produce waxy cerumen to keep eardrum moist and capture particles and insects entering external acoustic meatus
Nails	Fused keratinized epidermal cells formed within ingrowths of the epidermis	Protect the tips of fingers and toes; manipulating small objects

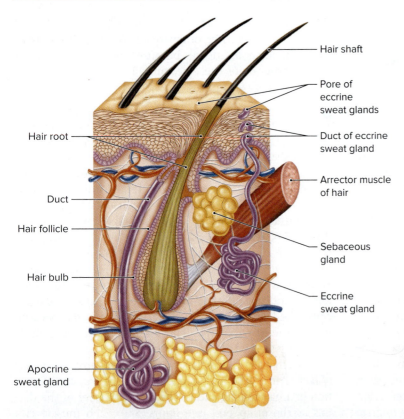

Figure 5.6 Accessory Structures of the Skin.
A section of skin showing the association of a hair follicle, arrector muscle of hair, sebaceous gland, apocrine sweat gland, and eccrine sweat gland. AP|R

erect hairs express rage to enemies or increase the thickness of the insulating layer of hair in cold weather in many mammals.

Hair is present over most of the body, but it is absent on the soles, palms, nipples, lips, and portions of the external genitalia. Their primary function is protection, although the tiny hairs over much of the body have little function. Eyelashes and eyebrows shield the eyes from sunlight and foreign particles. Hairs in the nostrils and external acoustic meatuses (ear canals) act as filters to protect against the entrance of foreign particles and insects. Hair protects the scalp from sunlight, mechanical injury, and heat loss. The nerve endings wrapping around the hair follicle detect movement of the hair shaft when skin contact is made (see figure 5.1).

Glands

Exocrine glands associated with the skin are of three types: sebaceous (oil), sweat, and ceruminous (wax) glands. Each type of gland is formed from either a hair follicle outpocketing or an inward growth from the epidermis during embryonic development.

Sebaceous Glands AP|R

Sebaceous (se-bā-shus) **glands** are oil-producing glands that usually empty their oily secretions into hair follicles (see figures 5.1 and 5.6). The oily

Clinical Insight

Normal hair loss from the scalp is about 75 to 100 hairs per day. Hair loss is increased and regeneration is decreased by poor diet, major illnesses, emotional stress, high fever, certain drugs, chemical therapy, radiation therapy, and aging. Baldness, an inherited trait, is much more common in males than in females.

secretion, called **sebum,** moves outward along the hair root until it reaches the skin surface. Sebum helps to protect the body by inhibiting the growth of some pathogens on the skin surface and by keeping the hair and skin pliable and soft. Moisturizing the skin keeps it from becoming dry and cracked, which creates passageways for pathogens to enter the body. It also aids in preventing dehydration by reducing evaporative water loss.

Sweat Glands

Sweat glands play an important role in maintaining homeostasis. The two types of sweat glands are *apocrine sweat glands* and *eccrine* (or *merocrine*) *sweat glands*. Both types of glands are tubular in shape, similar to a garden hose. The secretory portion of both types of sweat glands is coiled and located within the dermis and subcutaneous tissue, with the apocrine type located deeper than the eccrine type. In addition to being richly supplied with blood vessels, each sweat gland possesses a relatively straight, narrow duct that carries the glandular secretion to either the skin surface or a hair follicle (see figures 5.1 and 5.6).

Eccrine (ek´-rin) **sweat glands** are the most abundant skin glands, with over two million scattered across the body surface. The clear, watery sweat produced is delivered through a narrow duct directly to the skin surface. Eccrine sweat glands are active from birth and are stimulated to produce sweat when body temperature starts to rise during activities such as exercise. The watery nature of eccrine sweat assists in cooling the body through evaporation from the skin surface. Eccrine sweat is capable of protecting the body from environmental hazards. It contains a chemical called *lysozyme*, which has the ability to kill certain types of bacteria on the skin surface. The slightly acidic nature of eccrine sweat limits pathogen growth on the skin surface. Also the large volume of sweat that can be produced creates a flushing action that washes chemicals, pathogens, dirt, etc., from the skin surface. Although the kidneys and lungs are the primary organs of excretion, eccrine sweat does contain small amounts of salts and wastes, in addition to other substances that happen to be in excess within the blood. For example, glucose (blood sugar) can be detected in the sweat of individuals with diabetes mellitus.

Apocrine (ap´-ō-krin) **sweat glands** have ducts that deposit secretions into hair follicles (see figure 5.6). They occur primarily in the axillary, areolar, beard, and genital regions and become active at puberty under the influence of the sex hormones. These glands become stimulated during times of emotional stress and sexual excitement. Apocrine sweat is viscous and milky in color, due to the addition of lipid and proteins. Although the secretions are essentially odorless, decomposition by bacteria produces waste products that cause body odor. Apocrine sweat also contains *pheromones*, chemicals with the ability to alter physiological processes in nearby organisms. Studies have demonstrated the ability of human pheromones to affect reproductive function in males and females.

Ceruminous Glands

Ceruminous (se-rū´mi-nus) **glands** are modified apocrine sweat glands that are located in the external acoustic meatus and produce a waxy secretion called *cerumen* (earwax). The sticky, waxy nature of cerumen helps to keep foreign particles and insects out of the external acoustic meatus. Occasionally, excessive production of cerumen causes a buildup that becomes impacted in the external acoustic meatus. This condition may cause slight hearing loss, as well as pain. Once the impacted cerumen is removed by irrigation and/or mechanical means, hearing returns to normal.

Nails

Hard, hooflike *nails*, composed of dead keratinized epidermal cells, cover the tips of the fingers and toes (figure 5.7). A nail consists of a *nail body*, the portion that is visible, and a *nail root*, the inner portion that is inserted into the dermis. The *free edge* is the portion of the nail body that extends beyond the tip of the finger. Nails are colorless but they normally appear pinkish due to the blood vessels in the *nail bed*, which is the skin beneath the nail body. Nails appear bluish in persons suffering from severe anemia or oxygen deficiency. Near the nail root is a whitish, crescent-shaped area that is called the *lunula*. The *cuticle* is a band of epidermis attached to the proximal border of the nail body. The major function of nails is protection, but fingernails are also useful in manipulating small objects.

Clinical Insight

Sebaceous glands increase their production of sebum at puberty. Accumulated sebum in enlarged hair follicles may form blackheads, whose color comes from oxidized sebum and melanin and not from dirt as is commonly believed. Invasion of certain bacteria may result in pimples or boils.

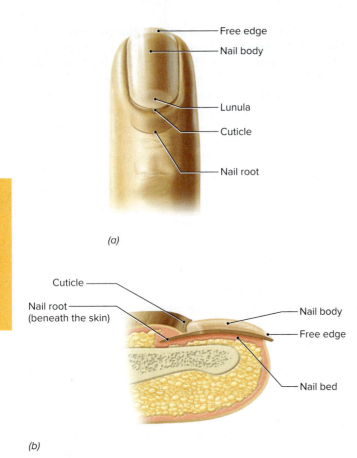

Figure 5.7 The Structure of a Fingernail. A fingernail is composed of dead, heavily keratinized epidermal cells that are formed and undergo keratinization in specialized tissue beneath the nail root. AP|R

Check My Understanding

5. Explain the functions of hair.
6. What are the functions of sebaceous and sweat glands?

5.5 Temperature Regulation

Learning Objectives

8. Describe how the skin aids in the regulation of body temperature.
9. Contrast hypothermia and hyperthermia, including the causes and bodily effects of each.

According to a 1992 study published in the *Journal of the American Medical Association*, humans are able to maintain an average healthy body temperature near 36.8°C (98.2°F), although the surrounding environmental temperature may vary widely. Variations in body temperature of 0.5° to 1.5°F during a 24-hour activity cycle are normal, but variations of more than a few degrees can be life threatening. The brain controls the regulation of body temperature, while the skin plays a key role in conserving or dissipating heat. Decomposition reactions (see chapter 2), especially in metabolically active tissues such as the liver and skeletal muscles, are the source of body heat. Figure 5.8 illustrates the major aspects of temperature regulation.

When body temperature begins to rise above an individual's set point, the brain triggers dilation (widening) of the blood vessels within the skin. The resulting increase in blood flow to the skin increases heat loss from the skin surface. When body temperature becomes excessively high, the brain also activates eccrine sweat glands. These glands release sweat onto the skin surface and its evaporation aids in the removal of excess heat. Once body temperature returns to homeostasis, these changes in blood flow and sweat production cease.

When body temperature begins to fall below an individual's set point, the brain triggers constriction (narrowing) of the blood vessels within the skin. The resulting decrease in blood flow to the skin decreases heat loss from the skin surface. Eccrine sweat glands are not activated, so heat is not lost through sweat evaporation. If heat loss becomes excessive, the brain stimulates small groups of skeletal muscles to produce involuntary, rapid, small contractions (shivering). The increase in skeletal muscle activity increases aerobic respiration and ATP hydrolysis, which in turn generates additional heat to raise body temperature. Once body temperature returns to homeostasis, the changes in blood flow and muscle activity return to normal.

Sometimes the temperature-regulating mechanism is insufficient to counter environmental extremes of temperature. *Hypothermia*, a body temperature below 35.0°C (95.0°F), can result from prolonged exposure to a cold environment, which overwhelms the body's temperature-regulating mechanism. Without treatment, an initial feeling of coldness and shivering can progress to mental confusion, lethargy, loss of consciousness, and death. Persons with little subcutaneous fat (e.g., elderly or thin persons) are more susceptible to hypothermia. In treating a person with hypothermia, the body temperature must be raised gradually to stabilize the cardiovascular and respiratory systems.

In *hyperthermia*, the temperature-regulating mechanism cannot prevent a dangerous increase in body temperature. A person with hyperthermia has a body temperature over 40°C (104°F). It can be caused by environmental factors, trauma, or drug exposure. Consider a person in an environment with both a high air temperature and high humidity level. The high humidity prevents sweat from evaporating and cooling the skin surface. The high air temperature also decreases heat loss and, in situations where the environmental temperature is higher than body temperature, heat is actually gained from surrounding air. In such an environment, excessive physical exertion is a common

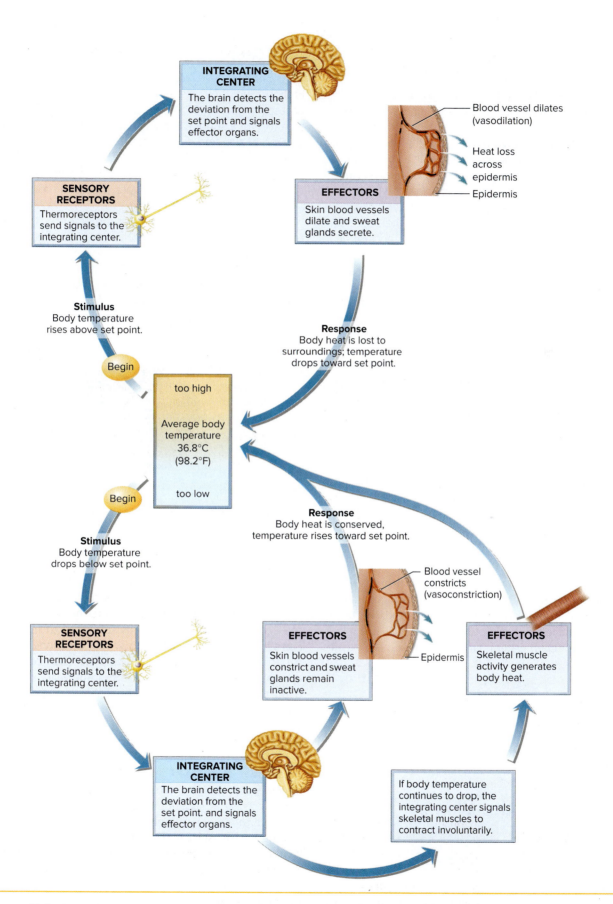

Figure 5.8 The Negative-Feedback Mechanism That Regulates Body Temperature.

⊕ Clinical Insight

Physicians have known for a long time that the UV radiation in sunlight produces damaging changes in the skin. The damage is cumulative, although a single day's exposure can produce noticeable changes. Short-term changes include sunburn and a tan as the body tries to protect itself from UV radiation by producing more melanin. Long-term UV damage ranges from increased wrinkling and loss of elasticity to liver spots and skin cancer.

Because the long-term effects of UV radiation do not appear for a number of years, the summer tans so important to some people will accelerate aging of the skin and may produce skin cancers in later life. The use of tanning salons only increases potential problems for unwary users.

Skin cancer is by far the most common type of cancer. Fortunately, most skin cancers are *carcinomas* involving the stratum basale (*basal cell carcinoma*) or stratum corneum (*squamous cell carcinoma*) and are usually curable by surgical removal. However, *melanoma,* cancer of melanocytes, tends to spread rapidly to other organs and can be lethal if it is not detected and removed in an early stage. Melanoma is fatal in about 45% of the cases.

The undesirable effects of overexposure to UV radiation can easily be prevented by reducing the exposure of the skin to sunlight and by the liberal use of sunblock. UV radiation is at its peak between 11:00 A.M. and 3:00 P.M., so avoiding exposure during these hours is especially helpful.

Sunblock is available with different levels of protection, and they are labeled according to the sun protection factor (SPF) provided. This allows for the selection of a sunblock that is appropriate for a particular type of skin and duration of exposure. For example, sunscreens with an SPF of 30 or higher are available for fair-skinned persons who burn easily. Sunblock with an SPF of 15 may give adequate protection for olive-skinned persons who rarely burn.

Because of its deadly nature, the American Cancer Society has devised the "ABCD rule" for distinguishing malignant melanoma from a mole: *A* for asymmetry (one side has a different shape than the other); *B* for border irregularity (the border is not uniform); *C* for color (color is not uniform, often a mixture of black, brown, tan, or red); *D* for diameter (more than ¼ inch).

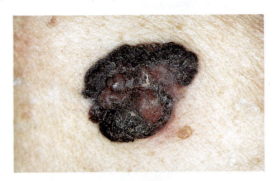

A Cutaneous Malignant Melanoma.
©James Stevenson/Science Source

---

trigger for hyperthermia. Persons who are dehydrated or overweight are also easily susceptible. Without treatment, progressive symptoms may include nausea, headache, dizziness, confusion, loss of consciousness, and death. Lying in a tub of cool (not cold) water is an effective treatment except in severe cases.

Check My Understanding

7. How is the skin involved in the regulation of body temperature?

5.6 Aging of the Skin

Learning Objective
10. Describe how aging affects the skin.

A newborn infant's skin is very thin, and there is not a lot of subcutaneous fat present. During infancy, an infant's skin thickens and more subcutaneous fat is deposited, producing the soft, smooth skin typical of infants.

As a child grows into adulthood, the skin continues to become thicker because it is subjected to numerous harmful conditions: sunlight, wind, abrasions, chemical irritants, and invasions of bacteria. The continued exposure of the skin through the adult years produces damaging effects. However, noticeable changes usually are not apparent until a person approaches 50 years of age. Thereafter, continued aging of the skin is more noticeable.

Typical changes in aging skin are as follows: (1) a breakdown of collagen and elastic fibers (hastened by exposure to sunlight) causes wrinkles and sagging skin; (2) a decrease in subcutaneous fat makes a person more sensitive to temperature changes; (3) a decrease in sebum production by sebaceous glands may cause dry, itchy skin; (4) a decrease in melanin production produces gray hair and sometimes a splotchy pattern of epidermal pigmentation; and (5) a decrease in hair replacement results in thinning hair or baldness, especially in males.

5.7 Disorders of the Skin

Learning Objective

11. Describe the common infectious and noninfectious disorders of the skin.

Because the skin is in contact with the environment, it is especially susceptible to injuries, such as abrasions (scraping), contusions (bruises), and cuts. Other common disorders of the skin may be subdivided into infectious and noninfectious disorders. Some inflammatory disorders may fall into either group, depending upon the specific cause of the disorder. Common childhood diseases, such as chicken pox and measles, are not listed here but produce skin lesions that characterize the particular disease.

Infectious Disorders

Acne (ak′-nē) is a chronic skin disorder characterized by plugged hair follicles that often form pimples (pustules) due to infection by certain bacteria. It often appears at puberty, when sex hormones stimulate increased sebum secretion.

Athlete's foot (tinea pedis) is a slightly contagious infection that is caused by a fungus growing on the skin. It produces reddish, flaky, and itchy patches of skin, especially between and under the toes, where moisture persists.

Boils are acute, painful *Staphylococcus* infections of hair follicles and their sebaceous glands, as well as the surrounding dermis and subcutaneous tissue. The union of several boils forms a *carbuncle*.

Fever blisters, or cold sores, are clusters of fluid-filled vesicles that occur on the lips or oral membranes. They are caused by a *Herpes simplex* virus (type 1) and are transmitted by oral or respiratory exposure. *Genital herpes*, which is caused by either *Herpes simplex* virus type 1 or *Herpes simplex* virus type 2, results in the formation of painful blisters on the genitals as a result of infection transmitted by sexual activity.

Impetigo (im-pe-tī′-gō) is a highly contagious skin infection caused by bacteria. It typically occurs in children and is characterized by fluid-filled pustules that rupture, forming a yellow crust over the infected area.

Noninfectious Disorders

Alopecia (al-ō-pāy′-shē-ah) is the loss of hair. It is most common in males who have inherited male pattern baldness, but it may result from noninherited causes, such as poor nutrition, sensitivity to drugs, and eczema.

Bed bugs (*Cimex lectulariur*) are microscopic parasitic insects that feed almost exclusively off human blood. Their preferred habitats are sleeping areas in hotels and homes, but they can also be found in office buildings, movie theaters, and public transportation vehicles. Bed bugs exhibit peak feeding activity at night, with a preference for exposed areas of skin, and leave behind itchy welts. Washing infested clothes and bedding at 115°F (46°C) will kill bed bugs. Insecticides, deep cleaning of infested areas, and discarding mattresses are recommended for living areas and other contaminated areas.

Bedsores (decubitus ulcers) result from a chronic deficiency of blood flow in the dermis and subcutaneous tissue. Bedsores form over bones that are subjected to prolonged pressure against a bed or cast. They are most common in bedridden patients. Frequent turning of bedridden patients helps to prevent bedsores.

Blisters, fluid-filled pockets, form when an abrasion, burn, or injury causes the epidermis to separate from the dermis.

Burns are damage to the skin caused by heat, chemicals, or radiation. Burns are classified according to the degree of damage. A *first-degree burn* involves only the epidermis. Healing is usually rapid. A *second-degree burn* produces damage to the epidermis and the outer portion of the dermis. Painful blisters form between the epidermis and dermis but usually no infection occurs. Healing requires two to several weeks, and scarring may occur. First- and second-degree burns are also referred to as *partial thickness burns* because they do not penetrate through the full thickness of the skin. A *third-degree burn* (or *full thickness burn*) destroys the epidermis, dermis, glands, hair follicles, and nerve endings. Most cases of third-degree burns are painless due to the complete destruction of nerve endings. Normal skin functions are lost, so care must be given to control fluid loss and bacterial infection. Skin grafting is often necessary, and scarring usually results.

Calluses and **corns** are thickened areas of skin that result from chronic pressure. Calluses are larger and often occur on the palms and on the balls of the feet. Corns are smaller and usually occur on the top of the toes. Improperly fitting shoes are frequent causes of corns on the feet.

A **common mole** (nevus) is a pink, tan, or brown growth usually appearing in childhood and continuing to develop into adulthood. Common moles result when melanocytes grow in clusters. Rarely do they develop into melanoma.

Dandruff (seborrheic dermatitis) is the excessive shedding of dead epidermal cells from the scalp as a result of excessive cell production. It is usually caused by seborrheic eczema of the scalp, a noninfectious dermatitis.

Eczema (ek-zē′-mah) (atopic dermatitis) is an inflammation producing redness, itching, scaling, and sometimes cracking of the skin. It is noninfectious and noncontagious, and it may result from exposure to irritants or from allergic reactions. *Seborrheic eczema* is characterized by hyperactivity of sebaceous glands and patches of red, scaling, and itching skin. It may occur at the corners of the mouth, in hairy areas, or in skin exposed to irritants.

Hives are red, itchy bumps or wheals that usually result from an allergic reaction to certain foods, drugs, or pollens.

Psoriasis (sō-rī′-ah-sis) is a chronic, noncontagious dermatitis that is characterized by reddish, raised patches of skin that are covered with whitish scales. It results from excessive cell production that may be triggered by emotional stress or poor health. The affected area may be slightly sore or may itch. Psoriasis occurs most often on the scalp, elbows, knees, buttocks, and lumbar areas.

Chapter Summary

5.1 Functions of the Skin

- The skin is also called the cutaneous membrane or integument.
- The functions of the skin are protection, excretion, temperature regulation, sensory perception, synthesis of vitamin D, and absorption.

5.2 Structure of the Skin and Subcutaneous Tissue

- The skin is composed of an outer epidermis that covers the inner dermis.
- The epidermis consists of keratinized stratified squamous epithelium, which lacks blood vessels.
- The epidermis is organized into five layers in thick skin and four layers in thin skin.
- New epithelial cells are constantly formed by the stratum basale. As the cells migrate toward the surface, they become keratinized, die, and finally are sloughed off.
- The epidermis contains four types of cells: keratinocytes, melanocytes, dendritic cells, and tactile epithelial cells.
- The dermis is divided into a papillary layer made of areolar connective tissue and a reticular layer made of dense irregular connective tissue containing both collagen and elastic fibers.
- The dermis contains blood vessels, nerves, and sensory receptors.
- Dermal papillae create dermal ridges that produce finger and toe print patterns.
- The dermal papillae and epidermal ridges help to interlock the dermis and epidermis.
- The subcutaneous tissue attaches the skin to underlying tissues and organs, absorbs impact, and stores fat.
- The subcutaneous tissue consists of areolar connective tissue and adipose tissue. It contains blood vessels and nerves.

5.3 Skin Color

- The color of the skin is inherited and results from the presence of three pigments: hemoglobin in dermal blood vessels, carotene in the epidermis and subcutaneous tissue, and melanin in the epidermis.

- Melanin is a brown-black pigment produced by melanocytes in the stratum basale of the epidermis and is incorporated into the keratinocytes.
- Melanin protects the body from UV radiation.

5.4 Accessory Structures

- Accessory structures are formed from the epidermis.
- A hair consists of keratinized epidermal cells that are formed at the base of a hair follicle.
- An arrector muscle of hair is attached to the side of each hair follicle at one end and to the papillary layer of the dermis at the other end. Its contraction pulls the hair into a more erect position.
- Hair occurs over most of the body.
- Glands associated with the skin are the sebaceous, sweat, and ceruminous glands.
- Sebaceous glands produce sebum, an oily secretion that is deposited into hair follicles.
- There are two types of sweat glands: apocrine and eccrine. Apocrine sweat glands occur in axillary and genital areas and secrete a relatively thick sweat that is deposited into hair follicles. Eccrine sweat glands occur all over the body and secrete a watery sweat that is deposited onto the surface of the skin.
- Eccrine sweat is used to cool the body, wash the skin surface, remove chemicals from blood, and protect against pathogens. Apocrine sweat contains pheromones.
- Ceruminous glands are located in the external acoustic meatus and secrete a waxy substance called cerumen.
- Nails protect the tips of fingers and toes.
- Nails are formed of layers of heavily keratinized and dead keratinocytes.

5.5 Temperature Regulation

- Average healthy human body temperature is 36.8°C (98.2°F).
- Body heat is produced by decomposition reactions.
- When body temperature rises above an individual's set point, blood vessels in the dermis dilate to increase heat loss and sweat is produced. The evaporation of sweat increases heat loss.
- When body temperature falls below an individual's set point, blood vessels in the dermis are constricted to reduce heat loss and arrector muscles of hair contract.

Under extreme heat loss, spontaneous skeletal muscle contractions (shivering) produce additional heat.
- When temperature regulation is overwhelmed, hypothermia and hyperthermia become medical emergencies.

5.6 Aging of the Skin

- After 50 years of age, wrinkles and sagging skin become noticeable.
- The effects of aging are caused by a breakdown of collagen and elastic fibers, a decrease in sebum production, a decrease in melanin production, and a decrease in subcutaneous fat.

5.7 Disorders of the Skin

- Infectious disorders of the skin include acne, athlete's foot, boils, fever blisters, and impetigo.
- Noninfectious disorders of the skin include alopecia, bed bugs, bedsores, blisters, burns, calluses and corns, common moles, dandruff, eczema, hives, and psoriasis.

Improve Your Grade

Connect Interactive Questions Reinforce your knowledge using multiple types of questions: interactive, animation, classification, labeling, sequencing, composition, and traditional multiple choice and true/false.

SmartBook Proven to help students improve grades and study more efficiently, SmartBook contains the same content within the print book but actively tailors that content to the needs of the individual.

Anatomy & Physiology REVEALED® Dive into the human body by peeling back layers of cadaver imaging. Utilize this world-class cadaver dissection tool for a closer look at the body anytime, from anywhere.

CHAPTER 6

Skeletal System

©Cultura Creative (RF)/Alamy Stock Photo RF

Steven and his friends are amateur skateboarders hanging out at their city skate park. At the urging of his friends, Steven decides to try a 360° spin for the very first time, even though he has forgotten his helmet and other safety gear. As he reaches the top of the ramp and begins his spin, his right foot slips off the skateboard, disturbing his balance. Steven throws his arms out to brace his fall but he is unable to keep his head from impacting the ramp as he rolls to the bottom. He sits up and rubs his head, stunned from the fall. His friends race down and begin to check Steven's limbs for fractures. Thankfully, the dense minerals in the bones of his arms and legs were able to resist fracturing. His skull was also hard enough to protect his brain from damage when it hit the ground. Covered in minor cuts and abrasions, Steven stands up and grabs his skateboard. Without even a thought, he fearlessly walks back to the top of the ramp, convinced that he will conquer the 360° spin today.

CHAPTER OUTLINE

6.1 Functions of the Skeletal System

6.2 Bone Structure
- Gross Structure of a Long Bone
- Microscopic Structure of a Long Bone

6.3 Bone Formation
- Intramembranous Ossification
- Endochondral Ossification
- Homeostasis of Bone

6.4 Divisions of the Skeleton

6.5 Axial Skeleton
- Skull
- Vertebral Column
- Thoracic Cage

6.6 Appendicular Skeleton
- Pectoral Girdle
- Upper Limb
- Pelvic Girdle
- Lower Limb

6.7 Joints
- Fibrous Joints
- Cartilaginous Joints
- Synovial Joints

6.8 Disorders of the Skeletal System
- Disorders of Bones
- Disorders of Joints

Chapter Summary

Module 5
Skeletal System

SELECTED KEY TERMS

Cartilaginous joint Joint with two bones joined by a thin band or pad of cartilage.
Compact bone Bone tissue composed of numerous tightly packed osteons.
Diaphysis (dia = through, apart; physis = to grow) The long shaft of a long bone.
Endochondral ossification (endo = inside; chondr = cartilage; oss = bone) The formation of bone tissue within cartilage tissue.
Epiphysial (growth) plate The hyaline cartilage between the epiphysis and diaphysis of immature long bones.
Epiphysis (epi = upon) The enlarged ends of a long bone.
Fibrous joint Joint with skeletal structures joined by a thin band of dense irregular connective tissue.
Intramembranous ossification (intra = inside) The formation of bone tissue within embryonic connective tissue.
Joint A junction between two bones or between a bone and a tooth.
Ligament A band or cord of dense regular connective tissue that joins bones together at joints.
Medullary cavity (medulla = marrow) The cavity within the shaft of a long bone that is filled with yellow bone marrow.
Spongy (trabecular) bone Bone tissue composed of trabeculae surrounded by red or yellow bone marrow.
Synovial joint (syn = with; ov = egg) Joint with two bones separated by a small fluid-filled cavity and surrounded by a connective tissue capsule.

THE SKELETAL SYSTEM SERVES as the supporting framework of the body, and performs several other important functions as well. The body shape, mechanisms of movement, and erect posture observed in humans would be impossible without the skeletal system. Two very strong tissues, bone and cartilage, compose the skeletal system.

6.1 Functions of the Skeletal System

Learning Objective
1. Describe the basic functions of the skeletal system.

The skeletal system performs five major functions:

1. **Support.** The skeleton serves as a rigid supporting framework for the soft tissues of the body.
2. **Protection.** The arrangement of bones in the skeleton provides protection for many internal organs. The thoracic cage provides protection for the internal thoracic organs, including the heart and lungs; the cranial bones form a protective case around the brain, ears, and all but the front of the eyes; the vertebrae protect the spinal cord; and the pelvic girdle protects some reproductive, urinary, and digestive organs.
3. **Attachment sites for skeletal muscles.** Skeletal muscles are attached to bones and extend across joints. Bones function as levers, enabling movement at joints when skeletal muscles contract.
4. **Production of formed elements.** The red bone marrow in spongy bone produces formed elements.
5. **Mineral storage.** The matrix of bones serves as a storage area for large amounts of calcium salts, which may be removed for use in other parts of the body when needed.

 Check My Understanding
1. What are the major functions of the skeletal system?
2. What are some examples of how the skeleton provides protection for internal organs?

6.2 Bone Structure

Learning Objectives
2. List the types of bones based on their shapes.
3. Describe the gross structure and microstructure of a long bone and a flat bone.

There are approximately 206 bones in an adult, and each bone is an organ composed of a number of tissues. Bone tissue forms the bulk of each bone and consists of both living cells and a nonliving matrix formed primarily of calcium salts. Other tissues include cartilage, blood, nerve tissue, adipose tissue, and dense irregular connective tissue.

There are six basic types of bones based on their shapes (figure 6.1). *Short bones* are most of the bones within the wrist and ankle, and they possess a small, boxy appearance. *Long bones* possess a long, skinny shape and include the clavicles (collarbones) and the bones in the upper and lower limbs, with the exception of the wrist and ankle bones and the patellae (kneecaps; singular, patella). *Sutural bones* are small bones that form within the sutures of the skull; they vary in number and location from person to person. Certain skull bones, the scapulae

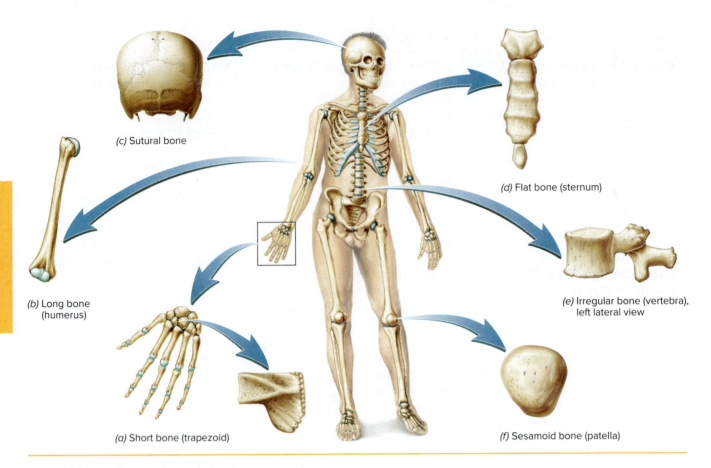

Figure 6.1 Basic Types of Bones.

(shoulder blades; singular, scapula), the ribs, and the sternum (breastbone) are classified as *flat bones*, meaning they are thin, flat, and slightly curved. *Irregular bones*, such as the vertebrae (backbones; singular, vertebra), coxal bones (hip bones), calcanei (heel bones; singular, calcaneus) of the ankles, and some skull bones, possess irregular shapes with numerous projections. *Sesamoid bones* are unique because of their sesame seed shape and the fact that they form inside muscle tendons. The patellae and pisiforms in the wrists are sesamoid bones.

Gross Structure of a Long Bone

The femur, the bone of the thigh, will be used as an example in considering the structure of a long bone. Refer to figure 6.2 as you study the following section.

At each end of the bone, there is an enlarged portion called an **epiphysis** (ē-pif′-e-sis; plural, epiphyses). The *epiphyses* articulate (articulo = joint) with adjacent bones to form joints. The **articular cartilage,** which is composed of hyaline cartilage, covers the articular surface of each epiphysis. Its purpose is to protect and cushion the end of the bone, in addition to providing a smooth surface for movement of joints. The long shaft of bone that extends between the two epiphyses is the **diaphysis** (dī-af′-e-sis). In immature bones, each epiphysis is joined to the diaphysis by an **epiphysial (growth) plate** of hyaline cartilage, which is used to lengthen the bone. The epiphysial plate is replaced by bone in mature bones, and the site of bone fusion is called an **epiphysial line.**

Except for the regions covered by articular cartilages, the entire bone is covered by the **periosteum** (per-ē-os′-tē-um), a dense irregular connective tissue membrane that is firmly attached to the underlying bone. The periosteum provides protection and also is involved in the formation and repair of bone. Tiny blood vessels from the periosteum help to nourish the bone.

The inner structure of a long bone is revealed by a longitudinal section. **Spongy (trabecular) bone** forms the inner region of the epiphyses and the inner surface of the diaphysis wall. It consists of thin rods or plates of bone tissue called **trabeculae** (trah-bek′-u-lē) that form a meshlike framework containing numerous spaces. The trabeculae are covered by a thin connective tissue membrane called **endosteum** (en-dos′-tē-um) that is involved in forming and repairing bone. Spongy bone reduces the weight of a bone without reducing its supportive strength. In an adult's long bones, **red bone marrow** fills the spaces between trabeculae within the proximal epiphyses of the humerus and femur. In other epiphyses

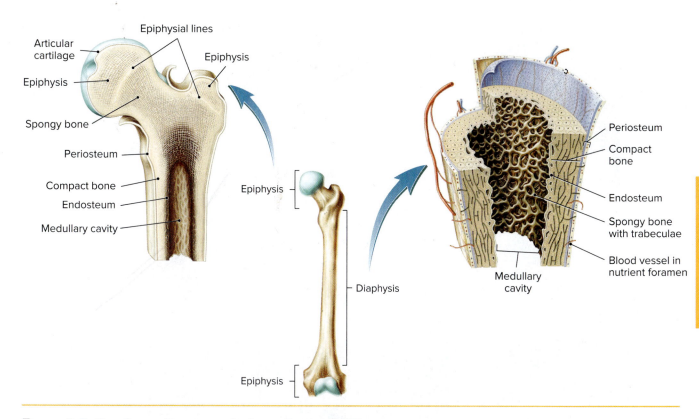

Figure 6.2 The Gross Structure of a Long Bone.

of the limbs, the spaces between trabeculae are filled with **yellow bone marrow,** which is composed of adipose tissue. The ability of red bone marrow to produce formed elements will be discussed in chapter 11.

Compact bone forms the wall of the diaphysis and a thin outer layer over the epiphyses. As the name implies, compact bone is formed of tightly packed bone tissue that lacks the spaces found in spongy bone. Compact bone is very strong, and it provides the supportive strength of long bones. The cavity that extends the length of the diaphysis is the **medullary cavity.** It is lined by the endosteum and is filled with yellow bone marrow.

The structure of the other bone types is like that of the epiphyses of long bones. Their outer surfaces are composed of a thin layer of compact bone covered with periosteum; the inner region is spongy bone covered with endosteum. In most of these bones, red bone marrow fills in the spaces between the trabeculae.

Microscopic Structure of a Long Bone

As noted earlier, there are two types of bone tissue: compact bone and spongy bone. When viewed microscopically, compact bone is formed of a number of subunits called osteons (figure 6.3). An **osteon** (os′-tē-on) is composed of a **central canal** containing blood vessels and nerves, surrounded by the **lamellae** (singular, *lamella*), concentric layers of bone matrix. Bone cells, the **osteocytes** (os′-tē-ō-sītz), are located between lamellae where they occupy tiny, interstitial-fluid filled spaces called **lacunae.**

Blood vessels and nerves enter a bone through a **nutrient foramen** (fō-rā′-men; plural, *foramina*), a channel entering or passing through a bone. The blood vessels form branches that pass through *perforating canals* and enter the central canals to supply nutrients to the osteocytes. **Canaliculi,** the tiny tunnels radiating from the lacunae, connect osteocytes with each other and the blood supply.

The trabeculae of spongy bone lack osteons, so osteocytes receive nutrients by diffusion of materials through canaliculi from blood vessels in the bone marrow surrounding the trabeculae (figure 6.3).

✓ Check My Understanding

3. What are the general functions of the skeletal system?
4. What are the major gross anatomical structures of a long bone?
5. How are compact and spongy bone histologically different?

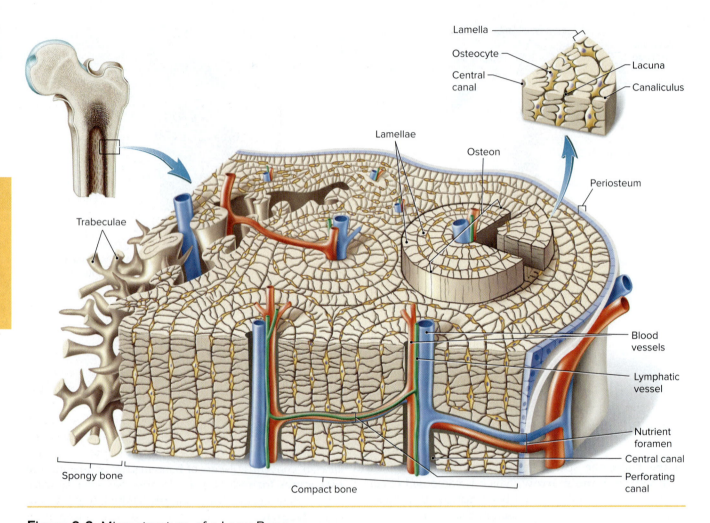

Figure 6.3 Microstructure of a Long Bone.

6.3 Bone Formation

Learning Objectives
4. Compare intramembranous and endochondral ossification.
5. Compare the functions of osteoblasts and osteoclasts.

The process of bone formation is called **ossification** (os-i-fi-kā′-shun). It begins during the sixth or seventh week of embryonic development. Bones are formed by the replacement of existing connective tissues with bone tissue (figure 6.4). There are two types of bone formation: intramembranous ossification and endochondral ossification. Table 6.1 summarizes these.

In both types of ossification, some primitive connective tissue cells are changed into bone-forming cells called **osteoblasts** (os′-tē-ō-blasts). Osteoblasts deposit bone matrix around themselves and soon become imprisoned in lacunae. Once this occurs, they are called osteocytes.

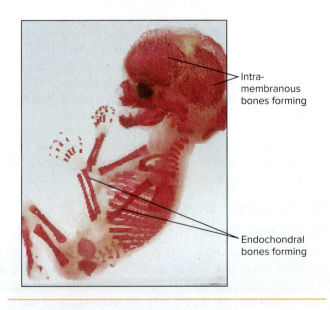

Figure 6.4 The Fetal Skeleton.
The stained, developing bones of a 14-week fetus.
©Biophoto Associates/Science Source

Table 6.1 Comparison of Intramembranous and Endochondral Ossification

Intramembranous	Endochondral
1. Membranes of embryonic connective tissue form at sites of future bones. 2. Some connective tissue cells become osteoblasts, which deposit spongy bone within the membrane. 3. Osteoblasts from the enclosing membrane, now called the periosteum, deposit a layer of compact bone over the spongy bone.	1. Bone is preformed in hyaline cartilage. 2. Osteoblasts of periosteum form a collar of compact bone that thickens and grows toward each end of the bone. 3. Cartilage is calcified, and osteoblasts derived from the periosteum form spongy bone, which replaces cartilage in ossification centers. The spongy bone is later removed in the diaphysis to form the medullary cavity.

Intramembranous Ossification

Most skull bones are formed by **intramembranous ossification**. Connective tissue membranes form early in embryonic development at sites of future intramembranous bones. Later, some connective tissue cells become osteoblasts and deposit spongy bone within the membranes starting in the center of the future bone. The membrane surrounding the developing bone becomes the periosteum. Osteoblasts from the periosteum then deposit a layer of compact bone over the spongy bone.

Some bone tissue must be removed and re-formed in order to produce the correct shape of the bone as it develops and grows. Cells that remove bone matrix are called **osteoclasts**. The opposing actions of osteoblasts and osteoclasts ultimately produce the shape of the mature bone.

Endochondral Ossification

Most bones of the body are formed by **endochondral** (en-dō-kon'-drul) **ossification**. Future endochondral bones are preformed in hyaline cartilage early in embryonic development. Figure 6.5 illustrates the ossification of a long bone.

In long bones, a new periosteum develops around the diaphysis of the hyaline cartilage template. Osteoblasts from the periosteum form a collar of compact bone around the diaphysis. A *primary ossification center* also forms in the middle of the cartilage shaft due to the enlargement of chondrocytes and a loss of cartilage matrix between lacunae. Calcification, which involves the depositing of calcium salts, occurs within the primary ossification center and leads to the death of chondrocytes. Blood vessels and nerves penetrate into the primary ossification center carrying along osteoblasts from the periosteum. The osteoblasts convert the calcified cartilage into spongy bone. As *secondary ossification centers* form in the epiphyses of the cartilage template, osteoclasts begin to remove spongy bone from the diaphysis to form the medullary cavity. The bone continues to grow as ossification progresses. As cartilage tissue continues to be replaced, the cartilage tissue between the primary and secondary ossification centers decreases until only a thin plate of cartilage, the epiphysial plate, separates the epiphyses from the diaphysis.

Subsequent growth in diameter results from continued formation of compact bone by osteoblasts from the periosteum. Growth in length occurs as bone tissue replaces cartilage tissue on the diaphysis side of each epiphysial plate, while new cartilage tissue is formed on the epiphysis sides. The opposing actions of osteoblasts and osteoclasts continually reshape the bone as it grows.

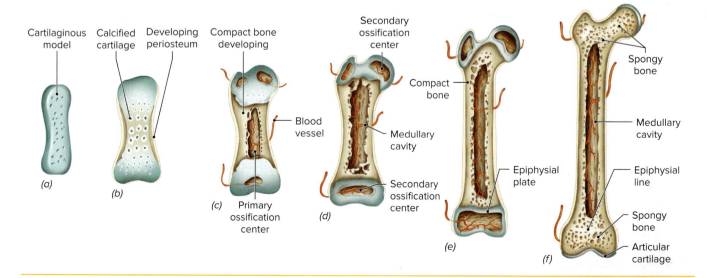

Figure 6.5 Major Stages (*a–f*) in the Development of an Endochondral Bone. (Bones are not shown to scale.)

Growth usually continues until about age 25, when the epiphysial plates are completely replaced by bone tissue. After this, growth in the length of a bone is not possible. The visible lines of fusion between the epiphyses and the diaphysis are called epiphysial lines.

Homeostasis of Bone

Bones are dynamic, living organs, and they are continually restructured throughout life. This occurs by the removal of bone matrix by osteoclasts and by the deposition of new bone matrix by osteoblasts. Physical activity causes the density and volume of bones to be maintained or increased, though inactivity results in a reduction in bone density and volume.

Calcium salts may be removed from bones to meet body needs anytime blood calcium levels are low, such as when dietary intake is inadequate. When dietary calcium intake increases blood calcium to a sufficient level, some calcium is used to form new bone matrix.

Children have a relatively large number of collagen fibers in their bone matrix, which makes their bones somewhat flexible. But as people age, the amount of collagen gradually decreases. This trend causes older people to have brittle bones that are prone to fractures. Older persons may also experience a gradual loss of bone matrix, which reduces the strength of the bones. A severe reduction in bone density, and therefore increased risk of fracture, is called *osteoporosis*.

Check My Understanding

6. How do intramembranous ossification and endochondral ossification differ?
7. How does physical activity affect the homeostasis of bones?

6.4 Divisions of the Skeleton

Learning Objectives

6. Name the two divisions of the skeleton.
7. Describe the major surface features of the bones and their importance.

The human adult skeleton is composed of two distinct divisions: the axial skeleton and the appendicular skeleton. The **axial** (ak'-sē-al) **skeleton** consists of the bones along the longitudinal axis of the body that support the head, neck, and trunk. The **appendicular** (ap-en-dik'-ū-lar) **skeleton** consists of the bones of the upper limbs and pectoral girdles and of the lower limbs and pelvic girdle (figure 6.6).

A study of the skeleton includes the various surface features of bones, such as projections, depressions, ridges, grooves, and holes. Specific names are given to each type of surface feature. Knowledge of surface bony features is essential for understanding the origins and insertions of skeletal muscles discussed in the muscular system, and is important in locating internal structures in clinical practice. The names of the major surface features are listed in table 6.2 and shown in figure 6.7 for easy reference as you study the bones of the skeleton.

Check My Understanding

8. What bones compose the axial skeleton?
9. What bones compose the appendicular skeleton?

6.5 Axial Skeleton

Learning Objectives

8. Identify the bones of the axial skeleton.
9. Compare the skulls of an infant and an adult.
10. Compare the bones of the vertebral column.
11. Compare true, false, and floating ribs.

The major components of the axial skeleton are the skull, vertebral column, and thoracic cage. Bones of the axial skeleton are shown in figure 6.6.

Skull

The **skull** is subdivided into the *cranium*, which is formed of eight bones encasing the brain, and 14 *facial bones*. With the exception of the mandible, all the skull bones are joined by joints that exhibit no movement, called **sutures** (sū'-churs) because they resemble stitches. The cranial cavity formed by the cranium protects the brain. The facial bones surround and support the openings of the digestive and respiratory systems. Several bones in the skull contain air-filled spaces called **paranasal sinuses** (figure 6.9) that are connected to the nasal cavity. These sinuses reduce the weight of the skull, add resonance to a person's voice, and produce mucus, which helps to moisten and purify the air within the sinuses and within the nasal cavity. The bones of the skull are shown in figures 6.8 to 6.12. Locate the bones on these figures as you study this section.

Cranium

The **cranium** is formed of one frontal bone, two parietal bones, one sphenoid, two temporal bones, one occipital bone, and one ethmoid.

The **frontal bone** forms the anterior part of the cranium, including the upper portion of the orbits (eye sockets), the forehead, and part of the roof of the nasal cavity. There are two large *frontal sinuses* in the frontal bone, one located above each eye.

The two **parietal** (pah-rī'-e-tal) **bones** form the sides and roof of the cranium. They are joined at the midline by the *sagittal suture* and to the frontal bone by the *coronal suture*.

The **occipital** (ok-sip'-i-tal) **bone** forms the posterior portion and floor of the cranium. It contains a large opening,

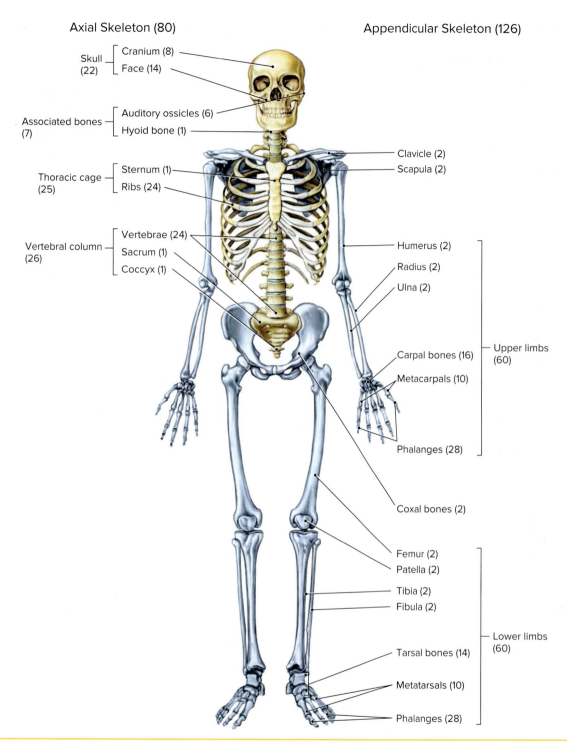

Figure 6.6 Bones of Axial Skeleton (colored gold) and Appendicular Skeleton (colored blue). AP|R

the *foramen magnum*, through which the spinal cord enters the skull to connect with the brainstem. On each side of the foramen magnum are the *occipital condyles* (kon′-dĭls), large knucklelike surfaces that articulate with the first vertebra of the vertebral column. The occipital bone is joined to the parietal bones by the *lambdoid* (lam′-doyd) *suture*.

The **temporal bones** are located below the parietal bones on each side of the cranium. They are joined to the parietal bones by *squamous* (skwā′-mus) *sutures* and to the occipital bone by the lambdoid suture. In each temporal bone, an **external acoustic meatus** leads inward to the eardrum. Just in front of the external acoustic meatus is the *mandibular fossa*, a depression that receives the mandibular condyle to form the *temporomandibular joint*.

Three processes are located on each temporal bone. The *zygomatic* (zī-gō-mat′-ic) *process* projects forward to join

Table 6.2 Surface Features of Bones

Feature	Description
Processes Forming Joints	
Condyle	A rounded or knucklelike process
Head	An enlarged rounded end of a bone supported by a constricted neck
Facet	A smooth, nearly flat surface
Processes for Attachment of Ligaments and Tendons	
Crest	A prominent ridge or border
Epicondyle	A prominence above a condyle
Spine	A sharp or slender process
Trochanter	A very large process found on the femur
Tubercle	A small, rounded process
Tuberosity	A large, roughed process
Depressions and Openings	
Alveolus	A deep pit or socket
Canal, Meatus	A tubelike passageway into or through a bone
Foramen	An opening or passageway through a bone
Fossa	A small depression
Groove	A furrowlike depression
Sinus	An air-filled cavity within a bone

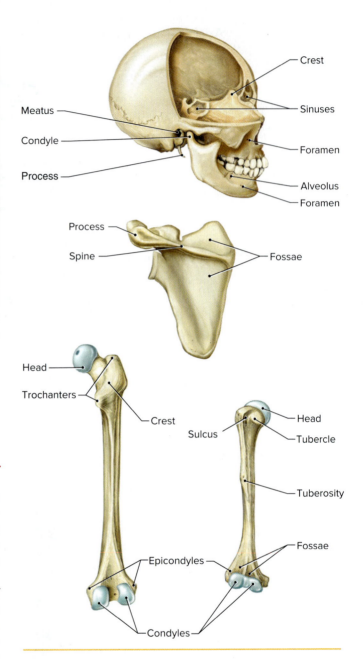

Figure 6.7 Surface Features of Bones.

with the zygomatic bone. The *mastoid* (mas′-toyd) *process* is a large, rounded projection that is located below the external acoustic meatus. It serves as an attachment site for some neck muscles. The *styloid process* lies just medial to the mastoid process. It is a long, spikelike process to which muscles and ligaments of the tongue and neck are attached.

The **sphenoid** (sfē′-noyd) forms part of the floor of the cranium, the posterior portions of the orbits, and the sides of the cranium just in front of the temporal bones. Because it articulates with all other cranial bones, the sphenoid is referred to as the "keystone" of the cranium. On its upper surface at the midline is a saddle-shaped structure called the *sella turcica* (ter′-si-ka), or "turkish saddle." It has a depression that contains the pituitary gland. Two *sphenoidal sinuses* are located just below the sella turcica.

The **ethmoid** (eth′-moyd) also contributes to the anterior portion of the cranium, including part of the medial surface of each orbit and part of the roof of the nasal cavity. The lateral portions contain several air-filled sinuses called *ethmoidal cells*. The *perpendicular plate* extends downward to form most of the nasal septum, which separates the right and left portions of the nasal cavity. It joins the sphenoid and vomer posteriorly and the nasal and frontal bones anteriorly.

The *superior* and *middle nasal conchae* (kong′-kē; singular, concha) extend from the lateral portions of the ethmoid toward the perpendicular plate. These delicate, scroll-like bones support the mucous membrane and increase the surface area of the nasal wall. The roof of the nasal cavity is formed by the *cribriform plates* of the ethmoid; the olfactory nerves enter the cranial cavity through foramina in the cribriform plate. On the upper surface where these plates join at the midline is a prominent projection called the *crista galli*, or cock's comb. The meninges that envelop the brain are attached to the crista galli.

Facial Bones

The paired bones of the face are the maxillae, palatine bones, zygomatic bones, lacrimal bones, nasal bones, and inferior nasal conchae. The single bones are the vomer and mandible.

Part 2 Covering, Support, and Movement of the Body   111

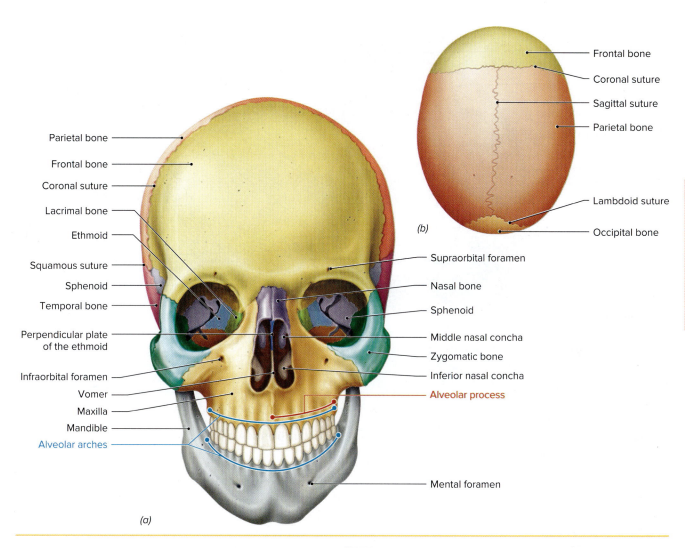

Figure 6.8 Anterior and Superior Views of the Skull. AP|R

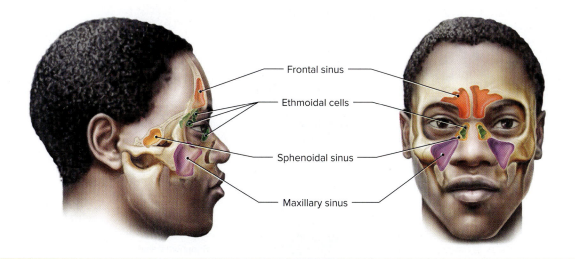

Figure 6.9 Paranasal Sinuses.
Paranasal sinuses are located in the frontal bone, the ethmoid, the sphenoid, and the maxillae. They are connected with the nasal cavity and increase the surface area of the nasal cavity.

112   Chapter 6   Skeletal System

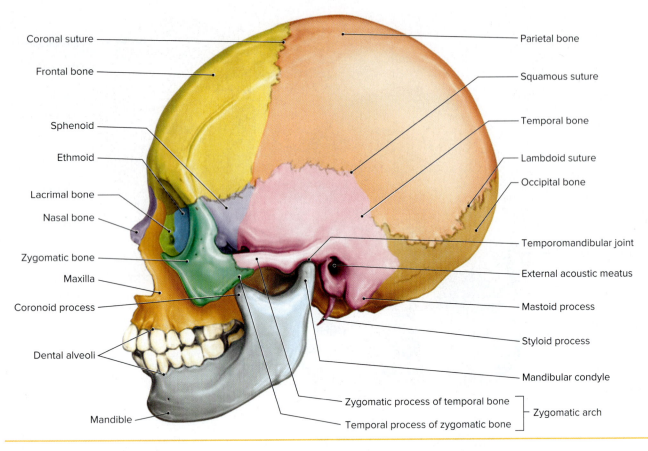

Figure 6.10 Lateral View of the Skull.

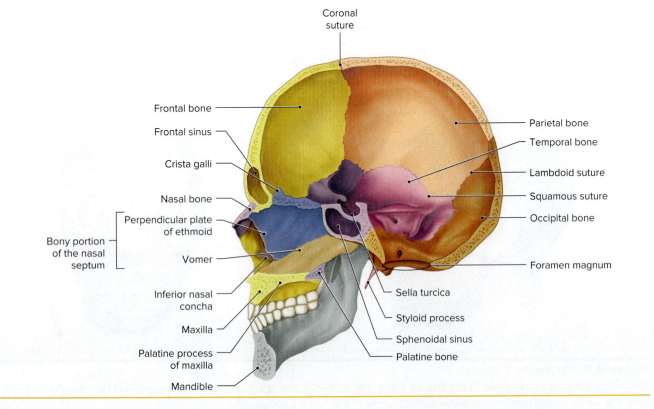

Figure 6.11 Median View of the Skull.

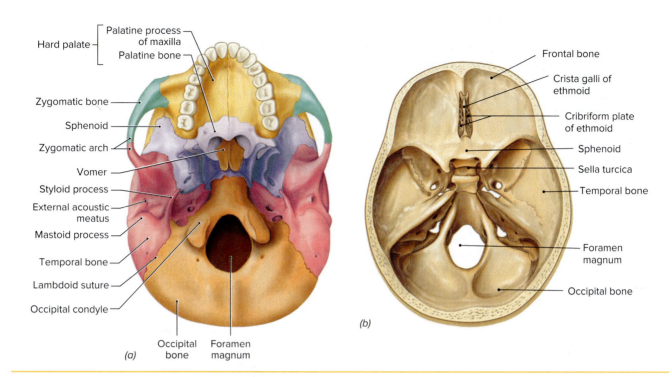

Figure 6.12 Inferior and Superior Views of the Skull.
(a) Inferior view of the skull. (b) Superior view of the transverse section of the skull. APR

The **maxillae** (mak-sil′-ē) form the upper jaw. Each maxilla is formed separately, but they are joined at the midline during embryonic development. The maxillae articulate with all of the other facial bones except the mandible. The *palatine processes* of the maxillae form the anterior portion of the *hard palate* (roof of the mouth and floor of the nasal cavity), part of the lateral walls of the nasal cavity, and the floors of the orbits.

Each maxilla possesses a downward-projecting, curved ridge of bone that contains the teeth. This ridge is the *alveolar process*, and the sockets containing the teeth are called *dental alveoli* (singular, alveolus). The alveolar processes unite at the midline to form the U-shaped maxillary *alveolar arch*. A large *maxillary sinus* is present in each maxilla just below the orbits.

The **palatine** (pal′-ah-tīn) **bones** are fused at the midline to form the posterior portion of the hard palate. Each bone has a lateral portion that projects upward to form part of a lateral wall of the nasal cavity.

The **zygomatic bones** (cheekbones) form the prominences of the cheeks and the floors and lateral walls of the orbits. Each zygomatic bone has a posteriorly projecting process, the *temporal process*, that extends to unite with the zygomatic process of the adjacent temporal bone. Together, they form the *zygomatic arch*.

The **lacrimal** (lak′-ri-mal) **bones** are small, thin bones that form part of the medial surfaces of the orbits. Each lacrimal bone is located between the ethmoid and maxilla.

The **nasal** (nā′-zal) **bones** are thin bones fused at the midline to form the bridge of the nose.

The **vomer** is a thin, flat bone located on the midline of the nasal cavity. It joins posteriorly with the perpendicular plate of the ethmoid, and these two bones form the bony part of the nasal septum.

Clinical Insight

The hard palate separates the nasal cavity from the oral cavity, which allows for chewing and breathing to occur at the same time. A *cleft palate* results when the palatine processes of the maxillae and the palatine bones fail to join before birth to form the hard palate. A *cleft lip,* a split upper lip, is often associated with a cleft palate. These congenital deformities can be corrected surgically after birth.

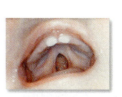

Cleft palate
©Biophoto Associates/
Science Source

Cleft lip
©Dr M.A. Ansary/
Science Source

The **inferior nasal conchae** are scroll-like bones attached to the lateral walls of the nasal cavity below the middle nasal conchae of the ethmoid. They project medially into the nasal cavity and serve the same function as the superior and middle nasal conchae of ethmoid.

The **mandible** (lower jaw) is the only movable bone of the skull. It consists of a U-shaped *body* with an upward-projecting portion, a *ramus*, extending from each end of the body. The upper portion of the body forms the mandibular *alveolar arch*, which contains the dental alveoli for the teeth. The upper part of each ramus is Y-shaped and forms two projections: an anterior *coronoid process* and a posterior *mandibular condyle*. The coronoid process is a site of attachment for muscles used in chewing. The mandibular condyle articulates with the mandibular fossa of the temporal bone to form a *temporomandibular joint*. These joints are sometimes involved in a variety of dental problems associated with an improper bite.

Associated Bones to the Skull

The **hyoid** (hī′-oyd) **bone** and **auditory ossicles** are known as associated bones to the skull because they are located in or near the skull but are not directly connected with any skull bones. The hyoid bone is a small, U-shaped bone located in the anterior portion of the neck, below the mandible. It does not articulate with any bone. Instead, it is suspended from the styloid processes of the temporal bones by muscles (figure 6.13). Muscles of the tongue are attached to the hyoid bone. The auditory ossicles (malleus, incus, stapes) are the smallest bones in the human body. They articulate with each other in the middle ears and assist in sound conduction and amplification (see chapter 9).

The Infant Skull

The skull of a newborn infant is incompletely developed. The face is relatively small with large orbits, and the bones are thin and incompletely ossified. The bones of the cranium are separated by dense connective tissue, with six rather large, nonossified areas called **fontanelles** (fon′-tah-nels) or soft spots (figure 6.14). The frontal bone is formed of two separate parts that fuse later in development. Incomplete ossification of the skull bones and the abundance of

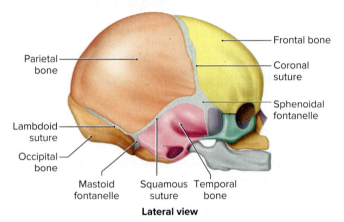

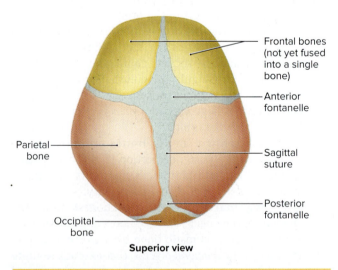

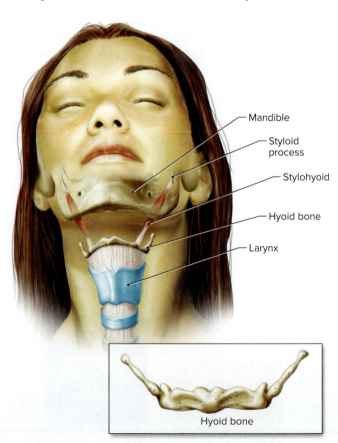

Figure 6.13 Anterior View of the Hyoid Bone.

Figure 6.14 The Fetal Skull.
There are one anterior, two mastoid, one posterior, and two sphenoidal fontanelles between the cranial bones in the fetus's skull. Note the fontanelles and the membranes between the cranial bones.

dense irregular connective tissue make the skull somewhat flexible and allow for partial compression of the skull to facilitate easier vaginal delivery. After birth, they allow for the skull to expand easily and accommodate the rapidly growing brain. Compare the infant skull in figure 6.14 with the adult skull in figures 6.8 and 6.10.

Check My Understanding

10. What bones form the cranium?
11. What bones form the face?

Vertebral Column

The vertebral column (spine or backbone) extends from the skull to the pelvis and forms a somewhat flexible but sturdy longitudinal support for the trunk. It is formed of 24 slightly movable vertebrae, the sacrum, and the coccyx. The vertebrae are separated from each other by **intervertebral discs** composed of fibrocartilage that serve as shock absorbers and allow bending of the vertebral column. Four distinct curvatures can be seen on the side view of the vertebral column (figure 6.15). From superior to inferior they are the *cervical, thoracic, lumbar,* and *sacral curvatures.* These curvatures provide flexibility and cushion, and allow the vertebral column to bear body weight more efficiently.

Structure of a Vertebra

Vertebrae are divided into three groups: cervical, thoracic, and lumbar vertebrae. Although each type has a distinctive anatomy, they have many features in common (figure 6.16).

The anterior, drum-shaped mass is the *vertebral body,* which serves as the major load-bearing portion of a vertebra. A bony *vertebral arch* surrounds the large *vertebral foramen* through which the spinal cord and nerve roots pass.

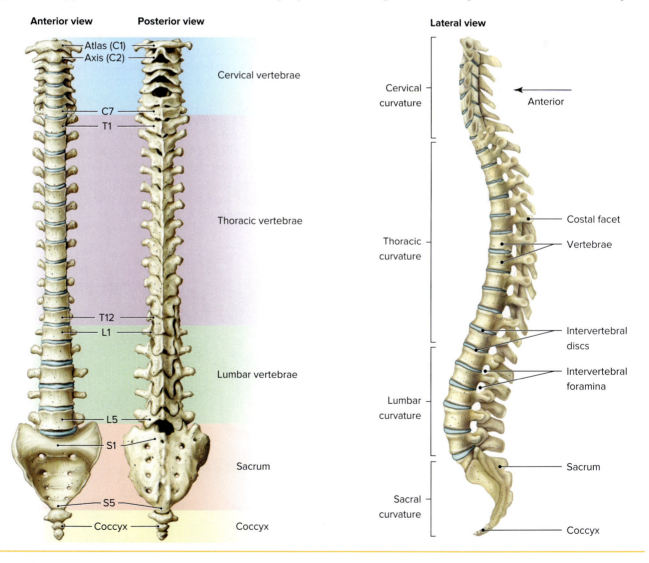

Figure 6.15 Structure of the Vertebral Column.
The vertebral column consists of 24 movable vertebrae, separated by intervertebral discs, sacrum, and coccyx.

A *spinous process* projects posteriorly and *transverse processes* project laterally from each vertebral arch.

A pair of *superior articular processes* project upward, while a pair of *inferior articular processes* project downward from the vertebral arch. The *articular facet* (fa′-set) of each superior articular process articulates with the articular facet of the inferior articular process of the adjacent vertebra above it. When joined by ligaments, the vertebrae form the *vertebral canal* that protects the spinal cord.

Small *intervertebral foramina* occur between adjacent vertebrae. They serve as side passageways for spinal nerves that exit the spinal cord (see figures 6.15 and 6.16).

Cervical Vertebrae

The first seven vertebrae are the **cervical** (ser′-vi-kul) **vertebrae** (C1–C7) that support the neck. They are unique in having a *transverse foramen* in each transverse process. It serves as a passageway for the vertebral arteries and veins, blood vessels involved in blood flow to and from the brain (figure 6.17a–d).

The first two cervical vertebrae are distinctly different from the rest. The first vertebra (C1), or **atlas,** whose superior articular facets articulate with the occipital condyles, supports the head. This joint allows for nodding of the head. The second vertebra (C2), which is called the **axis,** has a prominent *dens* that projects upward from the vertebral body, providing a pivot point for the atlas. When the head is turned, the atlas rotates on the axis (see figure 6.17a–c).

Thoracic Vertebrae

The 12 **thoracic vertebrae** (T1–T12) are larger than the cervical vertebrae, and their spinous processes are longer and slope downward. The ribs articulate with *costal facets* on the transverse processes and bodies of thoracic vertebrae (figures 6.17e and 6.19b).

Lumbar Vertebrae

The five **lumbar vertebrae** (L1–L5) have heavy, thick vertebral bodies to support the greater stress and weight that is placed on this region of the vertebral column. The spinous processes are blunt and provide a large surface area for the attachment of heavy back muscles (see figures 6.16 and 6.17f).

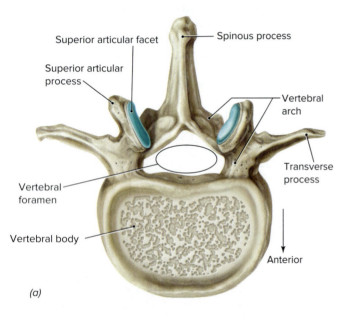

(a)

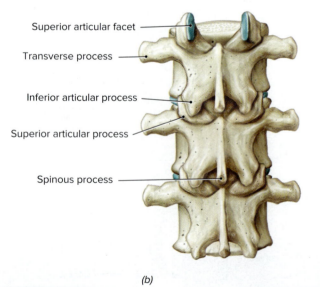

(b)

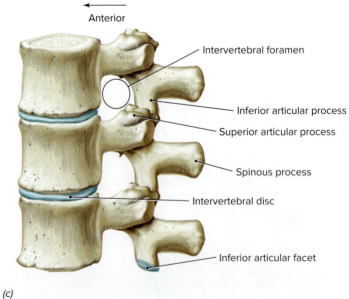

(c)

Figure 6.16 The Typical Vertebra.
(a) Superior view of a lumbar vertebra. *(b)* Posterior view of lumbar vertebrae. *(c)* Lateral view of lumbar vertebrae.

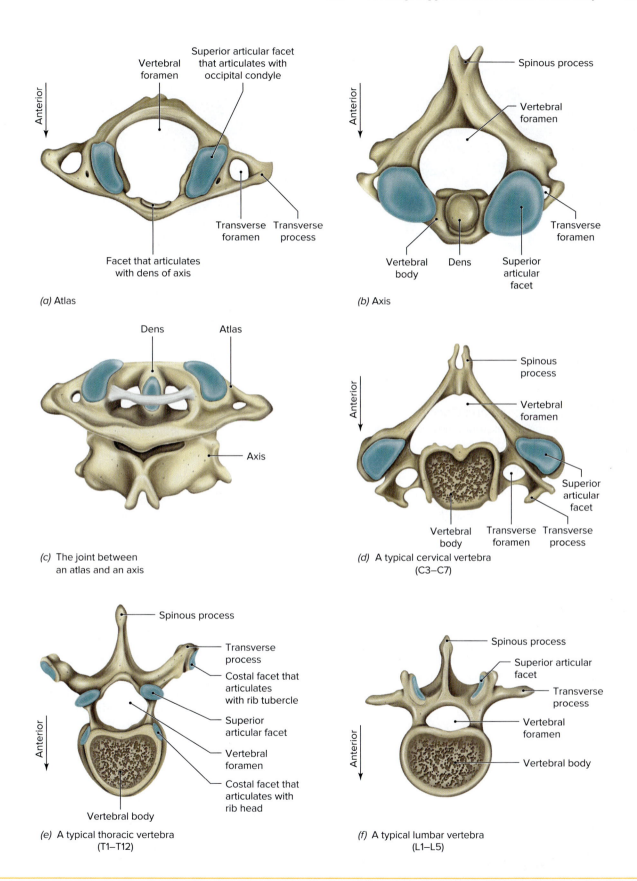

Figure 6.17 The Structures of Vertebrae.
(a), (b), (d), (e), and *(f)* are superior views. *(c)* is a posterior view. Bones are not shown to scale.

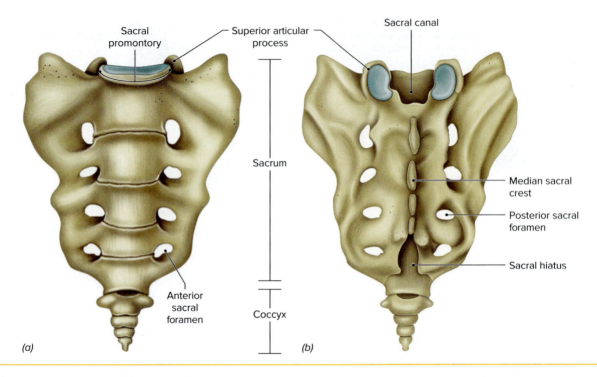

Figure 6.18 The Sacrum and Coccyx.
(a) Anterior view and (b) posterior view of the sacrum and coccyx. AP|R

Sacrum

The **sacrum** (sā-k′rum) is composed of five fused sacral vertebrae (S1–S5) (figure 6.18). It articulates with the fifth lumbar vertebra and coccyx and forms the posterior wall of the pelvis. The *sacral promontory* is the anterior, superior margin of the first sacral vertebra and serves as an important obstetric landmark. The spinous processes of the fused vertebrae form the *median sacral crest* on the posterior midline. On either side of the median sacral crest are the *posterior sacral foramina,* passageways for blood vessels and nerves. *Anterior sacral foramina* on the anterior surface serve a similar function. The *sacral canal* is a continuation of the vertebral canal that carries spinal nerve roots to the sacral foramina and the *sacral hiatus,* an inferior opening near the coccyx.

Coccyx

The most inferior part of the vertebral column is the **coccyx** (kok′-six), or tailbone, which is formed of three to five fused coccygeal vertebrae.

Thoracic Cage

The thoracic vertebrae, ribs, costal cartilages, and sternum form the **thoracic cage.** It provides protection for the internal organs of the thoracic cavity and supports the superior trunk, pectoral girdles, and upper limbs (figure 6.19).

Ribs

Twelve pairs of **ribs** are attached to the thoracic vertebrae. The *head* of each rib articulates with the costal facet on the vertebral body of its own vertebra and the costal facet on the vertebral body of the vertebra above it. A *tubercle* near the head articulates with the costal facet on the transverse processes of its own vertebra. The *shaft* of each rib curves around the thoracic cage and slopes slightly downward.

The upper seven pairs of ribs are attached directly to the sternum by the **costal** (kos′-tal) **cartilages,** which extend medially from the ends of the ribs. These ribs are the *true ribs.* The remaining five pairs are the *false ribs.* The first three pairs of false ribs are attached by cartilages to the costal cartilages of the ribs just above them. The last two pairs of false ribs are called *floating ribs* because they lack cartilages and are not attached anteriorly. The costal cartilages give some flexibility to the thoracic cage.

⊕ Clinical Insight

A biopsy of red bone marrow may be made by a *sternal puncture* because the sternum is covered only by skin and connective tissue. Under local anesthetic, a large-bore hypodermic needle is inserted into the sternum, and red bone marrow is drawn into a syringe.

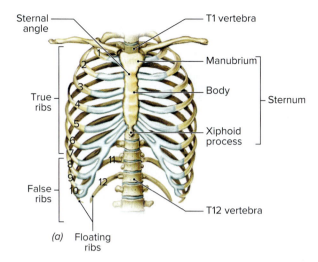

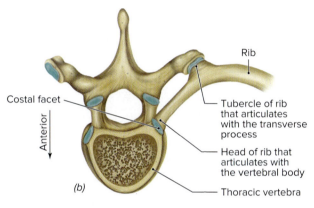

Figure 6.19 Thoracic Cage.
(a) The thoracic cage is formed by thoracic vertebrae, ribs, costal cartilages (colored light blue), and the sternum. *(b)* The joint between a rib and a thoracic vertebra.
AP|R

Sternum

The **sternum,** or breastbone, is a flat, elongated bone located at the midline in the anterior portion of the thoracic cage. It consists of three bones that are fused together. The *manubrium* (mah-nū′-brē-um) is the upper portion that articulates with the first two pairs of ribs; the *body* is the larger middle segment; and the *xiphoid* (zīf′-oyd) *process* is the small lower portion.

 Check My Understanding

12. How do cervical, thoracic, and lumbar vertebrae differ in structure and location?
13. How does the axial skeleton protect vital organs?

6.6 Appendicular Skeleton

Learning Objectives

12. Identify the bones of the appendicular skeleton.
13. Compare the structural and functional differences between the pectoral girdles and pelvic girdle.
14. Compare the structural and functional differences between the male and female pelves.
15. Describe how the appendicular skeleton is connected to the axial skeleton.

The appendicular skeleton consists of (1) the pectoral girdles and the bones of the upper limbs, and (2) the pelvic girdle and the bones of the lower limbs (see figure 6.6).

Pectoral Girdle

The human body possesses two **pectoral** (pek′-to-ral) **girdles,** or **shoulder girdles,** that assist in attaching the upper limbs to the torso. Each pectoral girdle consists of a clavicle and a scapula (figure 6.20). Each **S**-shaped **clavicle** (klav′-i-cul) articulates with the acromion of a scapula laterally and with the sternum medially. The **scapulae** (skap′-ū-le) are flat, triangular bones located on each side of the vertebral column, but they do not articulate with the axial skeleton. Instead, they are held in place by muscles, an arrangement that enables freedom of movement of the shoulder joints.

The anterior surface of each scapula is flat and smooth where it moves over the ribs. The *spine of the scapula* runs diagonally across the posterior surface from the *acromion* (ah-krōm′-ē-on) to the medial margin. On its lateral margin is the shallow *glenoid cavity,* which articulates with the head of the humerus. The *coracoid* (kor′-ah-koyd) *process* projects anteriorly from the upper margin of the glenoid cavity and extends below to the clavicle.

Upper Limb

The skeleton of each **upper limb** is composed of a humerus, an ulna, a radius, carpal bones, metacarpals, and phalanges (figure 6.21).

Humerus

The **humerus** (hū′-mer-us) articulates with the scapula at the shoulder joint, and the ulna and radius at the elbow joint. The rounded *head* of the humerus fits into the glenoid cavity of the scapula. Just below the head are two large tubercles where muscles attach. The *greater tubercle* (tū′-ber-cul) is on the lateral surface, and the *lesser tubercle* is on the anterior surface. An *intertubercular sulcus* lies between them. Just below these tubercles is the *surgical neck,* which gets its name from the frequent fractures that occur in this area. Near the midpoint on the lateral surface is the *deltoid tuberosity* (tū-be-ros′-i-tē), a rough, elevated area where the deltoid attaches.

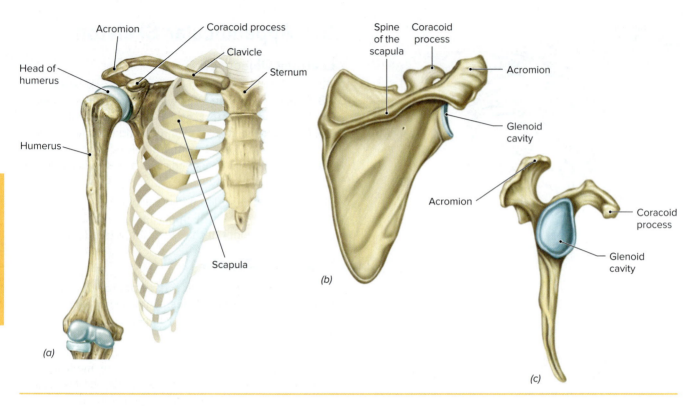

Figure 6.20 The Right Pectoral Girdle.
(a) The pectoral girdle consists of a scapula and a clavicle. Note how the head of the humerus articulates with the glenoid cavity of the scapula. The posterior view *(b)* and the lateral view *(c)* of the right scapula. AP|R

The distal end of the humerus has two condyles. The *trochlea* (trok′-lē-ah) is the medial condyle, which articulates with the trochlear notch of the ulna. The *capitulum* (kah-pit′-ū-lum) is the lateral condyle, which articulates with the head of the radius. Just above these condyles are two enlargements that project laterally and medially: the *lateral epicondyle* (ep-i-kon′-dīl) and the *medial epicondyle*. On the anterior surface between the epicondyles is a depression, the *coronoid* (kor′-o-noyd) *fossa*, that receives the coronoid process of the ulna whenever the upper limb is flexed at the elbow. The *olecranon* (o-lek′-rah-non) *fossa* is in a similar location on the posterior surface of the humerus, and it receives the olecranon of the ulna when the upper limb is extended at the elbow.

Ulna

The **ulna** (ul′-na) is the medial bone of the forearm. The proximal end of the ulna forms the *olecranon*, the bony point of the elbow. The large, half-circle depression just below the olecranon is the *trochlear notch*, which articulates with the trochlea of the humerus. This joint is secured by the *coronoid process* on the distal lip of the notch.

At the distal end, the knoblike *head* of the ulna articulates with the medial surface of the radius and with the wrist bones. The *styloid process* is a small medial projection to which ligaments of the wrist are attached.

Radius

The **radius** (rā′-dē-us) is the lateral bone of the forearm. The disclike *head* of the radius articulates with the lateral proximal surface of the ulna in a way that enables the head to rotate freely when the forearm is rotated. A short distance below the head is the *radial tuberosity*, an elevated, roughened area where the biceps brachii attaches. At its distal end, the radius articulates with the lateral distal end of the ulna and the carpal bones. A small lateral *styloid process* serves as an attachment site for ligaments of the wrist.

Carpal Bones, Metacarpals, and Phalanges

The skeleton of the hand consists of the carpal bones, metacarpals, and phalanges (figure 6.21d). The **carpal** (kar′-pul) **bones,** or wrist bones, consist of eight small bones that are arranged in two transverse rows of four bones each. They are joined by ligaments that allow limited gliding movement.

The **metacarpals,** bones of the palm, consist of five metacarpal bones that are numbered I to V starting with the metacarpal adjacent to the thumb. The bones of the fingers are the **phalanges** (fah-lan′-jēz; singular, *phalanx*).

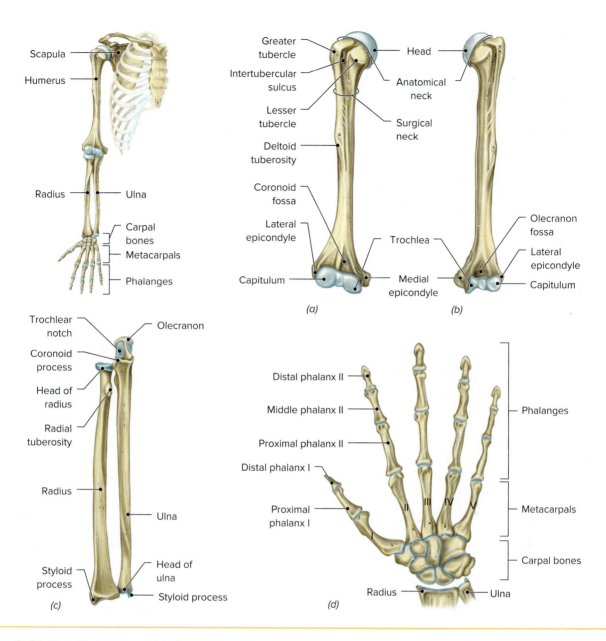

Figure 6.21 The Right Upper Limb.
(a) Anterior view of humerus; (b) Posterior view of humerus; (c) Anterior view of ulna and radius; (d) Posterior view of hand. AP|R

Each finger consists of three phalanges (proximal, middle, and distal), except for the thumb, which has only two (proximal and distal).

Pelvic Girdle

The **pelvic** (pel′-vik) **girdle** supports the attachment of the lower limbs to the torso and consists of the two **coxal** (kok′-sal) **bones,** or **hip bones,** and the sacrum (figure 6.22). The coxal bones articulate with the sacrum posteriorly and with each other anteriorly. The pelvic girdle is also often referred to as the **pelvis** (plural, *pelves*).

Coxal Bones

Each coxal bone is formed by three fused bones—ilium, ischium, and pubis—that join at the *acetabulum* (as-e-tab′-ū-lum), the cup-shaped socket on the lateral surface. The **ilium** (plural, *ilia*) is the broad superior portion whose upper margin forms the *iliac crest*, the prominence of the hip. Just below the *posterior inferior iliac spine* is the *greater sciatic* (sī-at′-ik) *notch*, which allows the passage of blood vessels and the sciatic nerve from the pelvis to the thigh. The *auricular surface* of each ilium joins with the sacrum to form a *sacroiliac joint*.

122 Chapter 6 Skeletal System

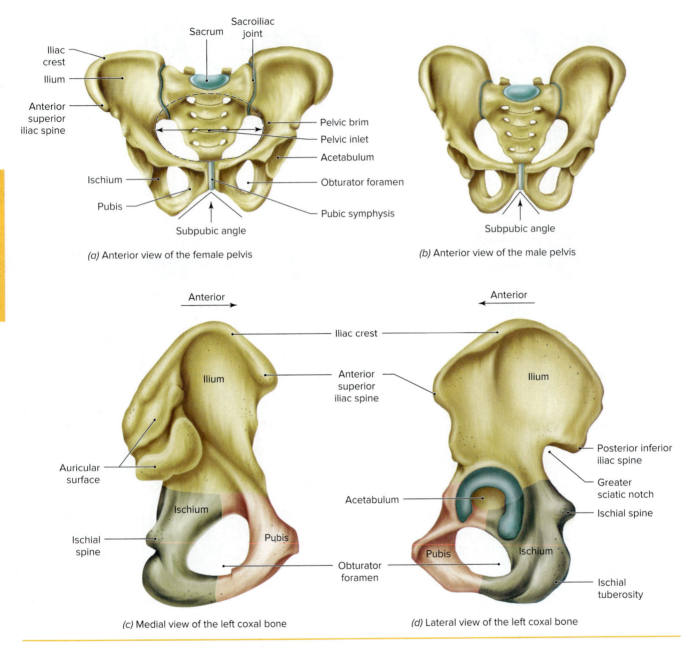

Figure 6.22 Pelvic Girdle and Coxal Bones. AP|R

The **ischium** (plural, *ischia*) forms the inferior, posterior portion of a coxal bone and supports the body when sitting. The roughened projection at the posterior, inferior angle of the ischium is the *ischial tuberosity*. Just above this tuberosity is the *ischial spine*, which projects medially. The distance between the left and right ischial spines in females is important during childbirth because it determines the diameter of the pelvic opening.

The **pubis** (plural, *pubes*) is the inferior, anterior portion of a coxal bone. A portion of the pubis extends posteriorly to fuse with the anterior extension of the ischium. The large opening created by this junction is the *obturator* (ob-tū-rā′-ter) *foramen*, through which blood vessels and nerves pass into the thigh. The pubes unite anteriorly to form the *pubic symphysis*, where the bones are joined by a pad of fibrocartilage.

Clinical Insight

When giving intramuscular injections in the hip, it is important to avoid the region near the greater sciatic notch to prevent possible injury to the large blood vessels and nerves in this area.

Table 6.3 Sexual Differences of the Pelves

Characteristic	Male	Female
General structure	Heavier; processes prominent	Lighter; processes not so prominent
Pelvic inlet	Narrower and heart-shaped	Wider and oval-shaped
Subpubic angle	Less than 90°	More than 90°
Relative width	Narrower	Wider
Acetabulum	Faces laterally	Faces laterally but more anteriorly

Table 6.3 lists the major differences between the male and the female pelves. Compare them with the male and female pelves in figure 6.22 and note the adaptations of the female pelvis for childbirth. The **pelvic inlet,** the opening at the top of the pelvic cavity, is encircled by the *pelvic brim,* a circular line passing along the sacral promontory and bony lines extending along the ilia and pubes. Its size and shape in females are critical to the success of the birth process.

 Clinical Insight

The fetus must pass through the pelvic inlet during birth. Physicians carefully measure this opening before delivery to be sure that it is of adequate size. If not, the baby is delivered via a *cesarean section.* In a cesarean section, a transverse incision is made through the pelvic and uterine walls to remove the infant.

Lower Limb

The bones of each **lower limb** consist of a femur, a patella, a tibia, a fibula, tarsal bones, metatarsals, and phalanges (figure 6.23).

Femur

The **femur,** or thigh bone, is the largest and strongest bone of the body (figure 6.23a, b). Structures at the proximal end include the rounded *head,* a short *neck,* and two large processes that are sites of muscle attachment: an upper, lateral *greater trochanter* (trō-kan′-ter) and a lower, medial *lesser trochanter.* The head of the femur fits into the acetabulum of the coxal bone. The neck is a common site of fractures in older people. At the enlarged distal end are the *lateral* and *medial condyles,* surfaces that articulate with the tibia.

Patella

The **patella,** or kneecap, is located anterior to the knee joint. It is embedded in the tendon of the quadriceps femoris. The dense regular connective tissue of the tendon continues from the patella to the tibial tuberosity as the *patellar ligament.* The patella offers protection to the structures within the knee joint during movement.

Tibia

The **tibia,** or shinbone, is the larger of the two bones of the leg (figure 6.23c). It bears the weight of the body. Its enlarged proximal portion consists of the *lateral* and *medial condyles,* which articulate with the femur to form the knee joint. The *tibial tuberosity,* a roughened area on the anterior surface just below the condyles, is the attachment site for the patellar ligament. The distal end of the tibia articulates with the talus, a tarsal bone, and laterally with the fibula. The *medial malleolus* (mah-lē-ō′-lus) forms the medial prominence of the ankle.

Fibula

The **fibula** is the slender, lateral bone in the leg (figure 6.23c). Both ends of the bone are enlarged. The proximal *head* articulates with the lateral surface of the tibia but is not involved in forming the knee joint. The distal end articulates with the tibia and talus. The *lateral malleolus* forms the lateral prominence of the ankle.

Tarsal Bones, Metatarsals, and Phalanges

The skeleton of the foot consists of the tarsal bones (ankle), metatarsals (instep), and phalanges (toes) (figure 6.23d, e). Seven bones compose the **tarsal bones.** The most prominent tarsal bones are the *talus,* which articulates with the tibia and fibula, and the *calcaneus* (kal-kā′-nē-us).

 Clinical Insight

Total hip replacement (THR) has become commonplace among older persons as a way to overcome the pain and immobility caused by osteoarthritis of the hip joint. This procedure utilizes two prostheses. A polyurethane cup replaces the damaged acetabulum, and a metal shaft and ball replace the diseased head of the femur. Surfaces of the prostheses in contact with bone are porous, allowing bone to grow into them to ensure a firm attachment. Patient recovery involves stabilization of the prostheses while bone grows into them, as well as normal healing from the surgery.

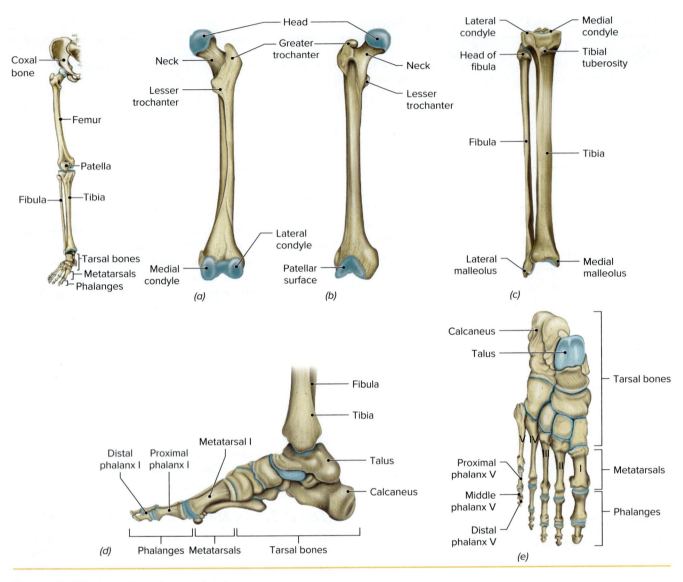

Figure 6.23 The Right Lower Limb.
(a) Posterior view of femur; (b) Anterior view of femur; (c) Anterior view of tibia and fibula; (d) Medial view of foot; (e) Superior view of foot. AP|R

Five **metatarsals** support the instep. They are numbered I to V, starting with the metatarsal adjacent to the great toe. The tarsal bones and metatarsals are bound together by ligaments to form strong, resilient arches of the foot. Each toe consists of three phalanges (proximal, middle, and distal), except for the great toe, which has only two (proximal and distal).

Check My Understanding

14. What bones form the pectoral girdles and upper limbs?
15. What bones form the pelvic girdle and lower limbs?

6.7 Joints

Learning Objectives

16. Compare the structures, functions, and locations of fibrous joints, cartilaginous joints, and synovial joints.
17. Compare the types of movements allowed by synovial joints.
18. Compare the six types of synovial joints.

A **joint,** or *articulation,* is a junction between two bones or between a bone and a tooth. Based upon its structure, a joint is classified as a fibrous joint, a cartilaginous joint, or a synovial joint. As you read the following descriptions, locate the different types of joints on the corresponding illustrations of skeletal parts in figures presented earlier in the chapter.

Fibrous Joints

A joint is classified as a **fibrous joint** when a thin band of dense irregular connective tissue joins the skeletal structures. These joints exhibit no movement because of how tightly the bones are connected. The sutures of the skull and the joints between the teeth and dental alveoli are examples of fibrous joints (figure 6.24a). The joint between the distal ends of the tibia and fibula is also a fibrous joint.

Cartilaginous Joints

A joint is classified as a **cartilaginous joint** when a thin band or pad of hyaline cartilage or fibrocartilage joins the bones. These joints permit no or only a slight degree of movement.

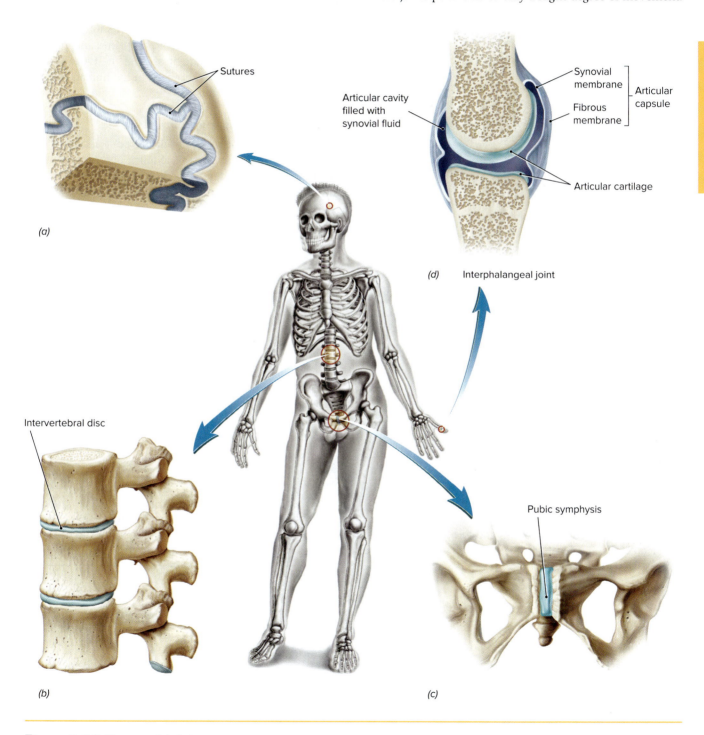

Figure 6.24 Types of Joints.
(a) Suture (fibrous joint); (b) Intervertebral disc (cartilaginous joint); (c) Pubic symphysis (cartilaginous joint); (d) Interphalangeal joint (synovial joint).

For example, an epiphysial plate is a thin pad of hyaline cartilage connecting an epiphysis to a diaphysis (see figure 6.5). This cartilaginous joint permits no movement because of how tightly the epiphysis and diaphysis are bound. However, the intervertebral discs joining vertebral bodies allow for slight movement between adjacent vertebrae (figure 6.24b), which accommodates movement of the vertebral column. Other examples include the pubic symphysis (figure 6.24c) and sacroiliac joints, both of which also allow for only slight movement between the joined structures.

Synovial Joints

The majority of the joints in the body are classified as **synovial** (si-nō′-ve-al) **joints.** The unique anatomy of synovial joints permits free movement of the joined bones. In a synovial joint, two bones are separated by a small fluid-filled cavity that is surrounded by a connective tissue capsule. The ends of the bones forming the joint are bound together by an *articular,* or *joint, capsule.* The thick outer layer of the capsule, called the *fibrous membrane,* is composed of dense irregular connective tissue. The thin inner layer of the capsule, called the *synovial membrane,* secretes synovial fluid into the *articular cavity.* The synovial fluid serves to lubricate the joint. The ends of the bones are covered with articular cartilage, which protects bones and reduces friction (figure 6.24d). **Ligaments,** the cords or bands of dense regular connective tissue that connect bones to each other, reinforce the joints. Synovial joints are categorized into several types based on their structure and types of movements.

Plane Joints

A *plane joint* occurs between two flat articular surfaces that slide over each other and allows for movement in one plane. Some examples of plane joints are the joints between carpal bones (figure 6.25a), between tarsal bones, and between clavicle and scapula.

Condylar Joints

A *condylar joint* is formed between an oval articular surface and an oval socket and allows for movements in two planes. The joints between carpal bones and radius and between metacarpals and proximal phalanges (figure 6.25b) are examples of condylar joints.

Saddle Joint

A *saddle joint* occurs where a saddle-like articular surface fits into a complementary depression, allowing movement in two planes. This type of joint occurs between the trapezium (a carpal bone) and metacarpal I (figure 6.25c).

Hinge Joints

A *hinge joint* involves a cylindrical articular surface and a complementary depression. It allows for movement similar to opening and closing a door. The elbow (figure 6.25d), knee, and joints between phalanges are all hinge joints.

Pivot Joints

A *pivot joint* involves a cylindrical articular surface and a complementary depression. It allows for rotation movements along a longitudinal axis. Examples of a pivot joint are the joint between atlas and axis (figure 6.25e) and the joint between the radius head and the ulna.

Ball-and-Socket Joints

In a *ball-and-socket joint,* a rounded head fits into a rounded socket. It allows for movements in all planes and provides the greatest range of movement of all of the types of synovial joints. The ball-and-socket joints in the human body are the shoulder and hip joints (figure 6.25f).

Movements at Synovial Joints

Movement at a joint results from the contraction of skeletal muscles that span across the joint. The type of movement that occurs is determined by the type of joint and the location of the muscle or muscles involved. The more common types of movements are listed in table 6.4 and illustrated in figure 6.26.

Clinical Insight

Older persons are prone to "breaking a hip," which means that a weakened femur breaks at the neck. This usually is a consequence of osteoporosis, the excessive loss of matrix from bones. Not only are older persons more prone to fractures, but healing of fractures takes much longer than in younger persons.

Check My Understanding

16. Where are fibrous joints, cartilaginous joints, and synovial joints found in the skeleton?
17. What types of synovial joints occur in the body, and where are they located?

6.8 Disorders of the Skeletal System

Learning Objectives

19. Describe common disorders of bones.
20. Describe common disorders of joints.

Common disorders of the skeletal system may be categorized as disorders of bones or disorders of joints.

Part 2 Covering, Support, and Movement of the Body 127

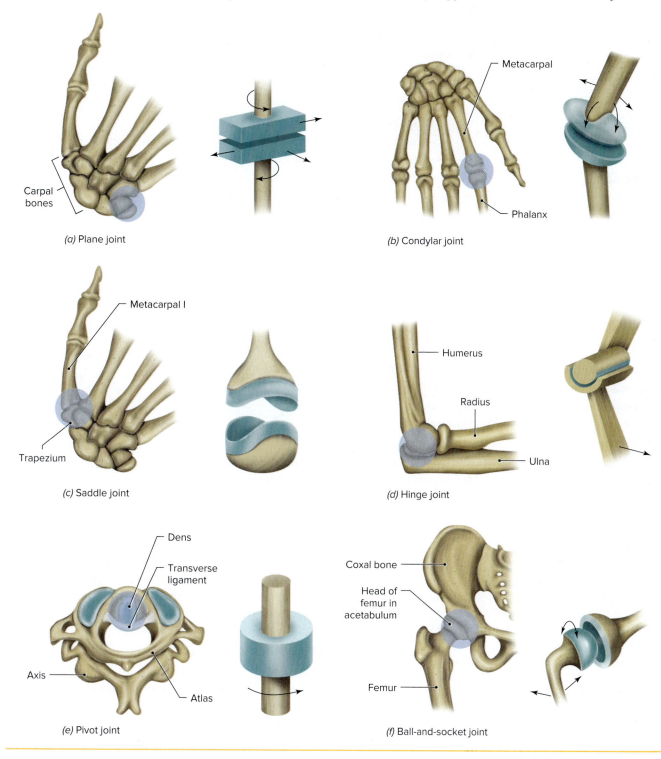

Figure 6.25 Types of Synovial Joints.

Orthopedics (or-thō-pē-diks) is the branch of medicine that specializes in treating diseases and abnormalities of the skeletal system.

Disorders of Bones

Fractures are broken bones. They are the most common type of bone injury and are categorized as either complete or incomplete. There are also several specific subtypes, such as the examples noted here and in figure 6.27.

- **Complete:** The break is completely through the bone.
- **Compound:** A broken bone pierces the skin.
- **Simple:** A bone does not pierce the skin.
- **Comminuted:** The bone is broken into several pieces.

Table 6.4 Movements at Synovial Joints

Movements	Description
Flexion	Decrease in the angle between bones forming the joint
Extension	Increase in the angle between bones forming the joint
Hyperextension	Increase in the angle between bones forming the joint beyond the anatomical position
Dorsiflexion	Flexion of the foot at the ankle
Plantar flexion	Extension of the foot at the ankle
Abduction	Movement of a bone away from the midline
Adduction	Movement of a bone toward the midline
Rotation	Movement of a bone around its longitudinal axis
Medial rotation	Rotation of a limb so its anterior surface turns medially
Lateral rotation	Rotation of a limb so its anterior surface turns laterally
Circumduction	Movement of the distal end of a bone in a circle while the proximal end forms the pivot joint
Eversion	Movement of the sole of the foot laterally
Inversion	Movement of the sole of the foot medially
Pronation	Rotation of the forearm when the palm is turned inferiorly or posteriorly
Supination	Rotation of the forearm when the palm is turned superiorly or anteriorly
Protraction	Movement of a body part anteriorly
Retraction	Movement of a body part posteriorly
Elevation	Movement of a body part superiorly
Depression	Movement of a body part inferiorly
Opposition	Movement of the thumb to touch the other four fingers
Reposition	Movement of the thumb back to the anatomical position

- **Segmental:** Only one piece is broken out of the bone.
- **Spiral:** The fracture line spirals around the bone.
- **Oblique:** The break angles across the bone.
- **Transverse:** The break is at right angles to the long axis of the bone.
- **Incomplete:** The bone is not broken completely through.
- **Greenstick:** The break is only on one side of the bone, and the other side of the bone is bowed.
- **Fissured:** The break is a lengthwise split in the bone.

Osteomyelitis is an inflammation of bone and bone marrow caused by bacterial infection. It is treatable with antibiotics but not easily cured.

Osteoporosis (os-tē-ō-pō-rō´-sis) is a weakening of bones due to the removal of bone matrix, which increases the risk of fractures. This is a common problem in older persons due to inactivity and a decrease in hormone production. It is more common in postmenopausal women because of the lack of estrogens. Exercise and calcium supplements retard the decline in bone density. Therapy includes drugs that reduce bone loss or those that promote bone formation. However, such drugs must be used with caution because they can have serious side effects.

Rickets is a disease of children that is characterized by a deficiency of calcium salts in the bones. Affected children have a bowlegged appearance due to the bending of weakened femurs, tibiae, and fibulae. Rickets results from a dietary deficiency of vitamin D and/or calcium. It is rare in industrialized nations.

Disorders of Joints

Arthritis (ar-thrī´-tis) is the general term for many different diseases of joints that are characterized by inflammation, swelling (edema), and pain. Rheumatoid arthritis and osteoarthritis are the most common types.

Rheumatoid (rū´-mah-toyd) *arthritis* is the most painful and crippling type. It is an autoimmune disorder, in which the joint tissues are attacked by the patient's own immune responses. The synovial membrane thickens, synovial fluid accumulates causing swelling, and articular cartilages are destroyed. The joint is invaded by dense irregular connective tissue that ultimately ossifies, making the joint immovable.

Osteoarthritis, the most common type, is a degenerative disease that results from aging and wear. The articular cartilages and the bone deep to the cartilages gradually disintegrate, which causes pain and restricts movement.

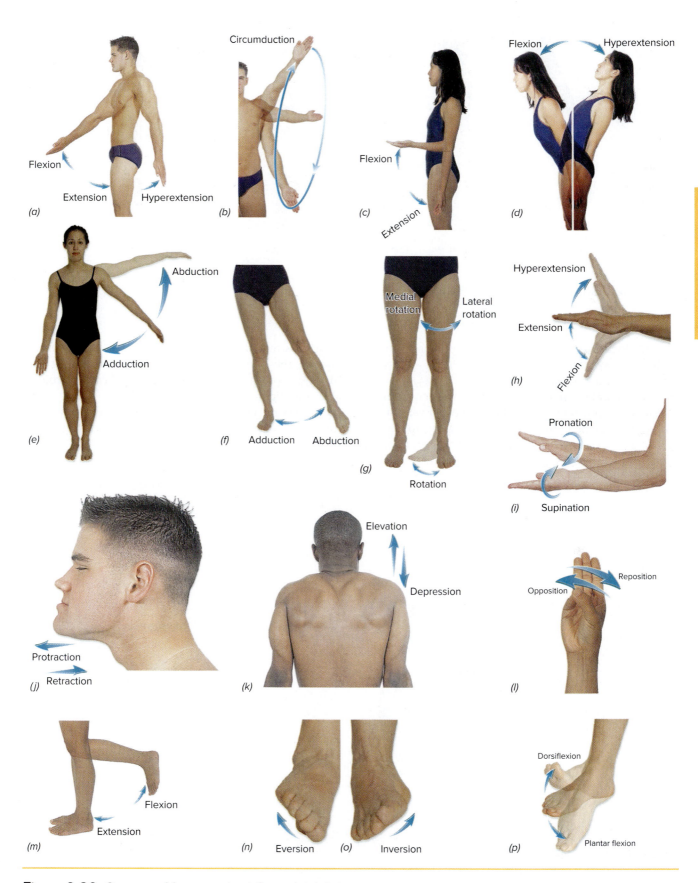

Figure 6.26 Common Movements at Synovial Joints.
(a-d, i-m) ©Eric Wise; (e-h, n-p) ©McGraw-Hill Education/Photo by JW Ramsey

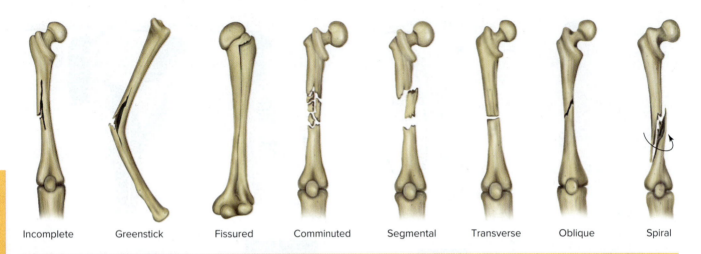

Figure 6.27 Some Types of Bone Fractures.

Dislocation is the displacement of bones forming a joint. Pain, swelling, and reduced movement are associated with a dislocation.

A **herniated disc** is a condition in which an intervertebral disc protrudes beyond the edge of a vertebra. A ruptured, or slipped, disc refers to the same problem. It is caused by excessive pressure on the vertebral column, which causes the *nucleus pulposus*, the centrally located gelatinous region of the disc, to protrude into the *anulus fibrosus*, the perimeter of the disc. The protruding disc may place pressure on a spinal nerve and cause considerable pain (figure 6.28).

Sprains result from tearing or excessive stretching of the ligaments and tendons at a joint without a dislocation.

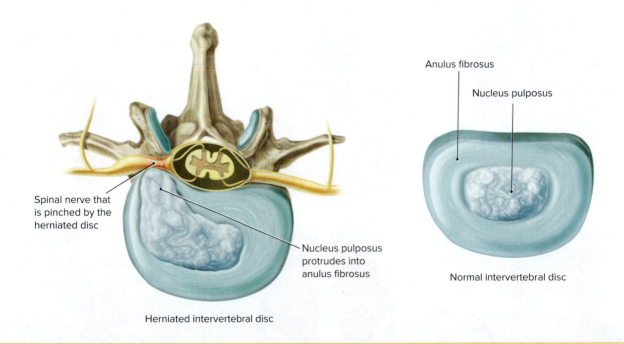

Figure 6.28 Herniated Disc.

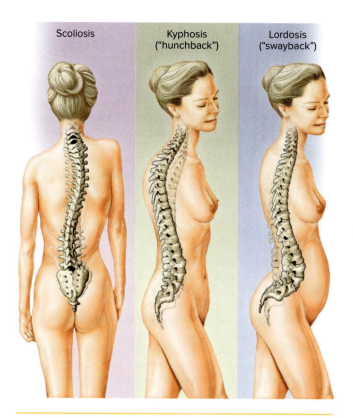

Figure 6.29 Abnormal Spinal Curvatures.

Abnormal spinal curvatures are usually congenital disorders. There are three major types (figure 6.29):

1. *Scoliosis* is an abnormal lateral curvature of the vertebral column. For some reason, it is more common in adolescent girls.
2. *Kyphosis* (kĭ-fō-sis) is an excessive thoracic curvature of the vertebral column, which produces a hunchback condition.
3. *Lordosis* is an excessive lumbar curvature of the vertebral column, which produces a swayback condition.

✓ Check My Understanding

18. What are some common types of fractures?
19. How do osteoarthritis and rheumatoid arthritis differ?

Chapter Summary

6.1 Functions of the Skeletal System

- The skeletal system provides support for the body and protection for internal organs.
- The bones of the skeleton serve as sites for the attachment of skeletal muscles.
- Formed elements are produced by red bone marrow.
- Bones serve as reservoirs for calcium salts.

6.2 Bone Structure

- Based on shapes, bones are classified into short, long, sutural, flat, irregular, and sesamoid bones.
- The diaphysis is the long shaft of a long bone that lies between the epiphyses, the enlarged ends of the bone.
- Each epiphysis is joined to the diaphysis by an epiphysial plate in immature bones, or by fusion at the epiphysial line in mature bones.
- Articular cartilages protect and cushion the articular surfaces of the epiphyses.
- The periosteum covers the bone surface except for the articular cartilages.
- Compact bone forms the wall of the diaphysis and the thin outer layer of the epiphyses.
- Spongy bone forms the inner region of the epiphyses and the inner wall of the diaphysis.
- The diaphysis contains a medullary cavity filled with yellow bone marrow.
- Compact bone is formed of numerous osteons.
- Central canals contain blood vessels and nerves.
- Spongy bone is composed of interconnected bony plates called trabeculae. The spaces between trabeculae are filled with red or yellow bone marrow.
- Flat, short, sutural, sesamoid, and irregular bones are composed of spongy bone covered by a thin layer of compact bone.

6.3 Bone Formation

- Intramembranous bones are formed when connective tissue membranes are replaced by bone.
- Connective tissue cells are transformed into osteoblasts, which deposit the spongy bone within the membrane.
- Osteoblasts from the periosteum form a layer of compact bone over the spongy bone.
- Endochondral bones are first formed of hyaline cartilage, which is later replaced by bone.
- In long bones, a primary ossification center forms in the center of the diaphysis and extends toward the epiphyses.
- Secondary ossification centers form in the epiphyses.
- An epiphysial plate of cartilage remains between the epiphyses and the diaphysis in immature bones.

- Growth in length occurs at the epiphysial plate, which is gradually replaced by bone.
- Compact bone is deposited on the outer bone surface by osteoblasts from the periosteum, which results in an increase in bone diameter.
- Osteoclasts hollow out the medullary cavity and reshape the bone.
- Bones are dynamic, living organs that are reshaped throughout life by the actions of osteoclasts and osteoblasts.
- Bone matrix may be removed from bones for other body needs and redeposited in bones later on.
- The number of collagen fibers decreases with age. The bones of older persons tend to be brittle and weak due to the loss of fibers and calcium salts, respectively.

6.4 Divisions of the Skeleton

- The skeleton is divided into the axial and appendicular divisions.
- The axial skeleton includes the bones that support the head, neck, and trunk.
- The appendicular skeleton includes the bones of the pectoral girdles and upper limbs and the bones of the pelvic girdle and the lower limbs.

6.5 Axial Skeleton

- The axial skeleton consists of the skull, vertebral column, and thoracic cage.
- The skull consists of cranial and facial bones; all are joined by joints that exhibit no movement except the mandible.
- The cranial bones are the frontal bone (1), parietal bones (2), sphenoid (1), temporal bones (2), occipital bone (1), and ethmoid (1).
- The facial bones are the maxillae (2), palatine bones (2), zygomatic bones (2), lacrimal bones (2), nasal bones (2), inferior nasal conchae (2), vomer (1), and mandible (1).
- The frontal bone, sphenoid, ethmoid, and maxillae contain paranasal sinuses.
- Cranial bones of an infant skull are separated by membranes and several fontanelles, which allow some flexibility of the skull during birth.
- Associated bones to the skull include a hyoid bone and six auditory ossicles.
- The vertebral column consists of 24 vertebrae, the sacrum, and the coccyx.
- Vertebrae are separated by intervertebral discs and are categorized as cervical (7), thoracic (12), and lumbar (5) vertebrae.
- The first two cervical vertebrae are unique. The joint between the atlas and the occipital condyles of the skull permits nodding of the head. The atlas rotates on the axis when the head is turned.
- Thoracic vertebrae have costal facets on the vertebral body and transverse processes for forming joints with the ribs.
- The vertebral bodies of lumbar vertebrae are heavy and strong.
- The sacrum is formed of five fused vertebrae and forms the posterior portion of the pelvis.
- The coccyx is formed of three to five fused vertebrae and forms the inferior end of the vertebral column.
- The thoracic cage consists of thoracic vertebrae, ribs, and sternum. It supports the upper trunk and protects internal thoracic organs.
- There are seven pairs of true ribs and five pairs of false ribs. The lowest two pairs of false ribs are floating ribs.
- The sternum is formed of three fused bones: manubrium, body, and xiphoid process.

6.6 Appendicular Skeleton

- The appendicular skeleton consists of two pectoral girdles, a pelvic girdle, and the bones of the limbs.
- The pectoral girdle consists of a clavicle and a scapula, and it supports the upper limbs.
- The bones of the upper limb are the humerus, the ulna, the radius, carpal bones, metacarpals, and phalanges.
- The humerus articulates with the glenoid cavity of the scapula to form the shoulder joint, and with the ulna and radius to form the elbow joint.
- The ulna is the innermost bone of the forearm. It articulates with the humerus at the elbow and with the radius and carpal bones at the wrist.
- The radius is the outermost bone of the forearm. It articulates with the humerus at the elbow and with the ulna and carpal bones at the wrist.
- The bones of the hand are the carpal bones (8), metacarpals (5), and phalanges (14).
- The carpal bones are joined by ligaments to form the wrist; metacarpal bones support the palm of the hand; and the phalanges are the bones of the fingers.
- The pelvic girdle consists of two coxal bones and the sacrum. The coxal bones are joined to each other anteriorly and to the sacrum posteriorly. The pelvic girdle supports the lower limbs.
- Each coxal bone is formed by the fusion of three bones: the ilium, ischium, and pubis.
- The ilium forms the superior portion of a coxal bone and joins with the sacrum to form a sacroiliac joint.
- The ischium forms the inferior, posterior portion of a coxal bone and supports the body when sitting.
- The pubis forms the inferior, anterior part of a coxal bone. The two pubes unite anteriorly at the pubic symphysis.
- The pelvic girdle is also referred to as the pelvis. There are structural and functional differences between male and female pelves.
- Each lower limb consists of a femur, a patella, a tibia, a fibula, tarsal bones, metatarsals, and phalanges.
- The head of the femur is inserted into the acetabulum of a coxal bone to form a hip joint. Distally, it articulates with the tibia at the knee joint.
- The patella is a sesamoid bone in the anterior portion of the knee joint.

- The tibia articulates with the femur at the knee joint and with the talus to form the ankle joint.
- The fibula lies lateral to the tibia. It articulates proximally with the tibia and distally with the talus.
- The skeleton of the foot consists of tarsal bones (7), metatarsals (5), and phalanges (14).
- Tarsal bones form the ankle, metatarsal bones support the instep, and phalanges are the bones of the toes.

6.7 Joints

- There are three types of joints: fibrous joints, cartilaginous joints, and synovial joints.
- In a fibrous joint, skeletal structures are joined by a thin band of dense irregular connective tissue. These joints exhibit no movement. Examples include the sutures, the joints between bones and teeth, and the joint between the distal ends of the tibia and fibula.
- In a cartilaginous joint, bones are joined by hyaline cartilage or fibrocartilage. These joints exhibit no movement or only a slight degree of movement. Examples include the epiphysial plates, joints between vertebral bodies, the pubic symphysis, and sacroiliac joints.
- In a synovial joint, two bones are separated by a small fluid-filled cavity that is surrounded by a connective tissue capsule. These joints exhibit free movement. The articular surfaces of the bones are covered by articular cartilages. The articular cavity is lubricated by synovial fluid secreted by the synovial membrane, the inner layer of articular capsule.
- There are six types of synovial joints: plane, condylar, hinge, saddle, pivot, and ball-and-socket.
- Movements at synovial joints include flexion, extension, hyperextension, dorsiflexion, plantar flexion, abduction, adduction, rotation, circumduction, inversion, eversion, protraction, retraction, elevation, depression, pronation, supination, opposition, and reposition.

6.8 Disorders of the Skeletal System

- Disorders of bones include fractures, osteomyelitis, osteoporosis, and rickets.
- Disorders of joints include arthritis, dislocation, herniated discs, abnormal spinal curvatures, and sprains.

Improve Your Grade

Connect Interactive Questions Reinforce your knowledge using multiple types of questions: interactive, animation, classification, labeling, sequencing, composition, and traditional multiple choice and true/false.

SmartBook Proven to help students improve grades and study more efficiently, SmartBook contains the same content within the print book but actively tailors that content to the needs of the individual.

Anatomy & Physiology REVEALED® Dive into the human body by peeling back layers of cadaver imaging. Utilize this world-class cadaver dissection tool for a closer look at the body anytime, from anywhere.

CHAPTER 7

Muscular System

©Olga_Danylenko/iStock/Getty Images RF

Melanie and a few of her friends head out early one morning for a short hike up a nearby mountain to a scenic overlook. As the wind gusts, forcing the temperature below freezing, they study a map and debate what trail to take. Melanie wonders if they made a good decision to hike today as her hands and feet begin to go numb despite her gloves and lined winter boots. Shivering violently, Melanie follows her friends up the mountain. The hike is strenuous because the trail they chose is both steep and rocky. The heat being created through the vigorous contractions of her skeletal muscles begins to gradually warm her body. In a short while, Melanie notices that she is no longer shivering or feeling the cold around her. By the time Melanie reaches the overlook, she is actually so warm that she begins to sweat. The friends sit on the edge of the overlook enjoying the view and each other's company. As the effects of the cold settle in once more, Melanie happily leads the way down the mountain to where a mug of hot chocolate and a roaring fire are waiting.

CHAPTER OUTLINE

7.1 Types of Muscle Tissue

7.2 Structure of Skeletal Muscle
- Skeletal Muscle Fibers
- Neuromuscular Interaction
- Motor Units
- Neuromuscular Junction

7.3 Physiology of Skeletal Muscle Contraction
- Mechanism of Contraction
- Energy for Contraction
- Contraction Characteristics

7.4 Actions of Skeletal Muscles
- Origin and Insertion
- Muscle Interactions

7.5 Naming of Muscles

7.6 Major Skeletal Muscles
- Muscles of Facial Expression and Mastication
- Muscles That Move the Head
- Muscles of the Abdominal Wall
- Muscles of Breathing
- Muscles That Move the Pectoral Girdle
- Muscles That Move the Arm and Forearm
- Muscles That Move the Wrist and Fingers
- Muscles That Move the Thigh and Leg
- Muscles That Move the Foot and Toes

7.7 Disorders of the Muscular System
- Muscular Disorders
- Neurological Disorders Affecting Muscles

Chapter Summary

Module 6
Muscular System

SELECTED KEY TERMS

Agonist (agogos = leader) A muscle whose contraction produces an action.
Antagonist (anti = against) A muscle whose contraction opposes the action of the agonist.
Aponeurosis (apo = from; neur = cord) A broad sheet of dense regular connective tissue that attaches a muscle to another muscle, bones, the dermis, or a ligament.
Creatine phosphate An energy storage molecule found in muscle cells.
Insertion The movable attachment of a muscle.
Motor unit A somatic motor neuron and the muscle fibers that it innervates.
Muscle fiber A single skeletal muscle cell.
Muscle tone The state of slight contraction in a skeletal muscle.
Myoglobin (myo = muscle) An oxygen-storage molecule in muscle cells.
Neurotransmitter (neuro = nerve; transmit = to send across) A chemical released by terminal boutons of neurons that can excite or inhibit a muscle cell, gland, or another neuron.
Origin The immovable attachment of a muscle.
Tendon A narrow band of dense regular connective tissue that attaches a muscle to a bone, other muscles, dermis, and ligaments.
Tetany (tetan = rigid, stiff) A sustained muscle contraction.

7.1 Types of Muscle Tissue

Learning Objective

1. Compare the types of muscle tissue.

Muscle tissue is the only tissue in the body that is specialized for contraction (shortening). The body contains three types of muscle tissue: skeletal, smooth, and cardiac. Each type of muscle tissue exhibits unique structural and functional characteristics. Contraction of skeletal muscle tissue produces locomotion, movement of body parts, and movement of the skin, as in making facial expressions. Contraction of skeletal muscles also produces heat, which is important in thermoregulation. Cardiac muscle tissue produces the driving force responsible for pumping blood through the cardiovascular system, as you will see in chapter 12. Smooth muscle tissue is responsible for various internal functions, such as controlling the movement of blood through blood vessels and air through respiratory passageways. It is also directly involved in vision and moving contents through hollow internal organs, as described in future chapters. Refresh your understanding of these tissues by referring to the discussion of muscle tissue in chapter 4. Table 7.1 summarizes the characteristics of muscle tissues.

Table 7.1 Types of Muscle Tissue

Characteristic	Skeletal	Smooth	Cardiac
Striations	Present	Absent	Present
Nucleus	Many peripherally located nuclei	Single centrally located nucleus	Usually a single centrally located nucleus
Cells	Long and parallel, called muscle fibers	Short; tapered ends; parallel	Short and branching; intercalated discs join cells end to end to form network
Neural control	Voluntary	Involuntary	Involuntary
Contractions	Fast, variable fatigability; slow; resistant to fatigue	Slow; resistant to fatigue	Rhythmic; resistant to fatigue
Location	Attached to bones, dermis, ligaments, and other muscles	Walls of hollow visceral organs and blood and lymphatic vessels, skin, and inside eyes	Wall of the heart
Photomicrograph	©Ed Reschke	©Ed Reschke/Getty Images	©Ed Reschke/Getty Images

7.2 Structure of Skeletal Muscle

Learning Objectives
2. Describe the structure of a skeletal muscle.
3. Explain how a skeletal muscle is attached to a bone or other tissues.
4. Describe the structure of a muscle fiber.
5. Describe a motor unit.
6. Describe the structure and function of a neuromuscular junction.

Skeletal muscles are the organs of the muscular system. They are called skeletal muscles because most of them are attached to bones. A skeletal muscle is composed mainly of skeletal muscle tissue bound together and electrically insulated by connective tissue layers. Individual skeletal muscle cells, called **muscle fibers** due to their long skinny shape, are wrapped in areolar connective tissue.

Muscle fibers extend most of the length of a whole muscle and are arranged in small bundles called *muscle fascicles* (fah′-si-kuls) that are each surrounded by a layer of dense irregular connective tissue. A muscle is formed when many muscle fascicles are packaged and held together by an outer layer of dense irregular connective tissue. Groups of whole muscles with similar functions are connected by a layer of dense irregular connective tissue called *fascia* (fash′-ē-ah). The fascia is beneath but connected to the subcutaneous tissue, which is how the muscles can produce skin movement. These muscle connective tissues extend beyond the end of the muscle tissue to form a tough, cordlike **tendon,** which attaches the muscle to a bone (figure 7.1). Fibers of the tendon and periosteum intermesh to form a secure attachment. A few muscles attach to other muscles, dermis, and ligaments, in addition to bones. In these muscles there is a broad, sheetlike attachment called an **aponeurosis** (ap″-ō-nū-rō′-sis) (plural, *aponeuroses*).

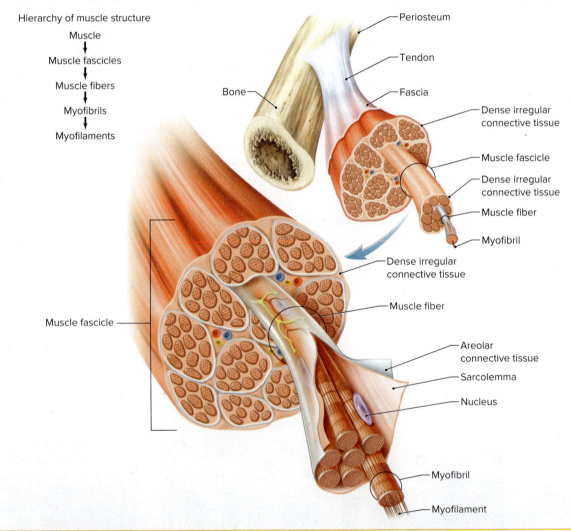

Figure 7.1 Hierarchy of Muscle Structure.
A skeletal muscle is primarily composed of skeletal muscle fibers supported and bound together in muscle fascicles by dense irregular connective tissue. Myofibrils are the contractile elements of a muscle fiber.

Skeletal Muscle Fibers

The microscopic structure of skeletal muscle tissue is so highly specialized that specific terminology is used to describe some muscle fiber structures. The prefixes *sarco-* (flesh) and *myo-* (muscle) are often used in renaming muscular structures. Therefore, the plasma membrane of a muscle fiber is called the **sarcolemma** (sar-kō-lem'-ah), and its cytoplasm is the **sarcoplasm.**

The sarcoplasm contains many threadlike **myofibrils,** which extend the length of the muscle fiber, as shown in figure 7.1. Myofibrils are the contractile elements of a muscle fiber. They consist of two kinds of myofilaments that interact to produce muscle contractions: (1) *thin myofilaments* composed mostly of the protein **actin** and (2) *thick myofilaments* composed of the protein **myosin** (table 7.2).

A thin myofilament consists of two twisted strands of actin molecules joined together like tiny strands of pearls. Two additional proteins, *troponin complex* and *tropomyosin,* are present in thin myofilaments and play a role in muscle contraction. Double strands of tropomyosin coil over each actin strand and cover the *myosin binding sites.* Troponin complex occurs at regular intervals on the tropomyosin strands. A thick myofilament is composed of hundreds of myosin molecules, each shaped like a double-headed golf club. The myosin heads are able to attach to the myosin binding sites on the actin molecules to form cross-bridges (figure 7.2). The organization of thin and thick myofilaments within a muscle fiber produces *striations*—the light and dark cross bands that are characteristic of skeletal muscle fibers when viewed microscopically.

As shown in figure 7.2, the arrangement of thin and thick myofilaments repeats itself throughout the length of a myofibril. These repeating units are called sarcomeres. A **sarcomere** is the functional unit of skeletal muscle—that is, it is the smallest portion of a myofibril capable of contraction. A sarcomere extends from a Z line to the next Z line. **Z lines** are composed of proteins arranged perpendicular to the longitudinal axis of the myofilament. Thin myofilaments are attached to each side of the Z lines and extend toward the middle of the sarcomeres. The *I band,* which is the light band in a photomicrograph, possesses thin myofilaments only and spans across the Z lines. The *A band,* which is the dark band in a photomicrograph, spans the length of the thick myofilaments. Note that the ends of the thin myofilaments do not meet, leaving a space at the center of the A band, which contains only thick myofilaments, called the *H band* (pale zone). Proteins that maintain the structure of the center of the sarcomere make up the *M line.*

Figure 7.3 illustrates the relationship of the sarcoplasmic reticulum and transverse (T) tubules to myofibrils in a muscle fiber. The **sarcoplasmic reticulum** is the name given to the smooth endoplasmic reticulum in a muscle fiber. It plays an important role in contraction by storing and releasing calcium (Ca^{2+}) ions. The **transverse (T) tubules** consist of extensions of the sarcolemma that penetrate into the sarcoplasm so that they lie alongside and contact the sarcoplasmic reticulum.

Neuromuscular Interaction

A muscle fiber must be stimulated by *action potentials* (electrochemical signals) in order to contract. Action potentials are carried from the brain or spinal cord to a muscle fiber by a long, thin process (an axon) of a motor neuron. A *motor neuron* is an action-causing neuron—its action potentials produce an action in the target cells. In muscle fibers, this action is contraction and the specific type of motor neuron is called a *somatic motor neuron.*

Motor Units

A somatic motor neuron and all of the muscle fibers to which it attaches, or innervates, form a **motor unit** (figure 7.4). Whereas a muscle fiber is attached to only one motor neuron, a single somatic motor neuron may innervate from 3 to 2,000 muscle fibers. Where precise muscle control rather than strength is needed, such as in the fingers, a motor unit contains very few muscle fibers. Large numbers of motor units are involved in the manipulative movements of the fingers. The area of the brain controlling finger movement is therefore proportionately large. In contrast, where strength rather than precise control is needed, such as in the postural muscles, a motor unit controls hundreds of muscle fibers. Whenever a motor neuron is activated, it stimulates contraction of all the muscle fibers that it innervates. Neighboring muscle fibers do not contract due to the insulation provided by the connective tissue coverings.

Table 7.2 Microscopic Anatomy of a Skeletal Muscle Fiber

Structure	Description/Function
Sarcolemma	Plasma membrane of a muscle fiber maintaining the integrity of the cell
Sarcoplasm	Cytoplasm of a muscle fiber that contains organelles
Nuclei	Contain DNA, which determines cell structure and function
Sarcoplasmic reticulum	Smooth ER in a muscle fiber that stores Ca^{2+}
Transverse tubules	Extensions of the sarcolemma that penetrate into the sarcoplasm carrying muscle impulses, which trigger the release of Ca^{2+} from the sarcoplasmic reticulum
Myofibril	A bundle of myofilaments
Myofilaments	Threadlike contractile proteins that interact to produce contractions

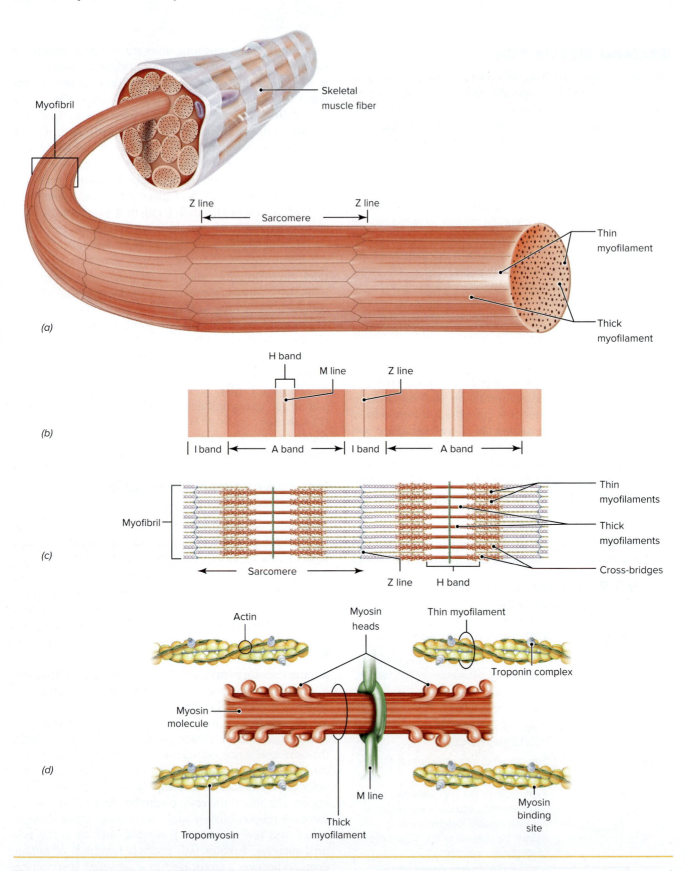

Figure 7.2 Structure of a Myofibril.
(a) A muscle fiber contains many myofibrils. Each myofibril consists of repeating functional units called sarcomeres. *(b)* The characteristic bands of sarcomeres. *(c)* The arrangement of thin and thick myofilaments within the sarcomeres. *(d)* Details of thin myofilaments and thick myofilaments. **AP|R**

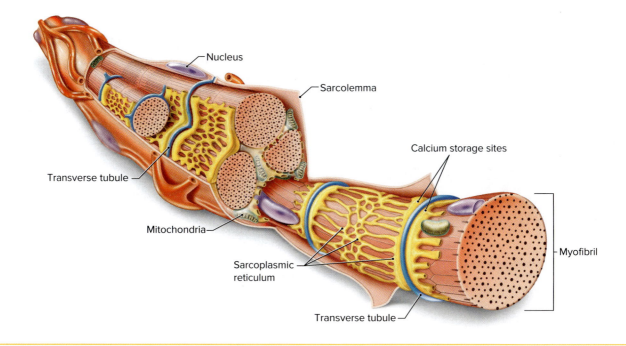

Figure 7.3 Special Structures Within Skeletal Muscle Fibers.
A portion of a muscle fiber showing the sarcoplasmic reticulum and the transverse (T) tubules associated with the myofibrils.

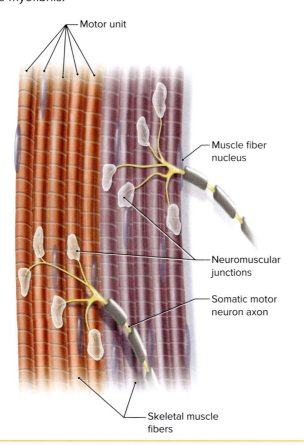

Figure 7.4 Motor Unit Structure.
A motor unit consists of one somatic motor neuron and all the muscle fibers that it innervates. Note the attachment of the terminal boutons to the muscle fibers. **AP|R**

Neuromuscular Junction

The part of a somatic motor neuron that leads to a muscle fiber is called an *axon*. The connection between the terminal branches of an axon and the sarcolemma of a muscle fiber is known as a **neuromuscular junction** (figure 7.4). As shown in figure 7.5, the *terminal boutons* of an axon fit into depressions, the *motor end plates*, in the sarcolemma. The tiny space between the terminal bouton and the motor end plate is the *synaptic cleft*. Numerous secretory vesicles in the terminal bouton contain the **neurotransmitter** (nū-rō-trans′-mit-er) *acetylcholine* (as″-ē-til-kō′-lēn), or ACh. When a somatic motor neuron is activated and an action potential reaches the terminal bouton, ACh is released from secretory vesicles into the synaptic cleft. The attachment of ACh to ACh receptors on the motor end plate triggers a series of reactions causing the muscle fiber to contract.

⊕ Clinical Insight

Anabolic steroids, substances similar to the male sex hormone testosterone, have been used by some athletes to promote muscle development and strength. However, physicians have warned that such use can produce a number of harmful side effects, including damage to kidneys, increased risk of heart disease and liver cancer, and increased irritability. Other side effects include decreased testosterone and sperm production in males and increased facial hair and deepening of the voice in females.

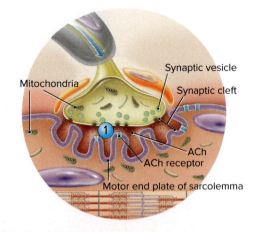

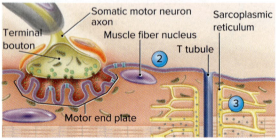

Figure 7.5 Structure of the Neuromuscular Junction. A neuromuscular junction is formed by the terminal bouton of a somatic motor neuron and the motor end plate of a muscle fiber. The detailed insert shows the synaptic vesicles, the synaptic cleft, and the folded surface of the motor end plate. **AP|R**

 Check My Understanding

1. How are muscle tissue and connective tissue arranged in a skeletal muscle?
2. What composes a muscle fiber?

7.3 Physiology of Skeletal Muscle Contraction

Learning Objectives
7. Describe the physiology of contraction.
8. Explain the cause of excess post-exercise oxygen consumption (EPOC).
9. Explain the all-or-none contraction of muscle fibers.
10. Discuss how graded contractions of whole muscles produce a variety of contraction strengths.

Contraction of a muscle fiber is a complex process that involves a number of rapid structural and chemical changes within the muscle fiber. The molecular mechanism of contraction is explained by the *sliding-filament model* described in the next section.

Mechanism of Contraction **AP|R**

As mentioned in the previous section, in order for a muscle fiber to contract it needs to first be stimulated or "excited" by a somatic motor neuron. The pairing of a muscle impulse and physical contraction of the muscle fiber is referred to as **excitation-contraction coupling.** Figure 7.5 shows the steps of excitation.

1. Contraction of a muscle fiber is initiated when the terminal bouton of an activated somatic motor neuron releases ACh into the synaptic cleft.
2. Acetylcholine binds to ACh receptors on the motor end plate causing the formation of a muscle impulse (similar to the action potential that will be described in chapter 8), that spreads over the sarcolemma and is carried into the sarcoplasm by the T tubules.
3. Stimulation of the sarcoplasmic reticulum from the nearby T tubules triggers the release of Ca^{2+} from the sarcoplasmic reticulum into the sarcoplasm.

Figure 7.6 shows the steps of the contraction cycle.

Step 1a: Ca^{2+} within the sarcoplasm binds to troponin complex, which then causes the tropomyosin strands to change position, exposing the myosin binding sites on actin molecules.

Step 1b: With the myosin binding sites exposed, each myosin head binds to a myosin binding site to form a cross-bridge with the actin molecule.

Step 2: While the cross-bridge is formed, the inorganic phosphate detaches, causing the myosin head to pivot and exert a power stroke that pulls the thin myofilaments toward the M line of the sarcomere. ADP detaches during the pivoting of the myosin head.

Step 3: The power stroke causes sliding of the myofilaments past one another, and the sarcomere shortens.

Step 4: A new molecule of ATP binds to the myosin head, causing myosin to release the actin molecule.

Step 5: The detached myosin head returns to its relaxed position and then becomes energized after hydrolyzing the ATP to ADP and P_i.

Step 6: This returns us to **Step 1b**, wherein the energized myosin head reattaches to a new binding site on actin, releases P_i, and uses its energy to repeat the power stroke in **Step 2**. This cycle rapidly repeats itself to maintain a contraction as long as ATP and Ca^{2+} are available.

When the somatic motor neuron stops stimulating the muscle fiber, an enzyme in the synaptic cleft called *acetylcholinesterase* begins decomposing ACh. The breakdown of ACh prevents continued stimulation of the muscle fiber. Consequently, Ca^{2+} is no longer released from the sarcoplasmic reticulum and is instead actively transported

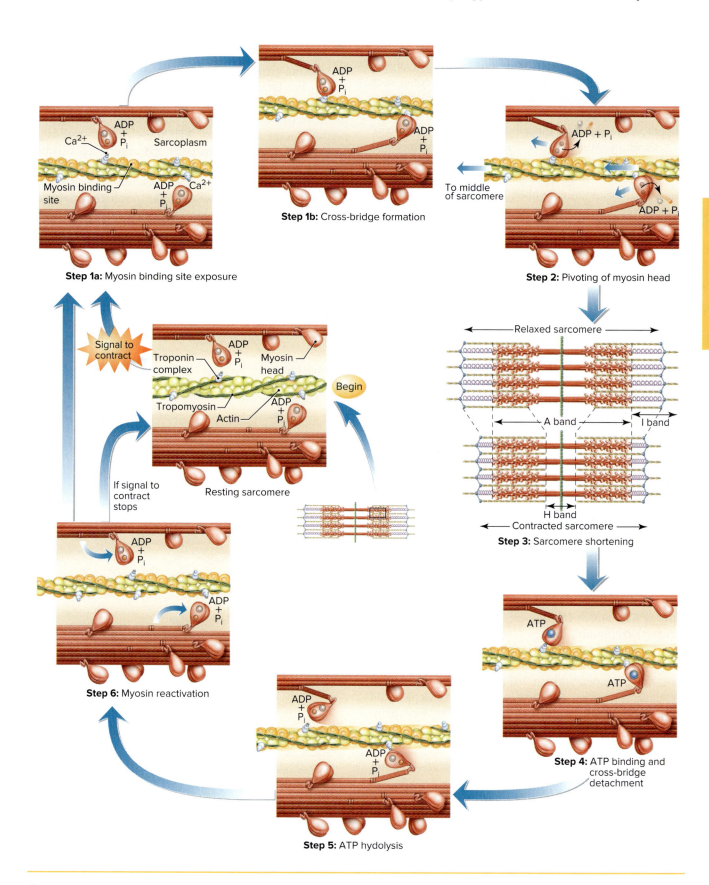

Figure 7.6 Sliding-Filament Model of Muscle Contraction.
The release of Ca^{2+} into sarcoplasm causes the exposure of myosin binding sites on actin molecules, enabling the contraction cycle to begin. ATP powers the contraction cycle. AP|R

from the sarcoplasm into the sarcoplasmic reticulum. This causes Ca^{2+} to unbind troponin complex, which allows tropomyosin to move back over the myosin binding sites and stop the contraction cycle. The thin and thick myofilaments then slide back to their original positions, moving the Z lines apart, lengthening the sarcomeres (muscle relaxation).

Carefully study figure 7.6, which illustrates the sliding-filament model of muscle contraction. Note the configuration of thin myofilaments and thick myofilaments in a relaxed muscle fiber, how they interact in the steps of the contraction cycle, and how contraction is powered by ATP. Although the sliding myofilaments produce contraction (i.e., the shortening of the sarcomeres), the lengths of the thin myofilaments and thick myofilaments remain unchanged (step 3, figure 7.6).

Energy for Contraction

The energy for muscle contraction comes from ATP molecules in the muscle fiber. Recall that ATP is a product of cellular respiration. However, there is only a small amount of ATP in each muscle fiber. Once it is used up, more ATP must be formed in order for additional contractions to occur. Figure 7.7 summarizes the processes involved in the replenishment of ATP.

While a muscle fiber is relaxed, it uses cellular respiration to release energy from nutrients and transfers that energy to the high-energy phosphate bonds of ATP. Once there are sufficient amounts of ATP available in the muscle fiber, the high-energy phosphate is transferred to creatine to form **creatine phosphate (CP),** which serves as a storage form of readily available energy. The resulting ADP is then reconverted to ATP using cellular respiration.

Muscle contraction quickly reduces ATP levels, resulting in the high-energy phosphate group being transferred back from the creatine phosphate to the ADP, forming ATP, which can then be used to power additional contractions (figure 7.7a).

There is four to six times more creatine phosphate than ATP in a muscle fiber, so it is an important source for immediate ATP formation without waiting for the slower process of cellular respiration. However, it can also be depleted in under 10 seconds in a muscle that is contracting repeatedly.

⊕ Clinical Insight

The reaction that transfers the phosphate between creatine phosphate and ADP is controlled by an enzyme unique to muscle tissue. When muscle tissue is damaged, this enzyme is released into the blood. Elevated levels of the cardiac version of this enzyme in blood tests suggest that a heart attack may have occurred. Blood levels of cardiac troponin complex can be used as an indicator of heart damage as well.

Oxygen and Cellular Respiration

Recall from chapter 3 that cellular respiration is the process of breaking down glucose in two steps: (1) anaerobic respiration in the cytosol and (2) aerobic respiration in the mitochondria. Due to the need of a constant supply for glucose to generate ATP, muscle fibers store large amounts of glucose as **muscle glycogen.** Recall from chapter 2 that glycogen is a polysaccharide of glucose.

Whether or not a muscle fiber uses just anaerobic respiration or also includes aerobic respiration depends on the availability of oxygen. During periods of strenuous exercise such as weight lifting, muscle fibers will employ mostly anaerobic respiration because the respiratory and

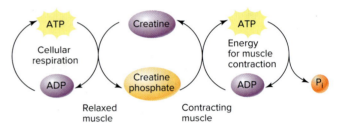
(a) ATP from creatine phosphate

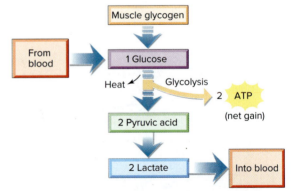
(b) ATP from anaerobic respiration

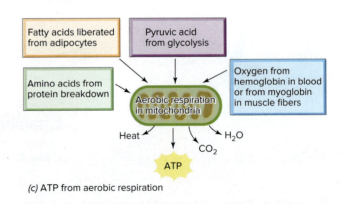
(c) ATP from aerobic respiration

Figure 7.7 Sources of ATP in Muscle Fibers.

cardiovascular systems cannot provide oxygen to muscle fibers quickly enough to maintain aerobic respiration. The muscle fibers will break down glycogen to glucose and glucose to pyruvic acid, in a process called *glycolysis*, forming only a small amount of ATP per molecule of glucose (see chapter 3 and figure 7.7b). In anaerobic respiration, pyruvic acid is further broken down into *lactate*.

Since anaerobic respiration is not favorable in muscle fibers, muscle tissue is adapted to facilitate aerobic respiration. Muscle tissue possesses a large number of blood vessels and obtains large amounts of oxygen from the blood via **hemoglobin,** the red pigment in red blood cells. Muscle fibers also have a similar pigment, **myoglobin,** which stores oxygen within the sarcoplasm and helps transfer oxygen to the mitochondria. In the same manner that creatine phosphate stores extra energy in times of muscle inactivity, some of the oxygen carried to muscle fibers is transferred from hemoglobin to myoglobin and stored for later use during periods of muscle activity. This function of myoglobin reduces the muscle fiber's dependence on oxygen carried to it by the blood at the onset of exercise. During inactivity or light to moderate physical activity (e.g., endurance training), muscle fibers receive sufficient oxygen to carry on the aerobic respiration. As shown in figure 7.7c, this process involves the breakdown of *pyruvic acid* produced in glycolysis, or other organic nutrients, into carbon dioxide and water. In contrast to anaerobic respiration, aerobic respiration provides a large amount of ATP per molecule of glucose (see chapter 3 and figure 7.7c).

Excess Post-Exercise Oxygen Consumption (EPOC)

When a muscle fiber utilizes anaerobic respiration, such as during strenuous exercise, it produces lactate and depletes its ATP, CP, and oxygen stores. To restore resting conditions within a muscle fiber after activity ceases, respiratory and heart rates remain elevated to support **excess post-exercise oxygen consumption** or **EPOC** (formerly oxygen debt). EPOC is the amount of oxygen required to replenish myoglobin and to produce the ATP needed for the metabolism of the lactate in the liver, heart, and skeletal muscles and the restoration of ATP and creatine phosphate in the muscle fibers.

Fatigue

If a muscle is stimulated to contract for a long period, its contractions will gradually decrease until it no longer responds to stimulation. This condition is called **fatigue.** Although the exact mechanism is not known, several factors seem to be responsible for muscle fatigue. The most likely cause of fatigue in long-term muscle activity is a lack of available nutrients, such as muscle glycogen and fatty acids, to utilize for ATP production.

Effects of Exercise on Muscles

Exercise has a profound effect on skeletal muscles. Strength training, which involves resistance exercise such as weight lifting, causes a muscle fiber to be repetitively stimulated to maximum contraction. Over time, the repetitive stimulation produces **hypertrophy**—an increase in muscle fiber size and strength. The number of muscle fibers cannot be increased after birth. Instead, hypertrophy results from an increase in the number of myofibrils in muscle fibers, which increases the diameter and strength of the muscle fibers and of the whole muscle itself. In comparison, lack of repetitive stimulation to maximum force causes muscular **atrophy,** which is the reduction in muscle size and strength due to loss of myofibrils. Atrophy can be caused by damage to the nerve stimulating the muscle or lack of use, such as when a limb is in a cast. Aerobic exercise, or endurance training, does not produce hypertrophy. Instead, it enhances the efficiency of aerobic respiration in muscle fibers by increasing (1) the number of mitochondria, (2) the efficiency of obtaining oxygen from the blood, and (3) the concentration of myoglobin.

Heat Production

Heat production by muscular activity is an important mechanism in maintaining average body temperature. Muscles are active organs that form a large proportion of the body weight. Heat produced by muscles results from cellular respiration and other chemical reactions within the muscle fibers. Recall that 60% of the energy released by cellular respiration is heat energy. Muscle generates so much heat that exercise leads to an increase in body temperature that requires sweating to help remove heat from the body. On the other hand, the major response to a decrease in body temperature is shivering, which is involuntary muscle contractions.

 Check My Understanding

3. What are the structure and function of a neuromuscular junction?
4. How do thin and thick myofilaments interact during muscle contraction?
5. What are the roles of ATP and creatine phosphate in muscle contraction?
6. What are the relationships among cellular respiration, lactate, and excess post-exercise oxygen consumption?

Contraction Characteristics

When studying muscle contraction, physiologists consider both single-fiber contraction and whole-muscle contraction.

Contraction of a Single Fiber

It is possible to remove a single muscle fiber in order to study its contraction in the laboratory. By using electrical stimuli to initiate contraction and by gradually increasing the strength (voltage) of each stimulus, it has been shown that the fiber will not contract until the stimulus reaches a certain minimal strength. This minimal stimulus is called the **threshold stimulus.**

Whenever a muscle fiber is stimulated by a threshold stimulus or by a stimulus of greater strength, it always contracts *completely*. Thus, a muscle fiber either contracts completely or not at all–contraction is *not* proportional to the strength of the stimulus. This characteristic of individual muscle fibers is known as the **all-or-none response.**

Contraction of Whole Muscles

Much information has been gained by studying the contraction of a whole muscle of an experimental animal. In such studies, electrical stimulation is used to cause contraction, and the contraction is recorded to produce a tracing called a *myogram*.

If a single threshold stimulus is applied, some of the muscle fibers will contract to produce a single, weak contraction (a muscle twitch) and then relax, all within a fraction of a second. The myogram will look like the one shown in figure 7.8. After the stimulus is applied, there is a brief interval before the muscle starts to contract. This interval is known as the *latent phase*. Then, the muscle contracts (shortens) during the *contraction phase* and relaxes (returns to its former length) during the *relaxation phase*. If a muscle is stimulated again after it has relaxed completely, it will contract and produce a similar myogram. A series of single stimuli applied in this manner will yield a myogram like the one in figure 7.9a.

If the interval between stimuli is shortened so that the muscle fibers cannot completely relax, the force of individual twitches combines by *summation*, which increases the force of contraction. Rapid summation produces incomplete tetany, a fluttering contraction (figure 7.9b). If stimuli are so frequent that relaxation is not possible, **tetany** results (figure 7.9c). Tetany is a state of sustained contraction without relaxation. In the body, tetany results from a rapid series of action potentials carried by somatic motor neurons to the muscle fibers that results in a prolonged state of contraction. Tetany for short time periods is the usual way in which muscles contract to produce body movements.

Graded Responses Unlike individual muscle fibers that exhibit all-or-none responses, whole muscles exhibit *graded responses*–that is, varying degrees of contraction. Graded responses enable the degree of muscle contraction to fit the task being performed. Obviously, more muscle fibers are required to lift a 14 kg (30 lb) weight than to lift a feather. Yet both activities can be performed by the same muscles.

Graded responses are possible because a muscle is composed of many different *motor units*, each responding to different thresholds of stimulation. In the laboratory, a weak stimulus that activates only low-threshold motor units produces a minimal contraction. As the strength of the stimulus is increased, the contractions get stronger as more motor units are activated until a **maximal stimulus** (one that activates all motor units) is applied, which produces a maximal contraction. Further increases in the strength of the stimulus (supramaximal) cannot produce a greater contraction. The same results occur in a normally functioning body. The nervous system provides the stimulation and controls the number of motor units activated in each muscle contraction. The activation of more and more motor units is known as *motor unit recruitment* (figure 7.9d).

Muscle Tone Even when a muscle is relaxed, some of its muscle fibers are contracting. At any given time, some of the muscle fibers in a muscle are involved in a sustained contraction that produces a constant partial, but slight, contraction of the muscle. This state of constant partial contraction, called **muscle tone,** keeps a muscle ready to respond. Muscle tone results from the alternating activation of different motor units by the nervous system so that some muscle fibers are always in sustained contraction, as seen in figure 7.10. Muscle tone of postural muscles plays an important role in maintaining erect posture.

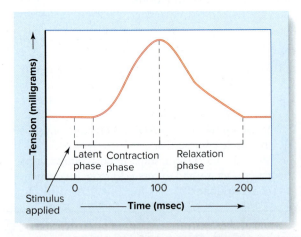

Figure 7.8 A Myogram of a Single Muscle Twitch. Note the brief latent phase, contraction phase, and longer relaxation phase.

 Check My Understanding

7. What is meant by the all-or-none response?
8. How are muscles able to make graded responses?

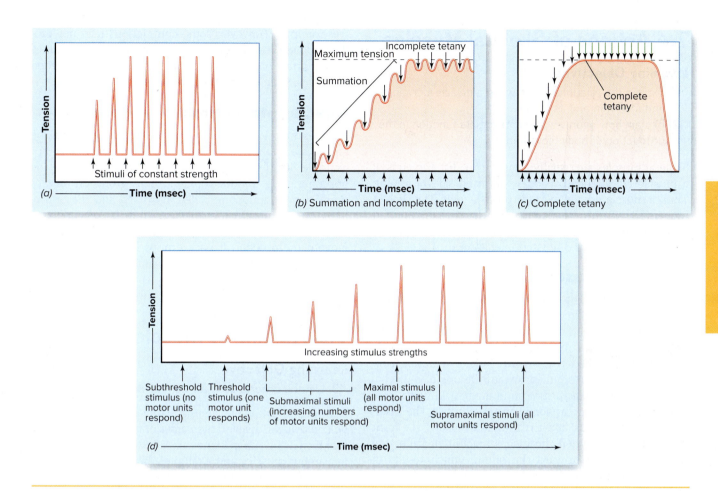

Figure 7.9 Muscular Responses.
Myograms of *(a)* a series of simple twitches, *(b)* summation caused by incomplete relaxation between stimuli, *(c)* tetany, and *(d)* motor unit recruitment.

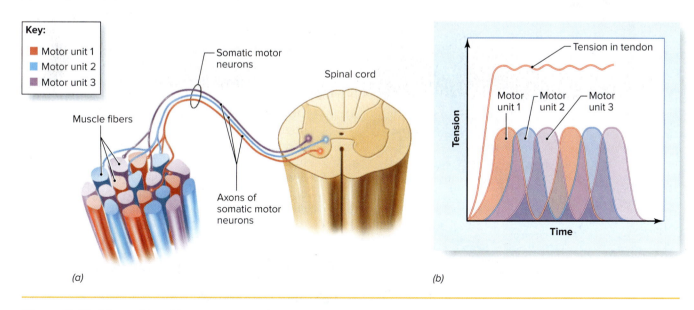

Figure 7.10 Motor Unit Alternation.
(a) Anatomy of motor units in a skeletal muscle. *(b)* Myogram showing mechanism of motor unit alternation in muscle tone.

7.4 Actions of Skeletal Muscles

Learning Objectives
11. Explain the relationship between a muscle's origin and insertion and its action.
12. Explain how agonists and antagonists function in the production of body movements.

Skeletal muscles are usually arranged so that the ends of a muscle are attached to bones on each side of a joint. Thus, a muscle usually extends across a joint. The type of movement produced depends upon the type of joint and the locations of the muscle attachments. Common movements at joints were discussed in chapter 6.

Origin and Insertion

During contraction, a bone to which one end of the muscle is attached moves, but the bone to which the other end is attached does not. The movable attachment of a muscle is called the **insertion,** and the immovable attachment is called the **origin.** When a muscle contracts, the insertion is pulled toward the origin.

Consider the *biceps brachii* in figure 7.11. It has two origins, and both are attached to the scapula. The insertion is on the radius, and the muscle lies along the front of the humerus. When the biceps brachii contracts, the insertion is pulled toward the origin, which results in flexion of the forearm at the elbow.

Most muscle contractions are *isotonic contractions,* which cause movement at a joint. Walking and breathing are examples. However, some contractions may not produce movement but only increase tension within a muscle. Contractions that maintain body posture are good examples. Such contractions are *isometric contractions.*

Muscle Interactions

Muscles function in groups rather than singly, and the groups are arranged to provide opposing movements. For example, if one group of muscles produces flexion, the opposing group produces extension. A group of muscles producing an action are called **agonists,** and the opposing group of muscles are called **antagonists.** When agonists contract, antagonists must relax, and vice versa, for movement to occur. If both groups contract simultaneously, the movable body part remains rigid. Figure 7.11 illustrates how the biceps brachii is the agonist of forearm flexion, while the triceps brachii is the antagonist.

7.5 Naming of Muscles

Learning Objective
13. List the criteria used for naming muscles.

Learning the complex names and functions of muscles can be confusing. However, the names of muscles are informative if their meaning is known. A few of the criteria used in naming muscles and examples of terms found in the names of muscles are listed below:

- **Function:** extensor, flexor, adductor, and pronator.
- **Shape:** trapezius (trapezoid), rhomboid (rhombus), deltoid (delta-shaped or triangular), biceps (two heads).
- **Relative position:** external, internal, abdominal, medial, lateral.
- **Location:** intercostal (between ribs), pectoralis (chest).
- **Site of attachment:** temporalis (temporal bone), zygomaticus (zygomatic bone).
- **Origin and insertion:** sternohyoid (sternum = origin; hyoid = insertion), sternocleidomastoid (sternum and clavicle = origins; mastoid process = insertion).
- **Size:** maximus (larger or largest), minimus (smaller or smallest), brevis (short), longus (long).
- **Orientation of fibers:** oblique (diagonal), rectus (straight), transversus (across).

7.6 Major Skeletal Muscles

Learning Objectives
14. Describe the location and action of the major muscles of the body.
15. Identify the major muscles on a diagram.

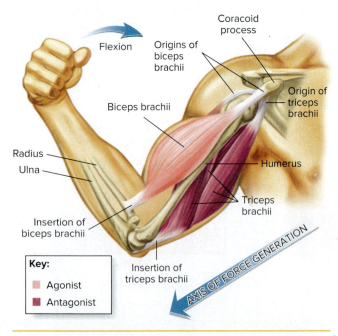

Figure 7.11 Muscle Interactions.
Demonstration of the actions of agonists and antagonists with origins and insertions labeled.

Part 2 Covering, Support, and Movement of the Body   147

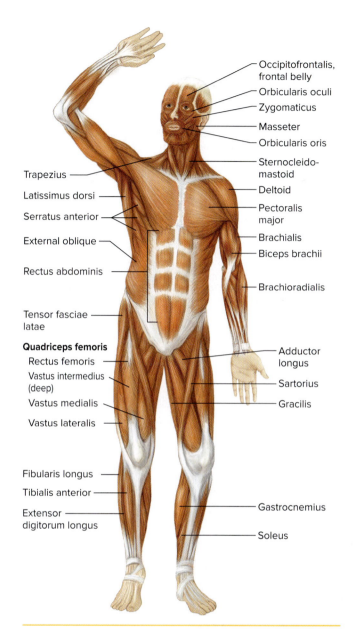

Figure 7.12 Anterior View of Superficial Skeletal Muscles.

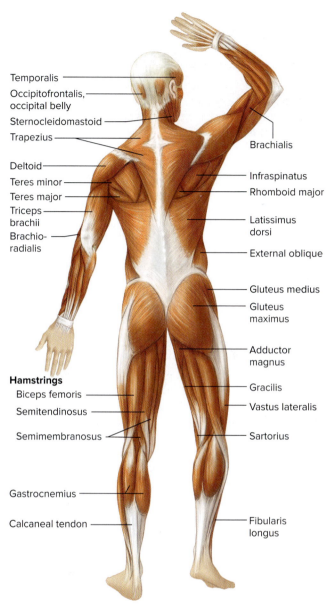

Figure 7.13 Posterior View of Superficial Skeletal Muscles.

This section is concerned with the name, location, attachment, and action of the major skeletal muscles. There are more than 600 muscles in the body, but only a few of the major muscles are considered here. Most of this information is presented in tables and figures to aid your learning. The tables are organized according to the primary actions of the muscles. The pronunciation of each muscle is included, because being able to pronounce the names correctly will help you learn the names of the muscles.

As you study this section, locate each muscle listed in the tables on the related figures 7.12 to 7.25. This will help you visualize the location and action of each muscle. Also, if you visualize the locations of the origin and insertion of a muscle, its action can be determined because contraction pulls the insertion toward the origin. It may help to refresh your understanding of the skeleton by referring to appropriate figures in chapter 6. Begin your study by examining figures 7.12 and 7.13 to learn the major superficial muscles that will be considered in more detail as you progress through the chapter.

Muscles of Facial Expression and Mastication

Muscles of the face and scalp produce the facial expressions that help communicate feelings, such as anger, sadness, happiness, fear, disgust, pain, and surprise. Most have origins on skull bones and insertions on the dermis of the skin (table 7.3 and figure 7.14).

Table 7.3 Muscles of Facial Expression

Muscle	Origin	Insertion	Action
Buccinator (buk′-si-nā-tor)	Lateral surfaces of maxilla and mandible	Orbicularis oris	Compresses cheeks inward
Occipitofrontalis (ok-sip-i-toe-fron-ta-lis)	This muscle consists of two parts: the frontal belly and the occipital belly. They are joined by the epicranial aponeurosis, which covers the top of the skull.		
Frontal belly	Epicranial aponeurosis	Skin and muscles above the eyes	Elevates eyebrows and wrinkles forehead
Occipital belly	Base of occipital bone	Epicranial aponeurosis	Pulls scalp backward
Orbicularis oculi (or-bik′-ū-lar-is ok′-ū-li)	Frontal bone and maxillae	Skin around eye	Closes eye
Orbicularis oris (or-bik′-ū-lar-is o′-ris)	Muscles around mouth	Skin around lips	Closes and puckers lips; shapes lips during speech
Platysma (plah-tiz′-mah)	Fascia of upper chest	Mandible and muscles around mouth	Draws angle of mouth downward (frowning)
Zygomaticus (zī-gō-mat′-ik-us)	Zygomatic bone	Orbicularis oris at angle of the mouth	Elevates corners of mouth (smiling)

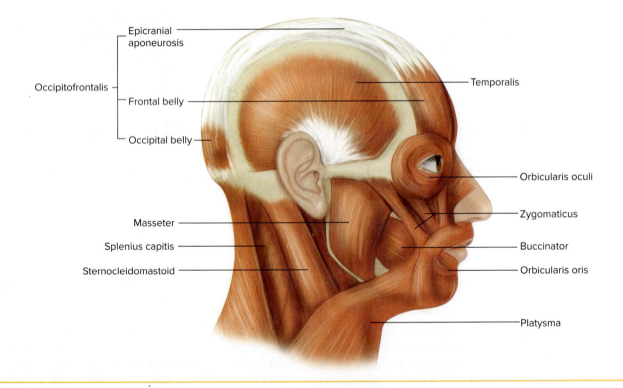

Figure 7.14 Muscles of Facial Expression and Mastication. **AP|R**

The *occipitofrontalis* is an unusual muscle. It has a large *epicranial aponeurosis* that covers the top of the skull and two contractile portions: the *frontal belly* over the frontal bone and the *occipital belly* over the occipital bone.

Two major pairs of muscles elevate the mandible in the process of mastication (chewing): the *masseter* and the *temporalis* (table 7.4 and figure 7.14).

Muscles That Move the Head

Several pairs of neck muscles are responsible for flexing, extending, and rotating the head. Table 7.5 lists two of the major muscles that perform this function: the *sternocleidomastoid* and the *splenius capitis*. As noted in table 7.8, the *trapezius* can also extend the head, although this is not its major function (figures 7.14 to 7.16).

Table 7.4 Muscles of Mastication

Muscle	Origin	Insertion	Action
Masseter (mas-se′-ter)	Zygomatic arch	Outer surface of mandible	Elevates mandible
Temporalis (tem-po-ra′-lis)	Temporal bone	Coronoid process of mandible	Elevates mandible

Table 7.5 Muscles That Move the Head

Muscle	Origin	Insertion	Action
Sternocleidomastoid (ster-nō-klī-dō-mas′-toid)	Clavicle and sternum	Mastoid process of temporal bone	Contraction of both muscles flexes head toward chest; contraction of one muscle turns head away from contracting muscle
Splenius capitis (splē′-nē-us kap′-i-tis)	Lower cervical and upper thoracic vertebrae	Mastoid process of temporal bone	Contraction of both muscles extends head; contraction of one muscle turns head toward same side as contracting muscle

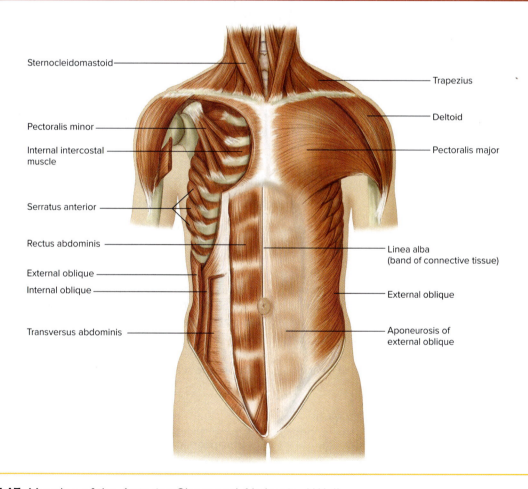

Figure 7.15 Muscles of the Anterior Chest and Abdominal Wall. The right pectoralis major is removed to show the underlying muscles. AP|R

Muscle of the Abdominal Wall

The abdominal muscles are paired muscles that provide support for the front and sides of the abdominal and pelvic regions, including support for the internal organs.

The muscles are named for the direction of their muscle fibers: *rectus abdominis, external oblique, internal oblique,* and *transversus abdominis.* They are arranged in overlapping layers and are attached by larger aponeuroses that

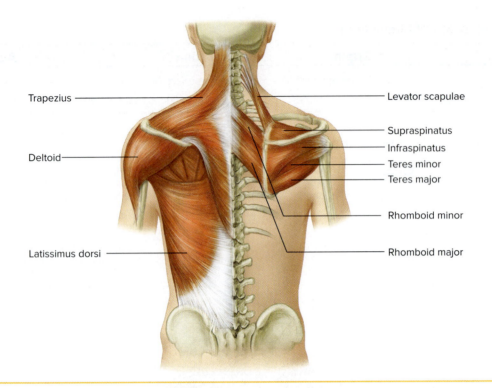

Figure 7.16 Muscles of the Posterior Shoulder.
The right trapezius is removed to show the underlying muscles. AP|R

merge at the anterior midline to form the *linea alba*, or white line (table 7.6 and figure 7.15).

Muscles of Breathing

Movement of the ribs occurs during breathing and is brought about by the contraction of two sets of muscles that are located between the ribs. The *external intercostal muscles* elevate and protract the ribs during inspiration, and the *internal intercostal muscles* depress and retract the ribs during expiration (table 7.7 and figure 7.15). The primary breathing muscle is the *diaphragm*, a thin sheet of muscle that separates the thoracic and abdominal cavities.

Muscles That Move the Pectoral Girdle

Pectoral girdle muscles originate on bones of the axial skeleton and insert on the scapula or clavicle. Because the scapula is supported mainly by muscles, it can be moved more freely than the clavicle. The *trapezius* is the outermost trapezoid-shaped muscle that covers much of the upper back. The *rhomboid major* and *minor* and the *levator scapulae* lie beneath the trapezius. Each *serratus anterior* is located on the sides of the ribs near the axillary region. The *pectoralis minor* lies beneath the pectoralis major. It protracts and depresses the scapula (table 7.8 and figures 7.15 to 7.18).

Table 7.6 Muscles of the Abdominal Wall

Muscle	Origin	Insertion	Action
Rectus abdominis (rek′-tus ab-dom′-i-nis)	Pubic symphysis and pubis	Xiphoid process of sternum and costal cartilages of ribs 5 to 7	Tightens abdominal wall; flexes the vertebral column
External oblique (eks-ter′-nal o-blēk′)	Anterior surface of ribs 5–12	Iliac crest and linea alba	Tightens abdominal wall; rotation and lateral flexion of the vertebral column
Internal oblique (in-ter′-nal o-blēk′)	Iliac crest and inguinal ligament	Cartilage of ribs 9–12, pubis, and linea alba	Same as above
Transversus abdominis (trans-ver′-sus ab-dom′-i-nis)	Iliac crest, cartilages of ribs 7–12, processes of lumbar vertebrae	Pubis and linea alba	Tightens abdominal wall

Table 7.7 Muscles of Breathing

Muscle	Origin	Insertion	Action
Diaphragm (dī-a-fram)	Lumbar vertebrae, costal cartilages of lower ribs, xiphoid process	Central tendon located at midpoint of muscle	Forms floor of thoracic cavity; depresses during contraction, causing inspiration
External intercostal muscle (eks-ter'-nal in-ter-kos'-tal)	Inferior border of rib above	Superior border of rib below	Elevates and protracts ribs during inspiration
Internal intercostal muscle (in-ter'-nal in-ter-kos'-tal)	Superior border of rib below	Inferior border of rib above	Depresses and retracts ribs during expiration

Table 7.8 Muscles That Move the Pectoral Girdle

Muscle	Origin	Insertion	Action
Trapezius (trah-pē-zē'-us)	Occipital bone; cervical and thoracic vertebrae	Clavicle; spine and acromion of scapula	Elevates clavicle; adducts and elevates scapula; extends head
Rhomboid major and minor (rom-boyd)	Upper thoracic vertebrae	Medial border of scapula	Adducts and elevates scapula
Levator scapulae (le-va'-tor skap'-ū-lē)	Cervical vertebrae	Superior medial margin of scapula	Elevates scapula
Serratus anterior (ser-ra'-tus)	Ribs 1–8	Medial border of scapula	Depresses, protracts, and rotates scapula
Pectoralis minor (pek-to-rah'-lis)	Front of the upper ribs	Coracoid process of scapula	Depresses and protracts scapula

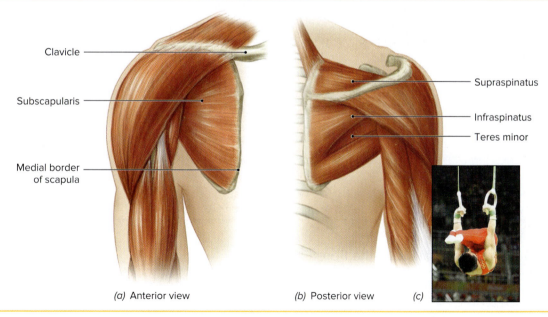

(a) Anterior view (b) Posterior view (c)

Figure 7.17 Muscles of the Right Rotator Cuff.
(a) Anterior view showing subscapularis. (b) Posterior view showing supraspinatus, infraspinatus, and teres minor. (c) A gymnast on the rings must have a strong rotator cuff. AP|R
(c) ©Leonard Zhukovsky/123RF RF

Muscles That Move the Arm and Forearm

Movement of the arm is enabled by the muscles that originate on the pectoral girdle, ribs, or vertebrae and insert on the humerus. The arrangement of these muscles and the ball-and-socket joint between the humerus and scapula enable great freedom of movement for the arm. The *pectoralis major* is the large outermost muscle of the chest. The *deltoid* is the thick muscle that caps the shoulder joint.

The *supraspinatus*, *infraspinatus*, and *teres minor* cover the back of the scapula. The front of each scapula is covered by the *subscapularis*. These four muscles and their tendons surround the head of the humerus at the shoulder joint, making up the **rotator cuff** (figure 7.17). The muscles and tendons of the rotator cuff are the only structures stabilizing the shoulder joint; thus, the joint is fairly unstable compared to other joints. However, this relative lack of stability is what allows the shoulder's mobility. The *latissimus dorsi* is a broad, sheetlike muscle that covers the lower back. The *teres major* assists the latissimus dorsi and is located just above it (table 7.9 and figures 7.15 to 7.17).

Muscles moving the forearm originate on either the humerus or the scapula and insert on either the radius or the ulna. Three flexors occur on the front of the arm: the *biceps brachii*, *brachialis*, and *brachioradialis*. One extensor, the *triceps brachii*, is located on the back of the arm (table 7.10 and figures 7.15, 7.18, and 7.19).

 Check My Understanding

9. What are the names and locations of the two parts of the occipitofrontalis?
10. What muscles are involved in chewing your food?
11. What muscles turn your head to the side?
12. What muscle separates the abdominal and thoracic cavities?
13. What are the names of the abdominal muscles from deep to superficial?
14. What three muscles elevate the scapula?

Table 7.9 Muscles That Move the Arm

Muscle	Origin	Insertion	Action
Pectoralis major (pek-tō-rah′-lis)	Clavicle, sternum, and cartilages of upper ribs	Greater tubercle of humerus	Adducts, flexes, and medially rotates arm
Deltoid (del′-toyd)	Clavicle and spine, and acromion of scapula	Deltoid tuberosity of humerus	Abducts, flexes, and extends arm
Latissimus dorsi (lah-tis′-i-mus dor′sī)	Lower thoracic and lumbar vertebrae; sacrum; lower ribs; iliac crest	Intertubercular sulcus of humerus	Adducts, extends, and medially rotates arm
Teres major (te′-rez)	Inferior angle of scapula	Distal to lesser tubercle of humerus	Same as above
Rotator cuff muscles	These four muscles stabilize the shoulder joint		
Supraspinatus (su-prah-spī′-na-tus)	Superior to spine of scapula	Greater tubercle of humerus	Abducts arm
Infraspinatus (in-frah-spī′-na-tus)	Inferior to spine of scapula	Greater tubercle of humerus	Laterally rotates arm
Teres minor	Lateral border of scapula	Greater tubercle of humerus	Laterally rotates arm
Subscapularis (sŭb-skap-ū-lār′-ris)	Front of scapula	Lesser tubercle of humerus	Medially rotates arm

Table 7.10 Muscles That Move the Forearm

Muscle	Origin	Insertion	Action
Biceps brachii (bī′-seps brā′-kē-ī)	Coracoid process and tubercle above glenoid cavity of scapula	Radial tuberosity of radius	Flexes forearm and supination, also flexes arm
Brachialis (brā′-kē-al-is)	Distal, front of humerus	Coronoid process of ulna	Flexes forearm
Brachioradialis (brā-kē-ō-rā-dē-a′-lis)	Lateral surface of distal end of humerus	Lateral surface of radius proximal to styloid process	Flexes forearm
Triceps brachii (trī′-seps brā′-kē-ī)	Lateral and medial surfaces of humerus and tubercle below glenoid cavity of scapula	Olecranon of ulna	Extends forearm, also extends arm

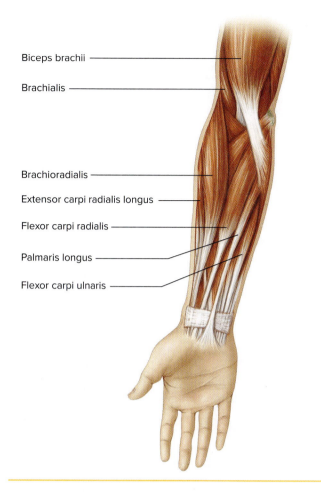

Figure 7.18 Muscles of the Anterior Right Forearm.

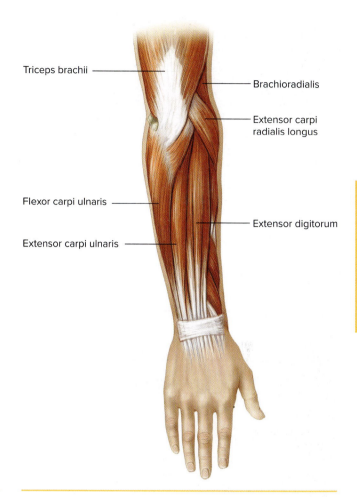

Figure 7.19 Muscles of the Posterior Right Forearm.

Muscles That Move the Wrist and Fingers

Many muscles that produce the various movements of the wrist and fingers are located in the forearm. Only a few of the larger superficial muscles are considered here. They originate from the distal end of the humerus and insert on carpal bones, metacarpals, or phalanges. Flexors on the front of the forearm include the *flexor carpi radialis, flexor carpi ulnaris,* and *palmaris longus.* Extensors on the back of the forearm include the *extensor carpi radialis longus, extensor carpi ulnaris,* and *extensor digitorum* (table 7.11; and figures 7.18 and 7.19). Note that the tendons of these muscles are held in position by a circular ligament at the wrist.

Check My Understanding

15. What muscle abducts and extends your arm?
16. What muscle extends your forearm?
17. What muscle extends your fingers?

Muscles That Move the Thigh and Leg

Muscles moving the thigh span the hip joint. They insert on the femur, and most originate on the pelvis. The *iliacus* and *psoas major* are located in the front, the *gluteus maximus* forms the buttocks, the *gluteus medius* is located beneath the gluteus maximus and extends laterally, and the *tensor fasciae latae* is located laterally. The *adductor longus* and *adductor magnus* are both located on the inside of the thigh (table 7.12 and figures 7.20 to 7.22).

The leg is moved by muscles located in the thigh. They span the knee joint and originate on the pelvis or femur and insert on the tibia or fibula. The **quadriceps femoris** is composed of four muscles with a common tendon that inserts on the patella. However, this tendon continues as the patellar ligament, which attaches to the tibial tuberosity–the functional insertion for these muscles. The *biceps femoris, semitendinosus,* and *semimembranosus* on the back of the thigh are collectively called the **hamstrings.** The medially located *gracilis* has two

Table 7.11 Muscles That Move the Wrist and Fingers

Muscle	Origin	Insertion	Action
Flexor carpi radialis (flek′-sor kar′-pī rā-dē-a′-lis)	Medial epicondyle of humerus	Metacarpals II and III	Flexes and abducts wrist
Flexor carpi ulnaris (flek′-sor kar′-pī ul-na′-ris)	Medial epicondyle of humerus and olecranon of ulna	Carpal bones and metacarpal V	Flexes and adducts wrist
Palmaris longus (pal-ma′-ris long′-gus)	Medial epicondyle of humerus	Fascia of palm	Flexes wrist
Extensor carpi radialis longus (eks-ten′-sor kar′-pī rā-dē-a′-lis long′-gus)	Lateral epicondyle of humerus	Metacarpal II	Extends and abducts wrist
Extensor carpi ulnaris (eks-ten′-sor kar′-pī ul-na′-ris)	Lateral epicondyle of humerus	Metacarpal V	Extends and adducts wrist
Extensor digitorum (eks-ten′-sor dij-i-to′-rum)	Lateral epicondyle of humerus	Back of phalanges II–V	Extends fingers

Table 7.12 Muscles That Move the Thigh

Muscle	Origin	Insertion	Action
Iliacus (il′-ē-ak-us)	Fossa of ilium	Lesser trochanter of femur	Flexes thigh
Psoas major (soh′-as)	Lumbar vertebrae	Lesser trochanter of femur	Flexes thigh
Gluteus maximus (glū′-tē-us mak′-si-mus)	Back of ilium, sacrum, and coccyx	Back of femur and iliotibial tract	Extends and laterally rotates thigh
Gluteus medius (glū′-tē-us mē′-dē-us)	Lateral surface of ilium	Greater trochanter of femur	Abducts and medially rotates thigh
Tensor fasciae latae (ten′-sor fash′-ē-ē lah-tē′)	Anterior iliac crest	Iliotibial tract	Flexes and abducts thigh
Adductor longus (ad-duk′-tor long′-gus)	Pubis near pubic symphysis	Back of femur	Adducts, flexes, and laterally rotates thigh
Adductor magnus (ad-duk′-tor mag′-nus)	Lower portion of ischium and pubis	Same as above	Same as above

Clinical Insight

Intramuscular injections are commonly used when quick absorption is desired. Such injections are given in three sites: (1) the side of the deltoid; (2) the gluteus medius in the upper, lateral portion of the buttock; and (3) the vastus lateralis near the midpoint of the outside of the thigh. These injection sites are chosen because there are no major nerves or blood vessels present that could be damaged, and the muscles have a good blood supply to aid absorption. The site chosen may vary with the age and condition of the patient.

insertions that give it dual actions. The long, straplike *sartorius* extends diagonally across the front of the thigh and spans both the hip and knee joints. Its contraction enables the legs to cross (tables 7.12 and 7.13, and figures 7.20 to 7.22).

Muscles That Move the Foot and Toes

Many muscles are involved in the movement of the foot and toes. They are located in the leg and originate on the femur, tibia, or fibula and insert on the tarsal bones, metatarsals, or phalanges. The muscles of the back of the leg include the *gastrocnemius* and *soleus*, which insert through a common tendon, the calcaneal (Achilles) tendon, which attaches to the calcaneus. The

Part 2 Covering, Support, and Movement of the Body   155

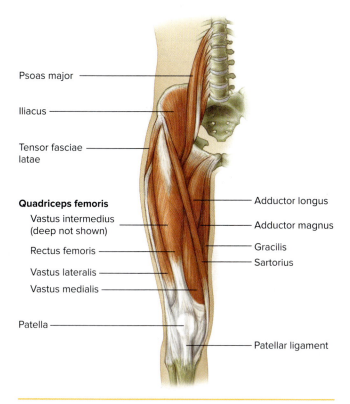

Figure 7.20 Muscles of the Anterior Right Thigh. (Note that the vastus intermedius is beneath the rectus femoris and is not visible in this view.) AP|R

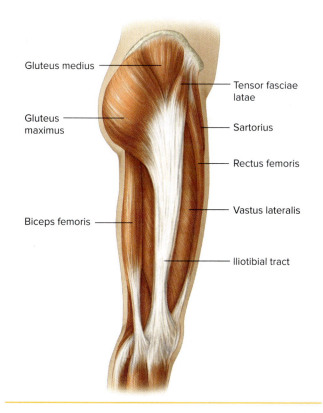

Figure 7.21 Muscles of the Lateral Right Thigh.

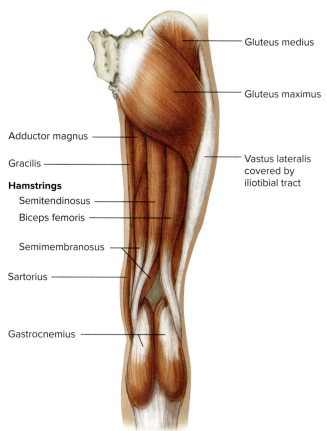

Figure 7.22 Muscles of the Posterior Right Thigh.

tibialis anterior is located adjacent to the shin, and the *extensor digitorum longus* lies lateral to it. Note that although the extensor digitorum extends the toes, as its name implies, it also dorsiflexes the foot. The *fibularis longus* is located on the outside of the leg (table 7.14 and figures 7.23 to 7.25).

Note how the tendons are held in position by the bands of ligaments at the ankle.

Clinical Insight

Repeated stress from athletic activities may cause inflammation of a tendon, a condition known as *tendonitis*. Tendons associated with the shoulder, elbow, hip, and knee joints are most commonly affected.

Check My Understanding

18. Which muscles flex the thigh?
19. What are the four parts of the quadriceps femoris?
20. What is the action of muscles inserting on the calcaneus?

Table 7.13 Muscles That Move the Leg

Muscle	Origin	Insertion	Action
Quadriceps femoris (quad'-ri-seps fem'-or-is)	Four muscles of the anterior thigh that extend the leg.		
Rectus femoris (rek'-tus fem'-or-is)	Anterior inferior iliac spine and superior margin of acetabulum	Patella; tendon continues as patellar ligament, which attaches to tibial tuberosity	Extends leg and flexes thigh
Vastus lateralis (vas'-tus lat-er-a'-lis)	Greater trochanter and back of femur	Same as above	Extends leg
Vastus medialis (vas'-tus me-de-a'-lis)	Medial and back of femur	Same as above	Extends leg
Vastus intermedius (vas'-tus in-ter-mē-'dē-us)	Anterior and lateral surfaces of femur	Same as above	Extends leg
Hamstrings	Three distinct muscles of the posterior thigh that flex leg and extend thigh.		
Biceps femoris (bī'-seps fem'-or-is)	Ischial tuberosity and back of femur	Head of fibula and lateral condyle of tibia	Flexes and laterally rotates leg; extends thigh
Semitendinosus (sem-ē-ten-di-nō'-sus)	Ischial tuberosity	Medial surface of tibia	Flexes and medially rotates leg; extends thigh
Semimembranosus (sem-ē-mem-brah-nō'-sus)	Ischial tuberosity	Medial condyle of tibia	Flexes and medially rotates leg; extends thigh
Gracilis (gras'-il-is)	Pubis near pubic symphysis	Medial surface of tibia	Adducts thigh; flexes leg and locks knee
Sartorius (sar-tor-e'-us)	Anterior superior iliac spine	Medial surface of tibia	Flexes thigh and leg; abducts and laterally rotates thigh

Table 7.14 Muscles That Move the Foot and Toes

Muscle	Origin	Insertion	Action
Gastrocnemius (gas-trōk-nē'-mē-us)	Medial and lateral condyles of femur	Calcaneus by the calcaneal tendon	Plantar flexes foot and flexes leg
Soleus (sō'-lē-us)	Back of tibia and fibula	Calcaneus by the calcaneal tendon	Plantar flexes foot
Fibularis longus (fib-yu-lar-ris long'-gus)	Lateral condyle of tibia and head and body of fibula	Metatarsal I and tarsal bones	Plantar flexes and everts foot; supports arch
Tibialis anterior (tib-ē-a'-lis an-te'-rē-or)	Lateral condyle and surface of tibia	Metatarsal I and tarsal bones	Dorsiflexes and inverts foot
Extensor digitorum longus (eks-ten'-sor dig-i-tor'-um long'-gus)	Lateral condyle of tibia and front of fibula	Phalanges of toes II–V	Dorsiflexes and everts foot; extends toes

7.7 Disorders of the Muscular System

Learning Objective

16. Describe the major disorders of the muscular system.

Some disorders of the muscular system may result from factors associated only with muscles, while others are caused by disorders of the nervous system. Certain neurological disorders are included here because of their obvious effect on muscle action.

Muscular Disorders

Cramps involve involuntary, painful tetany. The precise cause is unknown, but a cramp seems to result from chemical changes in the muscle, such as ionic imbalances or ATP deficiencies. Sometimes a severe blow to a muscle can produce a cramp.

Fibrosis (fī-brō'-sis) is an abnormal increase of connective tissue in a muscle. Usually, it results from connective tissue replacing dead muscle fibers following an injury.

Part 2 Covering, Support, and Movement of the Body   **157**

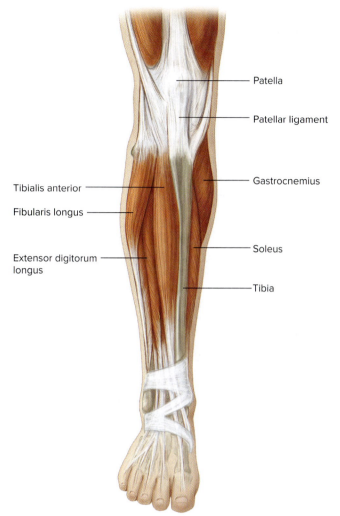

Figure 7.23 Muscles of the Anterior Right Leg. AP|R

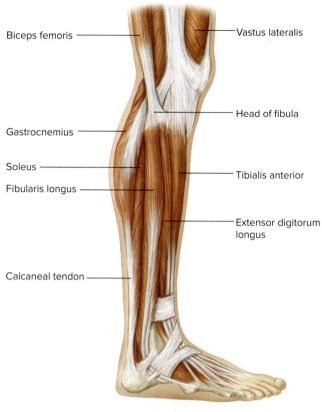

Figure 7.24 Lateral View of the Muscles of the Right Leg.

Fibromyalgia (fī-brō-mī-alj-a) is a painful condition of the muscles and joints with no known cause. Once thought to be a mental disorder, this is actually a musculoskeletal disorder that often leads to depression due to the helpless nature of the chronic symptoms.

Muscular dystrophy (dis'-trō-fē) is a general term for a number of inherited muscular disorders that are characterized by the progressive degeneration of muscles. The affected muscles gradually weaken and atrophy, producing a progressive crippling of the patient. There is no specific drug cure, but patients are encouraged to keep active and are given muscle-strengthening exercises.

Strains, or "pulled muscles," result when a muscle is stretched excessively. This usually occurs when an antagonist has not relaxed quickly enough as an agonist contracts. The hamstrings are a common site of muscle strains. In mild strains, only a few muscle fibers are damaged. In severe strains, both connective and muscle tissues are torn, and muscle function may be severely impaired.

Neurological Disorders Affecting Muscles

Botulism (boch'-ū-lizm) poisoning is caused by a neurotoxin produced by the bacterium *Clostridium botulinum*. The toxin prevents release of ACh from the terminal boutons of somatic motor axons. Without prompt treatment with an antitoxin, death may result from paralysis of breathing muscles. Poisoning results from eating improperly canned vegetables or meats that contain *C. botulinum* and the accumulated toxins.

Myasthenia gravis (mī-as-thē'-nē-ah gra'-vis) is characterized by extreme muscular weakness caused by improper functioning of the neuromuscular junctions. It is an autoimmune disease in which antibodies are produced that attach to the ACh receptors on the motor end plate and reduce or block the stimulatory effect of ACh. Myasthenia gravis occurs most frequently in women between 20 and 40 years of age. Usually, it first affects ocular muscles and other muscles of the face and neck, which may lead to difficulty in chewing, swallowing, and talking. Other muscles of the body may be involved later. Treatment typically involves the use of acetylcholinesterase inhibitors and immunosuppressive drugs, such as the steroid prednisone.

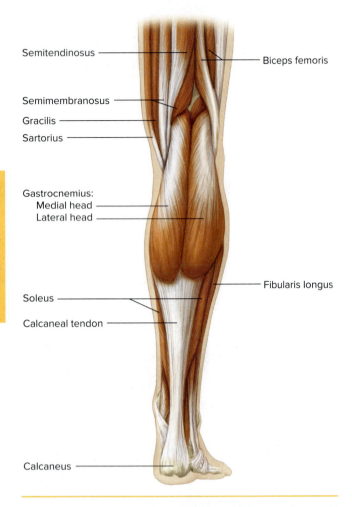

Figure 7.25 Muscles of the Posterior Right Leg. AP|R

Poliomyelitis (pō-lē-ō-mī-e-lī′-tis) is a viral disease of somatic motor neurons in the spinal cord. Destruction of the somatic motor neurons leads to paralysis of skeletal muscles. It is now rare in industrialized countries due to the availability of a polio vaccine. Virtually all children in the United States receive this vaccine, which protects them from polio.

Spasms are sudden, involuntary contractions of a muscle or a group of muscles. They may vary from simple twitches to severe convulsions and may be accompanied by pain. Spasms may be caused by irritation of the motor neurons supplying the muscle, emotional stress, or neurological disorders. Spasms of smooth muscle in the walls of the digestive and respiratory tracts, or in the walls of certain blood vessels, can be hazardous. Hiccupping is a spasm of the diaphragm.

Tetanus (tet′-ah-nus) is a disease caused by the anaerobic bacterium *Clostridium tetani*, which is common in soil. Infection usually results from puncture wounds. *C. tetani* produces a neurotoxin that affects somatic motor neurons in the spinal cord, resulting in continuous stimulation and tetany of certain muscles. Because the first muscles affected are those that move the mandible, this disease is often called "lockjaw." Without prompt treatment, mortality is high. Young children usually receive vaccinations of tetanus toxoid to stimulate production of antibodies against the neurotoxin. Booster injections are given at regular intervals to keep the concentration of antibodies at a high level in order to prevent the disease.

Chapter Summary

7.1 Types of Muscle Tissue
- The three types of muscle tissue in the body are skeletal, smooth, and cardiac.
- Each type of muscle tissue has unique structural and functional characteristics.

7.2 Structure of Skeletal Muscle
- Each skeletal muscle is formed of many muscle fibers that are arranged in muscle fascicles.
- Connective tissue envelops each muscle fiber, each muscle fascicle, and the entire muscle.
- Muscles are attached to bones or other tissues by either tendons or aponeuroses.
- The sarcolemma is the plasma membrane of a muscle fiber, and the sarcoplasm (cytoplasm) contains the myofibrils, the contractile elements.
- Myofibrils consist of thick and thin myofilaments. The arrangement of the myofilaments produces the microscopically visible striations that are characteristic of muscle fibers.
- Each myofibril consists of many sarcomeres joined end to end. A sarcomere is bounded by a Z line at each end.
- I bands are light areas in a muscle tissue photomicrograph, and A bands are dark areas.
- The H band is the center of a sarcomere and contains only thick myofilaments.
- The terminal bouton of a somatic motor neuron is adjacent to each muscle fiber at the neuromuscular junction. The terminal bouton fits into depressions in the sarcolemma, called motor end plates. The synaptic cleft is the small space between the terminal bouton and motor end plate. The neurotransmitter ACh is contained in tiny vesicles in the terminal bouton.
- Each muscle fiber is innervated and controlled by a somatic motor neuron.
- A motor unit consists of a somatic motor neuron and all of the muscle fibers it innervates.

7.3 Physiology of Skeletal Muscle Contraction

- An activated terminal bouton releases ACh into the synaptic cleft. ACh attaches to ACh receptors of the motor end plate, which leads to the release of Ca^{2+} within the sarcoplasm. This, in turn, leads to the formation of cross-bridges between the heads of myosin molecules and the myosin binding sites on actin molecules. A series of ratchetlike movements pulls the thin myofilaments toward the center of the sarcomere, producing contraction.
- Acetylcholinesterase quickly breaks down ACh to prevent continued stimulation and to prepare the muscle fiber for the next stimulus.
- Energy for contraction comes from high-energy phosphate bonds in ATP.
- After cellular respiration has formed a muscle fiber's normal supply of ATP, excess energy is transferred to creatine to form creatine phosphate, which serves as a reserve supply of energy.
- Small amounts of oxygen are stored in combination with myoglobin, which gives muscle fibers a reserve of oxygen for aerobic respiration.
- Vigorous muscular activity quickly exhausts available oxygen, leading to the production of lactate. The heavy breathing associated with excess post-exercise oxygen consumption provides the oxygen required to metabolize lactate and restore the pre-exercise state within the muscle fiber.
- Fatigue most likely results primarily from the lack of raw fuel in a muscle fiber.
- Large amounts of heat are produced by the chemical and physical processes of muscle contraction.
- When stimulated by a threshold stimulus, individual muscle fibers exhibit an all-or-none contraction response.
- A simple contraction consists of a latent phase, contraction phase, and relaxation phase.
- Whole muscles provide graded contraction responses, which are enabled by the number of motor units that are recruited.
- A sustained contraction of all motor units is tetany.
- Muscle tone is a state of partial contraction that results from alternating contractions of a few motor units.

7.4 Actions of Skeletal Muscles

- The origin is the immovable attachment, and the insertion is the movable attachment.
- Muscles are arranged in groups with opposing actions: agonists and antagonists.

7.5 Naming of Muscles

- Several criteria are used in naming muscles.
- These criteria include function, shape, relative position, location, site of attachment, origin and insertion, size, and orientation of fibers.

7.6 Major Skeletal Muscles

- Muscles of facial expression originate on skull bones and insert on the dermis of the skin. They include the occipitofrontalis, orbicularis oculi, orbicularis oris, buccinator, zygomaticus, and platysma.
- Muscles of mastication originate on fixed skull bones and insert on the mandible. They include the masseter and the temporalis.
- Muscles that move the head occur in the neck and upper back. They include the sternocleidomastoid and splenius capitis.
- Muscles of the abdominal wall connect the pelvis, thoracic cage, and vertebral column. They include the rectus abdominis, external oblique, internal oblique, and transversus abdominis.
- The diaphragm is the major muscle of breathing.
- External intercostal muscles and internal intercostal muscles move the ribs, helping breathing.
- Muscles that move the pectoral girdle originate on the thoracic cage or vertebrae and insert on the pectoral girdle. They include the trapezius, rhomboid major and minor, levator scapulae, pectoralis minor, and serratus anterior.
- Muscles that move the arm originate on the thoracic cage, vertebrae, or pectoral girdle and insert on the humerus. They include the pectoralis major, deltoid, subscapularis, supraspinatus, infraspinatus, latissimus dorsi, teres major, and teres minor.
- Supraspinatus, infraspinatus, teres minor, and subscapularis make up the rotator cuff.
- Muscles that move the forearm originate on the scapula or humerus and insert on the radius or ulna. They include the biceps brachii, brachialis, brachioradialis, and triceps brachii.
- Muscles that move the wrist and fingers are the muscles of the forearm. They include the flexor carpi radialis, flexor carpi ulnaris, palmaris longus, extensor carpi radialis longus, extensor carpi ulnaris, and extensor digitorum.
- Muscles that move the thigh originate on the pelvis and insert on the femur. They include the iliacus, psoas major, gluteus maximus, gluteus medius, tensor fasciae latae, adductor longus, and adductor magnus.
- Muscles that move the leg originate on the pelvis or femur and insert on the tibia or fibula. They include the quadriceps femoris, biceps femoris, semitendinosus, semimembranosus, gracilis, and sartorius.
- Muscles that move the foot and toes are the muscles of the leg. They include the gastrocnemius, soleus, fibularis longus, tibialis anterior, and extensor digitorum longus.

7.7 Disorders of the Muscular System

- Disorders of muscles include cramps, fibrosis, fibromyalgia, muscular dystrophy, and strains.
- Neurological disorders that directly affect muscle action include botulism, myasthenia gravis, poliomyelitis, spasms, and tetanus.

Improve Your Grade

Connect Interactive Questions Reinforce your knowledge using multiple types of questions: interactive, animation, classification, labeling, sequencing, composition, and traditional multiple choice and true/false.

SmartBook Proven to help students improve grades and study more efficiently, SmartBook contains the same content within the print book but actively tailors that content to the needs of the individual.

Anatomy & Physiology REVEALED® Dive into the human body by peeling back layers of cadaver imaging. Utilize this world-class cadaver dissection tool for a closer look at the body anytime, from anywhere.

CHAPTER 8

Nervous System

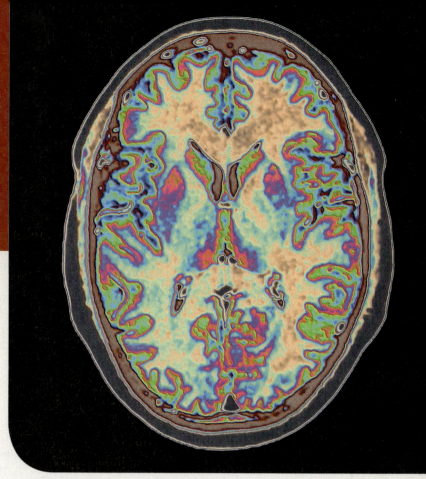

©Image Source/Getty Images RF

Have you ever wondered why you can handle multiple tasks at once? The answer is simple. You have a nervous system designed for rapid multitasking. Think about Bridgette and her typical commute to work. Bridgette is driving to work during the morning rush hour on Interstate 75 with her coworker Adam. The two are chatting about an important meeting later in the day that will outline the next quarter's objectives, while Bridgette continuously tracks the cars in all three lanes and adjusts her speed to match the flow of traffic. She subconsciously coordinates her use of turn signals, mirrors, and steering wheel to change lanes, while listening carefully to Adam's thoughts on an interoffice memo from the day before. Feeling a little bit tired, she begins to take a swig of her coffee but notices quite quickly that it is still too hot to drink. A few minutes later as Bridgette exits the highway, the two laugh hysterically and begin to sing when an old song comes on the radio. Clearly, the speed at which Bridgette's brain processes information and coordinates her body allows her morning commute to be productive, safe, and enjoyable.

CHAPTER OUTLINE

8.1 Introduction to the Nervous System

8.2 Divisions of the Nervous System
- Anatomical Divisions
- Functional Divisions of the Peripheral Nervous System

8.3 Nerve Tissue
- Neurons
- Types of Neurons
- Neuroglia

8.4 Neuron Physiology
- Membrane Potential
- Action Potential Formation
- Repolarization
- Action Potential Conduction
- Synaptic Transmission
- Neurotransmitters

8.5 Protection for the Central Nervous System
- Meninges

8.6 Brain
- Cerebrum
- Diencephalon
- Limbic System
- Brainstem
- Reticular Formation
- Cerebellum
- Ventricles and Cerebrospinal Fluid

8.7 Spinal Cord
- Structure
- Functions

8.8 Peripheral Nervous System (PNS)
- Cranial Nerves
- Spinal Nerves
- Reflexes

8.9 Autonomic Division
- Organization
- Autonomic Neurotransmitters
- Functions

8.10 Disorders of the Nervous System
- Inflammatory Disorders
- Noninflammatory Disorders

Chapter Summary

Module 7
Nervous System

SELECTED KEY TERMS

Action potential An electrochemical signal created by and conducted along the axon of a neuron.

Autonomic division (auto = self; nom = distribute) The division of the PNS involved in the involuntary control of cardiac muscle, smooth muscle, adipose tissue, and glands.

Axon (ax = axis, central) A neuronal process that carries action potentials away from the cell body.

Central nervous system The portion of the nervous system composed of the brain and spinal cord.

Dendrite (dendr = tree) A neuronal process that carries impulses toward the cell body or axon.

Ganglion (gangli = a swelling) A cluster of cell bodies located in the PNS.

Myelin sheath (myel = marrow) An insulating layer, formed by neuroglia, that surrounds an axon.

Nerve A bundle of axons in the peripheral nervous system.

Neuroglia Cells that support and protect neurons.

Neuron A cell capable of producing and transmitting an action potential.

Peripheral nervous system (peri = around) Portion of the nervous system composed of cranial and spinal nerves, ganglia, and sensory receptors.

Postsynaptic Pertaining to the cell that is activated by a signal at a synapse.

Presynaptic Pertaining to the neuron that releases a signal at a synapse.

Reflex An involuntary, rapid, and predictable response to a stimulus.

Somatic division The division of the PNS involved in the voluntary and involuntary control of skeletal muscles.

Synapse (syn = together) The junction between an axon and another neuron or effector cell.

8.1 Introduction to the Nervous System

Learning Objective

1. Describe the general functions of the nervous system.

The nervous system is the primary coordinating and controlling system of the body. Most of the activities of the nervous system occur below the level of consciousness and serve to maintain homeostasis. To maintain homeostasis, the nervous system requires almost instantaneous communication with the body. To achieve communication at this rate of speed, the nervous system uses **action potentials** that flow rapidly over and among neurons and between neurons and other body cells. An action potential is an electrochemical signal created and conducted along the axon of a neuron.

The general functions of the nervous system can be summarized as:

1. Detection of internal and external environmental changes
2. Analysis of the detected changes
3. Organization of the information for immediate and future use
4. Initiation of the appropriate actions in response to the changes

Check My Understanding

1. What are the functions of the nervous system?

8.2 Divisions of the Nervous System

Learning Objectives

2. Identify the anatomical divisions of the nervous system and their components.
3. Describe the functional divisions of the peripheral nervous system.

Although the nervous system functions as a coordinated whole, it is divided into anatomical and functional divisions as an aid in understanding this complex organ system.

Anatomical Divisions

The nervous system has two major anatomical divisions. The **central nervous system (CNS)** consists of the brain and spinal cord. The CNS is the body's neural integration center. It receives incoming information (action potentials), analyzes and organizes it, and initiates appropriate action. The **peripheral nervous system (PNS)** is located outside of the CNS and consists of cranial and spinal nerves, ganglia, and sensory receptors. The PNS carries action potentials formed by **sensory receptors,** such as pain and sound receptors, to the CNS. It also carries action potentials from the CNS to **effectors,** which are the muscles, glands, and adipose tissue.

Functional Divisions of the Peripheral Nervous System

The peripheral nervous system is divided into two major functional divisions. The **sensory division** carries action potentials from sensory receptors to the CNS. Somatic

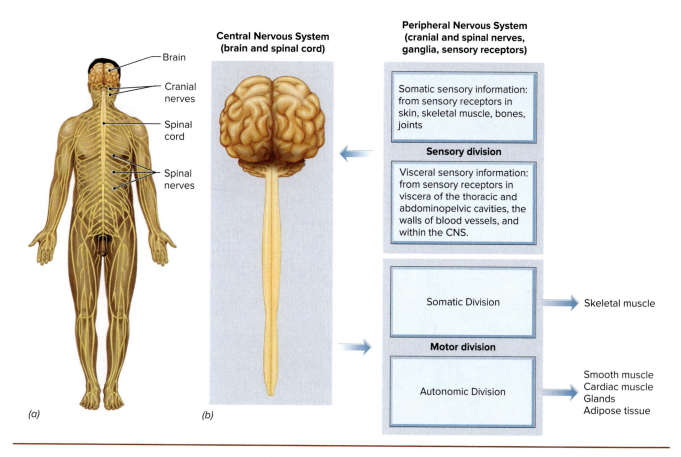

Figure 8.1 Components of the Nervous System.
(a) Anatomically, the nervous system consists of the central nervous system (brain and spinal cord) and the peripheral nervous system (cranial and spinal nerves, ganglia, sensory receptors). (b) Functionally, the peripheral nervous system consists of the sensory division and the motor division, which in turn is divided into the somatic division and autonomic division.

sensory information is collected by sensory receptors within the skin, skeletal muscles, bones, and joints. Visceral sensory information is collected by sensory receptors in the viscera in the thoracic and abdominopelvic cavities, in the walls of blood vessels, and within the CNS. The **motor division** carries action potentials from the CNS to effectors, which perform an action. The motor division is further divided into two subdivisions. The **somatic division** is involved in the *voluntary* (conscious) and *involuntary* (subconscious) control of skeletal muscles. The **autonomic** (aw-to-nom′-ik) **division** is involved in the involuntary control of cardiac muscle, smooth muscle, adipose tissue, and glands (figure 8.1).

 Check My Understanding

2. What structures are included in the central nervous system and in the peripheral nervous system?
3. What are the functional divisions of the peripheral nervous system?

8.3 Nerve Tissue

Learning Objectives
4. Describe the structure of a neuron.
5. Compare the three structural types of neurons.
6. Compare the three functional types of neurons.
7. Explain the functions of the five types of neuroglia.

The nervous system consists of organs composed primarily of nerve tissue supported and protected by connective tissues. As described in chapter 4, there are two types of cells that compose nerve tissue: neurons and neuroglia.

Neurons

Neurons (nū′-rahns), or nerve cells, are the structural and functional units of the nervous system. They are delicate cells that are specialized to generate and transmit action potentials. Neurons may vary in size and shape but they have many common features.

As shown in figures 8.2 through 8.4, the **cell body** is the portion of a neuron that contains the large, spherical

164 Chapter 8   Nervous System

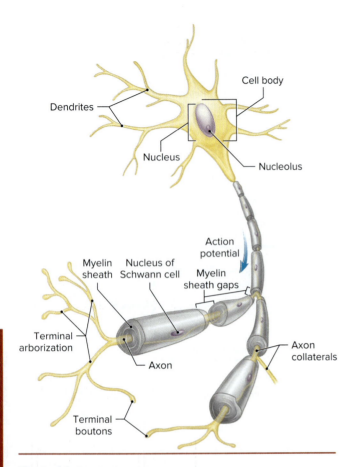

Figure 8.2 Neuron Anatomy.
The cell body contains the nucleus. One or more dendrites and a single axon are extensions from the cell body.

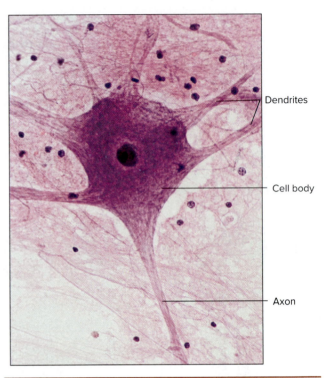

Figure 8.3 Photomicrograph of Nerve Tissue (50x).
Neurons are the structural and functional units of the nervous system. The dark spots in the area surrounding the neuron are nuclei of neuroglia. Note the location of the dendrites and axon. AP|R
©Ed Reschke/Getty Images

nucleus. The cell body also contains the usual cytoplasmic organelles. Two types of neuronal processes extend from the cell body: dendrites and axons. A neuron may have many dendrites but it has only one axon.

Dendrites (den′-drīts) are usually short, highly branched, tapering processes that create impulses (electrochemical signals) when stimulated by other neurons and sensory receptors. Dendrites carry the impulses toward the cell body or axon.

An **axon** (ak′-sahn), or *nerve fiber,* is a long, thin process of a neuron. It may have one or more side branches, called *axon collaterals,* that allow the neuron to make contact with more neurons or effectors. It also forms a number of short, fine branches, the *terminal arborization,* at its distal tip. The slightly enlarged tips of the terminal arborization are the **terminal boutons,** which form junctions (synapses) with other neurons, muscles, adipose tissue, or glands. An axon carries action potentials away from the cell body or dendrites.

Some axons are enclosed in an insulating **myelin sheath** formed by special neuroglia. Such axons are referred to as *myelinated* axons. The myelin sheath increases the speed of action potential transmission. The tiny spaces between adjacent myelin-forming cells, where the axon is exposed, are known as **myelin sheath gaps** (or nodes of Ranvier). Axons lacking a myelin sheath are referred to as *unmyelinated* axons and have a much slower speed of action potential transmission.

Types of Neurons

Neurons may be classified according to their anatomy or their function. Structurally, there are three basic types of neurons: multipolar, bipolar, and pseudounipolar neurons (figure 8.4).

Multipolar neurons have several dendrites and a single axon extending from the cell body. Most of the neurons whose cell bodies are located in the brain and spinal cord are multipolar neurons.

Bipolar neurons have only two processes: a dendrite and an axon extending from opposite ends of the cell body. Bipolar neurons occur in the sensory portions of the eyes, ears, and nose.

Pseudounipolar neurons have a single process extending from the cell body. This process quickly divides

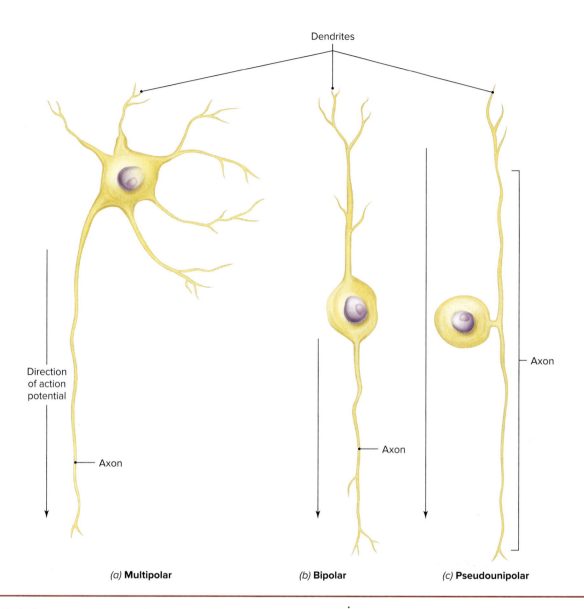

Figure 8.4 Structural Types of Neurons.

into two branches extending in opposite directions, with both branches functioning as a single axon. The end of the axon within the PNS ends in dendrites, while the end within the CNS ends in a terminal arborization. Pseudounipolar neurons carry action potentials from sensory receptors to the CNS. Clusters of cell bodies of pseudounipolar neurons often form **ganglia** (singular, *ganglion*), which are located in the PNS.

Functionally, there are three basic types of neurons: sensory neurons, interneurons, and motor neurons.

Sensory neurons carry action potentials from the peripheral parts of the body to the CNS. Their dendrites are associated with sensory receptors or are specialized to detect changes directly. Action potentials are carried over an axon within cranial or spinal nerves to the CNS. Cell bodies of sensory neurons are located external to the CNS in ganglia. Structurally, most sensory neurons are pseudounipolar neurons, although bipolar neurons are found in special sense organs.

Interneurons are located entirely within the CNS and synapse with other neurons. They are responsible for the processing and interpretation of action potentials by the CNS. Interneurons receive action potentials from sensory neurons and transmit them from place to place within the CNS. They also activate motor neurons, which results in a stimulation of effectors. Interneurons are multipolar neurons.

Motor neurons carry action potentials from the CNS to effectors to produce an action. Their cell bodies and dendrites are located within the CNS, while their axons are located in cranial and spinal nerves. Motor neurons are multipolar neurons (table 8.1).

Table 8.1 Functional Types of Neurons

Type	Structure	Function
Sensory neurons	Mostly pseudounipolar, some bipolar	Carry action potentials from peripheral sensory receptors to the CNS
Interneurons	Multipolar	Carry action potentials between neurons within the CNS
Motor neurons	Multipolar	Carry action potentials from the CNS to effectors (muscles, glands, and adipose tissue)

Neuroglia

The **neuroglia** (nū-rog′-lē-ah) provide support and protection for neurons. One type of neuroglia—Schwann cells—occurs in the PNS. Four types of neuroglia occur in the CNS, where they are even more numerous than neurons (figures 8.5 and 8.6).

Schwann cells form the myelin sheath around PNS myelinated axons. A Schwann cell wraps its plasma membrane tightly around an axon many times so that the nucleus and most of the cytoplasm become squeezed into the outer layer. The inner layers, formed by layers of plasma membrane, constitute the myelin sheath. The outermost layer forms the **neurilemma,** which is essential for axon regeneration after injury.

Oligodendrocytes (ōl-i-gō-den′-drō-sītz) form the myelin sheath of myelinated axons within the CNS but they do not form a neurilemma. Lack of a neurilemma is one factor that contributes to the inability of axons within the brain and spinal cord to regenerate after injury.

Astrocytes (as′-trō-sītz) are the primary supporting cells for neurons in the CNS. They stimulate the growth of neurons and influence synaptic transmission. Astrocytes also join with the epithelium of blood vessels to form the *blood-brain barrier,* which protects neurons by tightly regulating the exchange of materials between the blood and neurons.

Microglial cells are scattered throughout the CNS, where they keep the tissues clean by engulfing and digesting cellular debris and pathogens.

Ependymal (e-pen-dī′-mal) **cells** form the epithelial lining of cavities in the brain and spinal cord and aid in the production of cerebrospinal fluid, a unique fluid within the CNS that will be discussed later.

 Check My Understanding

4. What are the general functions of the nervous system?
5. What are the structural and functional types of neurons?
6. What are the roles of the five types of neuroglia?

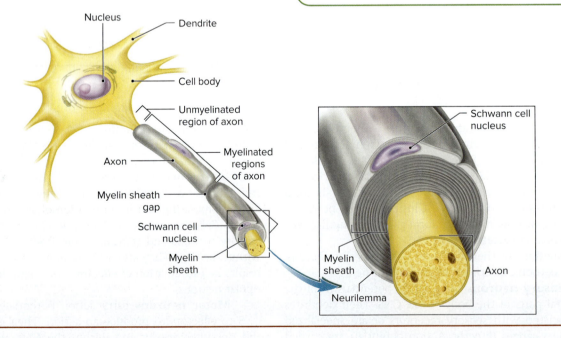

Figure 8.5 The Myelin Sheath.
The portion of a Schwann cell that winds tightly around an axon forms a myelin sheath, while the cytoplasm and nucleus of the Schwann cell remaining on the surface form the neurilemma. AP|R

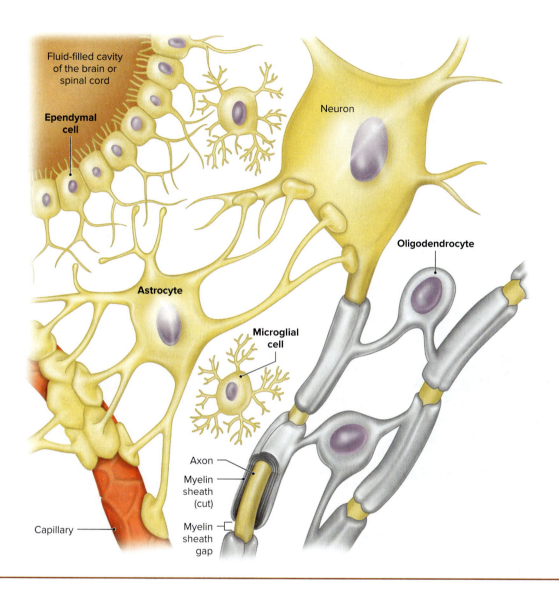

Figure 8.6 Types of Neuroglia in the Central Nervous System.

8.4 Neuron Physiology

Learning Objectives

8. Explain the formation and conduction of an action potential.
9. Describe how action potentials are transmitted across a synapse.

Neurons have two unique functional characteristics: irritability and conductivity. **Irritability** is the ability to respond to a stimulus by forming an action potential. **Conductivity** is the ability to transmit an action potential along an axon to other neurons or effector cells. These characteristics enable the functioning of the nervous system.

Membrane Potential

Most body cell plasma membranes are polarized, meaning there is an electrical charge difference across the plasma membrane. This difference creates a *voltage* that is called a **membrane potential.** In neurons and other cells with irritability that are inactive, this voltage is called a **resting membrane potential (RMP).** In neurons, the RMP is maintained at an average around −70mV. The reason for the difference in electrical charge is the unequal distribution of ions and proteins on either side of the plasma membrane (figure 8.7). In a resting neuron, sodium (Na^+) and chloride (Cl^-) ion concentrations are high in the ECF and low in the cytosol, whereas potassium (K^+) ions have the opposite distribution. There are also large, negatively charged proteins and ions, such as phosphates (PO_4^{3-}) and sulfates (SO_4^{2-}), in the cytosol that cannot cross the plasma membranes. These differences polarize the plasma membrane, meaning there are a net excess of positive charges on the ECF-side and a net excess of negative charges on the cytosol-side. The negative RMP indicates that the cytosol-side of the plasma membrane is more negative than the ECF-side.

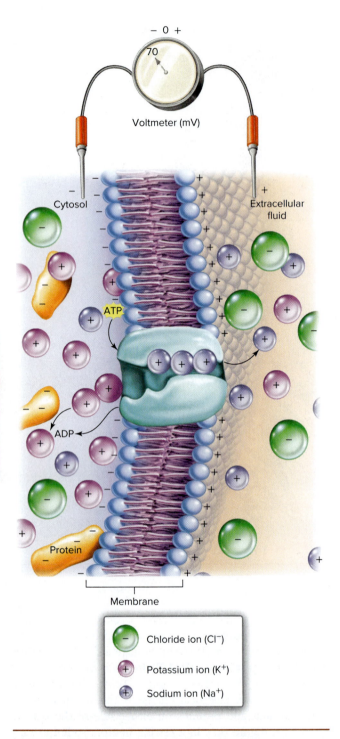

Figure 8.7 Resting Membrane Potential.
At rest, Na⁺ and Cl⁻ ions are in high concentration in the ECF and K⁺ ions are in high concentration in the cytosol. The plasma membrane possesses a net + charge on its ECF-side and a net − charge on its cytosol-side. The resulting voltage across the plasma membrane is −70mV. The Na⁺/K⁺ pump compensates for ion diffusion by moving 3 Na⁺ from the cytosol back into the ECF and 2 K⁺ from the ECF back into the cytosol.

To establish and maintain the RMP, neurons must be able to compensate for the diffusion of Na⁺ and K⁺ along their concentration gradients. The plasma membrane is more permeable to K⁺, but both ions exhibit movement that is capable of disrupting the RMP. The Na⁺/K⁺ pump is a carrier protein that uses ATP to move Na⁺ and K⁺ against their concentration gradient (see chapter 3). This carrier is continuously active to establish and maintain the RMP and to restore it after action potential formation.

Action Potential Formation

When stimulated, axons exhibit an all-or-none response. They either form an action potential that will travel along the axon or do not respond. The weakest stimulus that will activate a neuron to produce an action potential is called a *threshold stimulus*. Action potentials do not vary in their degree of electrical change, meaning every action potential is identical.

When a neuron is activated by a threshold stimulus, its plasma membrane becomes permeable to Na⁺ as Na⁺ channels open, which allows these ions to quickly diffuse into the neuron. The inward flow of Na⁺ for a brief instant causes the cytosol along the inside of the plasma membrane to become positively charged (an excess of positive charges) and the ECF along the outside to become negatively charged (an excess of negative charges) at the point of stimulation. The membrane potential changes to +30mV as a result of these changes. This switch in polarity is called *depolarization* and the plasma membrane is now referred to as *depolarized*. This sudden depolarization *is* the action potential (figure 8.8b). The wave of depolarization then flows along the axon. You will see how this happens momentarily.

Repolarization

Immediately after depolarization, K⁺ channels open and Na⁺ channels close, allowing K⁺ to diffuse into the ECF in order to *repolarize* or reestablish the RMP. The loss of K⁺ to the ECF creates an excess of positive charges along the ECF-side of the plasma membrane and an excess of negative charges along the cytosol-side. As a result, the membrane voltage changes from +30mV to −70mV (figure 8.8c). As described in the previous section, the Na⁺/K⁺ pump then reestablishes the resting-state distribution of ions (figure 8.8d). When this is accomplished, the neuron is ready to respond to another stimulus. Depolarization and repolarization are accomplished in about 1 millisecond.

Action Potential Conduction

When an action potential is formed at one point in an axon, it triggers the depolarization of adjacent portions of the plasma membrane, which, in turn, depolarizes still other regions of the plasma membrane. The result is a wave of depolarization that conducts an action potential along the axon. Repolarization immediately follows an action potential.

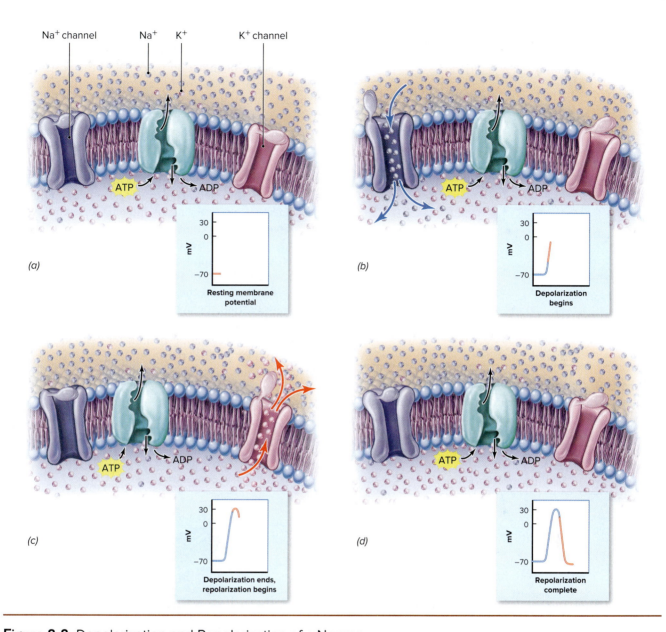

Figure 8.8 Depolarization and Repolarization of a Neuron.
(a) Neuron at rest. Both Na$^+$ and K$^+$ channels are closed. *(b)* Na$^+$ channels open and Na$^+$ flows into the neuron depolarizing the plasma membrane to +30mV. *(c)* Na$^+$ channels close. K$^+$ channels open and K$^+$ flows out of the neuron repolarizing the plasma membrane to −70mV. *(d)* K$^+$ channels close and Na$^+$/K$^+$ pumps reestablish resting ion distribution. APR

Conduction of action potentials is more rapid in myelinated axons than in unmyelinated axons. Recall that a myelinated axon is exposed only at myelin sheath gaps. Because of this, an action potential jumps from gap to gap and does not have to depolarize the intervening segments of the axon (figure 8.9).

Synaptic Transmission

A **synapse** (sin′-aps) is a junction of an axon with either another neuron or an effector cell. At a synapse, the terminal bouton of the neuron fits into a small depression on another neuron's dendrite or cell body or on a cell within a muscle, a gland, or adipose tissue. The neuron releasing the signal at a synapse is referred to as **presynaptic.** The neuron or cell being activated by the signal at the synapse is referred to as **postsynaptic.** There is a tiny space, the **synaptic cleft,** between the presynaptic and postsynaptic structures, so they are not in physical contact (figure 8.10).

In neuron-to-neuron synaptic transmission, when an action potential reaches the terminal bouton of the presynaptic neuron, it causes the terminal bouton to secrete **neurotransmitters** into the synaptic cleft. Then, the neurotransmitters bind to receptors on the postsynaptic

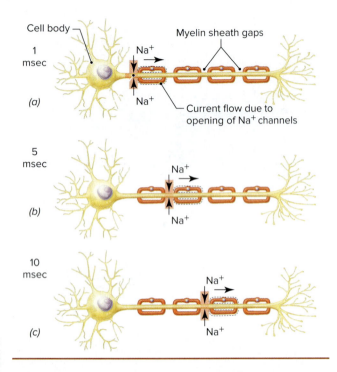

Figure 8.9 Action Potential Conduction. Movement of sodium ions during (a) action potential formation (depolarization), and (b) and (c) action potential conduction. Note that repolarization immediately follows depolarization. AP|R

neuron's plasma membrane, which triggers a response in the postsynaptic neuron. Some neurotransmitters stimulate formation of an action potential in the postsynaptic neuron, while others inhibit action potential formation. If an action potential is formed in the postsynaptic neuron, it is carried along the neuron's axon to the next synapse where synaptic transmission takes place again.

Because only terminal boutons can release neurotransmitters, action potentials can pass in only one direction across a synapse—from the presynaptic neuron to the postsynaptic neuron. Thus, action potentials always pass in the "correct" direction, which maintains order in the nervous system.

Some neurotransmitters are reabsorbed into the terminal bouton for reuse. Others diffuse out of the synaptic cleft or are decomposed by enzymes released into the synaptic cleft. Some of the decomposition products are then reabsorbed into the bouton for reuse, while others diffuse away from the synaptic cleft. Quick removal of a neurotransmitter prevents continuous stimulation or inhibition of the postsynaptic neuron (or cell) and prepares the synapse for another transmission. From start to finish, synaptic transmission takes only a fraction of a second.

Neurotransmitters

Neurotransmitters enable neurons to communicate with each other, as well as with other cells throughout the

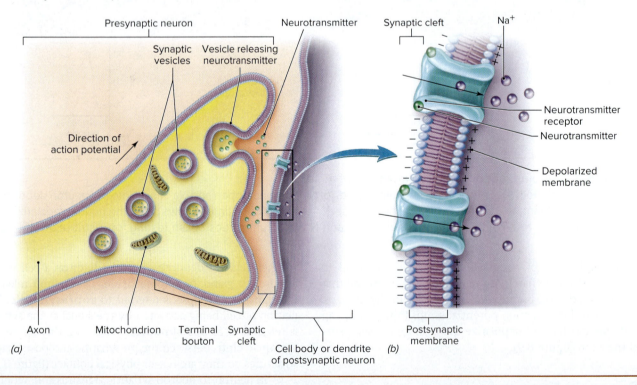

Figure 8.10 Synaptic Transmission from Neuron to Neuron.
(a) A terminal bouton of the presynaptic neuron fits into a depression on a dendrite or the cell body of the postsynaptic neuron. When an action potential reaches the terminal bouton, a neurotransmitter is released into the synaptic cleft. (b) The neurotransmitter binds with receptors on the plasma membrane of the postsynaptic neuron, causing depolarization of the membrane. AP|R

body. Scientific research has identified over 100 neurotransmitters at work within the human nervous system, and most likely more will be discovered in the future. When released, neurotransmitters create either excitatory or inhibitory effects on the postsynaptic cell. *Excitatory neurotransmitters* cause the formation of an impulse in the postsynaptic cell, which in turn promotes cell function. *Inhibitory neurotransmitters* inhibit the formation of an impulse in the postsynaptic cell, resulting in an inhibition of cell function. What makes the study of neurotransmitters intriguing is the fact that one neurotransmitter can create both excitatory and inhibitory effects depending upon the postsynaptic cell receiving the signal. For example, acetylcholine acts as an excitatory neurotransmitter in skeletal muscle by promoting contraction of skeletal muscle fibers (see chapter 7). However, acetylcholine acts as an inhibitory neurotransmitter in cardiac muscle by inhibiting contraction of cardiac muscle cells, resulting in a decrease in heart rate (see chapter 12).

The cell body and dendrites of a postsynaptic neuron synapse with hundreds of presynaptic neurons. Some of the neurotransmitters released in these synapses exert excitatory effects, while some exert inhibitory effects. Whether or not an action potential is formed in the postsynaptic neuron depends upon whether the excitatory or inhibitory effects are dominating at that time.

Check My Understanding

7. How are action potentials formed and conducted?
8. What is the mechanism of synaptic transmission?

Clinical Insight

Inhibitory and stimulatory drugs act by affecting synaptic transmission. Some *tranquilizers* and *anesthetics* inhibit synaptic transmission by increasing the threshold of postsynaptic neurons. *Nicotine, caffeine,* and *benzedrine* promote synaptic transmission by decreasing the threshold of postsynaptic neurons.

8.5 Protection for the Central Nervous System

Learning Objective
10. Describe how the brain and spinal cord are protected from injury.

Both the brain and the spinal cord are soft, delicate organs that would be easily damaged without adequate protection. Surrounding bones and fibrous membranes provide both protection and support. The brain occupies the cranial cavity formed by the cranial bones, and the spinal cord lies within the vertebral canal formed by the vertebrae. Three membranes are located between the CNS and the surrounding bones. These membranes are collectively called the meninges.

Meninges

The **meninges** (me-nin′-jēs; singular, *meninx*) consist of three membranes arranged in layers. From innermost to outermost, they are the pia mater, arachnoid mater, and dura mater (figures 8.11 and 8.12).

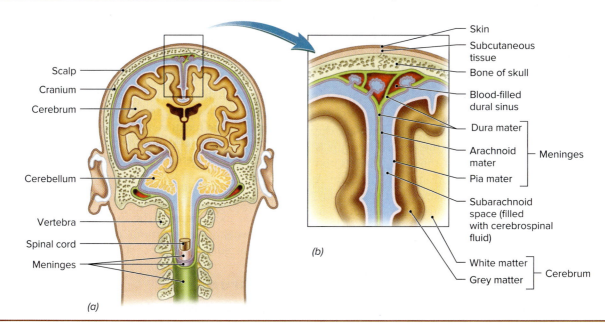

Figure 8.11 The Meninges of the Brain.
(a) Membranes called meninges enclose the brain and spinal cord. (b) The meninges include three layers: dura mater, arachnoid mater, and pia mater.

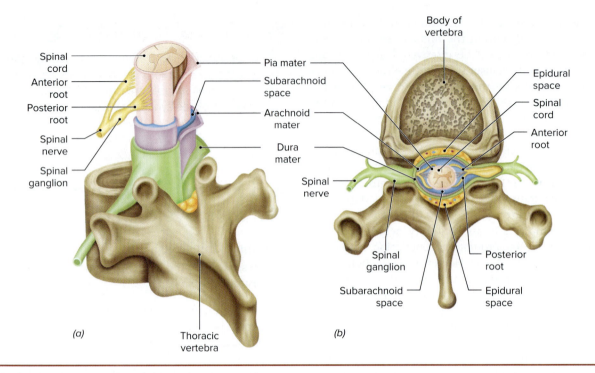

Figure 8.12 Meninges and Spinal Cord.
(a) The meninges support and protect the spinal cord. (b) Adipose tissue fills the epidural space, providing a protective cushion.

The **pia mater** (pee-uh mah-ter; "tender mother") is the very thin, innermost membrane. It tightly envelops both the brain and the spinal cord and penetrates into each groove and depression. It contains many blood vessels that nourish the underlying brain and spinal cord.

The **arachnoid** (ah-rak'-noyd) **mater** ("spider mother") is the middle membrane. It is a thin, weblike membrane without blood vessels that does not penetrate into the small depressions as does the pia mater. Between the pia mater and the arachnoid mater is the *subarachnoid space*, which contains *cerebrospinal fluid*. This clear, watery liquid serves as a shock absorber around the brain and spinal cord.

The **dura** (du'-rah) **mater** ("tough mother") is the tough, fibrous outermost layer. In the cranial cavity, it is attached to the inner surfaces of the cranial bones and penetrates into *fissures* between some parts of the brain. A fissure is a deep, wide groove that separates brain regions. In the vertebral canal, the dura mater forms a protective tube that extends to the sacrum. It does not attach to the bony surfaces of the vertebral canal but is separated from the bone by an *epidural space*. Adipose tissue fills the epidural space and serves as an additional protective cushion. Physical trauma can cause tearing of blood vessels extending between the dura and arachnoid maters. The pooling of blood between the two meninges, which is called a *subdural hematoma*, creates an artificial space called the *subdural space*.

 Check My Understanding

9. How is the CNS protected from mechanical injuries?

8.6 Brain

Learning Objectives
11. Describe the major parts of the brain in terms of structure, location, and function.
12. Identify the functions of the lobes of the cerebrum.
13. Describe the formation, circulation, absorption, and functions of cerebrospinal fluid.

The brain is a large, exceedingly complex organ. It contains about 100 billion neurons and innumerable neuronal processes and synapses. The brain consists of four major components: the cerebrum, cerebellum, diencephalon, and brainstem. Locate these structures in figure 8.13.

Cerebrum

The **cerebrum** (suh-ree-brum) is the largest portion of the brain. It performs the higher brain functions involved with sensations, voluntary actions, reasoning, planning, and problem solving.

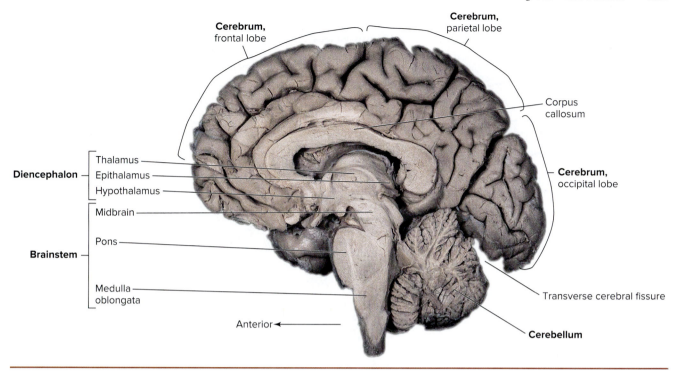

Figure 8.13 Cadaver Brain in Median Section.
The major components of the brain as shown in a median section. Note the unlabeled gyri and sulci of the cerebrum. **AP|R**
©McGraw-Hill Education/Dennis Strete, photographer

Structure

The cerebrum consists of the left and right **cerebral hemispheres,** which are joined by a mass of myelinated axons called the **corpus callosum.** The cerebral hemispheres are separated by the **longitudinal cerebral fissure,** which lies along the superior midline and extends downward to the corpus callosum.

The surface of the cerebrum has numerous folds or ridges, called *gyri* (jī′-rē; singular, gyrus). The shallow grooves between the gyri are called *sulci* (sul′-sē; singular, sulcus). The outer layer of the cerebrum is composed of *grey matter* (cell bodies, dendrites, terminal arborizations, and unmyelinated axons) and is called the **cerebral cortex.** *White matter,* composed of myelinated and unmyelinated axons, lies beneath the cortex and composes most of the cerebrum. These axons transmit action potentials between regions within the same cerebral hemisphere, between the cerebral hemispheres via the corpus callosum, and between the cerebral cortex and lower brain centers. Several masses of grey matter, called **nuclei,** are embedded deep within the white matter of each cerebral hemisphere.

Each cerebral hemisphere is divided into five lobes. Four lobes are named for the cranial bones under which they lie. Locate the cerebral lobes in figures 8.13 and 8.14.

1. The **frontal lobe** lies in front of the *central sulcus* and above the *lateral sulcus.*
2. The **parietal lobe** lies just behind the central sulcus, above the temporal lobe, and in front of the occipital lobe.
3. The **temporal lobe** lies below the frontal and parietal lobes and in front of the occipital lobe.
4. The **occipital lobe** lies behind the parietal and temporal lobes. The boundaries between the parietal, temporal, and occipital lobes are not distinct.
5. The **insula** lies underneath the lateral sulcus. It is the lobe that cannot be viewed superficially.

Functions

The cerebrum is involved in the interpretation of sensory action potentials as sensations and in controlling voluntary motor responses, intellectual processes, the will, and many personality traits. The cerebrum has three major types of functional areas: sensory, motor, and association areas (figure 8.14).

Sensory areas receive action potentials formed by sensory receptors and interpret them as sensations. These areas occur in several cerebral lobes. For example, the sensory areas for vision are in the occipital lobes and those for hearing are found in the temporal lobes. Areas identifying sensations from skin (cutaneous) stimulation lie along the *postcentral gyri* (gyri just behind the central sulci) of the parietal lobes. Sensory areas for taste are located at the lower end of the postcentral gyri. The sensory areas for smell are located in the inferior part of the frontal lobe and the medial aspect of the temporal lobe.

Ascending sensory axons carrying sensations from the skin cross over from one side to the other prior to reaching the thalamus. Thus, the postcentral gyrus in the left cerebral hemisphere receives action potentials from the skin on the right side of the body, and vice versa.

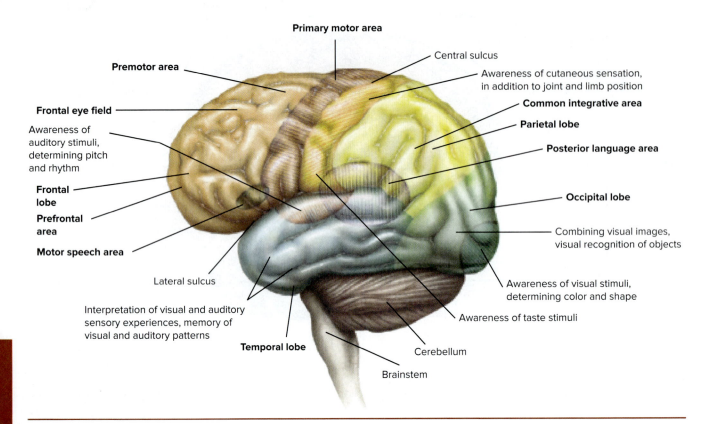

Figure 8.14 A Lateral View of the Left Brain.
Note the cerebral lobes and their functional areas. AP|R

Motor areas are located in the frontal lobe. The *primary motor areas* that control skeletal muscles lie along the *precentral gyri* (gyri just in front of the central sulci) of the frontal lobes. The region in front of the primary motor area is the *premotor area*.

The premotor area is involved in complex learned activities, such as writing, tying your shoes, and driving a car. Also in the premotor area is the *frontal eye field*, which controls voluntary eye movements. The *motor speech area (Broca area)*, which controls the ability to speak, is located near the lower end of the primary motor area. It is found in only one hemisphere: the left hemisphere in about 90% of people.

Descending motor axons cross over from one side to the other in the brainstem. Thus, the left side of the cerebrum controls skeletal muscles on the right side of the body, and vice versa.

Association areas occur in each cerebral lobe, where they interrelate sensory inputs and motor outputs. They play critical roles in the interrelationships of sensations, memory, will, and the coordination of motor responses. The *common integrative area* is a major association area that is located at the junction of the temporal, parietal, and occipital lobes. It is involved with the interpretation of complex sensory experiences and thought processes. The *posterior language area (Wernicke area)*, which is an association area located in the temporal and parietal lobes, is used to interpret the meaning of spoken and written language. Like the motor speech area, it is found in only one hemisphere: the left hemisphere in about 90% of people.

The *prefrontal area*, which is located in the frontal lobe just behind the frontal bone and just above the eyes, is involved with functions such as planning, complex behaviors, conscience, generating personality, and executive functions. Executive functions include distinguishing between good and bad, understanding future consequences, social control of urges, and working toward a goal. Portions of the prefrontal area are not fully developed until a person is in his or her 20s, which is why teenagers often have issues with impulse control and decision-making.

> **Clinical Insight**
>
> The transmission of action potentials by neurons in the brain produces electrical potentials that can be detected and recorded as brain waves. A recording of brain waves is called an *electroencephalogram (EEG)*. The patterns of brain waves are used in the diagnosis of certain brain disorders. The cessation of brain wave production is one criterion of brain death.

Hemisphere Specialization

The two cerebral hemispheres perform different functions in most people, although each performs basic functions of receiving sensory input and initiating voluntary motor output. In about 90% of the population, the left cerebral hemisphere controls analytical and verbal skills, such as mathematics, reading, writing, and speech. In these persons, the right hemisphere controls musical, artistic and spatial awareness, imagination, and insight. In some persons, this pattern is reversed; in a few, there seems to be no specialization. Men also have greater lateralization than women, which is why damage to a hemisphere can have greater effects in men.

Diencephalon

The **diencephalon** (dī-en-sef′-a-lon) is a small but important part of the brain. It lies between the brainstem and the cerebrum of the brain and consists of three major components: the thalamus, hypothalamus, and epithalamus (see figure 8.13).

Thalamus

The **thalamus** (thal′-ah-mus) consists of two lateral masses of nerve tissue that are joined by a narrow isthmus of nerve tissue called the *interthalamic adhesion*. Sensory action potentials (except those for smell) coming from lower regions of the brain and the spinal cord are first received by the thalamus before being relayed to the cerebral cortex. The thalamus provides a general but nonspecific awareness of sensations such as pain, pressure, touch, and temperature. It seems to associate sensations with emotions, but it is the cerebral cortex that interprets the precise sensation. The thalamus also serves as a relay station for communication between motor areas of the brain.

Hypothalamus

The **hypothalamus** (hī-pō-thal′-ah-mus) is located below the thalamus and in front of the midbrain. It communicates with the thalamus, cerebrum, and other parts of the brain. The hypothalamus is the major integration center for the autonomic division. In this role, it controls virtually all internal organs. The hypothalamus also is the connecting link between the brain and the endocrine system, which produces chemicals (hormones) that affect most cells in the body. This link results from hypothalamic control of the hypophysis, or pituitary gland, which is suspended from its inferior surface. Although it is small, the hypothalamus exerts a tremendous impact on body functions.

The primary function of the hypothalamus is the maintenance of homeostasis, and this is accomplished through its regulation of

- body temperature;
- mineral and water balance;
- appetite and digestive processes;
- heart rate and blood pressure;
- sleep and wakefulness;
- emotions; and
- secretion of hormones by the pituitary gland.

Epithalamus

The **epithalamus** (ep-i-thal′-ah-mus; epi = above) is a small mass of tissue located above and behind the thalamus, forming part of the roof of the third ventricle. The major structure within the epithalamus is the *pineal gland*. The pineal gland is stimulated to produce a hormone called *melatonin* when sunlight levels become low during the evening and overnight hours. This hormone induces sleepiness to initiate the night component of a person's day-night cycle and may assist in regulating the onset of puberty. This hormone will be discussed further in chapter 10.

Limbic System

The thalamus and hypothalamus are associated with parts of the cerebral cortex and nuclei deep within the cerebrum to form a complex known as the **limbic system.** The limbic system is involved in memory and in emotions such as sadness, happiness, anger, and fear. It seems to regulate emotional behavior, especially behavior that enhances survival. Mood disorders, such as depression, are usually a result of malfunctions of the limbic system. It also is referred to as the "motivational system" because it provides our desire to carry out the commands created by the cerebrum.

 Check My Understanding

10. What are roles of the functional areas of the cerebrum?
11. What are the functions of the thalamus, hypothalamus, and epithalamus?

Brainstem

The **brainstem** is the stalklike portion of the brain that joins higher brain centers to the spinal cord. It contains several nuclei that are surrounded by white matter. Ascending (sensory) and descending (motor) axons between higher brain centers and the spinal cord pass through the brainstem. The components of the brainstem include the midbrain, pons, and medulla oblongata (see figure 8.13).

Midbrain

The **midbrain** is the most superior portion of the brainstem. It is located behind the hypothalamus and above the pons. It contains reflex centers for head, eye, and body movements in response to visual and auditory stimuli. For example, reflexively turning the head to enable better vision or better hearing is activated by the midbrain.

Pons

The **pons** lies between the midbrain and the medulla oblongata and is recognizable by its bulblike anterior portion. It consists primarily of axons. Longitudinal axons connect lower and higher brain centers, and transverse axons connect with the cerebellum. The pons also works with the medulla oblongata by controlling the rate and depth of breathing (see chapter 14).

Medulla Oblongata

The **medulla oblongata** (me-dūl´-ah ob-lon-ga´-ta) is the most inferior portion of the brain, and it is the connecting link with the spinal cord. Descending (motor) axons extending between the brain and the spinal cord cross over to the opposite side of the brain within the medulla oblongata. The medulla oblongata contains three integration centers that are vital for homeostasis:

1. The **respiratory rhythmicity center** controls the basic rhythm of breathing by triggering each cycle of inhale and exhale. It is also involved in associated reflexes such as coughing and sneezing.
2. The **cardiac control center** regulates the rate and force of heart contractions.
3. The **vasomotor center** regulates blood pressure and blood flow by controlling the diameter of blood vessels.

Reticular Formation

The **reticular** (re-tik´-ū-lar) **formation** is a network of axons and small nuclei of grey matter that extends from the upper spinal cord, through the brainstem, into the diencephalon. This network generates and transmits action potentials that arouse the cerebrum to wakefulness. A decrease in activity results in sleep. Damage to the reticular formation may cause unconsciousness or a coma.

Cerebellum

The **cerebellum** (ser-e-bel´-um) is the second largest portion of the brain. The **transverse cerebral fissure** separates its upper surface from the occipital and temporal lobes of the cerebrum. It is also positioned behind the pons and medulla oblongata. It is divided into two lateral hemispheres by a medial constriction, the *vermis* (ver´-mis). Grey matter forms a thin outer layer covering the white matter underneath, which forms most of the cerebellum (see figure 8.13).

The cerebellum is a reflex center that controls and coordinates the interaction of skeletal muscles. It controls posture, balance, and muscle coordination during movement. Damage to the cerebellum may result in a loss of equilibrium, muscle coordination, and muscle tone.

Table 8.2 summarizes the major brain functions.

Table 8.2 Summary of Brain Functions

Part	Function
Cerebrum	Sensory areas interpret action potentials as sensations. Motor areas control voluntary skeletal muscle actions. Association areas interrelate various sensory and motor areas and are involved in intellectual processes, will, memory, emotions, and personality traits. The limbic system is involved with motivation and with emotions as they relate to survival behavior.
Diencephalon	
Thalamus	Receives and relays sensory action potentials (except smell) to the cerebrum and motor action potentials to lower brain centers. Provides a general awareness of pain, touch, pressure, and temperature.
Hypothalamus	Serves as the major integration center for the autonomic division. Controls water and mineral balance, heart rate and blood pressure, appetite and digestive activity, body temperature, and sexual response. Is involved in sleep and wakefulness and in emotions of anger and fear. Regulates functions of the pituitary gland.
Epithalamus	Production of the hormone melatonin
Brainstem	
Midbrain	Relays sensory action potentials from the spinal cord to the thalamus and motor action potentials from the cerebrum to the spinal cord. Contains reflex centers that move the eyeballs, head, and neck in response to visual and auditory stimuli.
Pons	Relays action potentials between the midbrain and the medulla oblongata and between the cerebellar hemispheres. Helps medulla oblongata regulate breathing.
Medulla oblongata	Relays action potentials between the brain and spinal cord. Reflex centers control heart rate and contraction force, blood vessel diameter, breathing, swallowing, vomiting, coughing, sneezing, and hiccupping. Motor axons cross over to the opposite side.
Cerebellum	Controls posture, balance, and the coordination of skeletal muscle contractions.

Check My Understanding

12. What are the functions of the medulla oblongata?
13. How is the cerebellum involved in skeletal muscle contractions?

Ventricles and Cerebrospinal Fluid

There are four interconnecting **ventricles,** or cavities, within the brain. Each ventricle is lined by ependymal cells and is filled with **cerebrospinal fluid (CSF).** The largest ventricles are the two *lateral ventricles* (first and second ventricles), which are located within the cerebral hemispheres. The *third ventricle* is a narrow space that lies on the midline between the lateral masses of the thalamus and above the hypothalamus. The *fourth ventricle* is located on the midline in the posterior portion of the brainstem just above the cerebellum. It is continuous with the central canal of the spinal cord. Observe the relative positions of the ventricles in figure 8.15.

Each ventricle contains a **choroid** (kō′-royd) **plexus,** a mass of special capillaries and ependymal cells that secrete CSF, but most of the CSF is produced in the lateral ventricles. The flow of CSF is shown in figure 8.16.

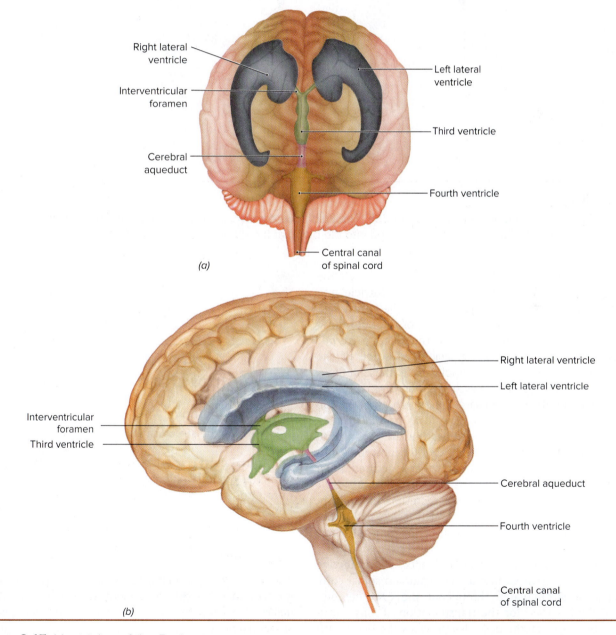

Figure 8.15 Ventricles of the Brain.
Anterior *(a)* and lateral *(b)* views of the ventricles of the brain. Note how they are interconnected.

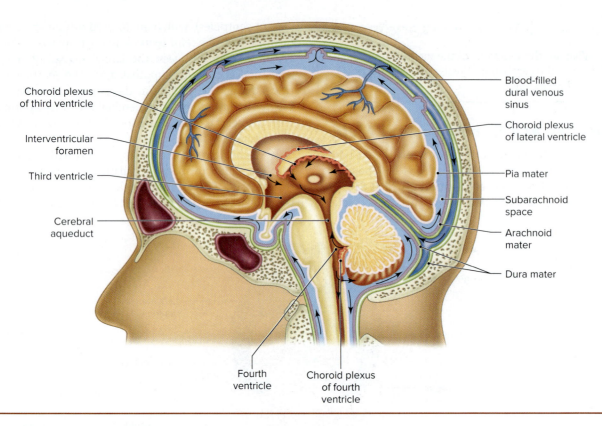

Figure 8.16 Circulation of CSF.
Choroid plexuses in ventricle walls secrete CSF. The fluid flows through the ventricles, central canal of the spinal cord, and subarachnoid space. It is reabsorbed into the blood at the dural venous sinus. AP|R

From the lateral ventricles, the CSF flows through the interventricular foramina into the third ventricle and then through the cerebral aqueduct into the fourth ventricle. From the fourth ventricle, some of the fluid flows down through the central canal of the spinal cord, but most of it passes into the subarachnoid space of the meninges. Within the subarachnoid space, the CSF flows in two directions. Some flows upward around the brain. The remainder flows downward along the back of the spinal cord, then turns to flow upward along the front of the spinal cord. The CSF continues upward around the brain in the subarachnoid space. CSF is reabsorbed into the blood-filled dural venous sinus that is located along the superior midline within the dura mater (figures 8.11 and 8.16). In a healthy state, the secretion and absorption of CSF occur at equal rates, which results in a rather constant hydrostatic pressure within the ventricles and subarachnoid space.

As mentioned previously, CSF acts as a protective shock absorber that surrounds the brain and spinal cord. Because it is circulated throughout the CNS, cerebrospinal fluid is used for the transportation of ions, nutrients, and waste products. It also provides the brain with buoyancy, which "floats" the brain within the skull and prevents damaging contact with the cranial floor.

 Check My Understanding
14. How does cerebrospinal fluid support healthy function of the CNS?

8.7 Spinal Cord

Learning Objective
14. Describe the structure and function of the spinal cord.

The **spinal cord** is continuous with the brain. It descends from the medulla oblongata through the foramen magnum into the vertebral canal and extends to the second lumbar vertebra. Beyond this point, only the roots of the lower spinal nerves occupy the vertebral canal.

Structure

The spinal cord is cylindrical in shape. It has two small grooves that extend throughout its length: the wider *anterior median fissure* and the narrower *posterior median sulcus*. These grooves divide the spinal cord into left and right portions. Thirty-one pairs of spinal nerves branch from the spinal cord. The spinal cord is divided into four segments–*cervical, thoracic, lumbar,* and *sacral*–based upon where the spinal nerves exit the vertebral column.

The cross-sectional structure of the spinal cord is shown in figures 8.12 and 8.17. Grey matter, shaped like the outstretched wings of a butterfly, is centrally located and is surrounded by white matter. The *central canal* extends the length of the spinal cord and contains CSF.

The pointed projections of the grey matter, as seen in cross section, are called horns. The *anterior horns* contain the cell bodies of somatic motor neurons whose axons enter spinal nerves and carry action potentials to skeletal muscles. The *posterior horns* contain interneurons that receive action potentials from sensory axons in the spinal nerves and carry them to sites within the CNS. *Lateral horns,* found only in the thoracic and lumbar segments of the spinal cord, contain the cell bodies of autonomic motor neurons whose axons follow autonomic pathways as they carry action potentials to cardiac and smooth muscle, glands, and adipose tissue. Interneurons form most of the grey matter in the CNS.

The horns of the grey matter divide the white matter into three regions: the *anterior, posterior,* and *lateral funiculi* (singular, funiculus). These funiculi contain **nerve tracts,** which are bundles of myelinated and unmyelinated axons of interneurons that extend up and down within the spinal cord.

Functions

The spinal cord has two basic functions. It transmits action potentials to and from the brain, and it serves as a reflex center for spinal reflexes. Action potentials are transmitted to and from the brain by axons composing

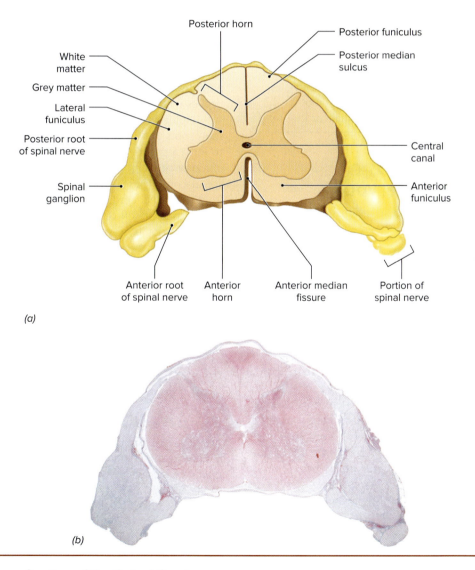

Figure 8.17 Cross Section of the Spinal Cord.
A drawing *(a)* and a photomicrograph *(b)* of the spinal cord in cross section show its basic structure. APR
(b) ©Per H. Kjeldsen, University of Michigan

the nerve tracts. *Ascending* (sensory) *tracts* carry sensory action potentials to the brain; *descending* (motor) *tracts* carry motor action potentials from the brain.

Clinical Insight

In *hydrocephalus*, a congenital defect restricts the movement of CSF from the ventricles into the subarachnoid space. In severe cases, the buildup of hydrostatic pressure within an infant's brain causes a marked enlargement of the ventricles and brain and widens the fontanelles of the cranium. Without treatment, death usually results within two to three years. Treatment involves surgical insertion of a small tube to drain the excess CSF from a ventricle into the peritoneal cavity, where it is reabsorbed.

Check My Understanding

15. What is the relationship between the ventricles, the meninges, and the cerebrospinal fluid?
16. What are the functions of the spinal cord?

8.8 Peripheral Nervous System (PNS)

Learning Objectives

15. Recall the name, type, and functions for each of the 12 pairs of cranial nerves.
16. Describe the classification of the spinal nerves and the plexuses they form.
17. Explain the functions of the components involved in a reflex.

The *peripheral nervous system (PNS)* includes the cranial and spinal nerves that connect the CNS to other portions of the body, along with sensory receptors and ganglia. A **nerve** consists of axons that are bound together by connective tissue. **Motor nerves** contain mostly axons of motor neurons; **sensory nerves** contain only axons of sensory neurons; and **mixed nerves** contain both motor axons and sensory axons. Most nerves are mixed. Nerves may contain both somatic axons and autonomic axons.

Cranial Nerves

Twelve pairs of **cranial nerves** arise from the brain and connect the brain with organs and tissues that are primarily located in the head and neck (table 8.3). Most cranial nerves arise from the brainstem. Cranial nerves are identified by both roman numerals and names. The numerals indicate the order in which the nerves arise from the inferior surface of the brain: CN I is closest to the front; CN XII is closest to the back (figure 8.18).

Five cranial nerves are primarily motor, three are sensory, and four are mixed.

Spinal Nerves

Arising from the spinal cord, there are 31 pairs of mixed nerves called **spinal nerves.** Each pair of spinal nerves is named based upon where it exits the vertebral column. The first pair of spinal nerves emerges from the spinal cord between the atlas and the occipital bone. The remaining 30 pairs of spinal nerves emerge through the *intervertebral foramina* between adjacent vertebrae, the *sacral foramina,* and the *sacral hiatus.* There are 8 pairs of *cervical nerves* (C1–C8), 12 pairs of *thoracic nerves* (T1–T12), 5 pairs of *lumbar nerves* (L1–L5), 5 pairs of *sacral nerves* (S1–S5), and 1 pair of *coccygeal nerves* (Co) (figure 8.19). Recall from chapter 6 that there are seven cervical vertebrae. Because the first pair of spinal nerves emerges above the atlas, there are eight pairs of cervical nerves instead of seven.

Spinal nerves branch from the spinal cord by two short roots that merge a short distance from the spinal cord to form a spinal nerve. The **anterior root** contains axons of motor neurons whose cell bodies are located within the spinal cord. These neurons carry motor action potentials from the spinal cord to effectors. The **posterior root** contains axons of sensory neurons. The swollen region in a posterior root is a **spinal ganglion,** which contains cell bodies of sensory neurons. The long axons of these neurons carry sensory action potentials to the spinal cord. Observe these structures and their relationships in figures 8.12, 8.17, and 8.20.

As shown in figure 8.19, the spinal cord ends at the second lumbar vertebra. The roots of lumbar, sacral, and coccygeal spinal nerves continue downward within the vertebral canal to exit between the appropriate vertebrae. These roots form the *cauda equina,* or "horse's tail," in the lower portion of the vertebral canal.

Spinal Plexuses

After a spinal nerve exits the vertebral canal, it divides into four major parts: the *anterior ramus* (plural, rami), *posterior ramus, meningeal branch,* and *rami communicantes.* The posterior ramus innervates the deep muscles and skin of the back. The meningeal branch innervates the vertebrae, meninges, and vertebral ligaments. The rami communicantes pass to the ganglia of the sympathetic trunk and are part of the autonomic division. The anterior rami of many spinal nerves merge to form **spinal plexuses,** networks of nerves, before continuing to the innervated structures. The anterior rami of most thoracic nerves do not form plexuses; rather, they form intercostal nerves.

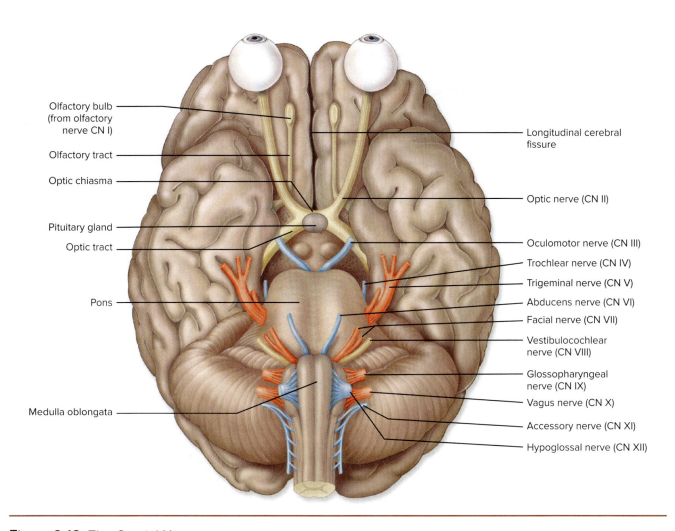

Figure 8.18 The Cranial Nerves.
Note the roots of the 12 pairs of cranial nerves in this inferior view of the brain. Cranial nerves are identified by both roman numerals and names. Most cranial nerves arise from the brainstem. Yellow = sensory nerve; Blue = motor nerve; Red = mixed nerve. APIR

The intercostal nerves innervate the intercostal and abdominal muscles, in addition to overlying skin.

In a plexus, the axons in the anterior rami are sorted and recombined so that axons going to a specific body part are carried in the same peripheral nerve, although they may originate in several different spinal nerves. There are four pairs of plexuses: cervical, brachial, lumbar, and sacral. Because many axons from the lumbar plexus contribute to the sacral plexus, these two plexuses are sometimes called the *lumbosacral plexus* (figure 8.19).

Cervical Plexus The upper cervical nerves merge to form a *cervical plexus* on each side of the neck. The nerves from these plexuses supply the muscles and skin of the neck and portions of the head and shoulders. The paired *phrenic* (fren′-ik) *nerves*, which stimulate the diaphragm to contract and begin inspiration, also arise from the cervical plexus.

Brachial Plexus The lower cervical nerves and perhaps nerves T1–T2 join to form a *brachial plexus* on each side of the vertebral column in the shoulder region. Nerves that serve skin and muscles of the pectoral girdle and upper limb emerge from the brachial plexuses. The *musculocutaneous, axillary, radial, median,* and *ulnar nerves* arise here.

Lumbar Plexus The last thoracic nerve (T12) and the upper lumbar nerves unite to form a *lumbar plexus* on each side of the vertebral column just above the coxal bones. Nerves from the lumbar plexuses supply the skin and muscles of the lower trunk, external genitalia, and the anterior and medial thighs. The *femoral* and *obturator nerves* arise here.

Sacral Plexus The lower lumbar nerves and the sacral nerves merge to form a *sacral plexus* on each side of the sacrum within the pelvis. Nerves from the sacral plexuses supply the skin and muscles of the buttocks and lower

Table 8.3 Summary of the Cranial Nerves

Nerve Number	Nerve Name	Type	Function
CN I	Olfactory	Sensory	Transmits sensory action potentials from olfactory receptors in olfactory epithelium to the brain.
CN II	Optic	Sensory	Transmits sensory action potentials for vision from the retina of the eye to the brain.
CN III	Oculomotor	Motor	Transmits motor action potentials to muscles that move the eyes superiorly, inferiorly, and medially; control the eyelids; adjust pupil size; and control the shape of the lens.
CN IV	Trochlear	Motor	Transmits motor action potentials to muscles that rotate the eyes.
CN V	Trigeminal	Mixed	Transmits sensory action potentials from scalp, forehead, face, teeth, and gums to the brain. Transmits motor action potentials to chewing muscles and muscles in floor of mouth.
CN VI	Abducens	Motor	Transmits motor action potentials to muscles that move the eyes laterally.
CN VII	Facial	Mixed	Transmits sensory action potentials from the front of the tongue to the brain. Transmits motor action potentials to facial muscles, salivary glands, and tear glands.
CN VIII	Vestibulocochlear	Sensory	Transmits sensory action potentials from the internal ear associated with hearing and equilibrium.
CN IX	Glossopharyngeal	Mixed	Transmits sensory action potentials from the back of the tongue, tonsils, pharynx, and carotid arteries to the brain. Transmits motor action potentials to salivary glands and pharyngeal muscles used in swallowing.
CN X	Vagus	Mixed	Transmits sensory action potentials from thoracic and abdominal organs, esophagus, larynx, and pharynx to the brain. Transmits motor action potentials to these organs and to muscles of speech and swallowing.
CN XI	Accessory	Motor	Transmits motor action potentials to muscles of the palate, pharynx, and larynx and to the trapezius and sternocleidomastoid muscles.
CN XII	Hypoglossal	Motor	Transmits motor action potentials to the muscles of the tongue.

limbs. The *sciatic nerves,* which emerge from the sacral plexuses, are the largest nerves in the body.

Check My Understanding

17. What composes the peripheral nervous system?
18. Identify and describe the functions of the 12 cranial nerves.
19. Name and locate the major spinal plexuses.

Reflexes

Reflexes are rapid, involuntary, and predictable responses to internal and external stimuli. Reflexes maintain homeostasis and enhance chances of survival. A reflex involves either the brain or the spinal cord, a sensory receptor, sensory and motor neurons, and an effector.

Most pathways of action potential transmission within the nervous system are complex and involve many neurons. In contrast, reflexes require few neurons in their pathways and therefore produce very rapid responses to stimuli. Reflex pathways are called **reflex arcs.**

Reflexes are divided into two types—autonomic and somatic—based on the effector(s) involved in the reflex. *Autonomic reflexes* act on smooth muscle, cardiac muscle, adipose tissue, and glands. They are involved in controlling homeostatic processes such as heart rate, blood pressure, and digestion. Autonomic reflexes maintain homeostasis and healthy body functions at the unconscious level, which frees the mind to deal with those actions that require conscious decisions. *Somatic reflexes* act on skeletal muscles. They enable quick movements such as moving the hand away from a painful stimulus. A person is usually unaware of autonomic reflexes but is aware of somatic reflexes. Reflexes are also divided into *cranial reflexes* and *spinal reflexes,* depending upon whether the brain or the spinal cord is involved in the reflex.

Figure 8.20 illustrates a somatic spinal reflex, which withdraws the hand after sticking a finger with a tack. Three neurons are involved in this reflex. Pain receptors are stimulated by the sharp pin and form action potentials that are carried by a sensory neuron to an interneuron in the spinal cord. Action potentials pass along the interneuron to a motor neuron, which carries the action potentials to a muscle that contracts to move the hand. Although the brain is not involved in this reflex, it does

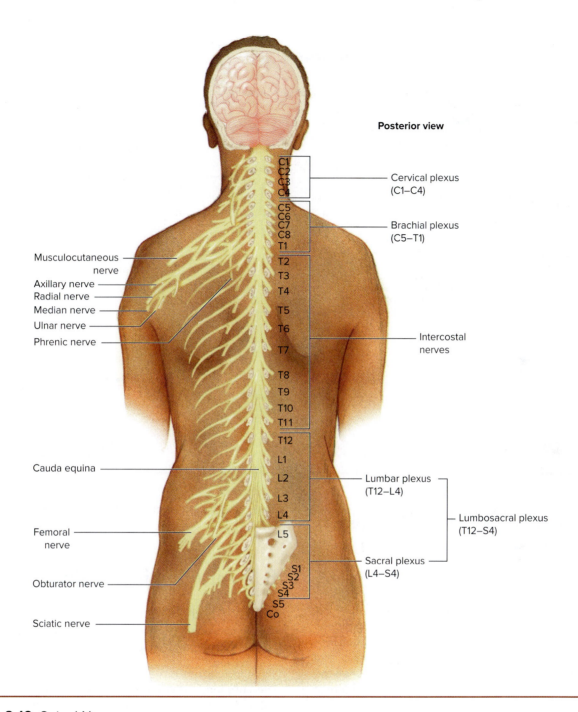

Figure 8.19 Spinal Nerves.
Thirty-one pairs of spinal nerves arise from the spinal cord. Anterior rami of spinal nerves in the thoracic region form the intercostal nerves. Those in other segments form nerve networks called spinal plexuses before continuing on to their target tissues. AP|R

receive sensory action potentials that make a person aware of a painful stimulus.

 Check My Understanding

20. What is a reflex?
21. What are the components of a spinal reflex?

 **Clinical Insight**

Because the responses of reflexes are predictable, physicians usually test a patient's reflexes in order to determine the health of the nervous system. Exaggerated, diminished, or distorted reflexes may indicate a neurological disorder.

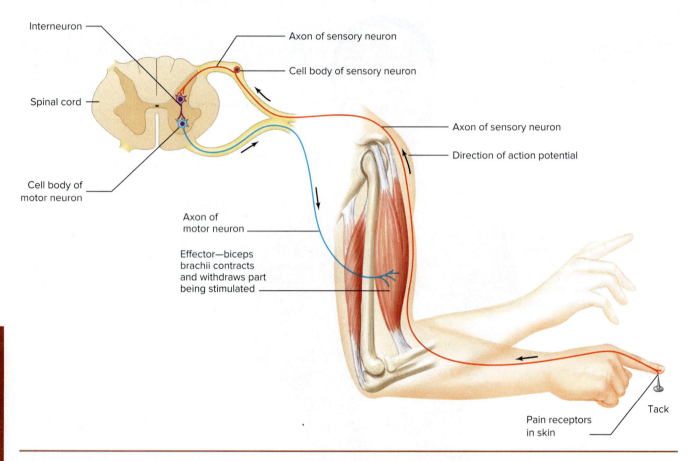

Figure 8.20 Typical Reflex Arc.
A somatic spinal reflex involving a sensory neuron, an interneuron, and a motor neuron. Red = sensory neuron; Purple = interneuron; Blue = motor neuron. AP|R

Clinical Insight

Because the spinal cord ends at the level of the second lumbar vertebra, spinal taps and epidural anesthetics are administered below this point. For these procedures, a patient is placed in a fetal position in order to open the spaces between the posterior margins of the vertebrae. A hypodermic needle is inserted into the vertebral canal either between the third and fourth lumbar vertebrae or between the fourth and fifth lumbar vertebrae. In a *spinal tap* (lumbar puncture), a hypodermic needle is inserted into the subarachnoid space to remove cerebrospinal fluid for diagnostic purposes. An *epidural anesthetic* is given by injecting an anesthetic into the epidural space with a hypodermic syringe. The anesthetic prevents sensory action potentials from reaching the spinal cord via posterior roots inferior to the injection. Epidurals are sometimes used to ease pain during childbirth.

8.9 Autonomic Division

Learning Objective

18. Compare the structure and functions of the sympathetic and parasympathetic parts of the autonomic division.

The autonomic division consists of portions of the central and peripheral nervous systems and functions without conscious control. Its role is to maintain homeostasis in response to changing internal conditions. The effectors under autonomic control are cardiac muscle, smooth muscle, adipose tissue, and glands. The autonomic division functions mostly by involuntary reflexes. Visceral sensory action potentials carried to the autonomic reflex centers in the hypothalamus, brainstem, or spinal cord cause visceral motor action potentials to be carried to effectors via cranial or spinal nerves. Higher brain centers, such as the limbic system and cerebral cortex, influence the autonomic division during times of emotional stress.

Table 8.4 compares the somatic and autonomic divisions..

Table 8.4 Comparison of Somatic and Autonomic Divisions

	Somatic	Autonomic
Control	Voluntary	Involuntary
Neural Pathway	One motor neuron extends an axon from the CNS to an effector	A preganglionic neuron extends an axon from the CNS to an autonomic ganglion and synapses with a postganglionic neuron that extends an axon to an effector
Neurotransmitters	Acetylcholine	Acetylcholine or norepinephrine
Effectors	Skeletal muscles	Smooth muscle, cardiac muscle, adipose tissue, and glands
Action	Excitatory	Excitatory or inhibitory

Organization

Unlike the somatic division, in which a single motor neuron extends from the CNS to a skeletal muscle, the autonomic division uses two motor neurons in sequence to carry motor action potentials to an effector. The cell body of the first neuron, or *preganglionic neuron*, is located within the brain or spinal cord. It extends an axon from the CNS to an **autonomic ganglion**. The cell body of the second neuron, or *postganglionic neuron*, is located within the autonomic ganglion and it extends an axon from the ganglion to the visceral effector (figure 8.21).

The autonomic division is subdivided into the **sympathetic part** and the **parasympathetic part**. The origin of their motor neurons and the organs innervated are shown in figure 8.22.

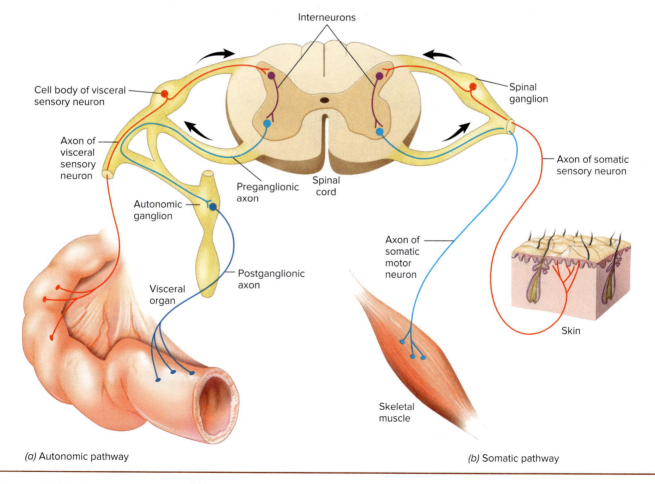

Figure 8.21 Comparison of Autonomic and Somatic Motor Pathways in Spinal Nerves.
(a) An autonomic pathway involves a preganglionic neuron and a postganglionic neuron that synapse at a ganglion outside of the CNS. *(b)* A somatic pathway involves a single motor neuron. Red = sensory neuron; Purple = interneuron; Blue = motor neuron.

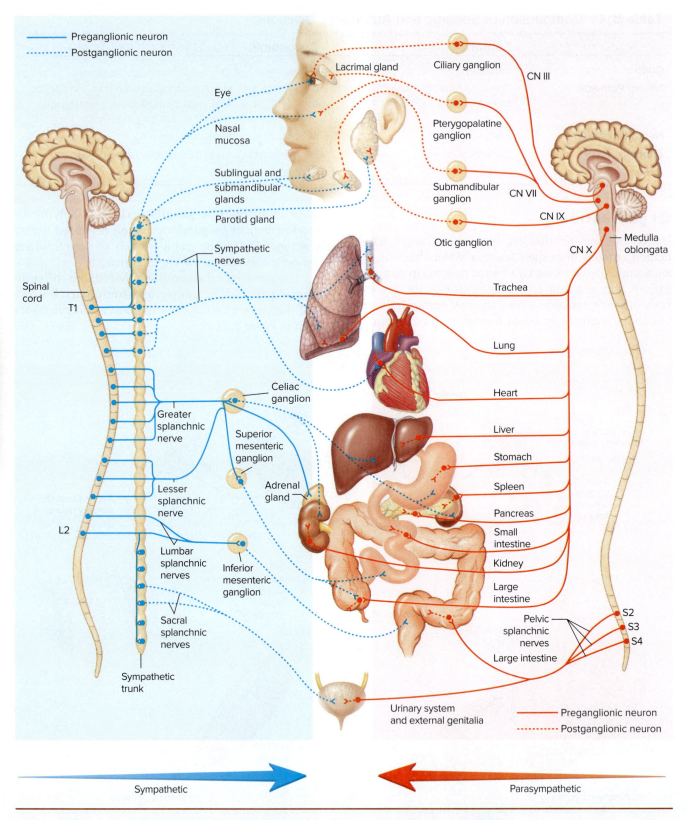

Figure 8.22 Innervation of Visceral Organs by the Autonomic Division.

Preganglionic axons of the sympathetic part arise from the thoracic and lumbar segments of the spinal cord–spinal nerves T1–L2. Some sympathetic preganglionic axons branch from the spinal nerves to synapse with postganglionic neurons in autonomic ganglia that are arranged in two chains, one on each side of the vertebral column. Each chain of ganglia is called a *sympathetic trunk*. Other sympathetic preganglionic axons pass through a ganglion of the sympathetic trunk without synapsing and extend to another type of ganglion, a *collateral ganglion*, before synapsing with a postganglionic neuron. Both pathways are shown in figure 8.22.

Preganglionic axons of the parasympathetic part arise from the brainstem and sacral segment (S2–S4) of the spinal cord. They extend through cranial or sacral nerves to synapse with postganglionic neurons within ganglia that are located very near or within visceral organs (figure 8.22).

Most visceral organs receive postganglionic axons of both the sympathetic and the parasympathetic parts; but a few, such as sweat glands and most blood vessels, receive only sympathetic axons.

Autonomic Neurotransmitters

Preganglionic axons of both the sympathetic and the parasympathetic parts secrete *acetylcholine* to initiate action potentials in postganglionic neurons, but their postganglionic axons secrete different neurotransmitters. Most sympathetic postganglionic axons secrete norepinephrine, a substance similar to adrenaline, which is why they are called *adrenergic axons*. Parasympathetic postganglionic axons secrete acetylcholine and thus are called *cholinergic axons* (figure 8.23).

Functions

Both sympathetic and parasympathetic parts stimulate some visceral organs and inhibit others. However, their effects on a given organ are opposite. For example, the sympathetic part increases heart rate whereas the parasympathetic part decreases heart rate. The contrasting effects are due to the different neurotransmitters secreted by sympathetic and parasympathetic postganglionic axons and the receptors of the receiving organs.

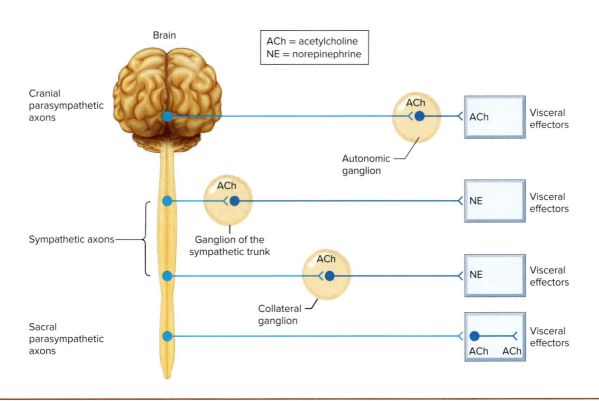

Figure 8.23 Comparison of Neurotransmitters Used by the Autonomic Division.

Table 8.5 Representative Actions of the Autonomic Division

Effector	Sympathetic Stimulation	Parasympathetic Stimulation
Eye	Dilation of pupil; changes lens shape for far vision	Constriction of pupil; changes lens shape for near vision
Heart	Increases rate and strength of contraction	Decreases rate of contraction
Arterioles	Constriction increases blood pressure	No innervation
Blood distribution	Increases supply to skeletal muscles; decreases supply to digestive organs	Decreases supply to skeletal muscles; increases supply to digestive organs
Lungs	Dilates bronchioles	Constricts bronchioles
Digestive tract	Inhibits motility and secretion by glands	Promotes motility and secretion by glands
Liver	Decreases bile production; increases blood glucose	Increases bile production; decreases blood glucose
Gallbladder	Relaxation	Contraction
Kidneys	Decreases urine production	No innervation
Pancreas	Decreases secretion of insulin and digestive enzymes	Increases secretion of insulin and digestive enzymes
Spleen	Constriction injects stored blood into circulation	No innervation
Urinary bladder	Contraction of internal urethral sphincter; relaxation of bladder wall	Relaxation of internal urethral sphincter; contraction of bladder wall
Reproductive organs	Vasoconstriction; ejaculation in males; reverse uterine contractions in females; stimulates uterine contractions in labor	Vasodilation; erection in males; vaginal secretion in females

The sympathetic part prepares the body for physical action to meet emergencies. Its actions have been summarized as preparing the body for *fight or flight*. The parasympathetic part is dominant under healthy, nonstressful conditions of everyday life. Because its actions are usually opposite those of the sympathetic part, it is often viewed as preparing the body for *resting and digesting*. Table 8.5 compares some of the effects of the sympathetic and parasympathetic parts on visceral organs.

Clinical Insight

Cocaine exerts major effects on the autonomic division. It not only stimulates the sympathetic part but also inhibits the parasympathetic part. In an overdose, this double-barreled action produces an erratic, uncontrollable heartbeat that may result in sudden death.

Check My Understanding

22. How do the somatic and autonomic divisions differ in terms of structure?
23. How does the autonomic division maintain homeostasis?

8.10 Disorders of the Nervous System

Learning Objective
19. Describe the common disorders of the nervous system.

Inflammatory Disorders

Meningitis (men-in-ji´-tis) results from a bacterial, fungal, or viral infection of the meninges. Bacterial meningitis cases are the most serious, with about 20% being fatal. If the brain is also involved, the disease is called *encephalitis*. Some viruses causing encephalitis are transmitted by bites of certain mosquitoes.

Neuritis is the inflammation of a nerve or nerves. It may be caused by several factors, such as infection, compression, or trauma. Associated pain may be moderate or severe.

Sciatica (si-at´-i-kah) is neuritis involving the sciatic nerve. The pain may be severe and often radiates down through the thigh and leg to the sole of the foot.

Shingles is an infection of one or more nerves. It is caused by the reactivation of the chicken pox virus, which, until that time, has been dormant in the nerve roots. The virus causes painful blisters on the skin at the sensory nerve endings, followed by prolonged pain (figure 8.24a).

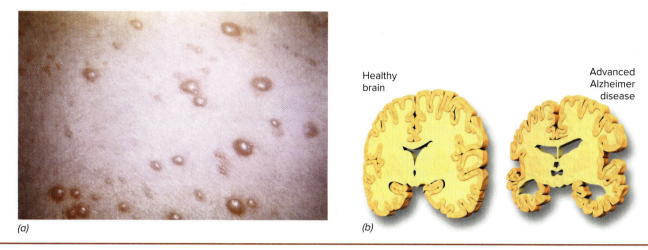

Figure 8.24 Nervous System Disorders.
(a) Shingles. *(b)* Alzheimer disease.
(a) Source: Dr. K.L. Hermann/CDC

Noninflammatory Disorders

Alzheimer (alts'-hī-mer) **disease (AD)** is a progressively disabling disease affecting older persons. It is associated with a loss of certain cholinergic neurons in the brain and a reduced ability of neurons to secrete acetylcholine. AD is characterized by a progressive loss of memory, disorientation, and mood swings (figure 8.24*b*).

Cerebral palsy (ser-ē'-bral pawl-zē) is characterized by partial paralysis and sometimes a degree of mental retardation. It may result from damage to the brain during prenatal development, often from viral infections caused by German measles or from trauma during delivery.

Cerebrovascular accidents (CVAs) are disorders of blood vessels serving the brain. They result from blood clots, aneurysms (an'-ū-rizms), or hemorrhage. Often called *strokes*, CVAs cause severe damage to the brain due to the loss of oxygen. Response time is crucial after the CVA in order to limit the amount of neural damage. They are a major cause of disability and the third highest cause of death in the United States.

Comas are states of unconsciousness in which the patient cannot be aroused even with vigorous stimulation. Illness or trauma to the brain may alter the functioning of the reticular formation, resulting in a coma.

Concussion results from a severe jarring of the brain caused by a blow to the head. Unconsciousness, confusion, and amnesia may result in severe cases.

Dyslexia (dis-lek'-sē-ah) causes the afflicted person to reverse letters or syllables in words and transpose words within sentences. It results from malfunction within the language center of the cerebrum.

Epilepsy (ep'-i-lep"-sē) may have a hereditary basis, or it may be triggered by injuries, infections, or tumors. There are two types of epilepsy. *Grand mal epilepsy* is the more serious form and is characterized by convulsive seizures. *Petit mal epilepsy* is the less serious form and is characterized by momentary loss of contact with reality without unconsciousness or convulsions.

Fainting is a brief loss of consciousness due to a sudden reduction in blood supply to the brain. It may result from either physical or psychological causes.

Headaches are triggered by various physical or psychological factors, but often result from a dilation of blood vessels within the meninges of the brain. Migraine headaches may have visual or digestive side effects and may be triggered by stress, allergies, or fatigue. Sinus headaches may result from inflammation that causes increased pressure within the paranasal sinuses. Some headaches result from tension in muscles of the head and neck.

Mental illnesses may be broadly categorized as either neuroses or psychoses. *Neuroses* are mild maladjustments to life situations that may produce anxiety and interfere with healthy behavior. *Psychoses* are serious mental disorders that sometimes cause delusions, hallucinations, or withdrawal from reality.

Multiple sclerosis (MS) is a progressive degeneration of the myelin sheath around axons in the CNS, accompanied by the formation of plaques of scar tissue called *scleroses*. This destruction results in a short-circuiting of neural pathways and an impairment of motor functions.

Neuralgia (nū-ral'-jē-ah) is pain arising from a nerve regardless of the cause of the pain.

Paralysis is the permanent loss of motor control of body parts. It most commonly results from accidental injury to the CNS.

Parkinson disease is caused by an insufficient delivery of the neurotransmitter *dopamine* to neurons in certain nuclei within the cerebrum. It produces tremors and impairs healthy skeletal muscle contractions. Parkinson disease is more common among older persons.

Chapter Summary

8.1 Introduction to the Nervous System

- The nervous system is the primary coordinating and controlling system of the body.
- An action potential is an electrochemical signal created and conducted along the axon of a neuron.
- The nervous system detects internal and external changes, analyzes the changes, organizes the information for immediate and future use, and initiates actions in response.

8.2 Divisions of the Nervous System

- Anatomical divisions are the central nervous system (CNS), composed of the brain and spinal cord, and the peripheral nervous system (PNS), composed of cranial and spinal nerves, ganglia, and sensory receptors.
- Functional divisions are the sensory and motor divisions. The motor division is subdivided into the somatic division and autonomic division.
- The somatic division is involved in the voluntary and involuntary control of skeletal muscles.
- The autonomic division is involved in the involuntary control of cardiac muscle, smooth muscle, glands, and adipose tissue.

8.3 Nerve Tissue

- Nerve tissue consists of neurons and neuroglia.
- A neuron is composed of a cell body, which contains the nucleus; one or more dendrites that conduct impulses toward the cell body or axon; and one axon that conducts action potentials away from the cell body or dendrites.
- Myelinated axons are covered by a myelin sheath. Schwann cells form the myelin sheath and neurilemma of peripheral myelinated axons. Oligodendrocytes form the myelin sheath of myelinated axons in the CNS; these axons lack a neurilemma.
- There are three structural types of neurons: multipolar, bipolar, and pseudounipolar.
- There are three functional types of neurons. Sensory neurons carry action potentials toward the CNS. Interneurons carry action potentials within the CNS. Motor neurons carry action potentials from the CNS.
- Schwann cells are neuroglia in the PNS. Four types of neuroglia occur in the CNS: oligodendrocytes, astrocytes, microglial cells, and ependymal cells.

8.4 Neuron Physiology

- Neurons are specialized to form and conduct action potentials.
- The plasma membrane of a resting neuron is polarized with an excess of positive charges on the ECF-side and negative charges on the cytosol-side. This difference creates a voltage called the resting membrane potential.
- When a threshold stimulus is applied, the neuron plasma membrane becomes permeable to sodium ions (Na^+), which quickly move into the neuron and cause depolarization of the membrane. This depolarization is the formation of an action potential.
- The depolarized portion of the plasma membrane causes the depolarization of adjacent portions so that a depolarization wave flows along the axon.
- Depolarization makes the neuron plasma membrane permeable to potassium ions (K^+), allowing them to quickly diffuse into the ECF and repolarize the plasma membrane.
- In neuron-to-neuron synaptic transmission, the terminal bouton secretes a neurotransmitter into the synaptic cleft. The neurotransmitter binds to receptors on the postsynaptic neuron, causing either the formation of an action potential or the inhibition of action potential formation. Then, the neurotransmitter is quickly removed by reabsorption into the terminal bouton, an enzymatic reaction or diffusion out of the cleft.
- The most common peripheral neurotransmitters are acetylcholine and norepinephrine. Some neurotransmitters are excitatory, while others are inhibitory.

8.5 Protection for the Central Nervous System

- The brain is encased by the cranial bones, and the spinal cord is surrounded by vertebrae.
- Both the brain and the spinal cord are covered by the meninges: the pia mater, arachnoid mater, and dura mater.
- Cerebrospinal fluid in the subarachnoid space provides buoyancy and serves as a fluid shock absorber surrounding the brain and spinal cord.

8.6 Brain

- The brain consists of the cerebrum, diencephalon, brainstem, and cerebellum.
- The cerebrum consists of two cerebral hemispheres joined by the corpus callosum. Gyri and sulci increase the surface area of the cerebral cortex. Each hemisphere is subdivided into five lobes: frontal, parietal, temporal, occipital, and insula.
- The cerebrum interprets sensations; initiates voluntary motor responses; and is involved in will, personality traits, and intellectual processes. The left cerebral hemisphere is dominant in most people.
- Sensory areas occur in the parietal, temporal, and occipital lobes. Motor areas occur in the frontal lobe. Association areas occur in all lobes of the cerebrum.
- The diencephalon consists of the thalamus, the hypothalamus, and the epithalamus.
- The thalamus is formed of two lateral masses connected by the interthalamic adhesion. It is a relay station for sensory and motor action potentials going to and from the cerebrum and provides an uncritical awareness of sensations.
- The hypothalamus is located below the thalamus and forms the floor of the third ventricle. It is a major integration center for the autonomic division. It also regulates several homeostatic processes such as body temperature, mineral and water balance, appetite, digestive processes, and secretion of pituitary gland hormones.
- The epithalamus possesses the pineal gland, which produces the hormone melatonin. Melatonin induces sleepiness in the evenings.

- The limbic system is associated with emotional behavior, memory, and motivation.
- The brainstem consists of the midbrain, pons, and medulla oblongata. Ascending and descending axons between higher brain centers and the spinal cord pass through the brainstem.
- The midbrain is a small, superior portion of the brainstem. It contains reflex centers for movements associated with visual and auditory stimuli.
- The pons is the middle portion of the brainstem. It works with the medulla oblongata to control breathing.
- The medulla oblongata is the most inferior portion of the brainstem and is continuous with the spinal cord. It contains reflexive integration centers that control breathing, heart rate and force of contraction, and blood pressure.
- The reticular formation consists of nuclei and axons that extend from the upper spinal cord and into the diencephalon. It is involved with wakefulness.
- The cerebellum lies behind the fourth ventricle. It is composed of two hemispheres separated by the vermis and coordinates skeletal muscle contractions.
- The ventricles of the brain, the central canal of the spinal cord, and the subarachnoid space around the brain and spinal cord are filled with cerebrospinal fluid. Cerebrospinal fluid is secreted by a choroid plexus in each ventricle.
- Cerebrospinal fluid is absorbed into blood of the dural venous sinus in the dura mater.

8.7 Spinal Cord

- The spinal cord extends from the medulla oblongata down through the vertebral canal to the second lumbar vertebra.
- Grey matter is centrally located and is surrounded by white matter. Anterior horns of grey matter contain cell bodies of somatic motor neurons; posterior horns contain interneuron cell bodies that receive incoming sensory action potentials; lateral horns contain cell bodies of autonomic motor neurons. White matter contains ascending and descending tracts of myelinated and unmyelinated axons.
- The spinal cord serves as a reflex center and conducting pathway for action potentials between the brain and spinal nerves.

8.8 Peripheral Nervous System (PNS)

- The PNS consists of cranial and spinal nerves, in addition to sensory receptors and ganglia. Most nerves are mixed nerves; a few cranial nerves are motor or sensory only. A nerve contains bundles of axons supported by connective tissue.
- The 12 pairs of cranial nerves are identified by roman numeral and name. The 31 pairs of spinal nerves are divided into 8 cervical, 12 thoracic, 5 lumbar, 5 sacral, and 1 coccygeal nerve.
- Anterior rami of many spinal nerves form spinal plexuses where axons are sorted and recombined so that all axons to a particular organ are carried in the same nerve. The four pairs of spinal plexuses are cervical, brachial, lumbar, and sacral plexuses.
- Reflexes are rapid, involuntary, and predictable responses to internal and external stimuli.
- Autonomic reflexes involve smooth muscle, cardiac muscle, adipose tissue, and glands. Somatic reflexes involve skeletal muscles.
- Cranial reflexes involve the brain, while spinal reflexes involve the spinal cord.

8.9 Autonomic Division

- The autonomic division involves portions of the central and peripheral nervous systems that are involved in involuntary maintenance of homeostasis.
- Two autonomic motor neurons are used to activate an effector. The axon of the preganglionic neuron arises from the CNS and ends in an autonomic ganglion, where it synapses with a postganglionic neuron. The axon of the postganglionic neuron extends from the ganglion to an effector.
- The autonomic division is divided into two subdivisions that generally have antagonistic effects. Nerves of the sympathetic part arise from the thoracic and lumbar segments of the spinal cord and prepare the body to meet emergencies. Nerves of the parasympathetic part arise from the brain and the sacral segment of the spinal cord and function mainly in nonstressful situations.

8.10 Disorders of the Nervous System

- Disorders may result from infectious diseases, degeneration from unknown causes, malfunctions, and physical injury.
- Inflammatory neurological disorders include meningitis, neuritis, sciatica, and shingles.
- Noninflammatory neurological disorders include Alzheimer disease, cerebral palsy, CVAs, comas, concussion, dyslexia, epilepsy, fainting, headaches, mental illness, multiple sclerosis, neuralgia, paralysis, and Parkinson disease.

Improve Your Grade

Connect Interactive Questions Reinforce your knowledge using multiple types of questions: interactive, animation, classification, labeling, sequencing, composition, and traditional multiple choice and true/false.

SmartBook Proven to help students improve grades and study more efficiently, SmartBook contains the same content within the print book but actively tailors that content to the needs of the individual.

Anatomy & Physiology REVEALED® Dive into the human body by peeling back layers of cadaver imaging. Utilize this world-class cadaver dissection tool for a closer look at the body anytime, from anywhere.

CHAPTER 9

Senses

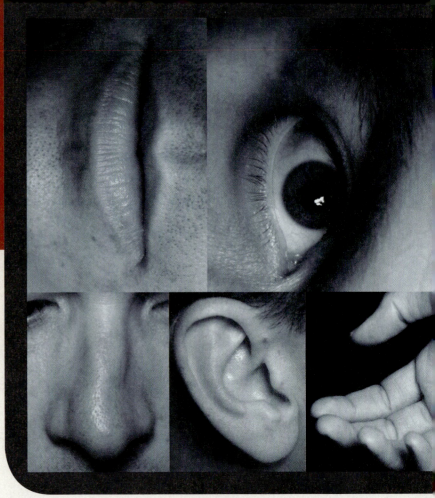

(Senses) ©Maartje van Caspel/Getty Images RF

Jeremy, age 14, was born blind and deaf. He is unable to see the world around him. He cannot see the sky, the earth, or his family. He cannot see a car heading toward him, and so he would not know to get out of the way. Jeremy cannot hear warning alarms or people yelling at him when danger nears. He cannot hear the spoken words used for quick, easy communication between people, and because he cannot hear words, he did not develop the auditory memories needed to produce speech. Nor can he verbally express his thoughts, opinions, or desires to those around him. To survive in the world and communicate with those around him, Jeremy has had to learn to use his other senses. He uses his sense of touch to identify people and objects around him and to learn about the world by reading in Braille. By feeling vibrations through his skin, he can detect the rhythm in music that is playing. His sense of smell is heightened, which allows him to detect certain types of hazards and aid in the identification of people and objects. Jeremy's life is the perfect example of just how important the senses are in maintaining health and wellness for each of us.

CHAPTER OUTLINE

9.1 Introduction to the Senses

9.2 Sensations
- Projection
- Adaptation

9.3 General Senses
- Temperature
- Pressure, Touch, and Stretch
- Chemoreceptors
- Pain

9.4 Special Senses
- Taste
- Smell
- Hearing
- Equilibrium
- Vision

9.5 Disorders of the Special Senses
- Disorders of Taste and Smell
- Disorders of the Ear
- Disorders of the Eye

Chapter Summary

Module 7
Nervous System

SELECTED KEY TERMS

Accommodation The focusing of light rays on the retina by the lens.
Chemoreceptor Sensory receptor stimulated by certain chemicals.
Cochlear hair cells Sensory receptors used in hearing.
Cones Photoreceptors for color vision.
Dynamic equilibrium Maintenance of balance during linear acceleration and rotational movement of the head.
Mechanoreceptor Sensory receptor stimulated by mechanical forces such as pressure or touch.
Nociceptor Sensory receptor stimulated by tissue damage.
Olfactory receptor Sensory receptor used to detect odors in inhaled air.

Photoreceptor (photo = light) Sensory receptor stimulated by light energy.
Projection The process by which the brain makes a sensation seem to come from the body part being stimulated.
Proprioceptor Sensory receptor stimulated by changes in body position or movements of the body or its parts.
Retina (retin = net) The inner layer of the eye, which contains the photoreceptors.
Rods Photoreceptors for black and white vision.
Semicircular canals The portion of the internal ear containing the sensory receptors detecting rotational head movement.

Sensory adaptation The decrease in the formation of action potentials by a sensory receptor when repeatedly stimulated by the same stimulus.
Spiral organ (organ of Corti) The sense organ in the internal ear containing the sensory receptors for hearing.
Static equilibrium The maintenance of balance during movement of the head with respect to gravitational force.
Taste bud Tongue organ that contains taste receptors.
Thermoreceptor Sensory receptor stimulated by changes in temperature.

9.1 Introduction to the Senses

Learning Objectives
1. Describe the purpose of a sensory receptor.
2. Identify the general senses and the special senses.

Our senses constantly inform us of what is going on in our internal and external environments so that our body can take appropriate voluntary or involuntary action and maintain homeostasis. Several different types of **sensory receptors** are involved in sending action potentials to the CNS, which then initiates the appropriate response.

The senses may be subdivided into two broad categories: general senses and special senses. *General senses* include pain, touch, pressure, stretching, chemical changes, cold, and heat. *Special senses* are taste, smell, vision, hearing, and equilibrium. Each of the senses depends upon (1) sensory receptors, which detect environmental changes and form action potentials; (2) sensory neurons, which carry the action potentials to the CNS; and (3) the brain, which interprets the action potentials.

✓ Check My Understanding
1. What are the general senses and special senses?
2. What parts of the nervous system are involved in the creation of a sense?

9.2 Sensations

Learning Objectives
3. Differentiate between sense, sensation, and perception.
4. Recall the five basic types of sensory receptors.
5. Compare the mechanisms of projection and adaptation of sensations.

Each type of sensory receptor is sensitive to a particular type of stimulus that causes the sensory receptor to form action potentials. The five types of sensory receptors, based on the specific stimuli to which they respond, are listed in table 9.1. These action potentials are carried by cranial or spinal nerves to the CNS. A **sensation** is a conscious or subconscious awareness of a change in the internal or external environment. The conscious awareness of a sensation, or **perception,** results from the interpretation of action potentials reaching sensory areas of the cerebral

Table 9.1 Types of Sensory Receptors

Type	Stimulus Detected
Thermoreceptors	Temperature changes
Mechanoreceptors	Mechanical forces
Nociceptors	Tissue damage
Chemoreceptors	Concentration of chemicals
Photoreceptors	Light energy

cortex. The sensation that is created is determined by the area of the brain receiving the action potentials rather than by the type of sensory receptor forming the action potentials. For example, all action potentials reaching the visual area of the occipital lobe are interpreted as visual sensations. A strong blow to an eye or to the back of the head may produce a visual sensation (flashes of light), although the stimulus is mechanical.

The perceived intensity of a sensation is dependent upon the frequency of action potentials reaching the cerebral cortex. The greater the frequency of action potentials, the greater is the intensity of the sensation. The frequency of action potentials sent to the brain is, in turn, dependent upon the action of sensory receptors. The greater the intensity of a stimulus, the greater the frequency of action potential formation by sensory receptors.

Projection

Whenever a sensation occurs, the cerebral cortex projects the sensation back to the body region where the action potentials originated so that the sensation seems to come from that region. This phenomenon is called **projection.** For example, if your thumb is injured, the pain is projected back to your thumb so that you are aware that your thumb hurts. Similarly, projection of visual and auditory sensations gives the feeling that eyes see and ears hear. The projection of sensations has obvious survival value in pinpointing the source of a sensation because it allows for corrective action to remove harmful stimuli.

Adaptation

If a sensory receptor is repeatedly stimulated by the same stimulus, the rate of action potential formation may decline until action potentials may not be formed at all. This phenomenon is called **sensory adaptation.** For example, when the odor of perfume is first encountered, it is very noticeable. But as the olfactory receptors become adapted to the stimulus, the strength of the sensation rapidly declines until the odor is hardly noticeable. Adaptation occurs within most sensory receptors, with the exception of those involved in pain and proprioception. Its purpose is to prevent overloading the nervous system with unimportant stimuli, such as clothes touching the body. Once a sensory receptor is adapted, a stronger stimulus is needed to form action potentials.

Check My Understanding

3. How are sensation and perception different?
4. Why are projection and sensory adaptation important for survival?

9.3 General Senses

Learning Objectives

6. Contrast the structures, locations, and functions of the sensory receptors involved in sensations of warm, cold, touch, pressure, stretch, chemical change, and pain.
7. Explain the mechanism of referred pain.

Sensory receptors for the general senses are widely distributed in the skin, muscles, tendons, ligaments, and visceral organs.

Temperature

Two types of **thermoreceptors** are located in the skin. *Warm receptors* are **free nerve endings,** which are sensory neuron dendrites, in the innermost part of the dermis that are most sensitive to temperatures above 25°C (77°F). *Cold receptors* are free nerve endings in the outermost part of the dermis that are most sensitive to temperatures below 20°C (68°F). Temperatures below 10°C (50°F) or above 45°C (113°F) stimulate pain receptors, which results in painful sensations. Thermoreceptors adapt very quickly to constant stimulation.

Pressure, Touch, and Stretch

Pressure, touch, and stretch receptors are **mechanoreceptors** (mek-ah-nō-re-cep′tors), which are sensitive to mechanical stimuli displacing the tissue in which they are located. *Lamellar (Pacinian) corpuscles* are rapidly adapting receptors used to detect deep pressure and stretch. They are located in the innermost part of the dermis, as well as in the ligaments and tendons associated with joints (figure 9.1).

There are several types of receptors that function in the skin as touch receptors (figure 9.1). Free nerve endings extend outward from the dermis into the spaces between the epidermal cells. These endings function primarily as pain receptors but also serve to detect touch, itch, and temperature. The free nerve endings surrounding a hair follicle function to detect hair displacement, such as when a bug lands on the forearm. *Tactile (Meissner) corpuscles* in the outer dermis are most abundant in hairless areas such as fingertips, palms, and lips. These rapidly adapting receptors are useful in detecting the onset of light touch to the skin. Some nerve endings in the outer dermis are associated with *tactile epithelial cells* in the stratum basale of the epidermis in areas such as the fingertips, hands, lips, and external genitalia. Together these slowly adapting structures function in detecting light touch and pressure, such as when reading Braille.

Baroreceptors are free nerve endings that monitor stretching within distensible internal organs such as blood vessels, the stomach, and the bladder. Signals from these receptors are used to help regulate visceral reflexes such as those used to regulate blood pressure, digestion, and urination. For example, baroreceptors within the urinary

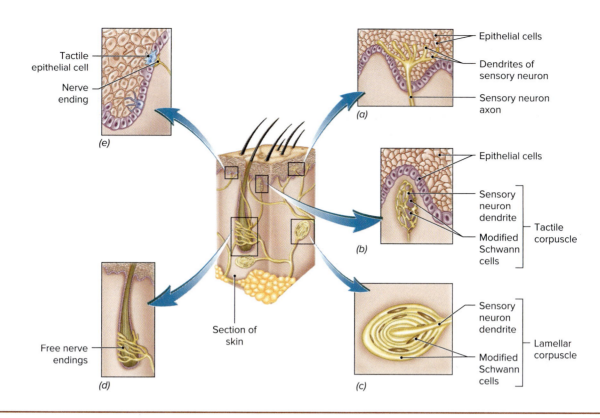

Figure 9.1 Touch and Pressure Receptors.
(a) Free nerve endings between epidermal cells detect touch, itch, pain, and temperature sensations. (b) Tactile corpuscles are light touch receptors located in the superifical dermis. (c) Lamellar corpuscles are pressure receptors located deep in the dermis, in addition to certain ligaments and tendons. (d) Free nerve endings around hair follicle detects movement of the hair shaft; (e) Tactile epithelial cell in stratum basale of the epidermis and a nerve ending in the adjacent dermis detect light touch and pressure.

bladder will trigger the urination reflex as the bladder fills and stretches. These receptors do not exhibit sensory adaptation owing to their role in regulating visceral reflexes.

Proprioceptors, such as *muscle spindles* and *tendon organs,* are used to monitor changes in skeletal muscle stretching and tendon tension during skeletal muscle contraction and relaxation (figure 9.2). These receptors keep us informed about the positioning of our body or body parts while stationary or moving. These receptors do not exhibit sensory adaptation owing to their role in maintaining posture, equilibrium, and muscle tone.

Chemoreceptors

The **chemoreceptors** that are part of the general senses are specialized neurons used to monitor body fluids for chemical changes. For example, chemoreceptors monitor changes in ion concentrations, pH, blood glucose level, and dissolved gases. The signals created by these chemoreceptors are not processed within the cerebral cortex; this means that the sensation created within the brain cannot be consciously detected.

Pain

Nociceptors, also referred to as pain receptors, are free nerve endings, which are widespread in body tissues, except within the nerve tissue of the brain. They are especially abundant in the skin, the organ that is in direct contact with the external environment. Nociceptors are stimulated whenever tissues are damaged, and the pain sensation initiates actions by the CNS to remove the source of the stimulation. Further, nociceptors do not easily adapt like many other sensory receptors. The lack of

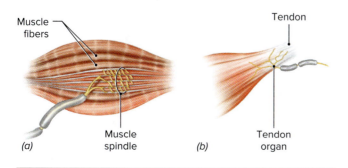

Figure 9.2 Proprioceptors.
(a) Muscle spindle. *(b)* Tendon organ.

adaptation is a protective mechanism that allows the person to be aware of a harmful stimulus until it is removed.

Referred Pain

Projection by the cerebral cortex is not always accurate when the action potentials originate from nociceptors in visceral organs. When damage to visceral organs occurs, pain sensations are often projected or referred to an undamaged part of the body wall or limb. This type of pain is called **referred pain.**

 Clinical Insight

Pain management in the United States costs billions of dollars each year. *Analgesia* (pain reduction) and *anesthesia* (complete loss of sensation) control pain by decreasing nociceptor sensitivity, blocking action potential formation, preventing action potential transmission to the CNS, or interfering with pain perception within the brain.

Referred pain is consistent from person to person and is important in the diagnosis of many disorders. For example, pain caused by a heart attack is referred to the left chest wall, left shoulder, and left upper limb in both genders, while women also commonly experience pain in the abdomen and jaw and between the scapulae. Referred pain is due to communication between neurons within the same nerve that are carrying action potentials from both visceral organs and the body wall or a limb. For example, neurons carrying action potentials from the heart use the same nerves as those from the left shoulder and upper limb (figure 9.3).

 Check My Understanding

5. What sensory receptors are involved in the general senses?
6. What are the roles of these sensory receptors in monitoring the external and internal environments?

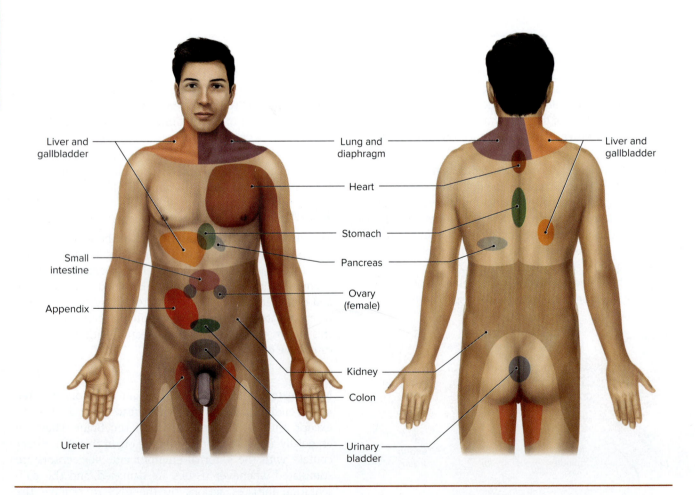

Figure 9.3 Referred Pain.
Surface regions to which visceral pain originating from various internal organs may be referred.

9.4 Special Senses

Learning Objectives

8. Contrast the location, structure, and function of olfactory and taste receptors.
9. Recall the location, structure, and function of the sensory receptors involved in hearing.
10. Distinguish the location, structure, and function of the sensory receptors involved in static equilibrium and dynamic equilibrium.
11. Identify the structures of the eye and the functions of these structures.
12. Describe the location, structure, and function of the sensory receptors involved in vision.

The sensory receptors for special senses are localized rather than widely distributed, and they, like all sensory receptors, are specialized to respond to only certain types of stimuli.

There are three different kinds of sensory receptors for the special senses. Taste and olfactory receptors are chemoreceptors, which are sensitive to chemical substances. Sensory receptors for hearing and equilibrium are mechanoreceptors, which are sensitive to vibrations formed by sound waves and movement of the head. Sensory receptors for vision are **photoreceptors,** which are sensitive to light energy.

Taste

The chemoreceptors for taste are located in specialized microscopic organs called **taste buds.** Most taste buds are located on the tongue in small, raised structures called *lingual papillae* (figure 9.4), though some can be found in areas such as the soft palate, pharynx, and esophagus.

A taste bud consists of a bulblike arrangement of rapidly adapting **taste receptors,** called *gustatory epithelial cells,* located within the epithelium of the lingual papillae. The taste bud possesses an opening called a *taste pore.* Taste

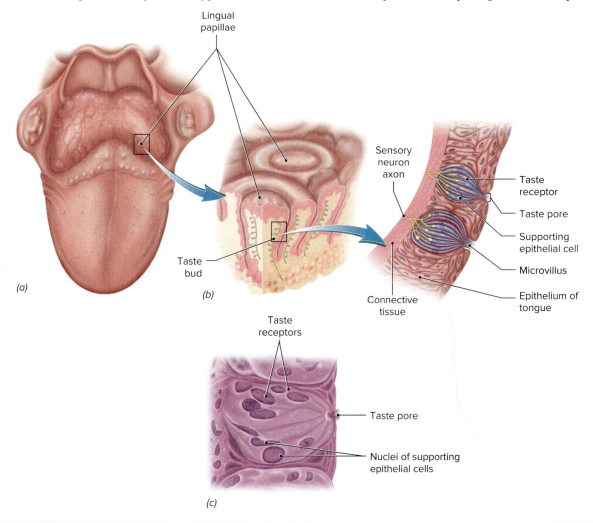

Figure 9.4 Taste Buds.
(a) Taste buds are located on lingual papillae of the tongue. *(b)* A taste bud contains taste receptors whose microvilli protrude through the taste pore. *(c)* Photomicrograph of a taste bud.

receptors have microvilli that extend through the pore and are exposed to chemicals on the tongue. Sensory axons leading to the brain are connected to the opposite end of the taste receptors. In order to activate the taste receptors, a substance must be dissolved in a liquid such as saliva.

There are five confirmed basic tastes that can be detected by the tongue: sweet, sour, salty, bitter, and umami (savory). The receptors for each basic taste are located across the tongue surface, which disproves the earlier belief that the basic tastes were mapped to specific regions of the tongue. It is probable that other substances, such as fats and Ca^{2+}, will be added as basic tastes in the near future as a result of ongoing taste research. It has been suggested that water is also a basic taste; however, not enough experimental data has been produced to support this claim. The many flavor sensations of food result from the stimulation of one or more taste receptors and, more importantly, the activation of olfactory receptors discussed in the next section.

The pathway of action potentials from taste receptors to the brain depends on where the taste receptors are located. Action potentials created by taste receptors on the front two-thirds of the tongue are carried by the facial nerve (CN VII), while those created on the back one-third travel over the glossopharyngeal nerve (CN IX). Action potentials created at the base of the tongue are carried by the vagus nerve (CN X). These cranial nerves carry the action potentials to the medulla oblongata, from which the action potentials travel to the thalamus and on to the taste areas in the parietal lobes of the cerebrum. AP|R

Smell

The **olfactory** (ōl-fak′-tō-rē) **receptors** are located in the upper portion of the nasal cavity, including the superior nasal conchae and nasal septum. The olfactory receptors, also called *olfactory sensory neurons*, are surrounded by the supporting epithelial cells of the olfactory epithelium. The outer ends of the olfactory receptors are covered with cilia that project into the nasal cavity, where they can contact airborne molecules. Chemicals in inhaled air are in a gaseous state and must dissolve in the mucus layer covering the olfactory epithelium in order to stimulate action potential formation (figure 9.5). The action potentials are carried by axons of the olfactory receptors, which form the olfactory nerves (CN I), to the olfactory bulbs. Here they synapse with neurons that form the olfactory tract and relay the action potentials to the olfactory areas deep within the temporal lobes and at the bases of the frontal lobes of the cerebrum.

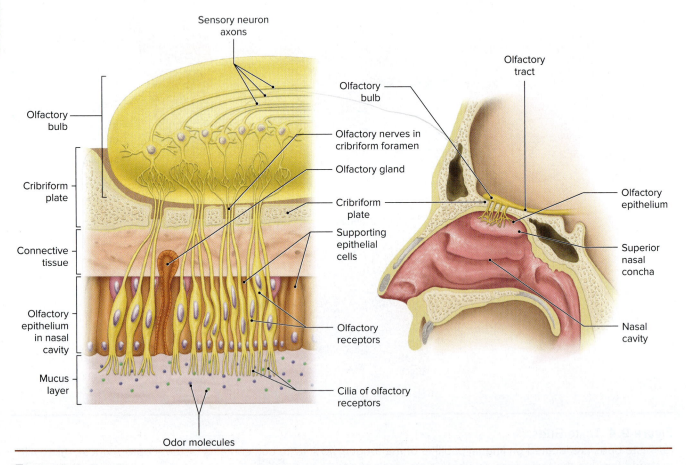

Figure 9.5 Smell.
Olfactory receptors are located between supporting epithelial cells in the upper portion of the nasal cavity. AP|R

It is common for a person to sniff the air when trying to detect faint odors. This is because the olfactory receptors are located above the usual path of inhaled air and additional force is needed to send larger amounts of air over the olfactory epithelium. Like taste receptors, olfactory receptors rapidly adapt to a particular stimulus.

The human olfactory epithelium possesses approximately 350 functional types of olfactory receptors. However, the average person can distinguish between 2,000 and 4,000 different odors. The ability to detect so many types of odors largely depends upon how the temporal lobes process the action potentials from various combinations of olfactory receptors. Studies have shown that women can detect, discern, and identify a wider range of odors than men. It is also possible with training to enhance your olfactory ability and potentially discern up to 10,000 different odors, an ability important for those in the wine industry. The decrease in odor 2 detection that occurs with age, which is why the elderly tend to use more cologne and perfume, is a result of receptor loss and desensitization rather than temporal lobe dysfunction. Research suggests that the olfactory epithelium is capable of detecting human pheromones. Human pheromones, which have been found in apocrine sweat and vaginal secretions, have been shown to have influence over reproductive functions. For example, pheromones from one female have been shown to lengthen or shorten the menstrual cycle of exposed females. The olfactory epithelium is also highly regenerative, owing to its direct exposure to the external environment. On average, an olfactory receptor lives only approximately 60 days before being replaced.

Check My Understanding

7. Where are taste and olfactory receptors located?
8. How are the five basic taste sensations produced?

Clinical Insight

The ability to distinguish various foods relies predominantly on the sense of smell. This explains why foods seem to have little taste for a person who is suffering from a head cold. The taste and smell of appetizing foods prepare the digestive tract for digestion by stimulating the flow of saliva in the mouth and gastric juice in the stomach.

Hearing

The *ear* is the organ of hearing. It is also the organ of equilibrium. The ear is subdivided into three major parts: the external ear, middle ear, and internal ear (figure 9.6). Table 9.2 summarizes the structures of the ear and their functions.

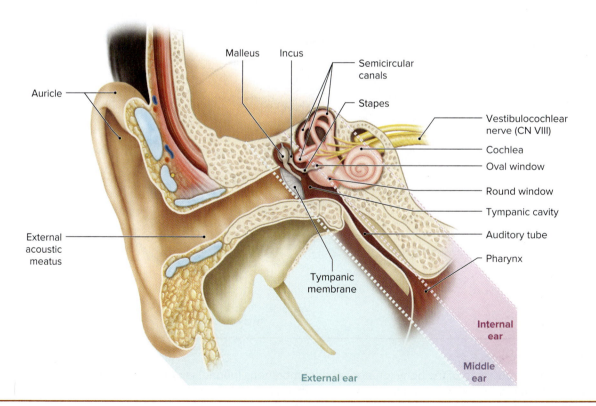

Figure 9.6 Anatomy of the Ear. AP|R

External Ear

The external ear consists of two parts: the auricle and the external acoustic meatus. The *auricle* (pinna) is the funnellike structure composed primarily of cartilage tissue and skin that is attached to the side of the head. The **external acoustic meatus** is a short tube that extends from the auricle through the temporal bone to the eardrum. Sound waves striking the auricle are channeled into the external acoustic meatus. **Cerumen** (earwax) and hairs in the external acoustic meatus help to prevent foreign particles from reaching the eardrum.

Middle Ear [AP|R]

The middle ear, or **tympanic** (tim-pan'-ik) **cavity,** is an air-filled space within the temporal bone. The tympanic membrane, auditory tube, and auditory ossicles are parts of the middle ear. The **tympanic membrane,** or eardrum, separates the tympanic cavity from the external acoustic meatus. The outer surface of the tympanic membrane is covered by skin, while the inner surface is covered by a mucous membrane. Sound waves, or air pressure waves, entering the external acoustic meatus cause the tympanic membrane to vibrate in and out at the same frequency as the sound waves.

The **auditory** (eustachian) **tube** connects the tympanic cavity with the pharynx. Its function is to keep the air pressure within the tympanic cavity the same as the external air pressure by allowing air to enter or exit the tympanic cavity. Equal air pressure on each side of the tympanic membrane is essential for the tympanic membrane to function properly. A valve at the pharyngeal end of the tube is usually closed, but it opens when a person swallows or yawns to allow air pressure to equalize. If you have experienced a rapid change in air pressure, you probably have noticed your ears "popping" as the air pressure is equalized and the tympanic membrane snaps back into place.

The **auditory ossicles** (os'-si-kulz) are three tiny bones that articulate to form a lever system from the tympanic membrane, across the tympanic cavity, to the internal ear. Each ossicle is named for its shape. The tip of the "handle" of the clubshaped *malleus* (mal'-ē-us), or hammer, is attached to the tympanic membrane, and its head articulates with the *incus* (ing'-kus), or anvil. The base of the incus articulates with the *stapes* (stā'-pēz), or stirrup, whose foot plate is inserted into the oval window of the internal ear.

The vibrations of the tympanic membrane cause corresponding movements of the ossicles, which result in the stapes vibrating in the oval window. In this way, vibrations of the tympanic membrane are transmitted to the fluid-filled internal ear. Due to the size difference between the larger tympanic membrane and the smaller oval window, vibrations are amplified by the ossicles.

Internal Ear

The internal ear is embedded in the temporal bone. It consists of two series of connecting tubes and chambers, one within the other: an outer *bony labyrinth* (lab'-i-rinth) and an inner *membranous labyrinth*. The two **labyrinths** are similar in shape (figure 9.7). The space between the bony and membranous labyrinths is filled with **perilymph,**

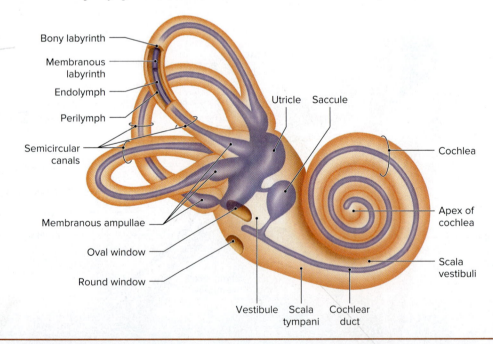

Figure 9.7 The Internal Ear.
The internal ear is composed of the bony (orange) and membranous (purple) labyrinths. Perilymph fills the space between the membranous labyrinth and the bony labyrinth. Endolymph fills the membranous labyrinth. Note that the ampullae of the semicircular canals, utricle, saccule, and cochlear duct are portions of the membranous labyrinth.

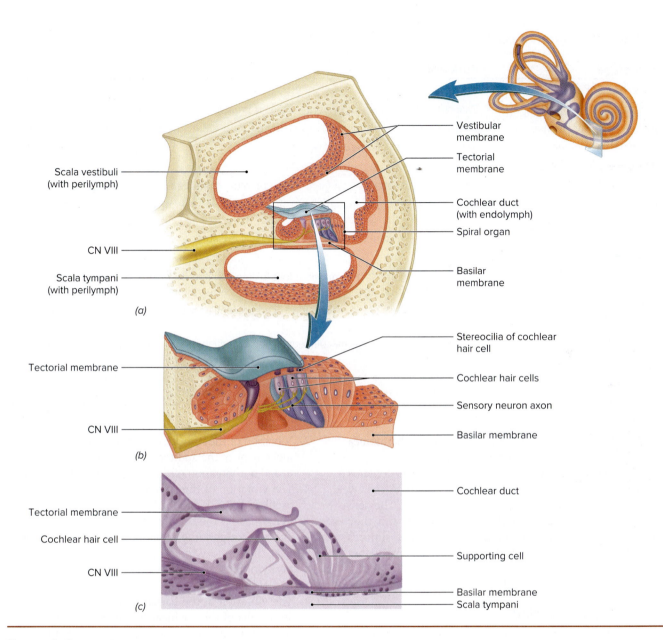

Figure 9.8 The Cochlea.
(a) A cross section of the cochlea shows the cochlear duct located between the scala vestibuli and the scala tympani and the spiral organ resting on the basilar membrane. *(b)* Detail of the spiral organ shows the tectorial membrane overlying the cochlear hair cells. *(c)* Photomicrograph of spiral organ. AP|R

whereas the membranous labyrinth contains **endolymph.** These fluids play important roles in the functions of the internal ear. The internal ear has three major parts: the cochlea, vestibule, and semicircular canals.

The **cochlea** (kok′-lē-ah) is the coiled portion of the internal ear. When viewed in cross section, as in figure 9.8, it can be seen that the cochlea is composed of three chambers that are separated from each other by membranes. The *scala vestibuli* (skā-la ves-tib′-ū-lī) and the *scala tympani,* both components of the bony labyrinth, extend the length of the cochlea and are continuous with each other at the apex of the cochlea. The scala vestibuli continues into the vestibule, which houses the membrane-covered *oval window*. The scala tympani extends toward the vestibule, ending at the membrane-covered *round window*.

The *cochlear duct*, which is part of the membranous labyrinth, extends nearly to the apex of the cochlea (see figure 9.7). As shown in figure 9.8, it is separated from the scala vestibuli by the *vestibular membrane* and from the scala tympani by the **basilar membrane.** The basilar membrane contains about 20,000 cross fibers that gradually increase in length from the base to the apex of the cochlea. The attachment of the basilar membrane to the bony center of the

Table 9.2 Summary of Ear Function

Structure	Function
External Ear	
Auricle	Channels sound waves into external acoustic meatus
External acoustic meatus	Directs sound waves to tympanic membrane
Tympanic membrane	Vibrates when struck by sound waves
Middle Ear	
Tympanic cavity	Air-filled space that allows tympanic membrane to vibrate freely when struck by sound waves
Auditory ossicles	Transmit and amplify vibrations produced by sound waves from the tympanic membrane to the perilymph within the cochlea
Auditory tube	Equalizes air pressure on each side of tympanic membrane
Internal Ear	
Cochlea	Fluids and membranes transmit vibrations initiated by sound waves to the spiral organ, whose cochlear hair cells generate action potentials associated with hearing
Saccule	Vestibular hair cells of the macula form action potentials associated with static and dynamic equilibrium
Utricle	Vestibular hair cells of the macula form action potentials associated with static and dynamic equilibrium
Semicircular canals	Vestibular hair cells of the cristae ampullares form action potentials associated with rotational movement of the head

cochlea allows it to vibrate like the reeds of a harmonica when activated by vibrations generated by sound.

The **spiral organ** (organ of Corti), which contains the sensory receptors for sound stimuli, is supported by the basilar membrane within the cochlear duct. The sensory receptors are called **cochlear hair cells,** and they have hairlike stereocilia extending from their free surfaces toward the overlying *tectorial* (tek-to′-rē-al) *membrane.* Axons of the vestibulocochlear nerve (CN VIII) exit the cochlear hair cells and extend to the brain.

Physiology of Hearing

The human ear is able to detect sound waves with frequencies ranging from near 20 to 20,000 Hertz (Hz; vibrations per second), but hearing is most acute between 2,000 and 3,000 Hz. For hearing to occur, vibrations formed by sound waves must be transmitted to the cochlear hair cells of the spiral organ. Then, the cochlear hair cells form action potentials that are transmitted to the hearing areas of the cerebrum for interpretation as sound sensations.

Figure 9.9 shows the structure of the internal ear with the cochlea uncoiled to show more clearly the relationships of its parts. Refer to this figure as you study the following outline of hearing physiology.

1. Sound waves enter the external acoustic meatus and strike the tympanic membrane, causing it to vibrate in and out at the same frequency and comparable intensity to the sound waves. Loud sounds cause a greater displacement of the tympanic membrane than do soft sounds.
2. Vibration of the tympanic membrane causes movement of the auditory ossicles, resulting in the in-and-out vibration of the stapes in the oval window.
3. The vibration of the stapes causes a corresponding oscillatory (back-and-forth) movement of the perilymph in the scala vestibuli and scala tympani and a corresponding movement of the membrane over the round window. This movement of the perilymph causes vibrations in the vestibular and basilar membranes.
4. The vibration of the basilar membrane causes the stereocilia of the cochlear hair cells to contact the tectorial membrane, which stimulates the formation of action potentials by the cochlear hair cells.
5. Action potentials formed by the cochlear hair cells are carried by CN VIII to the hearing areas of the temporal lobes of the cerebrum, where the sensation is interpreted. Some of the axons cross over to the opposite side of the brain so that the hearing areas in each temporal lobe interpret action potentials originating in each ear. AP|R

Pitch and Loudness

Because of the gradually increasing length of the fibers in the basilar membrane, different portions of the basilar membrane vibrate in accordance with the different frequencies (pitch) of sound waves. Low-pitched sounds cause the longer fibers of the membrane near the apex of the cochlea to vibrate, and high-pitched sounds activate the shorter fibers of the membrane near the base of the cochlea. The pitch

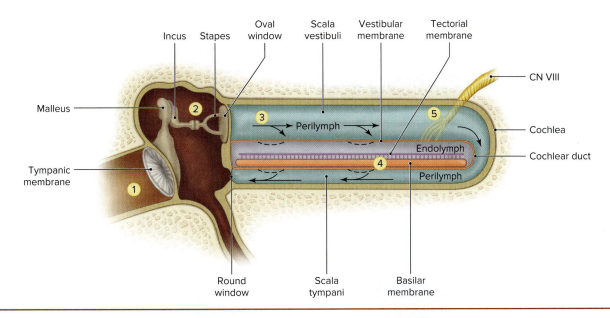

Figure 9.9 The Physiology of Hearing.
Sound waves entering the external acoustic meatus cause the tympanic membrane to vibrate. Vibrations of the tympanic membrane are carried by the auditory ossicles to the perilymph. Oscillating movements of the perilymph cause the vibration of the basilar membrane and spiral organ, which, in turn, results in the formation of action potentials by the cochlear hair cells. AP|R

of a sound sensation is determined by the portion of the basilar membrane and the spiral organ that are activated by the specific sound frequency and by the parts of the hearing areas that receive the action potentials. Action potentials from different regions of the spiral organ go to slightly different portions of the hearing areas in the brain, which causes them to be interpreted as different pitches.

The loudness of the sound is dependent upon the intensity of the vibration of the basilar membrane and spiral organ, which, in turn, determines the frequency of action potential formation. The greater the frequency of action potentials sent to the brain, the louder the sound sensation.

 Check My Understanding

9. How do sound waves stimulate the formation of action potentials?
10. How are pitch and loudness of a sound determined?

Equilibrium

Several types of sensory receptors provide information to the brain for the maintenance of equilibrium. The eyes and proprioceptors in joints, tendons, and muscles are important in informing the brain about equilibrium and the position and movement of body parts. However, unique receptors in the internal ear are crucial in monitoring two types of equilibrium. **Static equilibrium** involves the movement of the head with respect to gravitational force. **Dynamic equilibrium** involves linear acceleration in both horizontal and vertical directions, in addition to the rotational movement of the head.

Static Equilibrium

The **macula** (mak′-ū-lah; plural, *maculae*), an organ of static equilibrium, is located within the **utricle** (ū′-tri-kul) and the **saccule** (sak′-ūl), enlarged portions of the membranous labyrinth within the vestibule (see figure 9.7). Each macula contains thousands of sensory receptors called *vestibular hair cells* that possess hairlike stereocilia embedded in a gelatinous layer. *Otoliths* (ō′-tō-liths), crystals of calcium carbonate, are also embedded on the outer surface of the gelatinous layer. The otoliths increase the weight of the gelatinous layer and make it more responsive to the pull of gravity (figure 9.10).

The mechanism of static equilibrium may be summarized as follows:

1. Changes in head position cause gravity to pull on the gelatinous layer, which bends the stereocilia. This change stimulates the vestibular hair cells to form action potentials that are carried by CN VIII to the brain. No matter the position of the head, action potentials are formed that inform the brain of the head's position.
2. The cerebellum uses this information to maintain static equilibrium subconsciously.
3. Our awareness of static equilibrium results when the action potentials are interpreted by the cerebrum.

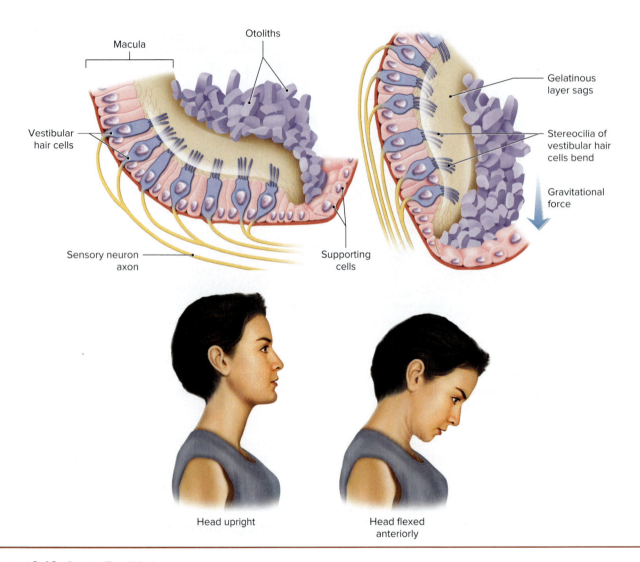

Figure 9.10 Static Equilibrium.
A macula is a sensory receptor for static equilibrium. Note how flexion of the head causes bending of the stereocilia of vestibular hair cells.

Dynamic Equilibrium

The maculae in the utricle and saccule also sense linear acceleration in both the horizontal and vertical directions. The mechanism is similar to that used to detect changes in static equilibrium. When the head accelerates either vertically or horizontally, inertia of the gelatinous layer causes the stereocilia of the vestibular hair cells to bend. When motion ends, the gelatinous layer continues to move for a moment, which bends the stereocilia in the opposite direction. The changes in stereocilia movement alert the brain to changes in velocity.

The membranous labyrinths of the **semicircular canals** contain the sensory receptors that detect rotational motion of the head. Examine figure 9.7 and note that the semicircular canals are arranged at right angles to each other so that each occupies a different plane in space, roughly equal to the frontal, paramedian, and transverse planes.

Near the attachment of each membranous canal to the utricle is an enlarged region called the **ampulla** (am-pūl′-lah). Each ampulla contains a sensory organ for dynamic equilibrium called the **crista ampullaris** (kris′-ta am-pūl-lar′-is; plural, *cristae ampullares*). Each crista ampullaris contains a number of vestibular hair cells, whose stereocilia extend into a dome-shaped gelatinous mass called the *ampullary cupula*. Axons of CN VIII lead from the vestibular hair cells to the brain (figure 9.11).

The mechanism for detecting rotational movement of the head may be described as follows:

1. When the head is turned, the endolymph pushes against the ampullary cupula, bending the stereocilia of vestibular hair cells, which stimulates the formation of action potentials. The action potentials are carried to the brain via CN VIII.

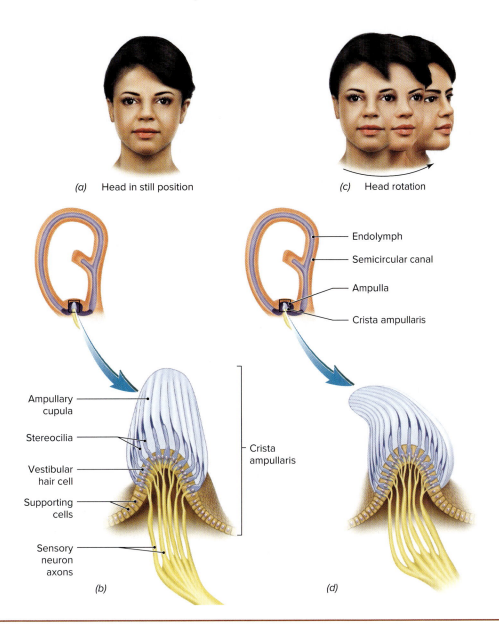

Figure 9.11 Dynamic Equilibrium and the Crista Ampullaris.
(a) When the head is upright and stationary, (b) the crista ampullaris is upright. (c) When the head is rotated, (d) the endolymph bends the cupula in the opposite direction, stimulating the vestibular hair cells to form action potentials.

2. Because each semicircular canal is oriented in a different plane, the vestibular hair cells of the cristae are not stimulated equally with a given head movement. Thus, the brain receives a different pattern of action potentials for each type of head movement.
3. The cerebellum uses the action potentials to make adjustments below the conscious level to maintain dynamic equilibrium.
4. Awareness of rotational movement, or lack of it, results from the cerebrum interpreting the pattern of action potentials it receives (figure 9.11).

Check My Understanding

11. What structures are involved in static and dynamic equilibrium?
12. What are the mechanisms of static and dynamic equilibrium?

Vision

Vision is one of the most important senses supplying information to the brain. The sensory receptors for light stimuli are located within the *eyes* (or *eyeballs*), the organs of vision. The eyes are located within the *orbits*, where they are protected by seven skull bones (see chapter 6). Connective tissues provide support and protective cushioning for the eyes.

Eyelids, Eyelashes, and Eyebrows

The exposed front of the eye is protected by the *eyelids*. Blinking spreads tears and mucus over the front of the eye to keep it moist. The inner surface of each eyelid is lined by a mucous membrane called the **conjunctiva** (kon-junk-tī'-vah), which continues across the front of the eye. Only its transparent outer epithelium covers the cornea. Mucus from the conjunctiva helps to lubricate the eye and keep it moist. The conjunctiva also contains many blood vessels and nociceptors.

Eyelashes help to keep airborne particles from reaching the eye surface and provide some protection from excessive light. *Eyebrows*, located on the brow ridges, also shield the eyes from overhead light and divert sweat from the eyes. Observe the accessory structures in figure 9.12.

Lacrimal Apparatus

The **lacrimal** (lak'-ri-mal) **apparatus,** shown in figure 9.13, is involved in the production and removal of tears. Tears are secreted continuously by the **lacrimal gland,** which is located in the upper, outer part of each orbit. Tears are carried to the surface of the eye by a series of tiny *excretory ducts*. The tears flow downward and inward across the eye surface as they are spread by blinking. Once collected at the inner corner of the eye by the *lacrimal canaliculi*, tears flow into the *lacrimal sac* and flow on through the *nasolacrimal duct* into the nasal cavity.

Tears perform an important function in keeping the anterior surface of the eye moist and in washing away foreign particles. An antibacterial enzyme (lysozyme) in tears helps to reduce the chance of eye infections.

Extrinsic Muscles

Movement of the eyes must be precise and in unison to enable good vision. Each eye is moved by six **extrinsic muscles of the eyeball** that originate from the posterior of the orbit and insert on the surface of the eye. Four muscles exert a direct pull on the eye, but two muscles pass through cartilaginous loops, enabling them to exert an oblique pull on the eyeball. Although each muscle has its own action, these muscles function as a coordinated group to enable eye movements. The locations and functions of these muscles are shown in figure 9.14. The extrinsic muscles are innervated by cranial nerves. The abducens nerve (CN VI) innervates the lateral rectus. The trochlear nerve (CN IV) innervates the superior oblique. The remaining extrinsic muscles are innervated by the oculomotor nerve (CN III).

Check My Understanding

13. What are the functions of the accessory structures of the eye?
14. Why does crying lead to a runny nose?

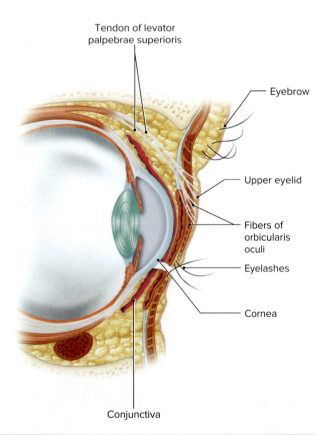

Figure 9.12 Sagittal Section of the Orbit.

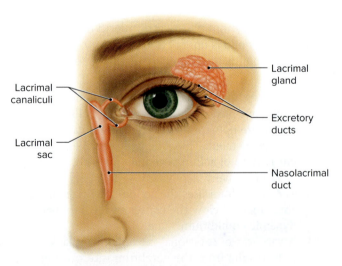

Figure 9.13 The Lacrimal Apparatus.

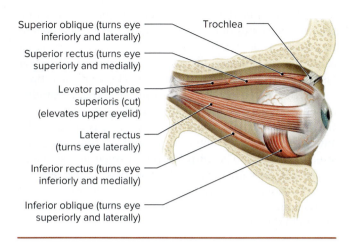

Figure 9.14 Extrinsic Eye Muscles.
The medial rectus, which is not shown in the image, turns the eye medially. AP|R

Structure of the Eye

The eye is a hollow, spherical organ about 2.5 cm (1 in) in diameter. It has a wall composed of three layers and internal spaces filled with fluids that support the walls and maintain the shape of the eye. The major parts of the eye are shown in figure 9.15.

Fibrous Layer The fibrous layer of the eye is the outermost layer and consists of two parts: the sclera and the cornea. The **sclera** (skle′-rah) is the opaque, white portion of the eye that forms most of the fibrous layer. The sclera is a tough, fibrous structure that provides protection for the delicate inner portions of the eye and for the optic nerve (CN II), which emerges from the back of the eye. The anterior portion of the sclera is covered by the conjunctiva. The **cornea** (kor′-nē-ah) is the anterior clear window of the eye. It has a greater convex curvature than the rest of the eyeball so that it can bend light rays as they pass through it. It lacks blood vessels and nerves that would block light rays from entering the eye.

Vascular Layer The vascular layer is the middle layer and includes the choroid, ciliary body, and iris. The **choroid** (kō′-royd), which is found in all but the anteriormost portion of the layer, contains both blood vessels that nourish the eye and large amounts of melanin. The absorption of light rays by melanin prevents back-scattering of

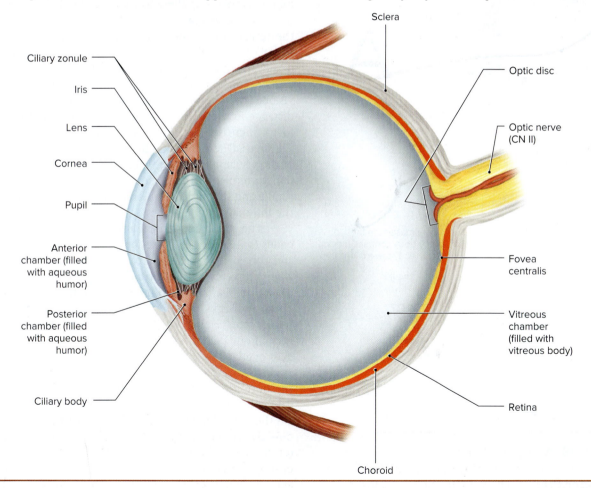

Figure 9.15 Transverse Section of the Left Eye. AP|R

light rays, which would impair vision. The anterior **ciliary** (sil′-ē-ar-ē) **body** contains the ciliary muscles and forms a ring around the **lens** just in front of the choroid. The *ciliary zonule* contains fibrous strands that extend from the ciliary body to the lens and hold the lens in place. Contraction and relaxation of ciliary muscles change the shape of the lens.

Although entering light rays are bent by the cornea, it is the lens that focuses light rays precisely on the retina. The transparent, somewhat elastic lens is composed of protein fibers and lacks blood vessels and nerves that would block the passage of light rays. Contraction and relaxation of the ciliary muscles change the shape of the lens during **accommodation**, the process of focusing light rays on the retina using the lens (figure 9.16). Contraction of the ciliary muscles relaxes the fibrous strands in the ciliary zonule and allows the lens to become more spherical in shape.

The relaxation of the ciliary muscles increases tension on the fibrous strands of the ciliary zonule and causes the lens to take on a more flattened shape. In this way, the shape of the lens is adjusted for distant, intermediate, and near vision so that the image is focused precisely on the retina.

The colored portion of the eye is the **iris**, a thin disc of connective tissue and smooth muscle that extends from the ciliary body in front of the lens. The iris controls the amount of light rays entering the eye by controlling the size of the pupil. The **pupil** is the opening in the center of the iris through which light rays pass to the lens. Its size is constantly adjusted by the iris as lighting conditions change. The pupil is constricted in bright light and is dilated in dim light.

Inner Layer The filmlike **retina** (ret′-i-nah) lines the inner surface of the eye behind the ciliary body. The retina contains two types of photoreceptor cells: rods and cones. Rod and cone anatomy can be seen in figure 9.17. The thin, elongate **rods** are photoreceptors for black and white vision because they are sensitive only to the presence of light rays. The shorter and thicker **cones** are photoreceptors for color vision. Because cones require bright light to function, only rods allow us to see in dim light.

The **macula** (mak′-u-lah) is a yellowish disc on the retina directly behind the lens. In the center of the macula is a small depression called the **fovea centralis** (fō′-vē-ah sen-trah′-lis). The fovea centralis contains densely packed cones, making it the area for the sharpest color vision. The density of the cones decreases with increased distance from the fovea. Rods, which are absent from the fovea, increase in density with increased distance from the fovea. Therefore, dim-light vision is best at the edge of the visual field (figure 9.18; see figure 9.15). It is important to remember that the macula on the retina is structurally and functionally different from the maculae within the internal ear.

The retina contains neurons in addition to rods and cones (see figure 9.17). Action potentials formed by rods and cones are transmitted to *retinal ganglion cells*, whose axons converge at the **optic disc** to form the optic nerve. The optic disc is located medial to the fovea. Because the optic disc lacks photoreceptors, it is also known as the "blind spot." However, we usually do not notice a blind spot in our field of vision because the visual fields of our eyes overlap.

An artery enters the eye and a vein exits the eye via the optic disc. These blood vessels are continuous with capillaries that nourish the inner tissues of the eye and are the only blood vessels in the body that can be viewed directly. A special instrument called an *ophthalmoscope* (of-thal′-mō-skōp) is used to look through the lens and observe these vessels. Figure 9.18 shows the appearance of blood vessels and the retina as viewed with an ophthalmoscope.

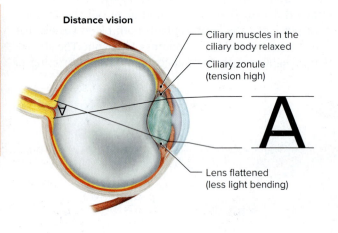

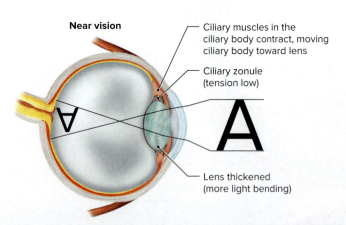

Figure 9.16 Accommodation.
Distance vision and near vision requires the focusing of light rays on the retina by the lens. Note how the lens changes shape in accommodation.

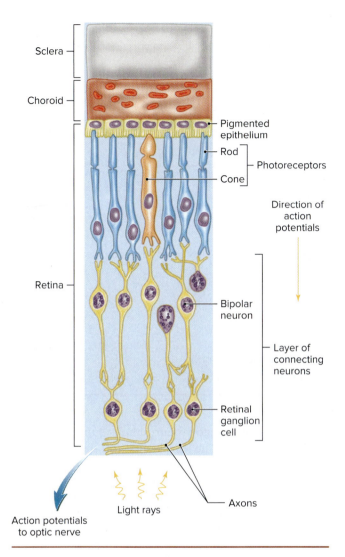

Figure 9.17 The Retina.
The retina consists of several cell layers. AP|R

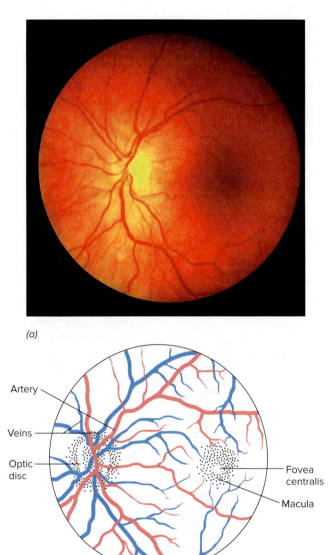

(a)

(b)

Figure 9.18 Ophthalmoscopy.
(a) A photo of the retina and *(b)* a diagram of the retina showing the optic disc and fovea centralis. Blood vessels enter and exit the eye at the optic disc. Axons exit the eye at the optic disc to form the optic nerve. The fovea centralis contains densely packed cones for direct color vision. AP|R

(a) ©Steve Allen/Getty Images RF

Check My Understanding

15. How are the components of the three layers of the eye involved in vision?

Chambers of the Eyeball The space between the cornea and the iris is known as the **anterior chamber,** which is filled with a watery fluid called **aqueous** (ā′-kwē-us) **humor.** The small **posterior chamber,** located between the iris and lens, is also filled with aqueous humor. The aqueous humor is filtered out of capillaries in the ciliary body, flows through the posterior chamber into the anterior chamber, and is reabsorbed into blood vessels located at the junction of the sclera and cornea. Aqueous humor is largely responsible for maintaining intraocular pressure and the healthy shape of the cornea. The aqueous humor also provides nourishment to the cornea and lens. Usually, it is secreted and absorbed at the same rate so that intraocular pressure is maintained at a constant level.

The large **vitreous chamber** is located behind the lens. It is filled with a clear, gellike substance called the **vitreous** (vit′-rē-us) **body.** The vitreous body, which forms during embryonic development, is not reabsorbed or regenerated. The vitreous body presses the retina firmly against the wall of the eye and helps to maintain the shape of the eye.

Table 9.3 summarizes the functions of eye structures.

Table 9.3 Functions of Eye Structures

Structure	Function
External Layer	
Sclera	Provides protection and shape for eye
Cornea	Allows entrance of light and bends light rays
Middle Layer	
Choroid	Contains both blood vessels that nourish deep structures and melanin that absorbs excessive light rays
Ciliary body	Supports and changes shape of lens in accommodation; secretes aqueous humor
Iris	Regulates amount of light entering eye by controlling the size of the pupil
Internal Layer	
Retina	Contains photoreceptors that convert light rays into action potentials; action potentials transmitted to brain via optic nerve (CN II)
Other Structures	
Lens	Bends light rays and focuses them on the retina
Anterior chamber	Contains the aqueous humor that controls intraocular pressure, maintains the shape of the cornea, provides nourishment to cornea and lens
Posterior chamber	Receives the aqueous humor produced by the ciliary body
Vitreous chamber	Contains the vitreous body that maintains shape of the eye and holds retina against choroid

Clinical Insight

Glaucoma results when the rate of absorption of aqueous humor is less than its rate of secretion. This causes a buildup of intraocular pressure that, without treatment, can compress and close the blood vessels nourishing the photoreceptors of the retina. If this occurs, the photoreceptors die and permanent blindness results.

Physiology of Vision

Light rays coming to the eye must be precisely bent so they are focused on the retina. This bending of the light rays is called *refraction* (rē-frak´-shun), and it is produced by the cornea and lens. The convex surface of the cornea produces the greatest refraction of light rays, while further bending (accommodation) by the lens provides a "fine adjustment" so that the image is focused precisely on the retina.

The optics of the eye cause the image to be inverted on the retina, as shown in figure 9.16. However, the visual areas of the cerebral cortex correct for this inversion so that objects are seen in their correct orientation. When images are incorrectly focused on the retina, poor vision results. Figure 9.19 shows common optical disorders and how they may be corrected with glasses, contact lenses, or Lasik surgery.

When light rays strike the retina, the light stimuli must be converted into action potentials that are sent to the brain. Both rods and cones contain light-sensitive pigments that break down into simpler substances when light is absorbed. The breakdown of these pigments results in the formation of action potentials.

Rods contain a light-sensitive pigment called **rhodopsin** that breaks down into *opsin*, a protein, and *retinal*, which is derived from vitamin A. This breakdown triggers the formation of action potentials that are carried via the optic nerve to the brain. Rhodopsin is resynthesized from opsin and retinal to prepare the rods for receiving subsequent stimuli. A deficiency of vitamin A may result in an insufficient amount of rhodopsin in the rods, which, in turn, may lead to *night blindness*, the inability to see in dim light.

Although the light-sensitive pigments are different in cones, they function in a similar way to rhodopsin. There are three different types of cones, and each has a pigment that responds best to a different color (wavelength) of light. One type responds best to red light, another type responds best to green light, and the third type responds to blue light. The perceived color of objects results from the combination of the cones that are stimulated and the interpretation of the action potentials that they form by the cerebral cortex.

Nerve Pathway Action potentials formed by the photoreceptors are transmitted via axons of the optic nerve to the brain. The optic nerves merge just in front of the pituitary gland to form an X-shaped pattern called the **optic chiasma** (kī-as´-mah) (figure 9.20). Within the optic chiasma, the axons from the medial half of the retina in each eye cross over to the opposite side. Thus, the medial axons of the left eye and the lateral axons of the right eye form the right optic tract leaving the optic chiasma.

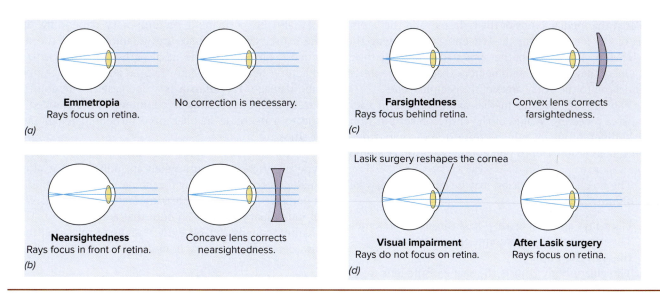

Figure 9.19 Common Optical Disorders.
Comparison of *(a)* emmetropia, or "normal" sight, *(b)* nearsightedness, *(c)* farsightedness, and *(d)* eyesight after Lasik surgery.

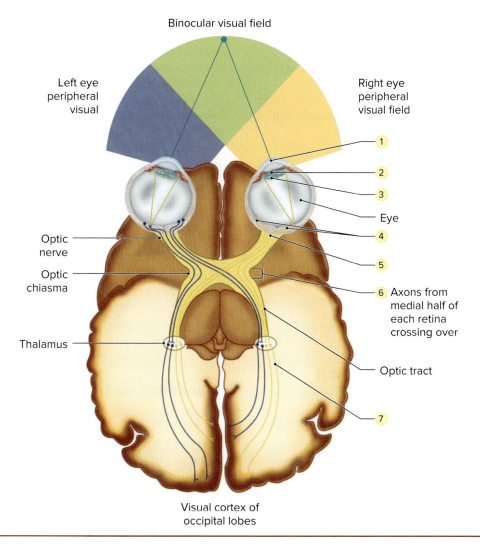

Figure 9.20 The Mechanism of Vision.

Similarly, the medial axons of the right eye and the lateral axons of the left eye form the left optic tract leaving the optic chiasma. The axons of the optic tracts enter the thalamus, where they synapse with neurons that carry the action potentials on to the visual areas of the occipital lobes.

The crossing of the medial axons results in each visual area receiving images of the entire object but from slightly different views, which create stereoscopic (three-dimensional) vision. Depth perception is a result of stereoscopic vision.

The mechanism of vision may be summarized as follows (figure 9.20):

1. Light rays are bent as they pass through the cornea.
2. The iris controls the amount of light passing through the pupil.
3. The ciliary body adjusts the shape of the lens to focus the light rays (image) on the retina.
4. Light absorbed by the rods and cones causes the formation of action potentials.
5. The action potentials are transmitted to neurons whose axons converge at the optic disc to form the optic nerve.
6. The medial axons of the optic nerves cross over at the optic chiasma and merge with the lateral axons on that side to form the optic tracts, which continue to the thalamus.
7. The action potentials are then carried to the vision areas in the occipital lobes of the cerebrum, where they are interpreted as visual images.

Check My Understanding

16. How are light rays converted into visual sensations?

9.5 Disorders of the Special Senses

Learning Objective

13. Describe the common disorders of taste, smell, hearing, and vision.

Disorders of Taste and Smell

Ageusia (uh-gyoo′-zee-uh) is a loss of taste function, meaning there is no perception of the five basic tastes, and is rare. **Hypogeusia,** or a reduced ability to taste, is more common and can be caused by zinc deficiency and chemotherapy. **Dysgeusia,** which is a distortion or impaired perception of taste, can be caused by taste bud distortion, pregnancy, diabetes, allergy medications like albuterol, zinc deficiency, and chemotherapy.

Anosmia is the inability to detect odor. The loss can be for one or more odors, up to all odors. It may also be permanent or temporary depending upon the cause. Typical causes are inflammation of the nasal mucosa, blockage of the nasal pathways, damage to the olfactory nerve, or head trauma leading to temporal lobe damage. **Hyposmia** is a decrease in the ability to detect odors. Hyposmia is common with advanced age due to a decrease in olfactory epithelium regeneration or smoking.

Dysosmia is a distorted sense of smell. *Parosmia*, a type of dysosmia, occurs when an individual has altered smell perception, meaning that something usually pleasant is perceived as being unpleasant. *Phantosmia* occurs when an individual perceives an odor that is not present. These phantom smells can be clinical signs of migraine, mood disorders, schizophrenia, or epilepsy.

Disorders of the Ear

Deafness is a partial or total loss of hearing. The cochlear hair cells of the spiral organ are easily damaged by high-intensity sounds, such as loud music and the noise of jet airplanes. Such damage produces a form of *nerve deafness* that may be partial or total, and it is permanent. Disorders of sound transmission by the tympanic membrane or auditory ossicles cause *conduction deafness*, which may be repairable by surgical means or overcome by the use of hearing aids.

Labyrinthine disease is a term applied to disorders of the internal ear that produce symptoms of dizziness, nausea, ringing in the ears (tinnitus), and hearing loss. It may be caused by an excess of endolymph, or by an infection, allergy, trauma, circulation disorders, or aging.

Motion sickness is a functional disorder that is characterized by nausea and is produced by repetitive stimulation of the equilibrium receptors in the internal ear.

Otitis media (ō-tī′-tis mē′-dē-ah) is an acute infection of the tympanic cavity. It may cause severe pain and an outward bulging of the tympanic membrane due to accumulated fluids. Pathogens enter the middle ear from the pharynx via the auditory tube or through a perforated tympanic membrane. Young children are especially susceptible because their auditory tubes are short and horizontal, which aids the spread of bacteria from the pharynx to the tympanic cavity. APR

Disorders of the Eye

Astigmatism (a-stig′-mah-tizm) is the unequal focusing of light rays on the retina, which causes part of an image to appear blurred. It results from an unequal curvature of the cornea or lens.

Blindness is partial loss or lack of vision. It may be caused by a number of disorders such as cataract, glaucoma, and detachment or deterioration of the retina. It may also result from damage to the optic nerves or the visual centers in the occipital lobes of the cerebrum.

⊕ Clinical Insight

A Caucasian male, age 63, presented for a routine eye exam with complaints of a blurry spot in his central vision and difficulty reading in dim light. Visual observation of his retinas with pupil dilation showed several medium-sized spots of drusen (yellow-white fatty protein deposits) in the macula of both eyes. The patient's tentative diagnosis was intermediate stage dry **age-related macular degeneration (AMD).** AMD manifests in two forms: *wet (neovascular) macular degeneration* and *dry (non-neovascular) macular degeneration.* Dry AMD in the early stages involves the formation of small drusen in the maculae and is usually asymptomatic. Drusen number and size increases during the intermediate stage, which often leads to blurry central vision and a need for brighter light to perform visual tasks (figure 9A). Wet AMD involves the growth of new vessels into the maculae that break easily and allow for fluid leakage. The resulting inflammation damages the maculae and causes loss of central vision. A fluorescein angiogram, using a fluorescent dye injected into the blood to photograph retinal blood vessels, detected no unhealthy vessel growth in the maculae. Because dry AMD can spontaneously become wet AMD, the man was instructed to monitor his condition daily with an *Amsler Grid.* The presence of wavy or blurred lines when viewed is an indication of the development of wet AMD. There is no cure currently for dry AMD. However, the man was placed on an AREDS Formula Eye Vitamin regime. The high levels of vitamin A, vitamin C, vitamin E, zinc (zinc oxide), and copper (cupric oxide) in the supplement have been shown to decrease the risk of developing advanced AMP (dry or wet) by as much as 25%.

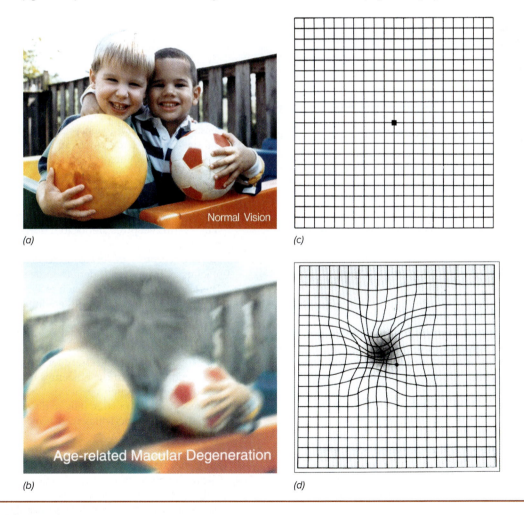

Figure 9A Macular Degeneration.
(a) Healthy vision. *(b)* Vision with macular degeneration. Appearance of Amsler grid with *(c)* healthy vision and *(d)* macular degeneration.
(a)–(b) Source: National Eye Institute, National Institutes of Health

A **cataract** is a cloudiness or opacity of the lens, which impairs or prevents vision. It is common in older people and is the leading cause of blindness. Surgical removal of the clouded lens and implantation of a plastic lens usually restores good vision.

Color blindness is the inability to perceive certain colors or, more rarely, all colors. Red-green color blindness, the most common type, is characterized by difficulty distinguishing reds and greens due to the absence of either red or green cones. Color blindness is inherited, and it occurs more often in males than in females because it is a sex-linked hereditary trait.

Conjunctivitis (con-junk-ti-vī′-tis) is inflammation of the conjunctiva. It may be caused by allergic reactions, physical or chemical causes, or infections. Inflammation that results from a bacterial or viral infection is commonly called *pink eye*. Such bacterial infections are highly contagious.

Farsightedness (hyperopia) is blurred vision caused by light rays being incorrectly focused behind the retina. Its causes include lens abnormalities and the eye being shorter than average.

Nearsightedness (myopia) is blurred vision caused by light rays being incorrectly focused in front of the retina. Its causes include lens abnormalities and the eye being longer than average.

Presbyopia (prez-bē-ō′-pē-ah) is the diminished ability of the lens to accommodate for near vision due to a decrease in its elasticity. It is a natural result of aging. At age 20, an object can be clearly observed about 10 cm (4 in) from the eye. At age 60, an object must be about 75 cm (30 in) from the eye to be clearly observed.

Retinoblastoma (ret-i-nō-blas-tō′-mah) is a cancer of immature retinal cells. It constitutes about 2% of the cancers in children.

Strabismus (strah-biz′-mus) is a disorder of the extrinsic eye muscles in which the eyes are not directed toward the same object simultaneously. Treatment may include eye exercises, corrective lenses, or corrective surgery.

Chapter Summary

9.1 Introduction to the Senses

- Sensory receptors are involved in sending action potentials to the CNS, which then initiates the appropriate response.
- The general senses include pain, touch, pressure, stretching, chemical changes, cold, and heat.
- The special senses are taste, smell, vision, hearing, and equilibrium.
- Each sense depends upon the actions of a sensory receptor, sensory neuron, and the brain.

9.2 Sensations

- Each type of sensory receptor is most sensitive to a particular type of stimulus.
- The five types of sensory receptors are thermoreceptors, mechanoreceptors, nociceptors, chemoreceptors, and photoreceptors.
- Sensations result from action potentials formed by sensory receptors that are carried to the brain for interpretation.
- Perception is the conscious awareness of sensation that is created by the cerebral cortex.
- A particular sensory area of the cerebral cortex interprets all action potentials that it receives as the same type of sensation.
- The brain projects the sensation back to the body region from which the action potentials seem to have originated.
- Sensory receptors for touch, pressure, warm, cold, taste, and smell adapt to repetitious stimulaton by decreasing the rate of action potential formation.

9.3 General Senses

- Thermoreceptors are located in the dermis of the skin. Warm receptors are located deeper than cold receptors.
- Pressure receptors are located in tendons and ligaments of joints and deep in the dermis of the skin.
- There are several types of touch receptors. Tactile corpuscles in the outer dermis are important for detecting the onset of light touch in hairless areas. Free nerve endings detect movement of a hair follicle. Nerve endings and tactile epithelial cells work together to detect light touch, which is necessary for reading Braille. Free nerve endings within the epidermis have a secondary role in detecting touch sensations.
- Baroreceptors detect stretching in distensible internal organs to regulate activities such as blood pressure and urination.
- Proprioceptors in skeletal muscles and tendons are involved in the maintenance of erect posture and muscle tone.
- Chemoreceptors are specialized neurons that monitor body fluids for chemical changes.
- Nociceptors are free nerve endings that detect painful stimuli. They are especially abundant within the skin and visceral organs and exhibit no adaptation as a protective mechanism.

- Referred pain is pain from visceral organs that is erroneously projected to the body wall or limbs. The pattern of referred pain is useful in diagnosing disorders of visceral organs.

9.4 Special Senses

Taste
- Taste receptors are located in taste buds, which are mostly located on lingual papillae of the tongue.
- There are five basic types of taste receptors: sour, sweet, salty, bitter, and umami.
- Chemicals must be in solution in order to stimulate the taste receptors.
- Action potentials from taste receptors are carried by the facial, glossopharyngeal, and vagus nerves to the brain.

Smell
- Olfactory receptors are located within the olfactory epithelium of the upper portion of the nasal cavity.
- Airborne molecules must be dissolved in the mucus layer covering the olfactory epithelium to stimulate the olfactory receptors.
- Action potentials are carried by the olfactory nerves to the olfactory bulbs and then by the olfactory tracts to the brain.

Hearing
- The ear is subdivided into the external, middle, and internal ear.
- The external ear consists of (a) the auricle, which directs sound waves into the external acoustic meatus; and (b) the external acoustic meatus, which channels sound waves to the tympanic membrane.
- The middle ear, or tympanic cavity, consists of (a) the tympanic membrane, which vibrates when struck by sound waves; (b) the auditory tube, which enables the equalization of air pressure on each side of the tympanic membrane; and (c) the auditory ossicles, which transmit vibrations of the tympanic membrane to the oval window of the internal ear.
- The internal ear is embedded in the temporal bone. It consists of (a) the membranous labyrinth, which is filled with endolymph, and lies within (b) the bony labyrinth, which is filled with perilymph. The major portions of the internal ear are the cochlea, vestibule, and semicircular canals.
- The cochlear duct, part of the membranous labyrinth, is bordered by the vestibular and basilar membranes. The spiral organ on the basilar membrane within the cochlear duct contains the cochlear hair cells, which are used to detect the vibrations produced by sound waves.
- Sound waves vibrate the tympanic membrane. These vibrations are transmitted by the auditory ossicles to the oval window of the internal ear, which then sets up oscillatory movements of the perilymph in the cochlea. The movements of the perilymph vibrate the basilar membrane and spiral organ, resulting in the formation of action potentials.
- Action potentials are carried to the brain by the vestibulocochlear nerve (CN VIII).

Equilibrium
- Sensory receptors in joints, muscles, eyes, and the internal ear send action potentials to the brain that are associated with equilibrium.
- The maculae within the saccule and utricle contain vestibular hair cells that are sensory receptors for static equilibrium and linear acceleration, which is a type of dynamic equilibrium.
- The cristae ampullares within the ampullae of the semicircular canals contain vestibular hair cells that are the sensory receptors for rotational movement of the head, which is a type of dynamic equilibrium.

Vision
- The eyes contain the sensory receptors for vision. Accessory organs include the extrinsic muscles of the eyeball, eyelids, eyelashes, eyebrows, and lacrimal apparatus.
- The wall of the eye is composed of three layers. The fibrous layer consists of (a) the sclera, which supports and protects inner structures; and (b) the cornea, which allows light to enter the eye. The vascular layer consists of (a) the darkly pigmented choroid, which contains both blood vessels to nourish internal structures and melanin to absorb excess light rays; (b) the ciliary body, which changes the shape of the lens to focus light rays on the retina; and (c) the iris, which controls the amount of light rays entering the eye via the pupil. The inner layer consists of the retina, which contains the photoreceptors.
- Fluids fill the chambers of the eye and give it shape. Aqueous humor fills the anterior and posterior chambers and is primarily responsible for maintaining intraocular pressure. The vitreous body fills the vitreous chamber and helps to hold the retina against the choroid.
- Light rays pass through the cornea, aqueous humor, lens, and vitreous body to reach the retina. Light rays are primarily refracted by the cornea, but the lens focuses them on the retina.
- The photoreceptors are rods and cones. Rods are adapted for dim light, black and white vision. Cones are adapted for bright light, color vision. There are three types of cones based on the color of light that they primarily absorb: red, green, and blue.
- Action potentials formed by the photoreceptors are carried to the brain by the optic nerve (CN II). Axons from the medial half of each eye cross over to the opposite side at the optic chiasma. The optic tracts continue from the optic chiasma and carry action potentials to the thalamus, which then relays the action potentials to the occipital lobes of the cerebrum.
- The slightly different retinal images of each eye enable stereoscopic vision when the two images are superimposed by the brain.

9.5 Disorders of the Special Senses
- Disorders of taste and smell include ageusia, hypogeusia, dysgeusia, anosmia, hyposmia, dysosmia, parosmia, and phantosmia.
- Disorders of the ears include deafness, labyrinthine disease, motion sickness, and otitis media.
- Disorders of the eyes include age-related macular degeneration, astigmatism, blindness, cataracts, color blindness, conjunctivitis, farsightedness, nearsightedness, presbyopia, retinoblastoma, and strabismus.

Improve Your Grade

Connect Interactive Questions Reinforce your knowledge using multiple types of questions: interactive, animation, classification, labeling, sequencing, composition, and traditional multiple choice and true/false.

SmartBook Proven to help students improve grades and study more efficiently, SmartBook contains the same content within the print book but actively tailors that content to the needs of the individual.

Anatomy & Physiology REVEALED® Dive into the human body by peeling back layers of cadaver imaging. Utilize this world-class cadaver dissection tool for a closer look at the body anytime, from anywhere.

CHAPTER 10

Endocrine System

©Trinette Reed/Blend Images LLC RF

Katherine, an endocrinologist in Los Angeles, has just finished with her last patient of the day and is headed off to her daily yoga class. An endocrinologist is a doctor who specializes in treating patients with hormonal imbalances. As she drives through traffic, she finds it amusing that the practice of yoga is a good metaphor for the endocrine system. That must be why she enjoys it so much. The ability to successfully maintain and change yoga positions requires focused control and coordination over muscle contraction and relaxation throughout every area of the body. If balance is lost at any time, the yoga position is lost and the person will fall, and even possibly become injured. The endocrine system functions in a similar fashion to maintain the body's homeostasis. Many glands work in concert, releasing hormones in precise amounts and with perfect timing, to maintain the health and balance of a human being. If even one of those hormones is produced incorrectly, the entire body can be thrust out of balance. Loss of balance within the body can be debilitating, which is why Katherine knows her medical practice provides such a valuable service.

CHAPTER OUTLINE

10.1 Introduction to the Endocrine System

10.2 The Chemical Nature of Hormones
- Mechanisms of Hormone Action
- Control of Hormone Production

10.3 Pituitary Gland
- Control of the Anterior Lobe
- Control of the Posterior Lobe
- Anterior Lobe Hormones
- Posterior Lobe Hormones

10.4 Thyroid Gland
- Thyroxine and Triiodothyronine
- Calcitonin

10.5 Parathyroid Glands
- Parathyroid Hormone

10.6 Adrenal Glands
- Hormones of the Adrenal Medulla
- Hormones of the Adrenal Cortex

10.7 Pancreas
- Glucagon
- Insulin

10.8 Gonads
- Female Sex Hormones
- Male Sex Hormone

10.9 Other Endocrine Glands and Tissues
- Pineal Gland
- Thymus

Chapter Summary

Module 8
Endocrine System

SELECTED KEY TERMS

Endocrine gland (endo = within; crin = secrete) A gland whose secretions diffuse into the blood for distribution.
Gene expression The use of DNA to promote protein synthesis.
Hormone (hormon = to set in motion) A chemical messenger secreted by an endocrine gland.
Hypersecretion (hyper = above) Production of an excessive amount of a secretion.

Hyposecretion (hypo = below) Production of a deficient amount of secretion.
Negative-feedback mechanism A mechanism used to keep a variable within its normal range, thereby maintaining homeostasis.
Paracrine signal (para = near) Local chemical signal that affects target cells within the same tissue from which it is produced.
Positive-feedback mechanism A mechanism used when the

originating stimulus needs to be amplified and continued in order for the desired result to occur.
Prostaglandin A type of paracrine signal.
Second messenger An intracellular substance that activates or inactivates enzymes to produce the characteristic effect for the hormone.
Target cell A cell that possesses receptors specific for that hormone.

10.1 Introduction to the Endocrine System

Learning Objectives
1. Distinguish between endocrine and exocrine glands.
2. Identify the major endocrine glands on a diagram or model.

Two interrelated regulatory systems coordinate body functions and maintain homeostasis: the nervous system and the *endocrine* (en′-do-krin) *system*. Unlike the almost instantaneous coordination by the nervous system, the endocrine system provides slower but longer-lasting coordination. The endocrine system consists of cells, tissues, and organs, collectively called **endocrine glands,** that secrete **hormones** (chemical messengers) into the interstitial fluid. The hormones then pass into the blood for transport to other tissues and organs, where they alter cellular functions (figure 10.1). In contrast, exocrine gland secretions are carried from the gland by a duct (tube) to an internal or external surface.

Many organs and tissues throughout the body release chemical signals. As the next section explains, only one of these types of chemical signals is released by the structures of the endocrine system. These structures include several major endocrine glands located throughout the body. Figure 10.2 illustrates the major endocrine glands of the body and the hormones those glands produce.

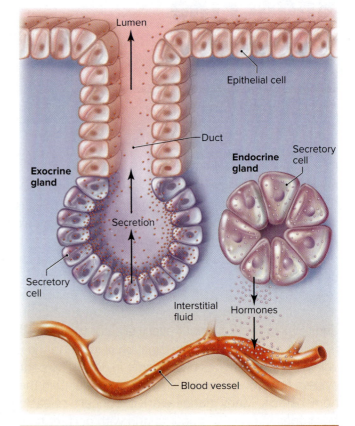

Figure 10.1 Exocrine Gland and Endocrine Gland Compared.

 Check My Understanding
1. What are the differences between exocrine and endocrine secretion?
2. What are the major endocrine glands of the body?

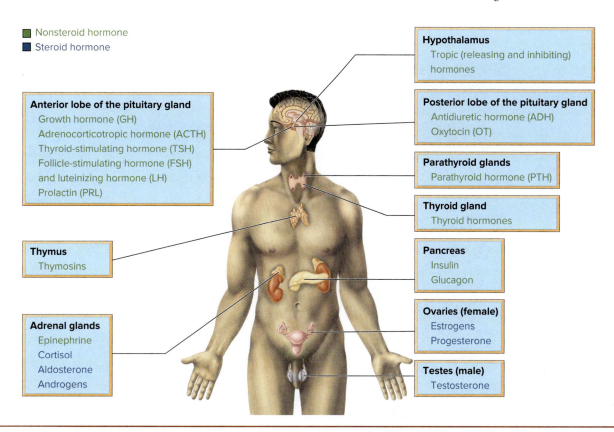

Figure 10.2 The Major Endocrine Glands and Their Hormones.

10.2 The Chemical Nature of Hormones

Learning Objectives

3. Distinguish between neurotransmitters, paracrine signals, and hormones.
4. Explain the three negative-feedback mechanisms that control hormone secretion.
5. Compare the mechanisms of action of steroid and nonsteroid hormones.

There are various modes of communication utilized within the human body (figure 10.3). Neural communication, which was described in chapter 8, uses the release of neurotransmitters at a synapse to transmit a signal from a neuron to another cell. Paracrine communication involves the release of **paracrine signals,** also referred to as "local hormones." Paracrine signals are released within a tissue and are used to affect the function of neighboring cells within that tissue. Endocrine communication involves the release of hormones into the blood for distribution throughout the body. Because hormones are transported in the blood, virtually all body cells are exposed to them. However, a hormone will create a response only in its **target cells,** which are cells that possess receptors specific for that hormone. Non-target cells lack these hormone-specific receptors and are unaffected by the hormone. Neuroendocrine communication is a hybrid mechanism in which a neuron releases a hormone that enters a blood vessel. The majority of this chapter focuses on the role of endocrine and neuroendocrine communication in homeostasis.

Eicosanoids (i-ko′-sa-noyds) are a major class of paracrine signals. Prostaglandins and leukotrienes are examples. **Prostaglandins** produce a variety of effects ranging from promoting inflammation and blood clotting to increasing uterine contraction in childbirth and raising blood pressure. **Leukotrienes** help regulate the immune response and promote inflammation and some allergic reactions.

Hormones are secreted in very small amounts, so their concentrations in the blood are extremely low. However, because they act on cells that have specific receptors for particular hormones, large quantities are not necessary to produce effects. Chemically, hormones may be classified in two broad groups: **steroids,** which are derived from cholesterol, and **nonsteroids,** which are derived from amino acids, peptides, or proteins.

 Clinical Insight

Aspirin and acetaminophen are widely used pain relievers. They function by inhibiting the synthesis of prostaglandins involved in the inflammatory response, which often is the basis of pain and fever.

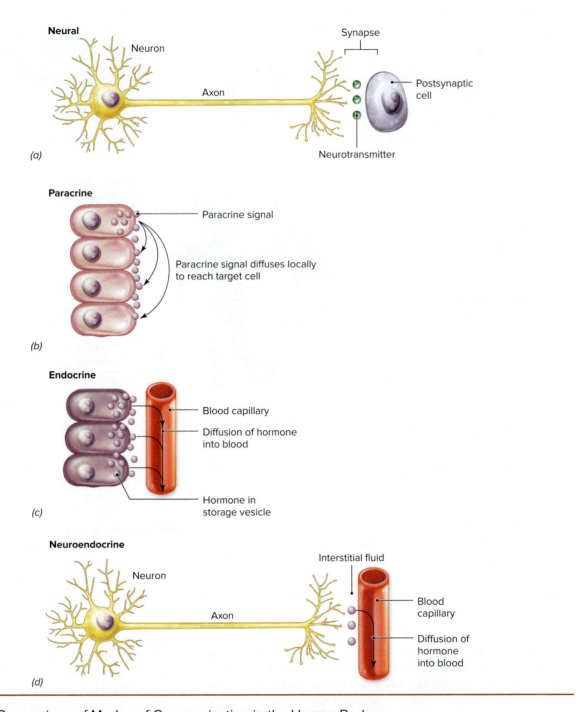

Figure 10.3 Comparison of Modes of Communication in the Human Body.

Mechanisms of Hormone Action

A hormone produces its effect by binding to a target cell's receptors for that hormone. The more receptors it binds to, the greater is the effect on the target cell. All hormones affect target cells by altering their metabolic activities. For example, they may change the rate of cellular processes in general, or they may promote or inhibit specific cellular processes. The end result is that homeostasis is maintained.

Steroid and Thyroid Hormones

Steroid hormones and thyroid hormones act on DNA in a cell's nucleus and affect **gene expression** causing protein synthesis (figure 10.4). ① Because they are lipid-soluble (see chapter 3), they can easily move through the phospholipid bilayers of plasma membranes to ② enter the nucleus. ③ After a hormone enters the nucleus, it combines with an intracellular receptor to form a hormone-receptor complex. ④ The hormone-receptor complex

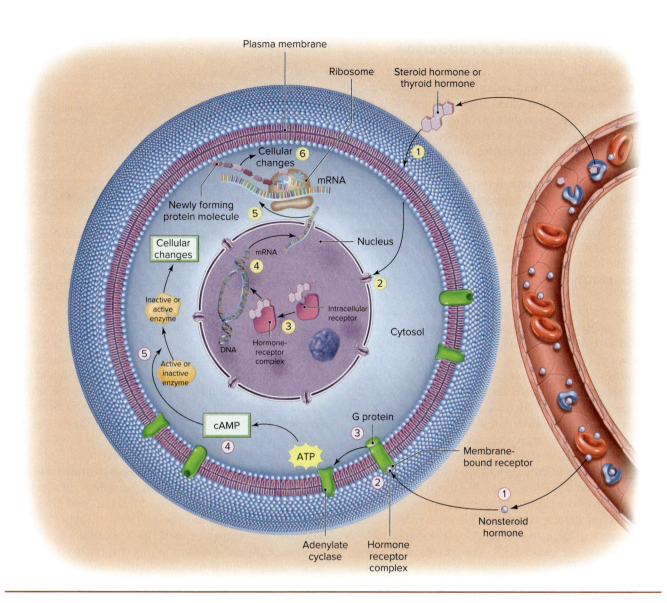

Figure 10.4 Comparison of Mechanisms of Hormone Action.
Mechanisms are numbered to match descriptions within text. AP|R

interacts with DNA, activating specific genes that synthesize messenger RNA (mRNA). ⑤ The mRNA exits the nucleus and interacts with ribosomes, which results in the synthesis of specific proteins, usually enzymes. ⑥ Then the newly formed proteins produce the specific effect that is characteristic of the particular hormone. Certain hormones, like thyroid hormones, enter the mitochondria to affect the DNA in them, thus altering metabolism.

Nonsteroid Hormones

Nonsteroid hormones are proteins, peptides, or modified amino acids that are not lipid-soluble, meaning they cannot pass across the phospholipid bilayer. Two messengers are required for these hormones to produce their effect on a target cell. The *first messenger* is the nonsteroid hormone bound to a receptor on the plasma membrane. The first messenger leads to the formation of a **second messenger** that is often, but not always, *cyclic adenosine monophosphate (cAMP)*. The second messenger is formed within the cell, and it activates or inactivates enzymes that produce the characteristic effect for the hormone (figure 10.4). When a cAMP is the second messenger, the sequence of events is as follows.

① A nonsteroid hormone binds to a receptor on the target cell's plasma membrane to ② form a hormone-receptor complex. ③ This complex activates a membrane protein (G protein), which, in turn, activates a membrane enzyme (adenylate cyclase), ④ which catalyzes the formation of cyclic adenosine monophosphate (cAMP) from ATP within the cytosol. ⑤ The cAMP activates enzymes that catalyze the activation or inactivation of cellular enzymes, which produce the cellular changes associated with the specific hormone.

Control of Hormone Production

Most hormone secretion is usually regulated by a **negative-feedback mechanism** that works to maintain homeostasis. When the blood concentration of a regulated substance begins to decrease, the endocrine gland is stimulated to increase the secretion of its hormone. The increased hormone concentration stimulates target cells to raise the blood level of the substance back to its set point. When the substance returns to its set point, the endocrine gland is no longer stimulated to secrete the hormone, and the secretion and concentration of the hormone decrease. Negative feedback keeps hormone levels in the blood relatively stable (figure 10.5). However, there are a few body processes that are hormonally regulated through **positive-feedback mechanisms.** An example that you will see later in this chapter and in chapter 18 is the production of oxytocin during labor and delivery.

As shown in figure 10.6, endocrine glands are controlled by these negative-feedback mechanisms in three ways. (1) In *hormonal control* (figure 10.6a), **tropic hormones** are used to regulate hormone secretion. A tropic hormone is released by one endocrine gland and regulates hormone secretion by another endocrine gland. For example, the hypothalamus secretes a tropic hormone that interacts with target cells in the anterior lobe of the pituitary gland. If stimulated, the anterior lobe of the pituitary gland will stimulate other endocrine glands to produce hormones. These hormones feed back and affect the function of the hypothalamus and anterior lobe. (2) In *neural control* (figure 10.6b), the nervous system stimulates an endocrine gland to produce a hormone, which affects target cells in the body. The actions of the target cells feed back to the nervous system to alter its activity. (3) In *humoral control* (figure 10.6c), a chemical change

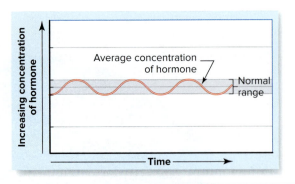

Figure 10.5 Homeostasis of Hormone Level. Negative-feedback mechanism controls the concentration of a hormone in the blood. The concentration may fluctuate slightly above and below the hormone's average concentration.

in the blood stimulates an endocrine gland to produce a hormone, which in turn affects target cells. The actions of the target cells then create a change in the blood level of a chemical, which feeds back to and alters the activity of the endocrine gland. These feedback mechanisms may have either stimulatory or inhibitory effects on the hormone production pathway.

The production of hormones is usually precisely regulated so that there is no **hypersecretion** (excessive production) or **hyposecretion** (deficient production). However, hormonal disorders do occur, and they usually result from severe hypersecretion or hyposecretion. Because endocrine disorders are specifically related to individual glands, disorders in this chapter are considered when each gland is discussed rather than at the end of the chapter.

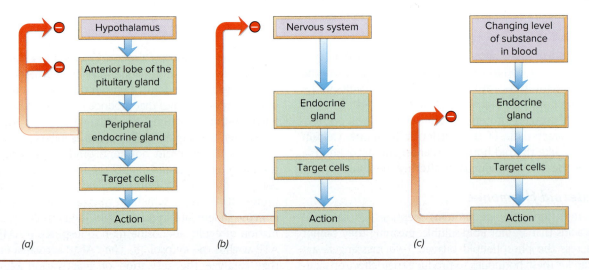

Figure 10.6 Negative-Feedback Mechanisms Used to Control the Release of Various Hormones.
(a) Hormonal control. *(b)* Neural control. *(c)* Humoral control.

Check My Understanding

3. How do steroid and nonsteroid hormones produce their effects on target cells?
4. How are hormones and prostaglandins similar but different?
5. How is the secretion of hormones regulated?

10.3 Pituitary Gland

Learning Objectives

6. Describe how the production of each of the anterior lobe hormones is controlled.
7. List the actions of hormones of the anterior lobe of the pituitary gland.
8. Describe how the production of each of the posterior lobe hormones is controlled.
9. List the actions of hormones of the posterior lobe of the pituitary gland.
10. Describe the major pituitary gland disorders.

The **pituitary** (pi-tū′-i-tar-ē) **gland,** or **hypophysis** (hī-pof′-i-sis), is attached to the hypothalamus by a short stalk. It rests in a depression of the sphenoid, the sella turcica, which provides protection. The pituitary gland consists of two major parts that have different functions: an *anterior lobe* and a *posterior lobe*. Although the pituitary gland is small, it regulates many body functions. The pituitary gland is controlled by neurons and hormones that originate in the **hypothalamus,** as shown in figure 10.7. The hypothalamus serves as a link between the brain and the endocrine system and is itself an endocrine gland. Table 10.1 summarizes the hormones of the pituitary gland and their functions.

Control of the Anterior Lobe

Special neurons (neurosecretory cells) in the hypothalamus regulate the secretion of hormones from the anterior lobe of the pituitary gland by secreting tropic hormones called *releasing* and *inhibiting hormones.* The hypothalamic hormones enter the portal veins of the hypophysis, which carry them directly into the anterior lobe without circulating throughout the body. In the anterior lobe, the hormones exert their effects on specific groups of cells. There is a releasing hormone for each hormone produced by the anterior lobe. There are inhibiting hormones for growth hormone and prolactin. As the names imply, releasing hormones stimulate the production and release of hormones from the anterior lobe, while inhibiting hormones have the opposite effect. The secretion of releasing and inhibiting hormones by the hypothalamus is regulated by various negative-feedback mechanisms.

Control of the Posterior Lobe

The posterior lobe of the pituitary gland is controlled by the neural control negative-feedback mechanism described previously and shown in figure 10.6b. Special neurons that originate in the hypothalamus have axons that extend into the posterior lobe of the pituitary gland. Nerve impulses passed along these neurosecretory axons cause the release of hormones from their terminal boutons within the posterior lobe, where they diffuse into the blood. Note the posterior lobe of the pituitary gland utilizes neuroendocrine communication (see figure 10.3d).

Anterior Lobe Hormones

The anterior lobe of the pituitary gland is sometimes called the "master gland" because it affects so many body functions. It produces and secretes six hormones: growth hormone (GH), thyroid-stimulating hormone (TSH), adrenocorticotropic hormone (ACTH), follicle-stimulating hormone (FSH), luteinizing hormone (LH), and prolactin (PRL).

Growth Hormone

As the name implies, **growth hormone (GH)** stimulates the division and growth of body cells. Increased growth results because GH promotes the synthesis of proteins and other complex organic molecules. GH also increases available energy for these synthesis reactions by promoting the release of fat from adipose tissue, the use of fat in aerobic respiration, and the conversion of glycogen to glucose. Although GH is more abundant during childhood and puberty, it is secreted throughout life.

Regulation of growth hormone secretion is by two hypothalamic hormones with antagonistic functions. GH-releasing hormone (GHRH) stimulates GH secretion, and GH-inhibiting hormone (GHIH) inhibits GH secretion. Whether the hypothalamus releases GHRH or GHIH depends upon changes in blood chemistry. For example, following strenuous exercise, a low level of blood glucose (hypoglycemia) and an excess of amino acids in the blood trigger the secretion of GHRH. Conversely, a high level of blood glucose (hyperglycemia) stimulates the secretion of GHIH.

Disorders If hypersecretion of GH occurs during the growing years, the individual becomes very tall–sometimes nearly 2.5 m (8 ft) in height. This condition is known as **gigantism.** If the hypersecretion of GH occurs in an adult after full growth in height has been attained, it produces a condition known as **acromegaly** (ak-rō-meg′-ah-lē). Because the growth of long bones has been completed, only the bones of the face, hands, and feet continue to grow. Over time, the individual develops heavy, protruding brow ridges, a jutting mandible, and enlarged hands and feet. Both gigantism and acromegaly may result from tumors of the anterior lobe of the pituitary gland. Affected

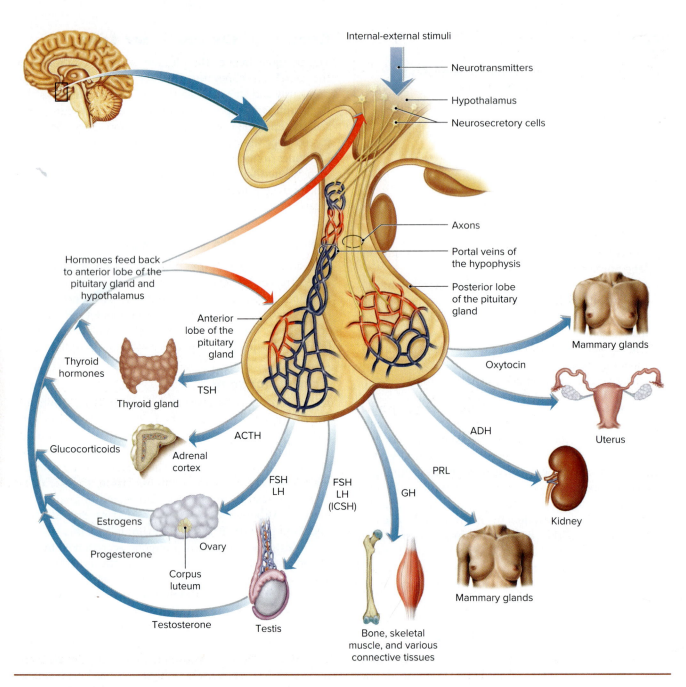

Figure 10.7 Control of Pituitary Gland Secretions.
Hypothalamic hormones are secreted by modified neurons and carried by the hypophyseal portal veins to the anterior lobe of the pituitary gland, where they either stimulate or inhibit the secretion of anterior lobe hormones. Nerve impulses stimulate modified neurons in the hypothalamus to secrete hormones that are released from their terminal boutons within the posterior lobe of the pituitary gland. AP|R

persons may have other health problems due to hypersecretion of other anterior lobe hormones.

If hyposecretion of GH occurs during childhood, body growth is limited. In extreme cases, this results in **pituitary dwarfism.** Affected persons have well-proportioned body parts but may be less than 1 m (3 ft) in height. They may suffer from other maladies due to a deficient supply of other anterior lobe hormones.

Thyroid-Stimulating Hormone

Thyroid-stimulating hormone (TSH) stimulates the thyroid gland to produce thyroid hormones. Blood concentrations of thyroid hormones control the negative-feedback mechanism for TSH production. Low levels of thyroid hormones activate the hypothalamus to secrete *thyrotropin-releasing hormone (TRH),* which stimulates release of TSH by the anterior lobe of the pituitary gland.

Table 10.1 Hormones of the Pituitary Gland

Hormone	Control	Action	Disorders
Anterior Lobe Hormones			
Growth hormone (GH)	Growth-hormone-releasing hormone (GHRH); growth-hormone-inhibiting hormone (GHIH)	Promotes growth of body cells and cell division; promotes protein synthesis; increases the use of fat and glucose for ATP	Hyposecretion in childhood causes pituitary dwarfism. Hypersecretion in childhood causes gigantism; in adults, it causes acromegaly.
Thyroid-stimulating hormone (TSH)	Thyrotropin-releasing hormone (TRH)	Stimulates thyroid gland to produce thyroid hormones	Hyposecretion leads to secondary hypothyroidism. Hypersecretion leads to secondary hyperthyroidism.
Adrenocorticotropic hormone (ACTH)	Corticotropin-releasing hormone (CRH)	Stimulates adrenal cortex to secrete glucocorticoids and androgens	
Follicle-stimulating hormone (FSH)	Gonadotropin-releasing hormone (GnRH)	In ovaries, stimulates development of ovarian follicles and secretion of estrogens; in testes, stimulates the production of sperm	
Luteinizing hormone (LH)	Gonadotropin-releasing hormone (GnRH)	In females, promotes ovulation and development of the corpus luteum, which leads to the production and secretion of progesterone, preparation of uterus to receive embryo, and preparation of mammary glands for milk secretion; in males, stimulates testes to secrete testosterone	
Prolactin (PRL)	Prolactin-releasing hormone (PRH); prolactin-inhibiting hormone (PIH)	Stimulates milk secretion and maintains milk production by mammary glands	
Posterior Lobe Hormones			
Antidiuretic hormone (ADH)	Concentration of water in body fluids	Promotes retention of water by kidneys	Hyposecretion causes diabetes insipidus.
Oxytocin (OT)	Stretching of uterus; stimulation of nipples	Stimulates contractions of uterus in childbirth and contraction of milk glands when nursing infant	
		In both sexes, promotes parental caretaking and feelings of sexual pleasure	

Conversely, high concentrations of thyroid hormones inhibit the secretion of TRH, which decreases production of TSH. Because TSH controls the thyroid gland, disorders of TSH secretion lead to thyroid disorders.

Adrenocorticotropic Hormone

Adrenocorticotropic (ad-re-nō-kor-ti-kō-trōp′-ik) **hormone (ACTH)** controls the secretion of hormones produced by the adrenal cortex (the outer portion of the adrenal gland). ACTH production is controlled by *corticotropin-releasing hormone (CRH)* from the hypothalamus. CRH release is controlled by blood levels of glucocorticoids (especially cortisol) from the adrenal cortex through a negative-feedback mechanism. A low level of cortisol from the adrenal cortex activates the hypothalamus to secrete CRH, which stimulates the release of ACTH from the anterior lobe of the pituitary gland. A high level of cortisol inhibits CRH secretion, and thus inhibits the production and secretion of ACTH. Excessive stress may stimulate the production of excessive amounts of ACTH by overriding the negative-feedback control.

Gonadotropins

Follicle-stimulating hormone (FSH) and **luteinizing** (lū-tē-in-īz-ing) **hormone (LH)** affect the gonads (testes and ovaries). Their release is stimulated by *gonadotropin-releasing hormone (GnRH)* from the hypothalamus. The onset of puberty in both sexes is caused by the start of FSH secretion. In females, FSH acts on the ovaries to promote the development of ovarian follicles, which contain ova and produce estrogens, the primary female sex hormones. In males, FSH acts on testes to promote sperm production. In females, LH stimulates ovulation and the development of the corpus luteum, a temporary gland in the ovary that produces progesterone, another female sex hormone. In males, LH is often referred to as *interstitial cell stimulating hormone (ICSH)* because it affects the interstitial cells of the testes, where it stimulates the secretion of testosterone. Further discussion of FSH and LH can be found in chapter 17.

Prolactin

Prolactin (prō-lak′-tin) **(PRL)** helps to initiate and maintain milk production by the mammary glands after the birth of an infant. Prolactin stimulates milk secretion after the mammary glands have been prepared for milk production by other hormones, including female sex hormones. Prolactin secretion is regulated by the antagonistic actions of *prolactin-releasing hormone (PRH)* and *prolactin-inhibiting hormone (PIH)* produced by the hypothalamus.

Posterior Lobe Hormones

The activities within the posterior lobe of the pituitary gland are examples of neuroendocrine communication (see figure 10.3*d*) that is regulated by the neural control mechanism (see figure 10.6*b*). The posterior lobe stores and releases two hormones: the antidiuretic hormone and oxytocin. Both of these hormones are secreted by neurons that originate in the hypothalamus and extend into the posterior lobe. The hormones are released into the blood within the posterior lobe and are distributed throughout the body (see figure 10.7).

Antidiuretic Hormone

The **antidiuretic** (an-ti-dī-ū-ret′-ik) **hormone (ADH)** promotes water retention by the kidneys to reduce the volume of water that is excreted in urine. ADH secretion is regulated by special neurons that detect changes in the water concentration of the blood. If water concentration decreases, secretion of ADH increases to promote water retention by the kidneys. If water concentration increases, secretion of ADH decreases, causing more water to be excreted in urine. By controlling the water concentration of blood, ADH helps to control blood volume and blood pressure. Further discussion of ADH can be found in chapter 16.

Disorders A severe hyposecretion of ADH results in the production of excessive quantities (20–30 liters per day) of dilute urine, a condition called **diabetes insipidus** (dī-ah-bē′-tēz in-sip′-i-dus). Diabetes means "overflow," and insipidus means "tasteless." Thus, diabetes insipidus essentially means to have overflow of tasteless urine. Conversely, mellitus means "sweet," so diabetes mellitus is an overflow of sweet urine. In diabetes insipidus, the affected person is always thirsty and must drink water almost constantly. This condition may be caused by injuries or tumors that affect any part of the ADH regulatory mechanism, such as the hypothalamus or posterior lobe of the pituitary gland, or nonfunctional ADH receptors in the kidneys. Diabetes insipidus must be treated with a drug called *desmopressin*, which mimics the action of ADH, or the patient will die of dehydration within a day due to excessive water loss in urine.

Clinical Insight

Pitocin, a synthetic oxytocin, is one of several drugs used to clinically induce labor. After delivery, these drugs may also be used to increase the muscle tone of the uterus and to control uterine bleeding.

Oxytocin

Oxytocin (ok-sē-tō′-sin) **(OT)** is released in large amounts during childbirth. It stimulates and strengthens contraction of the smooth muscles of the uterus, which culminates in the birth of the infant. It also has an effect on the mammary glands. Stimulation of a nipple by a suckling infant causes the release of OT, which, in turn, contracts the milk glands of the breast, forcing milk into the milk ducts, where it can be removed by the suckling infant.

Unlike other hormones, oxytocin secretion is controlled by a positive-feedback mechanism. For example, the greater the nipple stimulation by a suckling infant, the more OT released and the more milk available for the infant. When suckling ceases, OT production ceases.

OT is also produced in males and nonpregnant females, where it plays a role in promoting parental caretaking behaviors and feelings of sexual pleasure.

Check My Understanding

6. How does the hypothalamus control the secretions of the pituitary gland?
7. What are the functions of anterior lobe and posterior lobe hormones?

10.4 Thyroid Gland APR

Learning Objectives

11. Describe how the production of thyroid hormones is controlled.
12. List the actions of thyroid hormones.
13. Describe how the production of calcitonin is controlled.
14. List the actions of calcitonin.
15. Describe the major thyroid disorders.

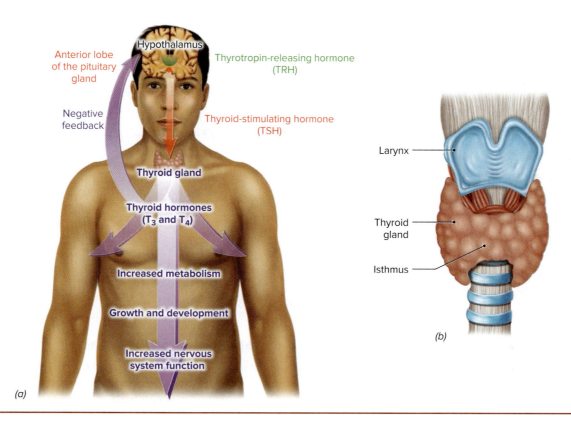

Figure 10.8 Anatomy and Physiology of the Thyroid Gland.
(a) Negative-feedback mechanism of thyroid control. (b) The thyroid gland consists of two lobes connected in front of the trachea at the isthmus. AP|R

The **thyroid gland** is located just below the larynx. It consists of two lobes, each one lateral to the trachea, that are connected by an isthmus in the front of the neck (figure 10.8). Table 10.2 summarizes the control, action, and disorders of the thyroid gland.

Thyroxine and Triiodothyronine

Iodide ions are essential for the formation and functioning of two similar thyroid hormones, produced by groups of cells forming *thyroid follicles* that respond to TSH.

Thyroxine is the primary hormone. It is also known as T_4 because each molecule contains four iodide ions. The other hormone, **triiodothyronine** (trī″-i-o″-dō-thī′-rō-nēn) or T_3, contains three iodide ions in each molecule. Both T_4 and T_3 exert their effect on body cells, and they have similar functions. They increase the metabolic rate, promote protein synthesis, and enhance neuron function. T_3 and T_4 are the primary factors that determine the basal metabolic rate (BMR), the number of calories required at rest to maintain life. Thyroid hormones are also important during

Table 10.2 Hormones of the Thyroid Gland

Hormone	Control	Action	Disorders
Thyroxine (T_4) and triiodothyronine (T_3)	TSH from anterior lobe of the pituitary gland	Increase metabolic rate; accelerate growth; stimulate neural activity	Hyposecretion in infants and children causes cretinism; in adults, it causes myxedema. Hypersecretion causes Graves disease. Iodine deficiency causes simple goiter.
Calcitonin (CT)	Blood Ca^{2+} level	Decreases blood Ca^{2+} level by promoting Ca^{2+} deposition in bones, inhibiting removal of Ca^{2+} from bones, promoting excretion of Ca^{2+} by kidneys	

infancy and childhood for healthy development of the nervous, skeletal, and muscular systems. Secretion of these hormones is stimulated by TSH from the anterior lobe of the pituitary gland, and TSH, in turn, is regulated by a negative-feedback mechanism as described in the discussion of the anterior lobe of the pituitary gland.

Disorders Hypersecretion, hyposecretion, and iodine deficiencies are involved in the thyroid disorders: Graves disease, simple goiter, cretinism, and myxedema.

Graves disease results from the hypersecretion of thyroid hormones. It is thought to be an autoimmune disorder in which antibodies bind to TSH receptors, stimulating excessive hormone production. It is characterized by restlessness and increased metabolic rate with possible weight loss. Usually, the thyroid gland is somewhat enlarged, which is called a **goiter** (goy-ter), and eyes bulge due to the swelling of tissues behind the eyes, producing what is called an *exophthalmic* (ek-sof-thal-mik) *goiter*.

Simple goiter is an enlargement of the thyroid gland that results from a deficiency of iodine in the diet. Without adequate iodine, the thyroid gland cannot produce thyroid hormones and hyposecretion of thyroid hormones occurs. The thyroid gland enlarges due to overstimulation with TSH in an attempt to produce more thyroid hormones. In some cases, the thyroid gland may become the size of an orange. Simple goiter can be prevented by including very small amounts of iodine in the diet. For this reason, salt manufacturers produce "iodized salt," which contains sufficient iodine to prevent simple goiter.

Cretinism (kre'-tin-izm) is caused by a severe hyposecretion of thyroid hormones in infants. Without treatment, it produces severe mental and physical retardation. Cretinism is characterized by stunted growth, abnormal bone formation, mental retardation, sluggishness, and goiter.

Myxedema (mik-se-de'-mah) is caused by severe hyposecretion of thyroid hormones in adults. It is characterized by sluggishness, weight gain, weakness, dry skin, goiter, and puffiness of the face.

Calcitonin

The thyroid gland produces a third hormone, **calcitonin** (kal-si-to'-nin) **(CT)**, from cells called *C cells* that are located between thyroid follicles. C cells do not respond to the hormonal mechanism the same as thyroid follicles do but respond by humoral control linked to the blood Ca^{2+} level. Calcitonin decreases blood Ca^{2+} by inhibiting the bone-resorbing action of osteoclasts, increasing the rate of Ca^{2+} deposition by osteoblasts, and promoting Ca^{2+} excretion by the kidneys. An excess of Ca^{2+} in the blood stimulates the thyroid gland to secrete calcitonin. The concentration of Ca^{2+} in the blood is important because it plays vital roles in metabolism, including maintenance of healthy bones, conduction of action potentials, muscle contraction, and clotting of blood. The function of calcitonin is antagonistic to parathyroid hormone, which is discussed in the next section.

10.5 Parathyroid Glands

Learning Objectives
16. Describe how the production of parathyroid hormone is controlled.
17. List the actions of parathyroid hormone.
18. Describe the major parathyroid disorders.

The **parathyroid glands** are small glands located on the back of the thyroid gland. There are usually four parathyroid glands, two glands on each lobe of the thyroid (figure 10.9).

Parathyroid Hormone

Parathyroid glands secrete **parathyroid hormone (PTH),** the most important regulator of the blood Ca^{2+} level. PTH increases the concentration of blood Ca^{2+} by promoting the removal of Ca^{2+} from bones by osteoclasts and by inhibiting Ca^{2+} deposition by osteoblasts. PTH acts in the kidneys to inhibit excretion of Ca^{2+} into urine and trigger the activation of vitamin D (also a hormone). Both PTH and

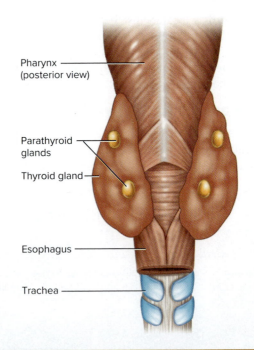

Figure 10.9 Parathyroid Glands.
Typically four parathyroid glands are attached to the back of the thyroid gland.

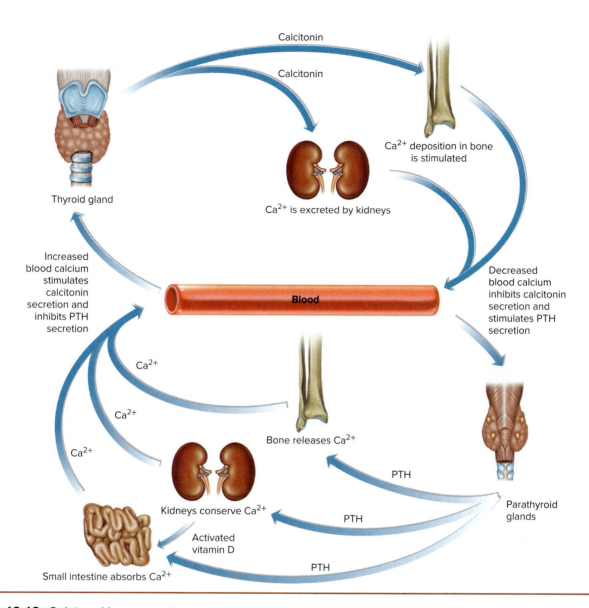

Figure 10.10 Calcium Homeostasis.
The concentration of Ca²⁺ in the blood controls the secretion of calcitonin and PTH.

vitamin D increase Ca²⁺ absorption by the small intestine. The antagonistic actions of PTH and calcitonin maintain blood Ca²⁺ homeostasis (figure 10.10 and table 10.3).

Disorders **Hypoparathyroidism,** the hyposecretion of PTH, can produce devastating effects. Without treatment, the concentration of blood Ca²⁺ may drop to a level that impairs neural and muscular activity. The effect on cardiac muscle may result in cardiac arrest and sudden death. Tetany of skeletal muscles may occur, and death may result from a lack of oxygen due to the inability of breathing muscles to function in a healthy way.

Hyperparathyroidism, the hypersecretion of PTH, causes too much Ca²⁺ to be removed from bones and raises

Table 10.3 Parathyroid Hormone

Hormone	Control	Action	Disorders
Parathyroid hormone (PTH)	Blood Ca²⁺ level	Increases blood Ca²⁺ level by promoting Ca²⁺ removal from bones and Ca²⁺ reabsorption by kidneys	Hyposecretion causes tetany, which may result in death. Hypersecretion causes weak, deformed bones that may fracture spontaneously.

blood Ca^{2+} to an abnormally high level. Without treatment, Ca^{2+} loss results in soft, weak bones that are prone to spontaneous fractures. The excess Ca^{2+} in the blood may lead to the formation of kidney stones or may be deposited in abnormal locations creating bone spurs (abnormal bony growths).

> **Check My Understanding**
> 8. What are the actions of thyroid hormones?
> 9. How is the level of blood Ca^{2+} regulated?

10.6 Adrenal Glands

Learning Objectives
19. Describe how the production of adrenal hormones is controlled.
20. List the actions of adrenal hormones.
21. Describe the major adrenal disorders.

There are two **adrenal glands;** one is located on top of each kidney. Each adrenal gland consists of two portions that are distinct endocrine glands: the inner adrenal medulla and the outer adrenal cortex (figure 10.11). Table 10.4 summarizes the control, action, and disorders of the adrenal gland.

Hormones of the Adrenal Medulla

The **adrenal medulla** secretes **epinephrine** (adrenaline) and **norepinephrine** (noradrenaline), two closely related hormones that have very similar actions on target cells. Epinephrine forms about 80% of the secretions.

The sympathetic part of the autonomic division of the nervous system regulates the secretion of adrenal medullary hormones. They are secreted whenever the body is under stress, and they duplicate the action of the sympathetic part on a bodywide scale. The medullary hormones have a stronger and longer-lasting effect in preparing the body for "fight or flight." The effects of epinephrine and norepinephrine include (1) a decrease in blood flow to the viscera and skin; (2) an increase in blood flow to the skeletal muscles, lungs, and nervous system; (3) conversion of glycogen to glucose to raise the glucose level in the blood; and (4) an increase in the rate of aerobic respiration. Epinephrine and norepinephrine are particularly important in short-term stress situations. In times of chronic stress, the adrenal cortex makes further adjustment, as will be discussed in the next section.

Hormones of the Adrenal Cortex

Several different steroid hormones are produced by the **adrenal cortex,** but the most important ones are aldosterone, cortisol, and the sex hormones.

Aldosterone (al-dō-ster'-ōn) is the most important mineralocorticoid secreted by the adrenal cortex. **Mineralocorticoids** regulate the concentration of electrolytes (mineral ions) in body fluids. Aldosterone stimulates the kidneys to retain sodium ions (Na^+) and to excrete potassium ions (K^+). This action not only maintains the healthy balance of Na^+ and K^+ in body fluids but also maintains blood volume and blood pressure. The reabsorption of Na^+ into the blood causes anions, such as chloride (Cl^-) and bicarbonate (HCO_3^-), to be reabsorbed due to their opposing charges. It also causes water to be reabsorbed by osmosis, which maintains blood volume and blood pressure. Aldosterone secretion is stimulated by several factors, including

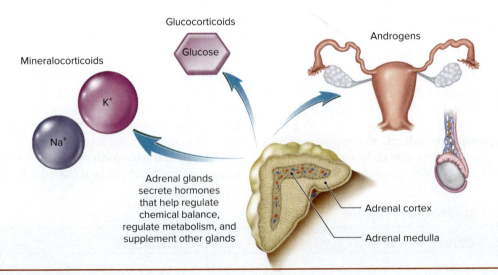

Figure 10.11 The Structure of the Adrenal Gland.
An adrenal gland consists of an inner adrenal cortex and an outer adrenal medulla.

Table 10.4 Hormones of the Adrenal Glands

Hormone	Control	Action	Disorders
Adrenal Medulla			
Epinephrine and norepinephrine	Sympathetic part of the autonomic division of the nervous system	Prepare body to meet emergencies; increase heart rate, cardiac output, blood pressure, and metabolic rate; increase blood glucose by converting glycogen to glucose; dilate respiratory passages	Hypersecretion causes prolonged responses. Hyposecretion causes no major disorders.
Adrenal Cortex			
Aldosterone	Blood electrolyte levels, angiotensin II	Increases blood levels of sodium and water, which decreases blood level of potassium; increases blood pressure	Hypersecretion inhibits neural and muscular activity, and also causes edema.
Cortisol	ACTH from anterior lobe of the pituitary gland	Promotes formation of glucose from noncarbohydrate nutrients; provides resistance to stress and inhibits inflammation	Hyposecretion causes Addison disease. Hypersecretion causes Cushing syndrome.
Androgens	ACTH from anterior lobe of the pituitary gland	Effects are insignificant in healthy adult males; contribute to the sex drive in females.	Hypersecretion as a result of tumors; causes masculinization in females.

 Clinical Insight

Everyone experiences stressful situations. Stress may be caused by physical or psychological stimuli that are perceived as threatening. Whereas mild stress can stimulate creativity and productivity, severe and prolonged stress can have serious consequences.

The hypothalamus is the initiator of the stress response. When stress occurs, the hypothalamus activates the sympathetic part of the autonomic division and the secretion of epinephrine and norepinephrine by the adrenal medulla. Thus, both neural and hormonal activity prepare the body to meet the stressful situation by increasing blood glucose, heart rate, breathing rate, blood pressure, and blood flow to the muscular and nervous systems.

Simultaneously, the hypothalamus stimulates the release of ACTH from the anterior lobe of the pituitary gland. ACTH, in turn, causes the secretion of cortisol by the adrenal cortex. Cortisol increases the levels of amino acids and fatty acids in the blood and promotes the formation of additional glucose from noncarbohydrate nutrients.

All of these responses prepare the body for an immediate response to cope with a stressful situation.

Prolonged stress may cause several undesirable side effects from the constant secretion of large amounts of epinephrine and glucocorticoids, such as decreased immunity and high blood pressure—problems that are common in our society.

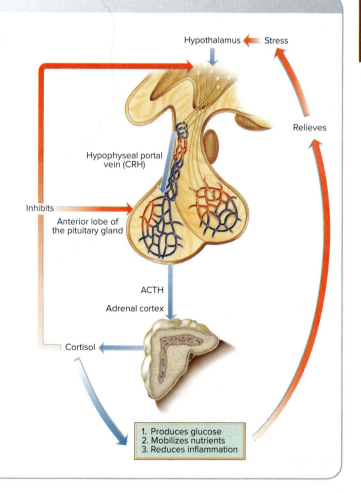

(1) a decrease in blood level of Na$^+$, (2) an increase in blood level of K$^+$, or (3) a decrease in blood pressure, which leads to angiotensin II production via the renin-angiotensin mechanism that is discussed in chapter 16.

Glucocorticoids are so named because they affect glucose metabolism. There are three major actions of glucocorticoids. (1) In response to chronic stress, glucocorticoids ensure a constant fuel supply by promoting the conversion of noncarbohydrate nutrients into glucose. This is important because carbohydrate sources, such as glycogen, may be exhausted after several hours without food or strenuous exercise. (2) They facilitate the utilization of glucose by cells. (3) They reduce inflammation.

Cortisol (kor'-ti-sol) is the most important of several glucocorticoids that are secreted by the adrenal cortex under the stimulation of ACTH. The blood levels of glucocorticoids are kept in balance because they exert a negative-feedback control on the secretion of CRH and ACTH, as described in the section of this chapter discussing the anterior lobe of the pituitary gland. The clinical insight box discussing stress illustrates the involvement of cortisol in the stress response.

The adrenal cortex also secretes small amounts of **androgens** (male sex hormones) and estrogens in response to ACTH from the anterior lobe of the pituitary gland. The estrogens have little significant function. The androgens promote the early development of male reproductive organs, but in adult males their effects are masked by sex hormones produced by testes. In females, adrenal androgens contribute to the female sex drive. In both sexes, excessive production results in exaggerated male characteristics. AP|R

Disorders **Cushing syndrome** results from hypersecretion by the adrenal cortex. It may be caused by an adrenal tumor or by excessive production of ACTH by the anterior lobe of the pituitary gland. This syndrome is characterized by high blood pressure, an abnormally high blood glucose level, protein loss, osteoporosis, fat accumulation on the trunk, fatigue, edema, and decreased immunity. A person with this condition tends to have a full, round face and an enlarged abdomen.

Addison disease results from a severe hyposecretion by the adrenal cortex. It is characterized by low blood pressure, low blood glucose and sodium levels, an increase in the blood potassium level, dehydration, muscle weakness, and increased skin pigmentation. Without treatment to control blood electrolytes, death may occur in a few days.

 Check My Understanding
10. How do secretions of the adrenal medulla prepare the body to react in emergencies?
11. How does the adrenal cortex help to maintain blood pressure?

10.7 Pancreas

Learning Objectives
22. Describe the control of pancreatic hormones.
23. List the actions of pancreatic hormones.
24. Describe the major pancreatic disorders.

The **pancreas** (pan'-krē-as) is an elongate organ that is located behind the stomach (figure 10.12). It is both an exocrine gland and an endocrine gland. Its exocrine functions are performed by secretory cells that secrete digestive enzymes into tiny ducts within the gland. These ducts merge to form the pancreatic duct, which carries the secretions into the small intestine. Its endocrine functions are performed by secretory cells that are arranged in clusters or clumps called the **pancreatic islets.** Their secretions diffuse into the blood. The islets contain alpha cells and beta cells. Alpha cells produce the hormone glucagon; beta cells form the hormone insulin. Table 10.5 summarizes the control, action, and disorders of the pancreas.

Glucagon

Glucagon (glū'-kah-gon) increases the concentration of glucose in the blood. It does this by activating the liver to convert glycogen and certain noncarbohydrates, such as amino acids, into glucose. Glucagon helps to maintain the blood level of glucose within healthy limits even when carbohydrates are depleted due to long intervals between meals. Epinephrine stimulates a similar action, but glucagon is more effective. Glucagon secretion is controlled by the blood level of glucose via a negative-feedback mechanism. A low level of blood glucose stimulates glucagon secretion, and a high level of blood glucose inhibits glucagon secretion.

Insulin

The effect of **insulin** on the level of blood glucose is opposite that of glucagon. Insulin decreases blood glucose by aiding the movement of glucose into body cells, where it can be used as a source of energy. Without insulin, glucose is not readily available to most

 Clinical Insight

Persons with inflamed joints often receive injections of *cortisone*, a glucocorticoid, to temporarily reduce inflammation and the associated pain. Such a procedure is fairly common in sports medicine.

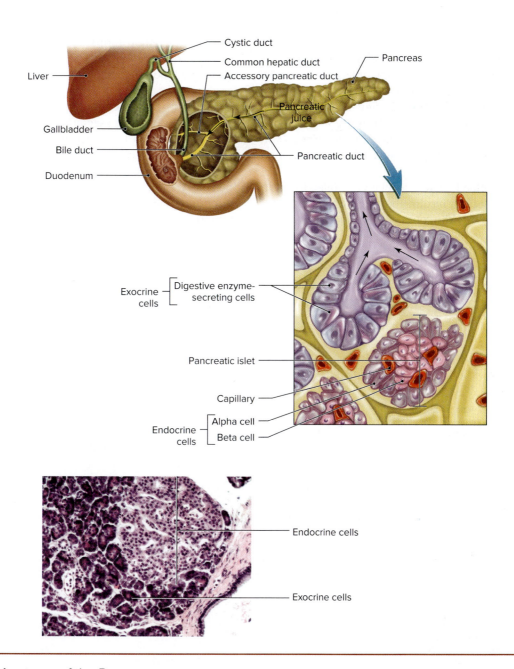

Figure 10.12 Anatomy of the Pancreas
The pancreas is both an endocrine and an exocrine gland. The hormone-secreting alpha and beta cells are grouped in clusters, called pancreatic islets. Other pancreatic cells secrete digestive enzymes. AP|R
©McGraw-Hill Education/Al Telser, photographer

Table 10.5 Hormones of the Pancreas

Hormone	Control	Action	Disorders
Glucagon	Blood glucose level	Increases blood glucose by stimulating the liver to convert glycogen and other nutrients into glucose	
Insulin	Blood glucose level	Decreases blood glucose by aiding movement of glucose into cells and promoting the conversion of glucose into glycogen	Hyposecretion causes type I diabetes mellitus. Hypersecretion may cause hypoglycemia.

cells for cellular respiration. Insulin also stimulates the liver to convert glucose into glycogen for storage. Figure 10.13 shows how the antagonistic functions of glucagon and insulin maintain the concentration of glucose in the blood at a healthy level. Like glucagon, the level of blood glucose regulates the secretion of insulin. A high blood glucose level stimulates insulin secretion; a low blood glucose level inhibits insulin secretion.

In reality, blood glucose level is controlled by multiple hormones, not just insulin and glucagon. Figure 10.14 illustrates how other hormones affect blood glucose level.

Epinephrine (short-term) and cortisol (long-term) raise blood glucose in times of stress. Growth hormone also raises the blood glucose level.

Disorders Diabetes mellitus (dī-ah-bē′-tēz mel-lī′-tus) is caused by the hyposecretion of insulin or the inability of target cells to recognize it due to a loss of insulin receptors. *Type I* or *insulin-dependent diabetes* is an autoimmune metabolic disorder that usually appears in persons less than 20 years of age. For this reason, it is sometimes called juvenile diabetes, although the condition persists for life. Type I diabetes results when the

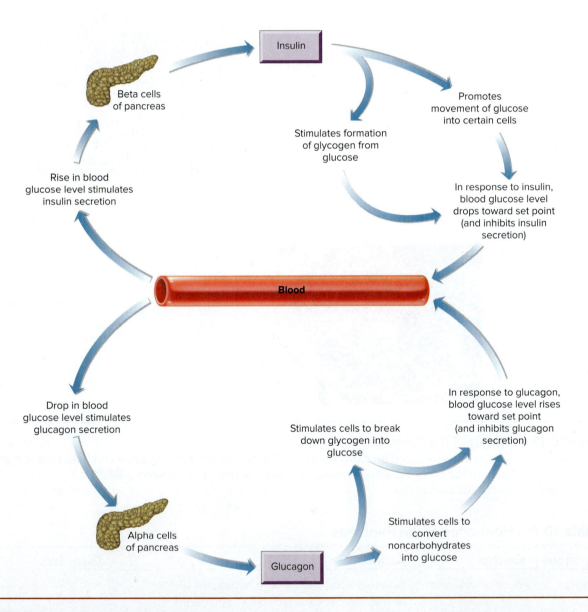

Figure 10.13 Blood Glucose Homeostasis.
Insulin and glucagon function together to help maintain a relatively stable blood glucose level. Negative-feedback mechanism responding to blood glucose level controls the secretion of both hormones. **AP|R**

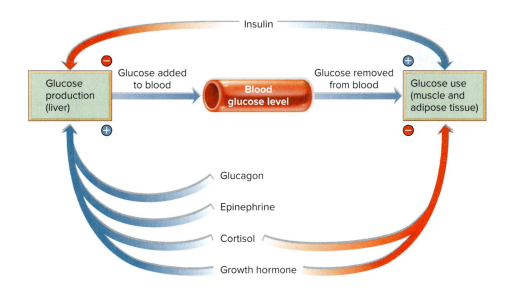

Figure 10.14 Blood Glucose Concentration.
Integrated control of blood glucose concentration. AP|R

immune response destroys the beta cells in pancreatic islets. Because the metabolism of carbohydrates, fats, and proteins is affected, persons with type I diabetes must follow a restrictive diet. They must also check their blood glucose level several times a day and inject themselves with insulin, or receive insulin from an implanted insulin pump to keep their blood glucose concentration within healthy limits.

The vast majority of diabetics have *type II* or *insulin-independent diabetes,* which is caused by a reduction of the insulin receptors on target cells. This form of diabetes, also called adult-onset diabetes, usually appears after 40 years of age in persons who are overweight. The symptoms are less severe than in type I diabetes and can be controlled by a careful diet and oral medications that help regulate the blood level of glucose. The current increase in obesity among children and young adults is of concern because it may lead to an increase in type II diabetes. In either case, the result is **hyperglycemia,** an excessively high level of glucose in the blood. With insufficient insulin or a reduction in target insulin receptors, glucose cannot get into cells easily, and cells must rely more heavily on triglycerides as an energy source for aerobic respiration. The products of this reaction tend to decrease blood pH (acidosis), which can inactivate vital enzymes and may lead to death.

An excessive production of insulin, or overdose of insulin, may lead to **hypoglycemia,** a condition characterized by an excessively low blood glucose level. Symptoms include acute fatigue, weakness, increased irritability, and restlessness. In extreme conditions, it may lead to an insulin-triggered coma.

Check My Understanding

12. How does the pancreas regulate the level of blood glucose?

10.8 Gonads

Learning Objectives
25. Describe how the production of female sex hormones is controlled.
26. List the actions of female sex hormones.
27. Describe how the production of male sex hormones is controlled.
28. List the actions of male sex hormones.

The gonads are the sex glands: the ovaries and testes. They not only produce oocytes and sperm, respectively, but also secrete the sex hormones. Table 10.6 summarizes the actions of the sex hormones. The gonads and their hormones are covered in more detail in chapter 17.

Female Sex Hormones

The **ovaries** are the female gonads. They are small, almond-shaped organs located in the pelvic cavity. The ovaries begin to function at the onset of puberty when the gonadotropins (FSH and LH) are released from the anterior lobe of the pituitary gland. Subsequently, ovarian hormones, FSH, and LH interact in an approximately 28-day *ovarian cycle* in which their concentrations increase and decrease in a rhythmic pattern.

Table 10.6 Hormones of Ovaries and Testes

Hormone	Control	Action
Ovaries		
Estrogens	FSH	Development of female reproductive organs, secondary sex characteristics, and sex drive; prepare uterus to receive a preembryo and help maintain pregnancy
Progesterone	LH	Prepares uterus to receive a preembryo and maintains pregnancy; prepares mammary glands for milk production
Testes		
Testosterone	LH (ICSH)	Development of male reproductive organs, secondary sex characteristics, and sex drive

Estrogens (es′-trō-jens), the primary female sex hormones, are several related steroid hormones that are secreted by developing ovarian follicles that also contain an oocyte (developing egg). Estrogens stimulate the development and maturation of the female reproductive organs and the secondary sex characteristics (e.g., female fat distribution, breasts, and broad hips). They also help to grow and maintain the uterine lining (endometrium) to support a pregnancy.

Progesterone (prō-jes′-te-rōn) is secreted by the corpus luteum, a gland that forms from the empty ovarian follicle after the oocyte has been released by ovulation. It helps prepare the uterus for receiving a preembryo and maintains the pregnancy. It also helps to prepare the mammary glands for milk production.

Male Sex Hormone

The **testes** are paired, ovoid organs located below the pelvic cavity in the scrotum, a sac of skin located behind the penis. The seminiferous tubules of the testes produce sperm, the male sex cell; and the interstitial cells (cells between the tubules) secrete the male hormone **testosterone** (tes-tos′-te-rōn). Testosterone stimulates the development and maturation of the male reproductive organs, the secondary sex characteristics (e.g., growth of facial and body hair, low voice, narrow hips, and heavy muscles and bones), the male sex drive, and it helps stimulate sperm production.

10.9 Other Endocrine Glands and Tissues

Learning Objectives
29. Describe the actions of melatonin.
30. Describe the action of the thymus.

There are a few other glands and tissues of the body that secrete hormones and are part of the endocrine system. These include the pineal gland, the thymus, the kidneys, the heart, and certain small glands in the lining of the stomach and small intestine. Hormones released from the kidneys, heart, and digestive system will be covered in their respective chapters. In addition, the placenta is an important temporary endocrine organ during pregnancy. It is considered in chapter 18.

Pineal Gland

The **pineal** (pin′-ē-al) **gland** is a small, cone-shaped nodule of endocrine tissue that is located in the epithalamus of the brain near the roof of the third ventricle. It secretes the hormone **melatonin** (mel-ah-tō′-nin), which seems to inhibit the secretion of gonadotropins and may help control the onset of puberty. Melatonin seems to regulate wake-sleep cycles and other biorhythms associated with the cycling of day and night. The secretion of melatonin is regulated by exposure to light and darkness. When exposed to light, action potentials from the retinas of the eyes are sent to the pineal gland, causing a decrease in melatonin production. During darkness, these action potentials decrease, and melatonin secretion is increased. Secretion is greatest at night and lowest in the day, which keeps our sleep-wakefulness cycle in harmony with the day-night cycle.

As frequent fliers know, jet lag results when the sleep-wakefulness cycle is out of sync with the day-night cycle. Jet lag can be more quickly reversed by exposure to bright light with wavelengths similar to sunlight, because the melatonin cycle is resynchronized to the new day-night cycle.

Thymus

The **thymus** is located in the mediastinum above the heart. It is large in infants and children, but it shrinks with age and is greatly reduced in adults. It plays a crucial role in the development of immunity, which is discussed in chapter 13. The thymus produces several hormones, collectively called **thymosins** (thi-mo′-sins), which are involved in the maturation of T lymphocytes, a type of white blood cell. Thymosins also seem to have some anti-aging effects. Hence, after the thymus shrinks, we age and are more prone to illness.

Chapter Summary

10.1 Introduction to the Endocrine System

- The endocrine system is composed of hormone-secreting cells, tissues, and organs.
- Exocrine glands have a duct; endocrine glands are ductless.
- Hormones are chemical messengers that are carried by the blood throughout the body, where they modify cellular functions of target cells.
- The major endocrine glands are the hypothalamus, pituitary gland, thyroid gland, parathyroid glands, pineal gland, thymus, adrenal glands, pancreas, ovaries (female), and testes (male).

10.2 The Chemical Nature of Hormones

- There are four major types of communication in the body: (1) neural, (2) paracrine, (3) endocrine, (4) neuroendocrine. All target cells have receptors for chemical messengers that affect them.
- Prostaglandins are not secreted by endocrine glands. They are formed by most body cells and have a distinctly local (paracrine) effect.
- Hormones may be classified chemically as either steroid hormones or nonsteroid hormones.
- Steroid hormones and thyroid hormones combine with a receptor within the target cell and interact with DNA to affect production of mRNA. All other nonsteroid hormones combine with a receptor in the plasma membrane of the target cell, which activates a membrane enzyme that promotes synthesis of cyclic AMP (cAMP), a second messenger. Cyclic AMP, in turn, activates other enzymes that bring about cellular changes.
- Production of most hormones is controlled by negative-feedback mechanisms.
- The negative-feedback mechanisms of hormone production work one of three ways: (1) hormonal, (2) neural, and (3) humoral.
- Endocrine disorders are associated with severe hyposecretion or hypersecretion of various hormones. Hyposecretion may result from injury. Hypersecretion is sometimes caused by a tumor.

10.3 Pituitary Gland

- The pituitary gland is attached to the hypothalamus by a short stalk. It consists of an anterior lobe and a posterior lobe.
- The hypothalamus secretes releasing hormones and inhibiting hormones that are carried to the anterior lobe by the portal veins of the hypophysis. The releasing and inhibiting hormones regulate the secretion of anterior lobe hormones.
- Anterior lobe hormones are
  a. growth hormone (GH), which stimulates growth and division of body cells;
  b. thyroid-stimulating hormone (TSH), which activates the thyroid gland to secrete thyroid hormones;
  c. adrenocorticotropic hormone (ACTH), which stimulates the secretion of hormones by the adrenal cortex;
  d. follicle-stimulating hormone (FSH) and luteinizing hormone (LH), which affect the gonads (in females, FSH stimulates production of estrogens by the ovaries, and the development of the ovarian follicles, leading to oocyte production; in males, it activates sperm production by the testes; in females, LH promotes ovulation and stimulates development of the corpus luteum, which produces progesterone; in males, it stimulates testosterone production); and
  e. prolactin (PRL), which initiates and maintains milk production by the mammary glands.
- Hyposecretion of GH in childhood causes pituitary dwarfism. Hypersecretion of GH in childhood causes gigantism, while during adulthood it causes acromegaly.
- Hyposecretion of TSH and hypersecretion of TSH lead to secondary thyroid disorders.
- Hormones of the posterior lobe are formed by neurons in the hypothalamus and are released within the posterior lobe.
- There are two posterior lobe hormones:
  a. antidiuretic hormone (ADH) promotes retention of water by the kidneys;
  b. oxytocin stimulates contraction of the uterus during childbirth, contractions of mammary glands in breast-feeding, and parental caretaking behaviors and sexual pleasure in both genders.
- Hyposecretion of ADH causes diabetes insipidus.

10.4 Thyroid Gland

- The thyroid gland is located just below the larynx, with two lobes lateral to the trachea.
- TSH stimulates the secretion of thyroxine (T_4) and triiodothyronine (T_3), which increase cellular metabolism, protein synthesis, and neural activity.
- Iodine is an essential component of the T_4 and T_3 molecules.
- Calcitonin decreases the level of blood Ca^{2+} by promoting Ca^{2+} deposition in bones. It also promotes the excretion of Ca^{2+} by the kidneys. Its secretion is controlled humorally by the level of Ca^{2+} in the blood.
- Hypersecretion of thyroid hormones causes Graves disease. Iodine deficiency causes simple goiter.
- Hyposecretion of thyroid hormones in infants and children causes cretinism; in adults, it causes myxedema.

10.5 Parathyroid Glands

- The parathyroid glands are embedded on the back of the thyroid gland.
- Parathyroid hormone increases the level of blood Ca^{2+} by promoting Ca^{2+} removal from bones, Ca^{2+} absorption from the small intestine, and Ca^{2+} retention by the kidneys. PTH also activates vitamin D, which helps stimulate Ca^{2+} absorption by the small intestine.
- Parathyroid secretion is controlled humorally by the level of blood Ca^{2+}.

- Parathyroid hormone and calcitonin work antagonistically to regulate blood Ca^{2+} level.
- Hyposecretion of PTH causes tetany, which may result in death. Hypersecretion causes weak, soft, deformed bones that may fracture spontaneously.

10.6 Adrenal Glands

- An adrenal gland is located on top of each kidney. Each gland consists of two parts: an inner adrenal medulla and an outer adrenal cortex.
- The adrenal medulla secretes epinephrine and norepinephrine, which prepare the body to deal with emergency situations. They increase the heart rate, circulation to nervous and muscular systems, and glucose level in the blood.
- The adrenal cortex secretes a number of hormones that are classified as mineralocorticoids, glucocorticoids, and androgens.
- Aldosterone is the most important mineralocorticoid. It helps to regulate the concentration of electrolytes in the blood, especially sodium and potassium ions, which increases blood pressure.
- Cortisol is the most important glucocorticoid. It promotes the formation of glucose from noncarbohydrate sources and inhibits inflammation. Its secretion is regulated by ACTH.
- Cortisol is involved in the response to chronic stress.
- Small amounts of androgens are secreted. They have little effect in adult males but contribute to the sex drive in adult females.
- Hyposecretion of cortisol causes Addison disease. Hypersecretion causes Cushing syndrome.

10.7 Pancreas

- The pancreas is both an exocrine and an endocrine gland. Its hormones are formed by the pancreatic islets, and their secretions are controlled by the level of blood glucose.
- Glucagon, from the alpha cells, increases the level of blood glucose by stimulating the liver to form glucose from glycogen and some noncarbohydrate sources.
- Insulin, from the beta cells, decreases the level of blood glucose by aiding the movement of glucose into cells.
- The antagonistic functions of glucagon and insulin keep the level of blood glucose within healthy limits.
- Hyposecretion of insulin or a decrease in the number of insulin receptors causes diabetes mellitus. Hypersecretion may cause hypoglycemia.

10.8 Gonads

- Gonads are the sex glands: the ovaries in females and the testes in males. They secrete sex hormones, in addition to producing sex cells. The secretion of these hormones is controlled by FSH and LH in both sexes.
- Estrogens are secreted by ovarian follicles and they stimulate development of female reproductive organs and secondary sex characteristics. Estrogens also help to prepare the uterus for a preembryo and help to maintain pregnancy.
- Progesterone is secreted mostly by the corpus luteum of the ovary after ovulation. It prepares the uterus for the preembryo, maintains pregnancy, and prepares the mammary glands for milk production.
- The testes secrete testosterone, the male sex hormone that stimulates the development of the male reproductive organs and secondary sex characteristics.

10.9 Other Endocrine Glands and Tissues

- The pineal gland is located near the roof of the third ventricle of the brain. It secretes melatonin, which seems to lead to the inhibition of secretion of FSH and LH by the anterior lobe of the pituitary gland. The pineal gland also seems to be involved in biorhythms.
- The thymus is located in the thoracic cavity above the heart. It secretes thymosins, which are involved in the maturation of white blood cells called T lymphocytes.
- Thymosins also seem to have anti-aging effects.

Improve Your Grade

Connect Interactive Questions Reinforce your knowledge using multiple types of questions: interactive, animation, classification, labeling, sequencing, composition, and traditional multiple choice and true/false.

SmartBook Proven to help students improve grades and study more efficiently, SmartBook contains the same content within the print book but actively tailors that content to the needs of the individual.

Anatomy & Physiology REVEALED® Dive into the human body by peeling back layers of cadaver imaging. Utilize this world-class cadaver dissection tool for a closer look at the body anytime, from anywhere.

CHAPTER 11

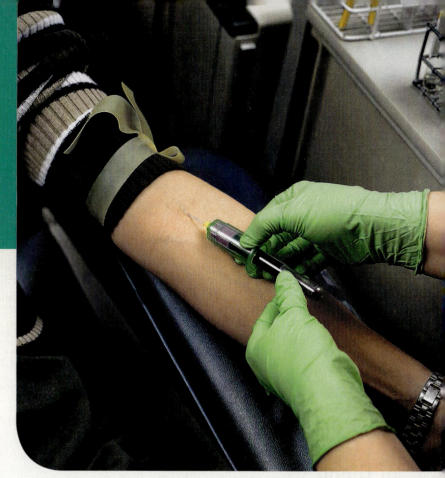

©Liquidlibrary/Getty Images RF

Blood

Phillip, at the age of 35, has been actively donating blood at the local Red Cross chapter for 10 years. Because he is type AB+, his whole blood donations can be used to help only type AB+ patients in need. However, at his last visit, Phillip learned that he had the ability to help more people by donating his platelets and plasma specifically. Cancer patients undergoing chemotherapy can suffer from platelet deficiency, which results in an increased risk of bleeding. These patients usually benefit from platelet transfusions to supplement what their own bodies cannot produce. Plasma, specifically the proteins within it, is frequently used to treat many rare diseases, such as bleeding disorders, immune deficiency disorders, and rabies. Because Phillip has type AB+ blood, his plasma lacks antibodies that are capable of creating adverse reactions in people with other blood types. Given that his plasma can be transfused into anyone with need safely, Phillip is considered a "universal plasma donor." Phillip's next appointment is in a few weeks and he is excited that, by donating specific blood components, he will be able to do so much for so many.

CHAPTER OUTLINE

11.1 General Characteristics of Blood

11.2 Red Blood Cells
- Hemoglobin
- Concentration of Red Blood Cells
- Production
- Life Span and Destruction

11.3 White Blood Cells
- Function
- Types of White Blood Cells

11.4 Platelets

11.5 Plasma
- Plasma Proteins
- Nitrogenous Wastes
- Electrolytes

11.6 Hemostasis
- Vascular Spasm
- Platelet Plug Formation
- Coagulation

11.7 Human Blood Types
- ABO Blood Group
- Rh Blood Group
- Compatibility of Blood Types for Transfusions

11.8 Disorders of the Blood
- Red Blood Cell Disorders
- White Blood Cell Disorders
- Disorders of Hemostasis

Chapter Summary

Module 9
Cardiovascular System

SELECTED KEY TERMS

Agglutination (agglutin = to stick together) The clumping of red blood cells in an antigen-antibody reaction.
Coagulation The formation of a blood clot.
Embolus A moving blood clot or foreign body in the blood.
Formed elements The solid components of blood: red blood cells, white blood cells, and platelets.

Hematopoiesis (hemato = blood; poiesis = to make) The production of formed elements.
Hemoglobin (hemo = blood) The pigmented protein in red blood cells, involved in transporting oxygen and carbon dioxide.
Hemostasis (hemo = blood; stasis = standing still) A positive-feedback mechanism initiated after vascular injury to stop or limit blood loss.

Plasma The liquid portion of blood.
Platelet A formed element that promotes coagulation.
Red blood cell A hemoglobin-containing formed element that transports respiratory gases; an erythrocyte.
Thrombus A stationary blood clot formed in an unbroken blood vessel.
White blood cell A formed element that has defensive and immune functions; a leukocyte.

BLOOD IS USUALLY CONFINED WITHIN THE HEART AND BLOOD VESSELS as it transports materials from place to place within the body.

11.1 General Characteristics of Blood

Learning Objective

1. Describe the general characteristics and functions of blood.

Blood is classified as a connective tissue that is composed of **formed elements** (the solid components, including blood cells and platelets) suspended in **plasma,** the liquid portion (matrix) of the blood. Blood is heavier and about four times more viscous than water. It is slightly alkaline, with a pH between 7.35 and 7.45. The volume of blood varies with the size of the individual, but it averages 5 to 6 liters in males and 4 to 5 liters in females. Blood comprises about 8% of the body weight.

About 55% of the blood volume consists of plasma, and 45% is made up of formed elements. Because the majority of the formed elements are red blood cells (RBCs), it can be said that almost 45% of the blood volume consists of red blood cells. White blood cells (WBCs) and platelets combined form less than 1% of the blood volume (figure 11.1).

The number of formed elements in blood is hard to imagine. There are approximately 5 million RBCs, 7,500 WBCs, and 300,000 platelets in one microliter (μl). A single drop of blood due to a finger stick (approximately 50 μl) contains 250 million RBCs!

The blood functions to transport various substances throughout the body. Substances carried by blood include oxygen, carbon dioxide, nutrients, waste products, hormones, electrolytes, and water. Blood also has several regulatory and protective functions that will be described in this chapter.

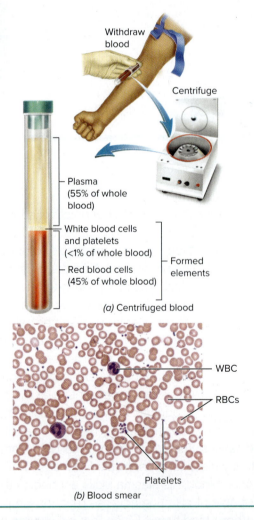

Figure 11.1 Blood Consists of Plasma and Formed Elements.
(a) If blood is centrifuged, the RBCs sink to the bottom of the tube and the liquid plasma forms the top layer. WBCs and platelets form a thin layer between the two. *(b)* The microscopic appearance of formed elements in a smear of blood. AP|R
(b) ©McGraw-Hill Education/Al Telser, photographer

> **Check My Understanding**
> 1. What are the functions of blood?

11.2 Red Blood Cells

Learning Objectives
2. Describe the appearance and healthy concentration of RBCs in blood.
3. Describe the structure of hemoglobin and its role.
4. Explain how the RBCs are produced and destroyed.

Red blood cells, or *erythrocytes* (eh-rith′-rō-sīts), are tiny, biconcave discs that are involved in respiratory gas transport throughout the body. The biconcave shape creates maximal surface area for the diffusion of these gases through the plasma membrane. Mature RBCs lack a nucleus and other organelles, although these are present in immature RBCs (figures 11.1, 11.2, and 11.4).

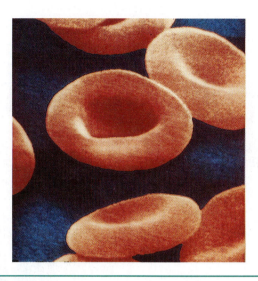

Figure 11.2 A false-color scanning electron photomicrograph of human red blood cells (5000×). AP|R
©Bill Longcore/Science Source

Hemoglobin

About 33% of the volume of each red blood cell consists of **hemoglobin** (hē′-mō-glō-bin). Hemoglobin gets its name because it consists of *heme*, an iron-containing pigment molecule, and a *globin*, a globelike protein. Blood is red because heme is a reddish pigment. Hemoglobin combines reversibly with oxygen and plays a vital role in the transport of oxygen by RBCs. It also plays a role in carbon dioxide transport.

When blood flows through the lungs, oxygen diffuses from air spaces in the lungs into the blood. Oxygen enters RBCs and combines with hemoglobin to form **oxyhemoglobin,** which gives a bright red color to blood. After the release of some oxygen from oxyhemoglobin to body cells, the resultant **deoxyhemoglobin** carries some of the carbon dioxide from body cells back to the lungs for removal. The reduced amount of oxygen carried by the deoxyhemoglobin gives a dark red color to blood. The mechanisms of transporting oxygen and carbon dioxide are covered in chapter 14.

Concentration of Red Blood Cells

Red blood cells are by far the most abundant formed elements. An *RBC count* is a routine clinical test to determine the number of RBCs in a μl of blood. For adult males, healthy values range from 4.7 to 6.1 million RBCs per μl. For adult females, healthy values range from 4.2 to 5.4 million RBCs per μl. Another common clinical test to determine the concentration of RBCs is the *hematocrit*, which is the percentage of blood volume composed of RBCs. This is shown as the "45% of whole blood" in figure 11.1. Average healthy values are 47% in adult males and 42% in adult females. The higher value in males results from the presence of testosterone. Testosterone increases the level of a hormone called erythropoietin, whose function will be discussed shortly. The lower value in females is partially due to the regular loss of blood through menstruation. Women also have higher average body fat, and hematocrit decreases as body fat increases.

The RBC count, hematocrit, and the *hemoglobin concentration* of the blood are commonly measured to determine the blood oxygen-carrying capacity. Hemoglobin concentration is the hemoglobin content expressed in grams per 100 ml of blood. Average healthy values are 13-18 g/100 ml for adult males and 12-16 g/100 ml for adult females.

Healthy values of RBC count, hematocrit, and hemoglobin concentration also vary with altitude. The concentration of RBCs is greater in persons living at higher altitudes because of the reduced oxygen content in air. This reduces the rate at which oxygen can enter the blood, causing a decline in the concentration of oxygen in the blood, which, in turn, stimulates RBC production, as discussed next.

Production

Prior to birth, red blood cells are produced largely by the liver and spleen. Following birth, production takes place primarily in the red bone marrow (myeloid tissue). In infants, RBCs are formed in the red bone marrow of all bones, but in adults RBC formation primarily occurs in the red bone marrow of the skull bones, ribs, sternum, vertebrae, and coxal bones as red bone marrow becomes restricted to these areas.

Red blood cell production varies with the oxygen concentration of the blood and is regulated by a negative-feedback mechanism. If the kidneys and liver sense

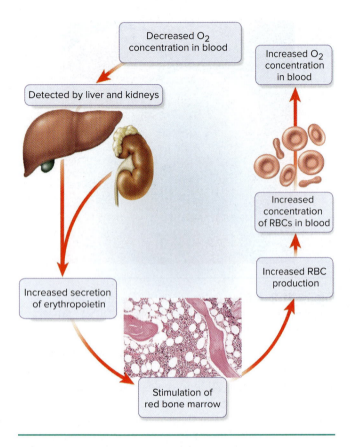

Figure 11.3 Regulation of Red Blood Cell Production. A negative-feedback mechanism corrects for a decreased O_2 concentration in blood. When blood O_2 concentration returns to set point, erythropoietin secretion declines to a basal level. APR

bone marrow in a process called **hematopoiesis.** Hemocytoblasts divide to form myeloid stem cells and lymphoid stem cells that divide to produce the precursor cells that develop into the different types of formed elements. The pattern of cell division and development is shown in figure 11.4. Note that RBCs lose their nuclei and other organelles as they mature.

Life Span and Destruction

The life span of red blood cells is about 120 days, and RBCs are destroyed and produced at a rate of about 2 million per second! Usually, destruction and production are kept in balance.

The plasma membranes of newly formed RBCs are flexible, which allows them to change shape as they pass through small blood vessels. However, with age the membranes lose their flexibility and become fragile and damaged because RBCs lack the organelles necessary to make membrane repairs. Worn-out RBCs are removed from circulation in the liver and spleen by phagocytic cells called **macrophages** (mak′-rō-fāj-es). Macrophages engulf and digest old and damaged RBCs in phagocytic vesicles. See chapter 3 to refresh your understanding of phagocytosis.

The globin portion of hemoglobin is broken down into amino acids, which are reused to form new hemoglobin and other proteins in the body. The heme portion of hemoglobin is broken down into an iron ion and a yellow pigment, *bilirubin* (bil-i-rū′-bin). The iron ion may be temporarily stored in the liver or spleen before being transported to the red bone marrow and used to form more hemoglobin. Bilirubin is secreted by the liver in bile, which is carried into the small intestine for disposal.

low blood oxygen concentration (hypoxemia), such as occurs with blood loss, they release **erythropoietin** (e-rith-ro-poi′-etin) **(EPO),** a hormone that stimulates red bone marrow to produce more RBCs. When the newly made RBCs restore blood oxygen homeostasis, production of EPO declines, causing a decrease in RBC production (figure 11.3). A small amount of EPO is always present to maintain RBC production at a basal rate. Note that the concentration of oxygen in blood triggers the negative-feedback mechanism, which regulates EPO secretion and, therefore, RBC production.

Iron, folic acid, and vitamin B12 are required for RBC production. Iron is required for hemoglobin synthesis because each hemoglobin molecule contains four iron ions. Folic acid and vitamin B12 are required for DNA synthesis during the early stages of RBC formation in red bone marrow. Vitamin B12 is sometimes called the *extrinsic factor* because it is obtained from a source external to the body, such as the diet or an injection. Effective absorption of vitamin B12 from food into the blood is facilitated by *intrinsic factor*, a glycoprotein secreted by the stomach.

All formed elements, including RBCs, are produced from *stem cells* called **hemocytoblasts** in red

 Clinical Insight

An elevated level of blood bilirubin leads to *jaundice*, a yellowing of the skin, mucous membranes, and sclera. It is commonly caused by impeding the removal of bilirubin from the blood due to malfunction of the liver or kidneys, or obstruction of the bile passageways. An elevated rate of RBC breakdown with certain disorders and diseases, such as sickle cell disease and malaria, directly increases the blood bilirubin level and the chance of developing jaundice. Newborns may experience jaundice because their livers are not mature enough to process the bilirubin resulting from the regular destruction of RBCs.

 Check My Understanding

2. How does hemoglobin contribute to the function of red blood cells?
3. How is RBC production regulated?

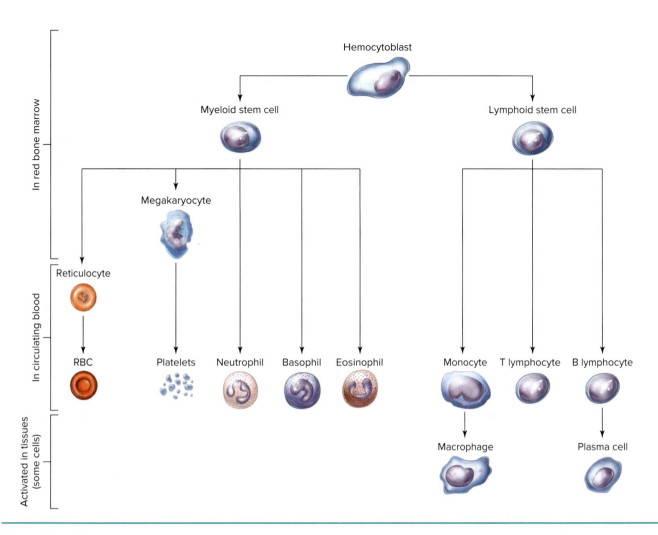

Figure 11.4 Hematopoiesis.
Formed elements develop from hemocytoblasts in red bone marrow. The color of the cells and platelets results from staining. AP|R

11.3 White Blood Cells

Learning Objectives
5. Describe the structure and functions of each type of WBC.
6. Describe the production of WBCs.
7. Indicate the healthy concentration of WBCs in blood and the percentage of each type of WBC.

White blood cells, or *leukocytes* (lu′-kō-sīts) get their name from the white color of pus. These spherical cells are the only formed elements with nuclei and other organelles. A healthy adult's WBC count is typically 4,500 to 10,000 per μl of blood. However, the number of a particular type of WBC increases whenever the body encounters pathogens (disease-causing organisms or chemicals) that it destroys.

Like other formed elements, WBCs are derived from the hemocytoblasts in the red bone marrow and their life span ranges from a few hours to many years. Their production is regulated by chemical signals released by red bone marrow cells, WBCs, and lymphoid tissues.

Function

White blood cells help provide defense against pathogens. Certain WBCs either promote or decrease inflammatory responses. Most of the functions of WBCs are performed within tissues located outside blood vessels. WBCs have the ability to move through capillary walls into tissues in response to chemicals released by damaged tissues or pathogens. They are able to follow a "chemical trail" through the tissue spaces to reach the source of the chemical, a behavior called *chemotaxis*. WBCs move by amoeboid movement, a motion characterized by flowing extensions of cytoplasm that pull the cell along. The WBCs then work to destroy dead cells, pathogens, and foreign substances.

Clinical Insight

Sickle-cell disease (sickle-cell anemia) is an inherited hemolytic disorder that affects about 0.2% of black Americans. Afflicted persons have inherited two abnormal forms of the gene responsible for hemoglobin formation, which causes their hemoglobin to differ from healthy hemoglobin by only a single amino acid. This small change is sufficient to cause RBCs to be sickle-shaped (C-shaped) or elongated and pointed. Such RBCs tend to clump together and block tiny arteries, depriving tissues of oxygen and causing intense pain and fatigue. This can lead to kidney disease, stroke, brain damage, and heart failure. The abnormal hemoglobin cannot transport oxygen efficiently, and the fragile RBCs rupture, further reducing the oxygen-carrying capacity of the blood. Without treatment, life expectancy is less than two years of age. With treatment, it is about age 50.

Persons who inherit only one abnormal form of the gene have a condition known as *sickle-cell trait*. They rarely have severe symptoms. About 8.3% of black Americans have sickle-cell trait. If a man and a woman, each with sickle-cell trait, reproduce, each of their children has a 25% chance of inheriting sickle-cell disease.

Sickle-cell disease is believed to have originated in tropical Africa where malaria was prevalent. Persons with sickle-cell trait have a natural resistance against the malarial parasite, which invades RBCs. This resistance to malaria is what has enabled the abnormal form of the gene to persist.

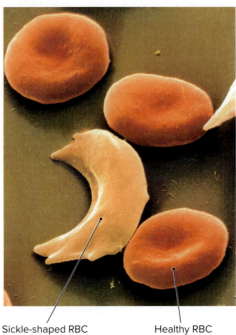

Sickle-shaped RBC   Healthy RBC

©Eye of Science/Science Source

Some WBCs destroy pathogens and cellular debris by phagocytosis. Others release chemicals that clump pathogens together, aiding phagocytosis, and still others release chemicals that kill pathogens. How WBCs fight disease is discussed in chapter 13.

Types of White Blood Cells

White blood cells may be distinguished from red blood cells by microscopic examination of fresh blood. However, WBCs must be stained in order to distinguish them from each other.

The five types of WBCs are neutrophils, eosinophils, basophils, lymphocytes, and monocytes. WBCs are classified by the presence or absence of visible cytoplasmic granules when stained. Neutrophils, eosinophils, and basophils are collectively known as **granulocytes** (gran′-ū-lō-sīts), because their cytoplasm contains small, colored granules. Lymphocytes and monocytes lack visible granules and are therefore called **agranulocytes**. Granulocytes are about 1.5 times larger than RBCs, and are distinguished from each other by the shapes of their nuclei and the color of their cytoplasmic granules. Agranulocytes are distinguished from each other by cell size and nuclear shape. Lymphocytes are only slightly larger than RBCs, while monocytes are two to three times larger than RBCs. See table 11.1 and figure 11.5.

Neutrophils

Neutrophils (nū′-trō-fils) are the most abundant white blood cells and form 40% to 60% of the total WBCs. They contain a nucleus with two to five lobes and inconspicuous lavender-staining granules. Neutrophils are attracted by chemicals released from damaged tissues and are the first WBCs to respond to tissue damage. They engulf bacteria and cellular debris by phagocytosis and release the enzyme *lysozyme*, which destroys some bacteria. Because their primary function is to destroy bacteria, the number of neutrophils increases dramatically in acute bacterial infections.

Eosinophils

Eosinophils (ē-ō-sin′-ō-fils) constitute 1% to 4% of the white blood cells. They are characterized by a bilobed nucleus and red-staining cytoplasmic granules. Eosinophils

Table 11.1 Formed Elements in Blood

Formed Elements	Description	Healthy Count*	Function
Red blood cells	Biconcave discs; no nucleus and other organelles; contain hemoglobin	4.2–5.4 million/μl in females; 4.7–6.1 million/μl in males	Transport O_2 and CO_2
White blood cells	Spherical shape; have nucleus and other organelles	4,500–10,000/μl	Help provide the body with defense and immunity
Granulocytes	Cytoplasmic granules present; 1.5 times larger than RBCs		
Neutrophils	Nucleus with two to five lobes; tiny cytoplasmic granules stain lavender	40–60% of total WBCs	Phagocytize bacteria and cellular debris
Eosinophils	Nucleus bilobed; cytoplasmic granules stain red	1–4% of total WBCs	Counteract histamine released in allergic reactions; destroy parasitic worms; phagocytize antigen–antibody complexes
Basophils	Nucleus U-shaped or bilobed; cytoplasmic granules stain blue	0.5–1% of total WBCs	Intensify inflammatory response in allergic reactions by releasing histamine and heparin
Agranulocytes	Cytoplasmic granules absent		
Lymphocytes	Very little cytoplasm around spherical nucleus; slightly larger than RBCs	20–40% of total WBCs	Provide immunity by producing antibodies and destroying pathogens and unhealthy cells
Monocytes	Nucleus usually U- to kidney-shaped; two to three times larger than RBCs	2–8% of total WBCs	Phagocytosis of bacteria and cellular debris
Platelets	Tiny cytoplasmic fragments	150,000–400,000/μl	Form platelet plugs and start clotting of the blood

*Values vary slightly depending on measurement procedures.

reduce inflammation by neutralizing histamine, a chemical released by basophils during allergic reactions. They also destroy parasitic worms and phagocytize antigen–antibody complexes.

Basophils

Basophils (bā′-sō-fĭls) are the least numerous of the white blood cells, forming only 0.5% to 1% of the WBCs. They are characterized by a nucleus that is U-shaped or bilobed and by large, blue-staining cytoplasmic granules. They release histamine and heparin when tissues are damaged and in allergic reactions. Histamine promotes inflammation by dilating blood vessels to increase blood flow in affected areas and making blood vessels more permeable, which allows other WBCs to enter the affected tissues. Heparin inhibits clot formation.

Lymphocytes

Lymphocytes (lĭm′-fō-sīts) form 20% to 40% of the circulating white blood cells. They are the smallest WBCs and are distinguished by a spherical nucleus that is enveloped by very little cytoplasm. Lymphocytes are especially abundant in lymphoid tissues and play a vital role in immunity, a defense mechanism that fights against specific pathogens and builds a memory of these encounters. There are two types of lymphocytes. **T lymphocytes** directly attack and destroy pathogens (bacteria and viruses), and **B lymphocytes** develop into antibody-producing *plasma cells* in response to foreign antigens. The details of lymphocytes and immunity are discussed in chapter 13.

 Clinical Insight

A *complete blood count (CBC)* is one of the most common and clinically useful blood tests. It consists of several different blood tests, some of which are RBC count, WBC count, platelet count, differential WBC count (the percentage of each type of WBC), hematocrit, and hemoglobin concentration. Unhealthy values for these tests are associated with infectious and inflammatory processes and with specific blood disorders.

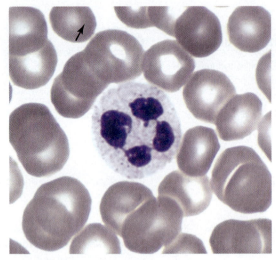
(a) Neutrophil (1,600×)

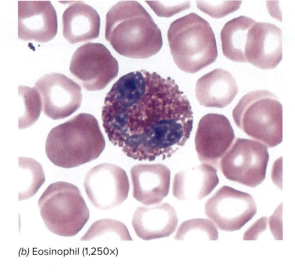
(b) Eosinophil (1,250×)

(c) Basophil (320×)

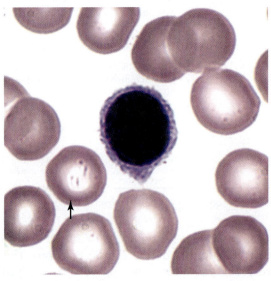
(d) Lymphocyte (1,600×)

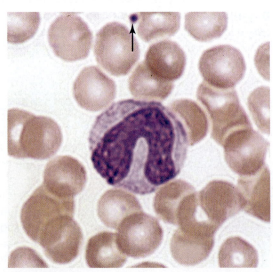
(e) Monocyte (410×)

Figure 11.5 White Blood Cells.
Note the platelets indicated by the arrows in (c) and (e). The various WBCs in these photomicrographs are shown at magnifications ranging from 320× to 1,600×. Therefore, the best way to understand their size is to compare them to the nearby RBCs. **AP|R**
(a–b, d) ©McGraw-Hill Education/Al Telser, photographer; (c–e) ©Victor P. Eroschenko RF

Monocytes

Monocytes (mon'-ō-sīts) are the largest white blood cells, and they comprise 2% to 8% of the WBCs. A U-shaped or kidney-shaped nucleus and abundant cytoplasm distinguish monocytes. Monocytes are active in phagocytosis. The number of monocytes in the blood increases during viral infections and inflammation of tissues. Monocytes in body tissues are called **macrophages.** They are very active phagocytic cells that join with neutrophils to clean up damaged tissues and pathogens. They carry out their functions of engulfing dead cells, cellular debris, and bacteria only after migrating into body tissues.

 Check My Understanding

4. What are the functions of each type of WBC?
5. What are the characteristics that differentiate each type of WBC?

11.4 Platelets AP|R

Learning Objectives
8. Describe the structure, production, and healthy concentration of platelets.
9. Describe the function of platelets.

Platelets are cytoplasmic fragments of *megakaryocytes*, large cells that develop from hemocytoblasts in red bone marrow (see figure 11.4). A platelet is much smaller than a red blood cell (see figure 11.5a, d). There are typically 150,000 to 400,000 platelets per μl of blood, and their life span is about one to two weeks. The primary role of platelets is to stop bleeding. When a blood vessel is injured, platelets clump together at the injured site and release chemicals that promote vascular spasm and coagulation, which are discussed later (figure 11.6).

 Check My Understanding

6. What is the function of platelets?

11.5 Plasma

Learning Objective
10. Explain the importance of the components of plasma.

Plasma is the fluid portion of the blood and consists of over 90% water. Water is the liquid carrier of plasma solutes (dissolved substances) and formed elements, in addition to being the solvent of all living systems. Plasma contains a great variety of solutes, such as nutrients, enzymes, hormones, antibodies, waste products, electrolytes, and respiratory gases. Table 11.2 lists the major types of solutes in plasma. Plasma solutes are constantly being added and removed, so the solutes are usually in a state of dynamic balance that is maintained by a variety of homeostatic mechanisms.

Plasma Proteins

Plasma proteins are the most abundant solutes. They are not used as an energy source but remain in the plasma. Less than 1% of plasma proteins are enzymes and hormones. The three major groups of plasma proteins are albumin, globulins, and fibrinogen. Except for gamma globulins, plasma proteins are produced by the liver and are released into the blood.

Albumins form about 60% of the plasma proteins. Albumins play an important role in transporting many hydrophobic substances, including lipids, lipid-soluble vitamins, some hormones, and certain ions. They also serve as buffers that help to keep the pH of the blood within narrow limits and play an important role in maintaining the osmotic pressure of the blood. Osmotic pressure determines the water balance between the blood and body cells. If the osmotic pressure of the blood declines, water moves into the body tissues and causes the tissues to swell (edema). This also decreases blood volume and, in severe cases, may decrease blood pressure as well. If the osmotic pressure of

Table 11.2 Major Solutes in Blood Plasma

Solute	Description
Albumins	Help transport hydrophobic substances, maintain osmotic pressure and pH of blood
Globulins	Alpha and beta types transport hydrophobic substances and some ions; gamma type is antibodies
Fibrinogen	Soluble protein that is converted to insoluble fibrin during formation of blood clot
Nitrogenous wastes	Breakdown products of proteins, nucleic acids, and creatine phosphate
Nutrients	Amino acids, fatty acids, glycerol, vitamins, and glucose
Enzymes and hormones	Help regulate metabolic processes
Electrolytes	Help regulate blood pH, osmotic pressure, and the ionic balance between blood and interstitial fluid
Respiratory gases	Approximately 1.5% of the oxygen and 7% of the carbon dioxide transported by blood is dissolved in plasma

Clinical Insight

A high level of blood cholesterol is associated with an increased risk of heart disease. Cholesterol occurs in the blood in combination with triglycerides and carrier proteins. These lipid-protein complexes are called *lipoproteins*. Considerable evidence links a high concentration of blood low-density lipoprotein (LDL), the so-called "bad" cholesterol, with heart disease. In contrast, a high level of blood high-density lipoprotein (HDL), the "good" cholesterol, reduces the risk of heart disease. The cholesterol level in the blood results from a combination of heredity, diet, and exercise.

A total blood cholesterol level less than 200 mg/dl (milligrams per deciliter) is a desirable goal. A blood LDL concentration of 100 to 130 mg/dl is near optimal. Persons at risk of coronary artery disease, such as smokers and the elderly, should strive for an LDL level less than 100. Reducing the amount of saturated fats (red meat, milk products, and egg yolks) and trans fats (present in hydrogenated oils) in the diet can decrease the LDL level.

The desired HDL level averages 40 to 50 mg/dl in men and 50 to 60 mg/dl in women. HDL level may be increased by exercise and maintaining a healthy weight.

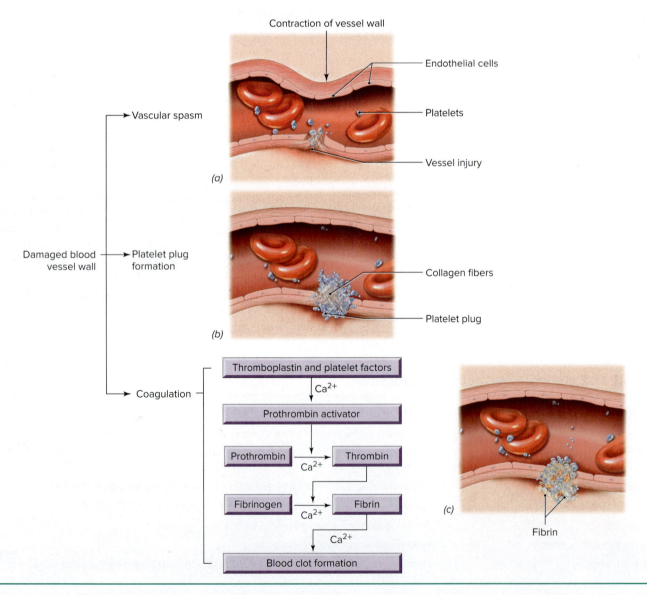

Figure 11.6 Processes of Hemostasis.
(a) Vascular spasm. *(b)* Platelet plug formation. *(c)* Coagulation.

the blood increases, water moves into the blood, causing an increase in blood volume and in blood pressure while reducing the amount of water available to body cells.

Globulins form about 36% of plasma proteins. The three types of globulins are alpha, beta, and gamma globulins. Many alpha and beta globulins play a role in carrying hydrophobic substances and some ions. Alpha and beta globulins make up the protein portion of low-density lipoproteins (LDLs) and high-density lipoproteins (HDLs), which function in transporting lipids. Gamma globulins are antibodies (immunoglobulins) that are produced by the B lymphocytes and are involved in immunity (see chapter 13 for details).

Fibrinogen forms only 4% of the plasma proteins. It plays a vital role in the blood-clotting process. Fibrinogen is a soluble protein that is converted to insoluble fibrin to form blood clots (figure 11.6).

Nitrogenous Wastes

Nitrogenous wastes are nitrogen-containing substances that include ammonia, urea, uric acid, and creatinine. Ammonia and urea are wastes produced during protein metabolism. Uric acid comes from the catabolism of nucleic acids. Creatinine is produced as a result of creatine phosphate breakdown in the muscle cells (see chapter 7). These wastes are carried in the blood to the kidneys, where they are excreted into urine. Plasma levels of these wastes are commonly used as indicators of kidney health.

Electrolytes

Most of the plasma electrolytes are ions of inorganic compounds that are either absorbed from food or released from body cells. The common electrolytes include sodium ions (Na^+), potassium ions (K^+), calcium ions (Ca^{2+}), chloride ions (Cl^-), bicarbonate ions (HCO_3^-), and phosphate ions (PO_4^{3-}). Electrolytes help to maintain the osmotic pressure and pH of the blood, and a healthy ionic balance between interstitial fluid and blood.

Check My Understanding

7. What are the major components of blood plasma?

11.6 Hemostasis

Learning Objective

11. Describe the sequence of events that occurs during hemostasis.

Blood contains both *procoagulants,* substances that promote clotting, and *anticoagulants,* substances that inhibit clotting. Healthy blood within a healthy blood vessel does not clot because the anticoagulants have a higher concentration than procoagulants. Vessel injury triggers an increase in procoagulant concentration that initiates the process of hemostasis. **Hemostasis** is a positive-feedback mechanism initiated after vascular injury to stop or limit blood loss. There are three separate but interrelated processes involved in hemostasis: vascular spasm, platelet plug formation, and coagulation (figure 11.6). As you will see next, vascular spasm, platelet plug formation, and coagulation increase once they begin, and the three processes activate each other, essentially occurring simultaneously.

Vascular Spasm

A *vascular spasm,* or constriction, of the blood vessel results from contraction of smooth muscle within the vessel wall at the damaged site (figure 11.6a). Physical damage to the vessel causes the release of chemicals that initiate the spasm. Narrowing of the blood vessel restricts blood loss from the damaged vessel and it lasts for several minutes, which allows time for formation of the platelet plug and clotting. As platelets accumulate at the site of the damage, they secrete *serotonin,* a chemical that continues the contraction of the smooth muscle in the damaged vessel.

Platelet Plug Formation

Platelets usually do not stick to each other or to the wall of the blood vessel because the vessel wall contains several substances that repel platelets. However, when a vessel is damaged, the collagen in the vessel wall is exposed. Platelets are attracted to the site and adhere to the collagen and to each other so that a cluster of platelets accumulates to plug the break (figure 11.6b). This process is enhanced by the chemicals released from both the damaged blood vessel wall and platelets aggregated at the damaged site. The formation of a *platelet plug* may not seal off the damaged blood vessel but it sets the stage for coagulation.

Coagulation

Coagulation (kō-ag-ū-lā′-shun), or blood clotting, is the most effective process of hemostasis. Clot formation is a complex process but it is completed within one minute after a blood vessel has been damaged. The clot is restricted to the site of damage because that is where procoagulants outnumber anticoagulants. The key steps in coagulation are summarized here and shown in figure 11.6c:

1. Damaged tissues release *thromboplastin* and aggregated platelets release *platelet factors,* which react with several clotting factors in the plasma to produce **prothrombin activator.**
2. In the presence of calcium ions, prothrombin activator stimulates the conversion of **prothrombin,** an inactive enzyme, into the active enzyme **thrombin.**
3. In the presence of calcium ions, thrombin converts molecules of fibrinogen, a soluble plasma

Clinical Insight

Sometimes unwanted blood clots (thrombi) form in unbroken blood vessels, where they may pose a serious health threat. Certain enzymes, such as streptokinase and urokinase, have been used for some time to help dissolve such clots. It is also common to use a form of tissue plasminogen activator (tPA) to dissolve thrombi. Because it is an engineered form of a clot-dissolving enzyme that naturally occurs in the body, unwanted side effects are minimal. tPA is less likely to trigger allergic reactions or antibody production.

Persons at risk for thrombus formation may be advised to take periodic low dosages of aspirin as a preventive measure. Aspirin inhibits platelets' release of thromboxanes, which are essential for all three processes of hemostasis. In this way, aspirin slows clotting and helps prevent thrombus formation.

protein, into threadlike, interconnected strands of insoluble **fibrin.** Fibrin strands crosslink to form a meshwork that entraps blood cells and platelets and sticks to the damaged tissue to form a **thrombus,** a blood clot.

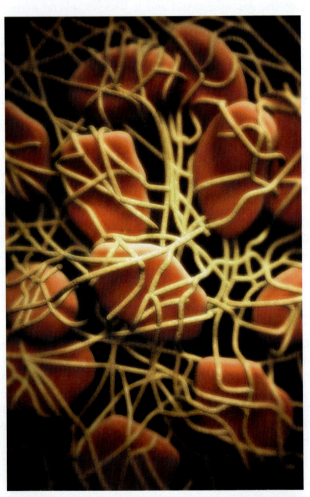

Figure 11.7 A Blood Clot.
Digitally generated illustration simulating a microscopic view of a blood clot, which consists of blood cells and platelets trapped in a meshwork of fibrin strands.
©MedicalRF.com RF

After a clot has formed, the platelets pull on the fibrin strands to bring the damaged edges closer together, which is important for vessel healing and the formation of a more compact clot that is harder to dislodge (figure 11.7). Simultaneously, fibroblasts migrate into the clot and form dense irregular connective tissue that repairs the damaged area. As healing occurs, *tissue plasminogen* (plaz-min′-o-jen) *activator (tPA),* released by the tissues of the damaged blood vessel, converts *plasminogen,* an inactive enzyme in blood plasma, into *plasmin,* its active form. Plasmin breaks down fibrin and dissolves the blood clot.

Check My Understanding

8. What are the three major processes in hemostasis?
9. How are blood clots formed?

11.7 Human Blood Types

Learning Objectives
12. Explain the basis of blood typing and why it is important.
13. Identify the blood typing antigens and antibodies in each ABO blood type and Rh blood type.

Several different blood types occur in humans. The most familiar ones involve the ABO blood group (types A, B, AB, and O) and the Rh blood group (Rh+ and Rh−).

Blood types are classified by the presence or absence of certain antigens, which are glycoproteins and glycolipids, located within the plasma membrane of the red blood cells. Each person has a unique set of RBC antigens that are inherited and remain unchanged throughout life. Within the plasma, an individual possesses antibodies against antigens that are not present on the RBCs. Remember, antibodies are defensive proteins produced by plasma cells. Whenever RBCs with one type of antigen are transfused into the blood of a person whose RBCs do not possess the antigen, the antigens on the transfused RBCs are recognized as foreign by the recipient's antibodies and agglutination occurs. During **agglutination,** the recipient's antibodies bind to the antigens on the transfused RBCs, which causes the RBCs to clump together. This reaction can be fatal because the clumps of RBCs block small vessels and deprive the tissues supplied by these vessels of nutrients and oxygen. Of the 600 potential antigens on human RBCs, only a few can cause significant agglutination in a blood transfusion. These antigens are the A antigen, B antigen, and Rh antigen.

ABO Blood Group

The ABO blood group includes types A, B, AB, and O blood, which are classified by the presence or absence of A and B antigens on red blood cells. Type A blood has A antigens on RBCs. Type B blood has B antigens on RBCs. Type AB blood has both A and B antigens on RBCs. In type O blood, neither the A antigen nor the B antigen is present (figure 11.8).

After birth, each person's plasma cells start producing antibodies against the A or B antigen that is not present on his or her RBCs. As a result, people with type A blood develop anti-B antibodies in their plasma. Those with type B blood develop anti-A antibodies in their plasma. Those with type O blood develop both anti-A and anti-B antibodies in their plasma. People with type AB blood have neither of these antibodies in their plasma (figure 11.8).

Rh Blood Group

Blood typing also routinely tests for the presence of the **Rh (D) antigen.** There are several Rh antigens, but it is the D antigen that is most important. The Rh antigen is named after *Rhesus* monkeys, in which the blood group was first discovered.

If the Rh antigen is present on the red blood cells, the blood is typed as Rh positive (Rh+). If the Rh antigen is absent, the blood is Rh negative (Rh−). Like the A and B antigens, the presence or absence of the Rh antigen is inherited.

A major difference between the ABO blood group and the Rh blood group is that antibodies are not automatically produced against the Rh antigen after birth.

Clinical Insight

The ABO blood type can be easily determined by placing two separate drops of blood to be tested on a glass slide. A drop of *serum* (the remaining fluid after blood has clotted) containing anti-A antibodies is added to one drop, while serum containing anti-B antibodies is added to the other. The pattern of agglutination that occurs in the separate drops of blood indicates the blood type.

The Rh blood type is determined by adding serum containing anti-Rh antibodies to a drop of blood on a glass slide. If agglutination occurs, the blood is Rh+. If agglutination does not occur, the blood is Rh−.

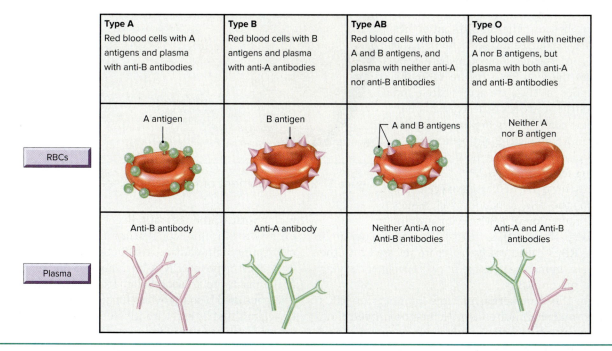

Figure 11.8 Antigen and Antibody Characteristics of the ABO Blood Group.

Instead, a person has to be exposed to the Rh antigen for the lymphoid system to respond by making anti-Rh antibodies. The first time this occurs, there is no agglutination reaction but the production of anti-Rh antibodies begins. The buildup of anti-Rh antibodies sensitizes the person to future introductions of Rh antigens. If a person with Rh− blood is sensitized and receives a subsequent transfusion of Rh+ RBCs, the anti-Rh antibodies will cause agglutination of the transfused Rh+ RBCs, usually with serious or fatal results. Anti-Rh antibodies are never found in individuals with Rh+ RBCs.

Hemolytic Disease of the Newborn

A similar kind of problem occurs in **hemolytic disease of the newborn (HDN),** a blood disorder of newborn infants that results from destruction of fetal red blood cells by maternal antibodies.

When a woman with Rh− blood is pregnant with her first Rh+ fetus, some of the fetal Rh+ RBCs may accidentally enter the maternal blood due to broken placental blood vessels. This occurs most often during the third trimester and childbirth. The introduction of fetal RBCs with Rh antigens triggers the buildup of anti-Rh antibodies in the woman's blood. The buildup is slow but the mother has become sensitized to the Rh antigen.

Hemolytic disease of the newborn may develop in a subsequent pregnancy with an Rh+ fetus because the anti-Rh antibodies in maternal blood readily pass through the placenta into the fetal blood, where they agglutinate the fetal RBCs (figure 11.9). If a large number of RBCs are agglutinated and destroyed, the fetus has a decreased ability to transport oxygen. It is important to note that the anti-A and anti-B antibodies cannot cross the placenta and pose no threat to the developing fetus.

In response to a decreased oxygen concentration, the fetal blood-forming tissues increase production of RBCs. In an attempt to rapidly produce RBCs, large numbers of nucleated, immature RBCs called *erythroblasts* are released into the blood. These immature cells are not as capable of carrying oxygen as mature RBCs.

Also, the destruction of large numbers of RBCs produces other harmful effects. Hemoglobin freed from RBCs may interfere with healthy kidney function and cause kidney failure. Blood flow to other vital organs could also be blocked. The breakdown of large amounts of hemoglobin forms an excess of bilirubin, a yellow

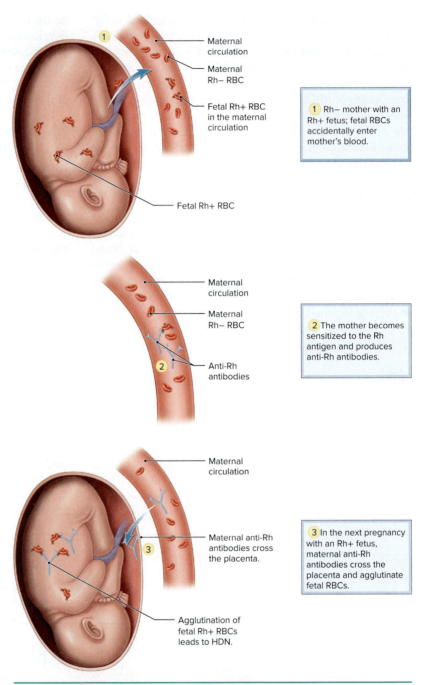

Figure 11.9 Development of Hemolytic Disease of the Newborn.

pigment that produces jaundice. Oxygen deficiency and excessive bilirubin concentrations in the fetal blood may cause brain damage in afflicted infants.

Treatment of HDN at birth involves the replacement of the infant's total blood volume slowly with Rh− blood. The transfused blood provides functional RBCs that cannot be agglutinated by anti-Rh antibodies that may still be present and reduces the bilirubin concentration to eliminate the jaundice. Subsequently, the infant's own RBC production will again produce Rh+ RBCs, but by then all anti-Rh antibodies will have been removed from the blood.

Compatibility of Blood Types for Transfusions

When blood loss is substantial, blood transfusions are routinely given to replace lost blood. A blood transfusion is prepared by separating whole blood into its separate components through centrifugation (spinning it at high speeds) (figure 11.1). Once the plasma layer is removed, the compacted red blood cells are suspended in a nutrient-rich additive and are ready for transfusion. The removal of the plasma removes donor antibodies that can cause an agglutination reaction in the recipient.

It is preferable to perfectly match the donor's blood type with that of the recipient's in blood transfusions. However, a compatible but different blood type may be used in an extreme emergency. If this is done, care must be taken to ensure that the antigens of the donor's blood are compatible with the antibodies of the recipient's blood. For example, RBCs with the A antigen can be given to recipients with type A or type AB blood because neither type contains anti-A antibodies. However, if RBCs with the A antigen were given to recipients with type B or type O blood, agglutination would occur because both types contain anti-A antibodies (figure 11.10). Individuals with Rh+ blood can be given both Rh+ and Rh− blood types in a transfusion, because an Rh+ individual will never produce anti-Rh antibodies. However, individuals with Rh− blood are given only Rh− blood types to prevent sensitization and the formation of anti-Rh antibodies.

Table 11.3 indicates the preferred ABO and Rh blood types that are used for transfusions. Blood types listed in this table are classified by combining the ABO and Rh

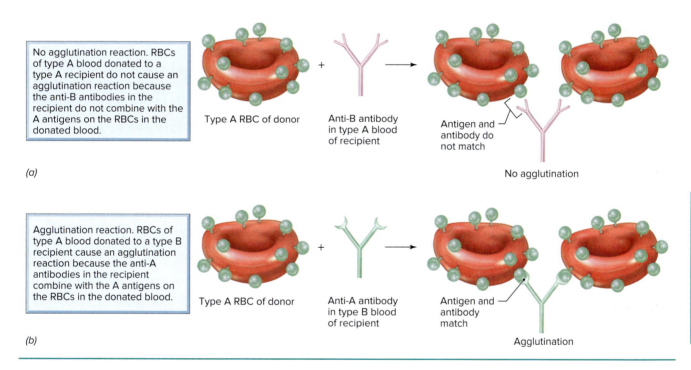

Figure 11.10 Compatible and Incompatible Transfusions.

Table 11.3 Preferred and Acceptable ABO and Rh Blood Types for Transfusions

Blood Type of Recipient	Preferred Blood Type of Donor	Acceptable Blood Types of Donor
A−	A−	O−
A+	A+	A−, O−, O+
B−	B−	O−
B+	B+	B−, O−, O+
AB−	AB−	A−, B−, O−
AB+	AB+	AB−, A−, A+, B−, B+, O−, O+
O−	O−	None
O+	O+	O−

groups; for example, type A− means the blood contains A antigens and no Rh antigens, type A+ means the blood contains both A and Rh antigens. Note that type AB+ blood may receive RBCs from all blood types and that the RBCs of type O− blood may be given to all blood types.

Clinical Insight

Hemolytic disease of the newborn is preventable by injecting serum containing anti-Rh antibodies (trade name RhoGAM) into the blood of Rh− females. The first dose is injected at 28 weeks of pregnancy, with a second dose given immediately after the birth of an Rh+ infant, or after miscarriage or abortion. The anti-Rh antibodies agglutinate and destroy any fetal Rh+ RBCs that may have entered the mother's blood before they can stimulate the production of anti-Rh antibodies and sensitize the mother. Further, pregnant Rh− mothers will be given an injection of RhoGAM near the fifth month of subsequent pregnancies as a safety precaution.

Check My Understanding

10. What determines an individual's ABO blood type?
11. Why is blood typing important in transfusions?
12. What is the cause of hemolytic disease of the newborn?

11.8 Disorders of the Blood

Learning Objective
14. Describe the major blood disorders.

Blood disorders may be grouped as red blood cell disorders, white blood cell disorders, and disorders of hemostasis. Healthy values for common blood tests are located in Appendix C. Blood tests are valuable in diagnosing a variety of disorders. Note that many of the disorders described in the next section are associated with unhealthy values of blood tests.

Red Blood Cell Disorders

Anemia (ah-nē′-mē-ah) is a decrease in the oxygen-carrying capacity of the blood and is the most common blood disorder. A decreased number of red blood cells or an insufficient amount of hemoglobin reduces the blood's capacity to carry oxygen. There are several different types of anemia:

- *Nutritional anemia* results from insufficient amounts of iron in the diet.
- *Hemorrhagic anemia* results from the excessive loss of RBCs through bleeding.
- *Pernicious anemia* results from a deficiency of intrinsic factor, which prevents absorption of sufficient vitamin B12 from the intestine to support adequate RBC production.
- *Hemolytic anemia* results from premature rupture of RBCs so that hemoglobin is released into the plasma.
- *Aplastic anemia* results from destruction of red bone marrow or its inability to produce a sufficient number of RBCs.
- *Sickle-cell disease* (see the clinical insight box earlier in this chapter)

Polycythemia (pol-ē-sī-thē-mē-ah) is a condition characterized by an excess of RBCs in the blood. The excess RBCs increase blood volume and viscosity, which impairs circulation. It also leads to an increase in blood pressure, which can cause the rupture of blood vessels. It may result from cancer of the RBC-forming cells, dehydration, hypersecretion of EPO, or from *blood doping*. Blood doping is any technique used to raise the number of RBCs in the blood to enhance athletic performance. Blood doping is considered cheating by various athletic authorities.

White Blood Cell Disorders

Infectious mononucleosis is a contagious disease of the lymphoid tissue caused by the Epstein–Barr virus (EBV). It occurs primarily in young adults and kissing is a common mode of transmission. Three times more females contract the disease than males. It infects B lymphocytes, which enlarge and resemble monocytes. Symptoms include fever, headache, fatigue, sore throat, and swollen lymph nodes. There is no cure, but infectious mononucleosis usually persists for about four weeks. However, in some persons it may linger for months or years, and relapses may be frequent.

Leukemia (lū-kē′-mē-ah) is a group of cancers of the red bone marrow cells that form WBCs. It is characterized by an excess production of WBCs and the crowding out of RBC- and platelet-forming cells. Acute forms affect primarily children or young adults; chronic forms occur more often in adults. The various types of leukemia are classified according to the predominant WBC involved. Treatment usually involves chemotherapy and sometimes a transplant of red bone marrow from a compatible donor.

Disorders of Hemostasis

Hemophilia (hē-mō-fil′-ē-ah) is a group of inherited disorders that occur more often in males because they are linked to the X chromosome (see chapter 18). Hemophilia is characterized by spontaneous bleeding and a reduced ability to form blood clots. It may be caused by a deficiency of any one of several plasma clotting factors. There is no cure for hemophilia, but it is treated by injection or transfusion of the missing clotting factors.

Thrombocytopenia (throm-bō-sī-tō-pē'-nē-ah) is a condition in which the number of platelets is low enough (<50,000/μl) that spontaneous bleeding cannot be prevented. Bleeding from many small vessels typically results in purplish blotches appearing on the skin.

Thrombosis is the condition resulting from the formation of a blood clot in an unbroken blood vessel. Such clots tend to form where the lining of a blood vessel is roughened or damaged. They can cause serious effects if they plug an artery and deprive vital tissues of blood. Blood clots form more frequently in veins than in arteries, causing a condition known as *thrombophlebitis*, which is inflammation of the veins due to a blood clot.

Sometimes, a clot formed in a vein breaks free and is carried by the blood only to lodge in an artery, often a branch of a pulmonary artery. A moving blood clot or foreign body in the blood is called an **embolus,** and when it blocks a blood vessel, the resulting condition is known as an **embolism.** An embolism can produce very serious and sometimes fatal results if it lodges in a vital organ and blocks the flow of blood.

Chapter Summary

11.1 General Characteristics of Blood

- Blood is composed of plasma (55%) and formed elements (45%). Red blood cells constitute nearly all of the formed elements.
- Blood is heavier and about four times more viscous than water, and it is slightly alkaline.
- About 8% of the body weight consists of blood. Blood volume ranges between 4 and 6 liters.

11.2 Red Blood Cells

- Red blood cells are biconcave discs that lack nuclei and other organelles, and contain a large amount of hemoglobin. Their primary function is the transport of respiratory gases.
- Hemoglobin is composed of heme, an iron-containing pigment, and globin, a protein. It plays a vital role in oxygen transport and participates in carbon dioxide transport.
- RBCs are very abundant in the blood. They number 4.7 to 6.1 million per μl in males and 4.2 to 5.4 million per μl in females.
- RBCs are formed from hemocytoblasts in the red bone marrow. The rate of production is controlled by the oxygen concentration of the blood via a negative-feedback mechanism. A decreased oxygen concentration stimulates kidney and liver cells to release erythropoietin, which stimulates increased production of RBCs by red bone marrow.
- Iron, amino acids, vitamin B12, and folic acid are essential for RBC production.
- RBCs live about 120 days before they are destroyed by macrophages in the spleen and liver. In hemoglobin breakdown, the iron ions are recycled for use in forming more hemoglobin. Bilirubin, a yellow pigment, is a waste product of hemoglobin breakdown. Amino acids from globin are recycled for use in making new proteins.

11.3 White Blood Cells

- White blood cells are also formed from hemocytoblasts in the red bone marrow. They retain their nuclei and other organelles, and number 4,500 to 10,000 per μl of blood.
- WBCs help to defend the body, and most of their activities occur within body tissues.
- The five types of WBCs are categorized into two groups. Granulocytes have visible cytoplasmic granules and include neutrophils, eosinophils, and basophils. Agranulocytes lack visible cytoplasmic granules and include lymphocytes and monocytes.
- Neutrophils and monocytes are phagocytes that destroy bacteria and clean up cellular debris.
- Eosinophils help to reduce inflammation and destroy parasitic worms.
- Basophils promote inflammation.
- Lymphocytes play vital roles in immunity.

11.4 Platelets

- Platelets are fragments of megakaryocytes in the red bone marrow. They number 150,000 to 400,000 per μl of blood.
- Platelets play a crucial role in hemostasis by forming platelet plugs and starting coagulation.

11.5 Plasma

- Plasma, the liquid portion of the blood, consists of over 90% water along with a variety of solutes, including nutrients, nitrogenous wastes, proteins, electrolytes, and respiratory gases.
- There are three major types of plasma proteins. Albumins are most numerous. Their major functions include the transport of hydrophobic substances, and helping to maintain the osmotic pressure and pH of the blood. Alpha and beta globulins transport hydrophobic substances and some ions. Gamma globulins are antibodies that are involved in immunity. Fibrinogen is a soluble protein that is converted into insoluble fibrin during coagulation.
- Less than 1% of plasma proteins are enzymes and hormones.
- Nitrogenous wastes in plasma include urea, uric acid, ammonia, and creatinine.
- Electrolytes include ions of sodium, potassium, calcium, bicarbonate, phosphate, and chloride. Electrolytes help to maintain the pH and osmotic pressure of the blood, in addition to the ionic balance between blood and interstitial fluid.

11.6 Hemostasis

- Hemostasis is a series of processes involved in the stoppage of bleeding. It consists of three processes: vascular spasm, platelet plug formation, and coagulation.
- Vascular spasm reduces blood loss until the other processes can occur.
- Platelets stick to the damaged tissue of the blood vessel wall and to each other to form a platelet plug.
- Platelets and the damaged blood vessel wall initiate clot formation by releasing platelet factors and thromboplastin, which cause the formation of prothrombin activator. Prothrombin activator converts prothrombin into thrombin, which, in turn, converts fibrinogen into fibrin. Fibrin strands form the clot.
- After clot formation, fibroblasts invade the clot and gradually replace it with dense irregular connective tissue as the clot is dissolved by enzymes.

11.7 Human Blood Types

- Blood types are determined by the presence or absence of specific antigens on the plasma membranes of red blood cells.
- The four ABO blood types, A, B, AB, and O, are based on the presence or absence of A antigen and B antigen.
- Anti-A and anti-B antibodies are spontaneously formed against the antigen(s) that is (are) not present on a person's RBCs.
- Blood with RBCs containing the Rh antigen is typed as Rh+. Blood without the Rh antigen is typed as Rh−.
- Anti-Rh antibodies are produced only after Rh+ RBCs are introduced into a person with Rh− blood. Once a person is sensitized in this way, a subsequent transfusion of Rh+ blood results in agglutination of the transfused RBCs.
- If incompatible blood is transferred, agglutination of the transfused RBCs occurs. The clumped RBCs plug small blood vessels, depriving tissues of nutrients and oxygen. The result may be fatal.
- Transfusions must be made using only compatible blood types. Types A, B, AB, and O blood recipients can only receive RBCs with antigens that will not trigger an agglutination reaction with antibodies present in plasma. Type Rh+ blood recipients can receive the RBCs of types Rh− and Rh+ blood. Type Rh− blood recipients can receive the RBCs of type Rh− blood only.
- Hemolytic disease of the newborn occurs in newborn infants when a sensitized Rh− woman is pregnant with an Rh+ fetus. Her anti-Rh antibodies pass through the placenta into the fetus and agglutinate the fetal RBCs, producing anemia and jaundice.

11.8 Disorders of the Blood

- Anemia is the most common disorder, and it may result from a variety of causes.
- Other disorders include polycythemia, infectious mononucleosis, leukemia, hemophilia, thrombocytopenia, thrombosis, and embolism.

Improve Your Grade

Connect Interactive Questions Reinforce your knowledge using multiple types of questions: interactive, animation, classification, labeling, sequencing, composition, and traditional multiple choice and true/false.

SmartBook Proven to help students improve grades and study more efficiently, SmartBook contains the same content within the print book but actively tailors that content to the needs of the individual.

Anatomy & Physiology REVEALED® Dive into the human body by peeling back layers of cadaver imaging. Utilize this world-class cadaver dissection tool for a closer look at the body anytime, from anywhere.

CHAPTER 12

©Jupiterimages/Getty Images RF

Cardiovascular System

A two-alarm fire is called in and the alarm begins to sound in the local fire station. Charlie, a veteran firefighter, begins to shout directions as he and the others in his unit don their gear. As they travel to the site of the blaze, Charlie is so focused on the task at hand that he is barely aware of the cardiovascular changes occurring within his body. His heart rate increases in order to increase his blood pressure, which in turn increases blood flow through his body. Changes within his blood vessels allow blood flow to be prioritized to organs that will be called upon once he arrives at the scene. Increasing activity in his skeletal muscle tissue, cardiac muscle tissue, and nervous tissue requires elevated rates of ATP production, which in turn require an increase in the delivery of oxygen, glucose, and fatty acids. Increased blood flow to the lungs, liver, and adipose tissue is needed to maintain sufficient levels of these vital chemicals. By the time the fire truck reaches the scene, Charlie is physically prepared to rush into the burning building to rescue trapped inhabitants, thanks in part to the actions of his cardiovascular system.

CHAPTER OUTLINE

12.1 Anatomy of the Heart
- Protective Coverings
- The Heart Wall
- Heart Chambers
- Heart Valves
- Flow of Blood Through the Heart
- Blood Supply to the Heart

12.2 Cardiac Cycle
- Heart Sounds

12.3 Conducting System of the Heart
- Electrocardiogram

12.4 Regulation of Heart Function
- Autonomic Regulation
- Other Factors Affecting Heart Function

12.5 Types of Blood Vessels
- Structure of Arteries and Veins
- Arteries
- Capillaries
- Veins

12.6 Blood Flow
- Velocity of Blood Flow

12.7 Blood Pressure
- Factors Affecting Blood Pressure
- Control of Peripheral Resistance

12.8 Circulation Pathways
- Pulmonary Circuit
- Systemic Circuit

12.9 Systemic Arteries
- Major Branches of the Aorta
- Arteries Supplying the Head and Neck
- Arteries Supplying the Shoulders and Upper Limbs
- Arteries Supplying the Pelvis and Lower Limbs

12.10 Systemic Veins
- Veins Draining the Head and Neck
- Veins Draining the Shoulders and Upper Limbs
- Veins Draining the Pelvis and Lower Limbs
- Veins Draining the Abdominal and Thoracic Walls
- Veins Draining the Abdominal Viscera

12.11 Disorders of the Heart and Blood Vessels
- Heart Disorders
- Blood Vessel Disorders

Chapter Summary

Module 9
Cardiovascular System

SELECTED KEY TERMS

Arteries Blood vessels that carry blood away from the heart.
Atrium (atrium = vestibule) A heart chamber that receives blood returned to the heart by veins.
Capillaries Smallest and most numerous blood vessels in tissues where exchange of materials between the blood and interstitial fluid occurs.
Cardiac cycle The sequence of events that occur during one heartbeat.
Cardiac output The volume of blood pumped from each ventricle in one minute.
Diastole The relaxation phase of the cardiac cycle.
Pulmonary circuit (pulmo = lung) The blood pathway that transports blood to and from the lungs.
Stroke volume The volume of blood pumped from each ventricle per heartbeat.
Systemic circuit The blood pathway that transports blood to and from all parts of the body except the lungs.
Systole The contraction phase of the cardiac cycle.
Vasoconstriction (vas = vessel) Contraction of vessel smooth muscle to decrease the diameter of the blood vessel.
Vasodilation Relaxation of vessel smooth muscle to increase the diameter of the blood vessel.
Veins Blood vessels that carry blood toward the heart.
Ventricle (ventr = underside) A heart chamber that pumps blood into an artery.

THE HEART AND BLOOD VESSELS form the *cardiovascular* (kar-dē-ō-vas′-kū-lar) *system*. The heart pumps blood through a closed system of blood vessels. Figure 12.1 shows the general scheme of circulation of blood in the body. Blood vessels colored blue carry deoxygenated (oxygen-poor) blood; those colored red carry oxygenated (oxygen-rich) blood. Large arteries carry blood away from the heart and branch into smaller and smaller arteries that open into capillaries, the smallest blood vessels, where materials are exchanged with interstitial fluid. Capillaries open into small veins that merge to form larger and larger veins, and the largest veins return blood to the heart. AP|R

12.1 Anatomy of the Heart

Learning Objectives

1. Identify the protective coverings of the heart.
2. Describe the parts of the heart and their functions.
3. Trace the flow of blood through the heart.
4. Describe the blood supply to the heart.

The heart is a four-chambered muscular pump that is located within the mediastinum in the thoracic cavity. It lies between the lungs and just above the diaphragm. The *apex* of the heart is the lower pointed end, which extends toward the left side of the thoracic cavity at the level of the fifth rib. The *base* of the heart is the upper portion, which is attached to several large blood vessels at the level of the second rib. The heart is about the size of a closed fist. Note the relationship of the heart with the surrounding organs in figure 12.2.

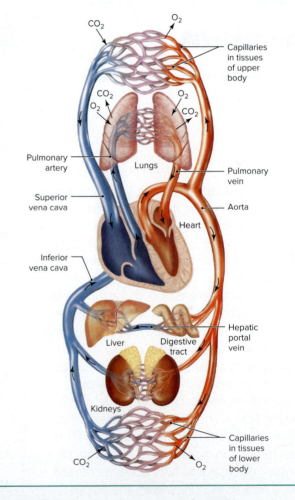

Figure 12.1 The Cardiovascular System.
Blood vessels carrying oxygenated blood are colored red; those carrying deoxygenated blood are colored blue. AP|R

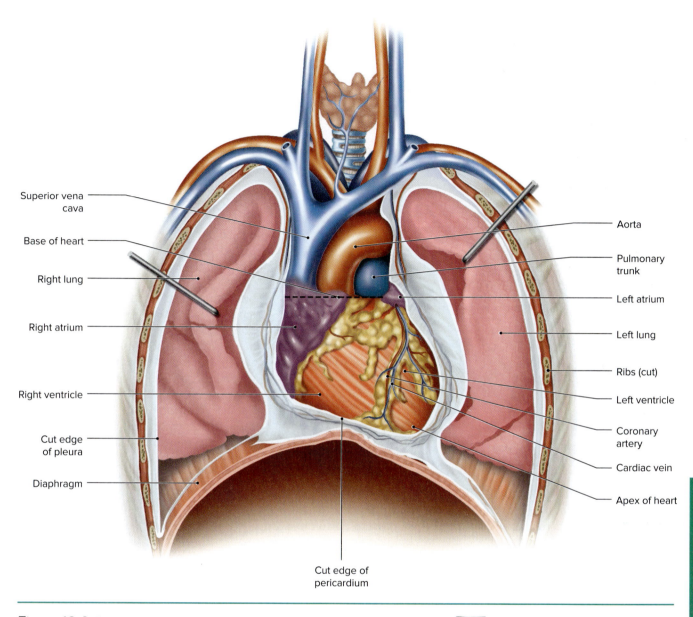

Figure 12.2 The Heart Within the Mediastinum of the Thoracic Cavity.

Protective Coverings

The heart and the bases of the attached blood vessels are enveloped by membranes that are collectively called the **pericardium** (per-i-kar′-dē-um). The pericardium is a loosely fitting sac that separates the heart from surrounding tissues and allows space for the heart to expand and contract as it pumps blood. The pericardium consists of two membranes: an outer *fibrous pericardium* and an inner **parietal layer of serous pericardium.** The fibrous pericardium is a tough, unyielding membrane composed of dense irregular connective tissue. It is attached to the diaphragm, inner surfaces of the sternum and thoracic vertebrae, and to adjacent connective tissues (figure 12.2). The delicate parietal pericardium lines the inner surface of the fibrous pericardium. At the bases of the large vessels (base of the heart), the parietal layer of serous pericardium folds back to form the **epicardium (visceral layer of serous pericardium)**, which forms the thin membrane that tightly adheres to the surface of the heart. The space between the parietal pericardium and the epicardium is the **pericardial cavity** (figure 12.3). This cavity is filled with pericardial fluid, which reduces the friction between the two layers of the serous pericardium when the heart contracts and expands.

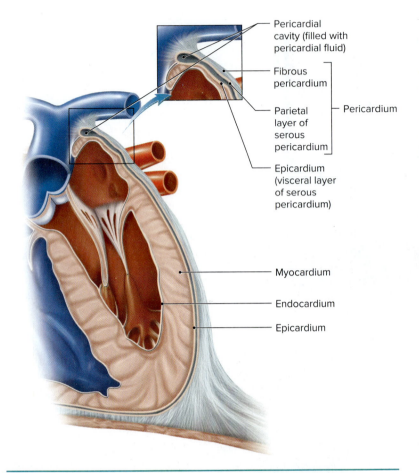

Figure 12.3 The Pericardium and Heart Wall.
The inset shows that the fibrous pericardium is lined by the parietal layer of serous pericardium, which folds back to form the epicardium.

arteries. There is no opening between the two atria or between the two ventricles. The atria are separated from each other by a partition called the *interatrial septum*. The ventricles are separated by the *interventricular septum*, a thick partition of cardiac muscle tissue (figure 12.4). The heart is a double pump. The right atrium and right ventricle compose the right pump. The left atrium and left ventricle compose the left pump.

The walls of the atria are much thinner than the walls of the ventricles. Differences in thickness are due to differences in the amount of cardiac muscle tissue that is present, which in turn reflects the work required of each chamber. Atrial walls possess less cardiac muscle tissue because blood movement from atria to ventricles is mostly passive, so that force from contraction is not as essential. The ventricles have more cardiac muscle tissue in order to create enough force to push blood up and out of the heart. The left ventricle has a thicker, more muscular wall than the right ventricle because it must pump blood throughout the entire body, except the lungs, whereas the right ventricle pumps blood only to the lungs. Locate the atria and ventricles in figure 12.4, and also in figures 12.2 and 12.5, which show views of the outer surface of the heart. Table 12.1 summarizes the functions of the heart chambers.

The Heart Wall

The wall of the heart consists of a thick layer of cardiac muscle tissue, the **myocardium** (mī-ō-kar′-dē-um), sandwiched between two thin membranes. Contractions of the myocardium provide the force that pumps the blood through the blood vessels. The epicardium, attached to the outside of the myocardium, contains the blood vessels that nourish the heart itself. The inside of the myocardium is covered with a simple squamous epithelium called the **endocardium.** The endocardium not only lines the chambers and valves of the heart, but also is continuous with the inner lining of the blood vessels attached to the heart (figure 12.3).

Heart Chambers

The two upper chambers are the **atria** (ā′-trē-ah) (singular, *atrium*), which receive blood being returned to the heart by the veins. The two lower chambers are the **ventricles** (ven′-tri-kuls), which pump blood into the

Heart Valves

Like all pumps, the heart contains valves that allow the blood to flow in only one direction through the heart. The two types of heart valves are atrioventricular valves (AV valves) and semilunar valves. Observe the location and structure of the heart valves in figures 12.4 and 12.6.

Atrioventricular Valves

The opening between each atrium and its corresponding ventricle is guarded by an **atrioventricular** (ā-trē-ō-ven-trik′-ū-lar) **valve** that is formed of dense irregular connective tissue. Each valve allows blood to flow from the atrium into the ventricle but prevents a backflow of blood from the ventricle into the atrium. The AV valve between the right atrium and the right ventricle is the **tricuspid** (trī-kus′-pid), or **right atrioventricular, valve.** Its name indicates that it is composed of three cusps, or flaps, of tissue. The **mitral** (mī′-tral), or **left atrioventricular, valve** consists of two cusps and is located between the left atrium and the left ventricle.

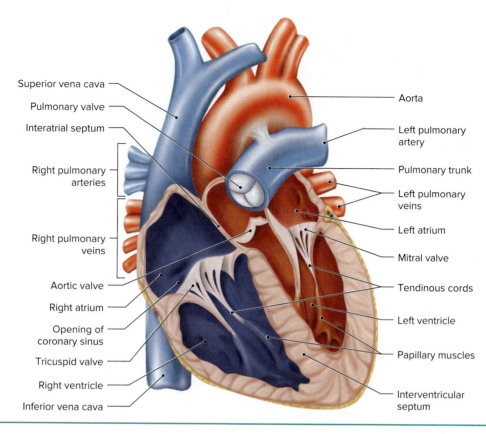

Figure 12.4 Frontal Section Showing the Internal Structure of the Heart. AP|R

Table 12.1 Functions of the Heart Chambers

Chamber	Function
Right atrium	Receives deoxygenated blood from the superior and inferior venae cavae and the coronary sinus, and passes this blood through the tricuspid valve to the right ventricle
Right ventricle	Receives deoxygenated blood from the right atrium and pumps this blood through the pulmonary valve into the pulmonary trunk
Left atrium	Receives oxygenated blood from the pulmonary veins and passes this blood through the mitral valve to the left ventricle
Left ventricle	Receives oxygenated blood from the left atrium and pumps this blood through the aortic valve into the aorta

The AV valves originate from rings of thick, dense irregular connective tissue that support the junction of the ventricles with the atria and the large arteries attached to the ventricles. This supporting dense irregular tissue is called the *fibrous skeleton* of the heart (figure 12.6). The fibrous skeleton not only provides structural support but also serves as insulation separating the electrical activity of the atria and ventricles. This insulation enables the atria and ventricles to contract independently.

Thin strands of dense regular connective tissue, the **tendinous cords,** extend from the valve cusps to the **papillary muscles,** small mounds of cardiac muscle tissue that project from the inner walls of the ventricles (see figure 12.4). The tendinous cords prevent the valve cusps from being forced into the atria during ventricular contraction. In fact, they are usually just the right length to allow the cusps to press against each other and tightly close the opening during ventricular contraction. Table 12.2 summarizes the functions of the heart valves.

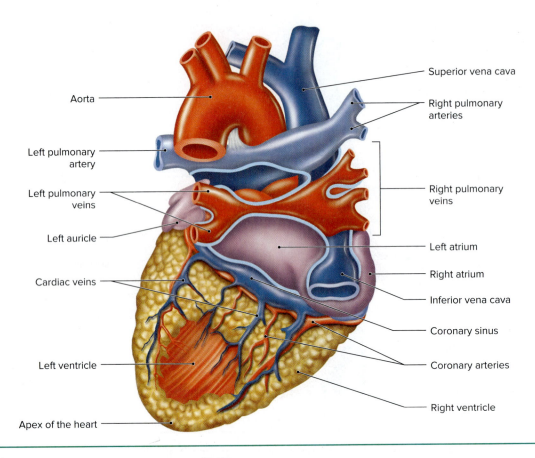

Figure 12.5 Posterior View of the Heart. AP|R

Semilunar Valves

The **semilunar valves** are located in the bases of the large arteries that carry blood from the ventricles. The **pulmonary valve** is located at the base of the pulmonary trunk, which extends from the right ventricle. The **aortic valve** is located at the base of the aorta, which extends from the left ventricle.

Each semilunar valve is composed of three pocket-like cusps of dense irregular connective tissue. They allow blood to be pumped from the ventricles into the arteries during ventricular contraction, but they prevent a backflow of blood from the arteries into the ventricles during ventricular relaxation.

Table 12.2 Heart Valves

Valve	Location	Function
Atrioventricular Valves		
Tricuspid valve	Opening between the right atrium and right ventricle	Prevents backflow of blood from the right ventricle into the right atrium
Mitral valve	Opening between the left atrium and left ventricle	Prevents the backflow of blood from the left ventricle into the left atrium
Semilunar Valves		
Pulmonary valve	Entrance to the pulmonary trunk	Prevents backflow of blood from the pulmonary trunk into the right ventricle
Aortic valve	Entrance to the aorta	Prevents backflow of blood from the aorta into the left ventricle

Flow of Blood Through the Heart

Figure 12.7 diagrammatically shows the flow of blood through the heart and the major vessels attached to the heart. Blood is oxygenated as it flows through the lungs and becomes deoxygenated as it releases oxygen to body tissues. Trace the flow of blood through the heart and major vessels in figure 12.7 as you read the following description.

The right atrium receives deoxygenated blood from all parts of the body except the lungs via three veins: the superior and inferior venae cavae and the coronary sinus. The **superior vena cava** (vē′-nah kā′-vah) returns blood from the head, neck, shoulders, upper limbs, and thoracic and lumbar regions. The **inferior vena cava** returns blood from the lower trunk and lower limbs. The **coronary sinus** drains deoxygenated blood from cardiac muscle tissue. Simultaneously, the left atrium receives oxygenated blood returning to the heart from the lungs via the **pulmonary veins.**

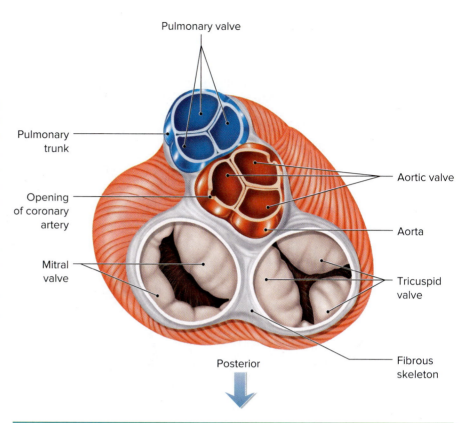

Figure 12.6 Superior View of the Heart Valves. Note the fibrous skeleton of the heart.

Blood flows from the left and right atria into the corresponding ventricles. About 70% of the blood flow into the ventricles is passive, and about 30% results from atrial contraction.

After blood has flowed from the atria into their respective ventricles, the ventricles contract. The right ventricle pumps deoxygenated blood into the **pulmonary trunk.** The pulmonary trunk branches to form the **left** and **right pulmonary arteries,** which carry blood to the lungs. The left ventricle pumps oxygenated blood into the **aorta** (ā-or′-tah). The aorta branches to form smaller arteries that carry blood to all parts of the body except the lungs. Locate these major blood vessels associated with the heart in figures 12.2, 12.4, 12.5, and 12.7.

Because the heart is a double pump, there are two basic pathways, or circuits, of blood flow as shown in figure 12.7. The **pulmonary circuit** carries deoxygenated blood from the right ventricle to the lungs and returns oxygenated blood from the lungs to the left atrium. The **systemic circuit** carries oxygenated blood from the left ventricle to all parts of the body except the lungs and returns deoxygenated blood to the right atrium.

Blood Supply to the Heart

The heart requires a constant supply of blood to nourish its own tissues. Blood is supplied by **left** and **right coronary** (kor′-ō-na-rē) **arteries,** which branch from the aorta just above the aortic valve (figures 12.6 and 12.18a). Blockage of a coronary artery may result in a heart attack. After passing through capillaries in cardiac muscle tissue, blood is returned via **cardiac** (kar′-dē-ak) **veins,** which lie next to the coronary arteries. These veins empty into the coronary sinus, which drains into the right atrium. Locate these blood vessels in figures 12.2 and 12.5 and note the adipose tissue that lies alongside the vessels. Also, study the relationships of the atria, ventricles, and large blood vessels associated with the heart.

Check My Understanding

1. What are the names and functions of the heart chambers?
2. What are the names and functions of the heart valves?
3. Trace a drop of blood as it flows through the heart and the pulmonary and systemic circuits.
4. Describe the flow of blood throughout the myocardium.

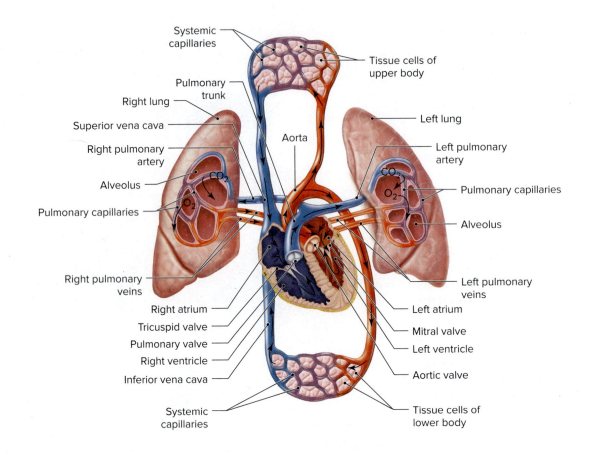

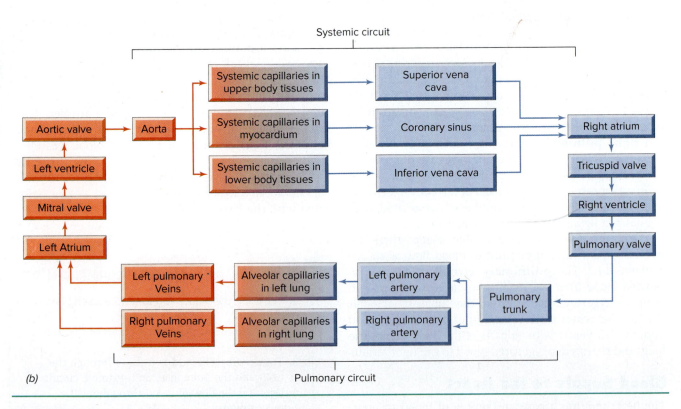

Figure 12.7 The Systemic and Pulmonary Circuits.
Heart chambers and vessels colored red carry oxygenated blood. Those colored blue carry deoxygenated blood.

Clinical Insight

If cusps of an AV valve collapse and open into the atrium, some blood may regurgitate (backflow) into the atrium during ventricular contractions. This is what happens in a disorder known as *mitral valve prolapse (MVP)*. In some cases, it causes no serious dysfunction. In others, fatigue and shortness of breath may occur. Persons with MVP are susceptible to *endocarditis*, inflammation of the endocardium, caused by some species of *Streptococcus* bacteria. Endocarditis can result in scarring of the valve cusps, which further decreases valve function. Persons with MVP are often advised to take antibiotics prior to dental work to prevent bacteria from entering the blood and being carried to the heart.

12.2 Cardiac Cycle

Learning Objectives
5. Describe the events of the cardiac cycle.
6. Describe the sounds of the heartbeat.

The **cardiac cycle** refers to the sequence of events that occur during one heartbeat. The contraction phase of a cardiac cycle is known as **systole** (sis′-to-lē); the relaxation phase is called **diastole** (dī′-as-to-lē). These phases are illustrated in figure 12.8. Note that the ventricles are relaxed when the atria contract, and the atria are relaxed when the ventricles contract. Systole increases blood pressure within a chamber, while diastole decreases blood pressure within a chamber.

When both the atria and ventricles are relaxed between beats, blood flows passively into the atria from the large veins leading to the heart and then passively into the ventricles. Then, the atria contract (atrial systole), forcing more blood into the ventricles so that they are filled. Immediately thereafter, the ventricles contract. Ventricular systole produces high blood pressure within the ventricles, which causes both AV valves to close and both semilunar valves to open. Opening of the semilunar valves allows blood to move into the arteries leading from the heart. Ventricular diastole immediately follows and the decrease in ventricle pressure allows the AV valves to open. Simultaneously, the semilunar valves close because of the greater blood pressure within the arteries. The cardiac cycle is then repeated. Study these relationships in figure 12.8.

Heart Sounds

The sounds of the heartbeat are usually described as *lub-dup* (pause) *lub-dup*, and so forth. These sounds are produced by the closing of the heart valves. The first sound results from the closing of the AV valves at the beginning of ventricular systole. The second sound results from the closing of the semilunar valves at the beginning of ventricular diastole. If any of the heart valves are defective and do not close properly, an additional sound, known as a heart murmur, may be heard.

 Check My Understanding

5. What are the events of a cardiac cycle?
6. What produces the heart sounds?

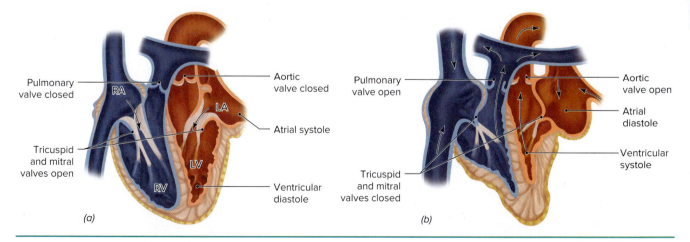

Figure 12.8 The Cardiac Cycle.
(a) Blood flows from the atria into the ventricles during ventricular diastole. *(b)* Blood is pumped from the ventricles during ventricular systole. **AP|R**

12.3 Conducting System of the Heart

Learning Objective
7. Describe the parts of the conducting system of the heart and their functions.

The heart is able to contract on its own because it contains specialized cardiac muscle tissue that spontaneously forms action potentials and transmits them to the myocardium to initiate contraction. This specialized tissue forms the *conducting system of the heart*, which consists of the sinoatrial node, atrioventricular node, atrioventricular bundle, bundle branches, and subendocardiac conducting network. Observe the location of the conducting system of the heart and its parts in figure 12.9.

The **sinoatrial** (sī-nō-ā'-trē-al) **node (SA node)** is located in the right atrium at the junction of the superior vena cava. It is known as the pacemaker of the heart because it rhythmically forms electrical action potentials to initiate each heartbeat. The action potentials are transmitted to the myocardium of the atria, where they produce a simultaneous contraction of the atria. The flow of action potentials causes downward contraction of the atria forcing blood into the ventricles. At the same time, the action potentials are carried to the **atrioventricular node (AV node)**, which is located in the right atrium near the lower portion of the interatrial septum.

There is a brief time delay as the action potentials pass slowly through the AV node, which allows time for the ventricles to fill with blood. From the AV node, the action potentials pass along the **atrioventricular bundle (AV bundle)**, a group of large fibers that divide into **left and right bundle branches** extending downward through the interventricular septum and upward to the lateral walls of the ventricles. The smaller **subendocardiac conducting network** arises from the bundle branches and carries the action potentials to the myocardium of the ventricles, where they stimulate ventricular contraction. The distribution of the ventricular fibers causes the ventricles to contract upward from the apex so that blood is forced into the pulmonary trunk and aorta.

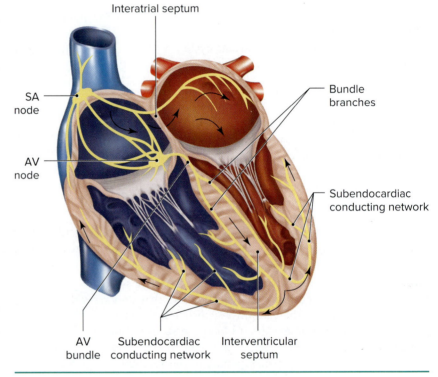

Figure 12.9 Conducting System of the Heart. Arrows indicate the flow of action potentials from the SA node. AP|R

⊕ Clinical Insight

If a coronary artery is partially obstructed by the fatty deposits of *atherosclerosis* (see the disorders section in this chapter for details), portions of the myocardium may be deprived of adequate blood. This produces chest pain known as *angina pectoris*. In severe cases, treatment may involve one of two approaches: coronary angioplasty or coronary bypass surgery.

In *coronary angioplasty*, a catheter that contains a balloon at its tip is inserted into an artery of an upper or lower limb and is threaded into the affected coronary artery. The balloon is positioned at the obstruction and is inflated for a few seconds to compress the fatty deposit and enlarge the lumen of the affected coronary artery. A meshlike metal tube called a stent is then inserted and positioned at the site of the obstruction to hold open the artery. The stent may be coated with a chemical that inhibits the growth of cells to minimize the chances that the artery will become obstructed again.

In *coronary bypass surgery*, a portion of an artery or a vein from elsewhere in the body is removed and is surgically grafted, providing a bypass around the obstruction to supply blood beyond the affected portion of the coronary artery.

Electrocardiogram

The origination and transmission of action potentials through the conducting system of the heart generate electrical currents that may be detected by electrodes placed on the body surface. An instrument called an *electrocardiograph* is used to transform the electrical currents picked up by the electrodes into a recording called an **electrocardiogram** (**ECG** or **EKG**).

Figure 12.10 shows a healthy ECG of five cardiac cycles and an enlargement of a healthy ECG of one cardiac cycle. Note that an ECG consists of several deflections, or waves. These waves correlate with the flow of action potentials during particular phases of the cardiac cycle.

An electrocardiogram has three distinct waves: the P wave, QRS complex, and T wave. The *P wave* is a small wave. It is produced by the depolarization of the atria. The *QRS complex* is produced by the depolarization of the ventricles. The greater size of the QRS complex is due to the greater muscle mass of the ventricles. The last wave is the *T wave*, which is produced by the repolarization of the ventricular myocardium. The repolarization of the atria is not detected because it is masked by the stronger QRS complex. An ECG provides important information in the diagnosis of heart disease and abnormalities. In reading an ECG, physicians pay close attention to the height of each wave and to the time required for an interval between each wave.

> ✅ **Check My Understanding**
> 7. What comprises the conducting system of the heart?
> 8. What events produce the waves of an electrocardiogram?

12.4 Regulation of Heart Function

Learning Objective
8. Explain how the heart rate and contraction strength are regulated.

Cardiac output is the volume of blood pumped from each ventricle in one minute, and it is an important measure of heart function. It is determined by two factors: **stroke volume** and **heart rate.** Stroke volume (SV) is the volume of blood pumped from each ventricle per heartbeat. Multiplying this volume by the heart rate (HR), heartbeats per minute, yields the cardiac output (CO).

$$CO = SV \times HR$$

At average resting values of stroke volume (70 ml/beat) and heart rate (72 beats/min), the cardiac output is 5,040 ml/min. This means that the total volume

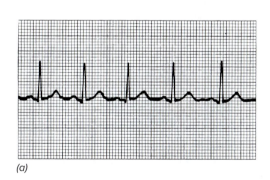
(a)

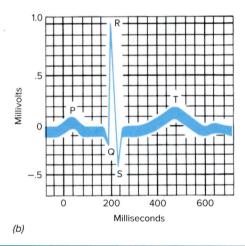
(b)

Figure 12.10 Electrocardiograms (ECGs).
(a) An ECG showing five cardiac cycles. (b) An ECG showing one cardiac cycle.

 Clinical Insight

Some irregularities in heart rhythms result from improper transmission of action potentials by the conducting system of the heart. In patients in whom the SA node or AV node malfunctions, a heartbeat may be obtained by implanting an artificial *pacemaker* in the chest wall. Wires (leads) are threaded through a vein to connect the pacemaker to the heart. This battery-operated device synchronizes heart contractions and controls the heart rate by sending weak electrical pulses to the heart to initiate contraction.

of blood, 4 to 6 liters, passes through each ventricle of the heart each minute. Cardiac output increases with exercise because both stroke volume and heart rate increase.

Heart function is regulated by both *intrinsic* and *extrinsic* factors. For example, *venous return*, the amount of blood returning to the heart during diastole, is an intrinsic factor that affects stroke volume. If venous return increases, more blood enters and is pumped from the ventricles, increasing the stroke volume and cardiac output. Heart rate is primarily controlled extrinsically by the autonomic division, although hormones and certain ions also affect it.

Autonomic Regulation

Heart rate regulation is primarily under the control of the **cardiac control center** located within the medulla oblongata of the brain. It receives sensory information about the level of blood pressure from baroreceptors located in the aortic arch and the carotid sinuses of the internal carotid arteries. It also receives sensory information from chemoreceptors in the aortic arch and the carotid bodies near the bifurcation of the common carotid arteries (figures 12.11 and 12.19). Baroreceptors are sensitive to changes in vessel wall stretching caused by both high and low blood pressure. Chemoreceptors are stimulated by low blood pH, high blood carbon dioxide level, and very low blood oxygen level. The cardiac control center is also affected by emotions, which are generated by the limbic system (see chapter 8).

The cardiac control center consists of both sympathetic and parasympathetic components. Action potentials transmitted to the heart via sympathetic axons cause an increase in heart rate and contraction strength, while action potentials transmitted by parasympathetic axons cause a decrease in heart rate. The cardiac control center constantly adjusts the frequency of sympathetic and parasympathetic action potentials to produce a heart rate and a contraction strength that meet the changing needs of tissue cells (figure 12.11).

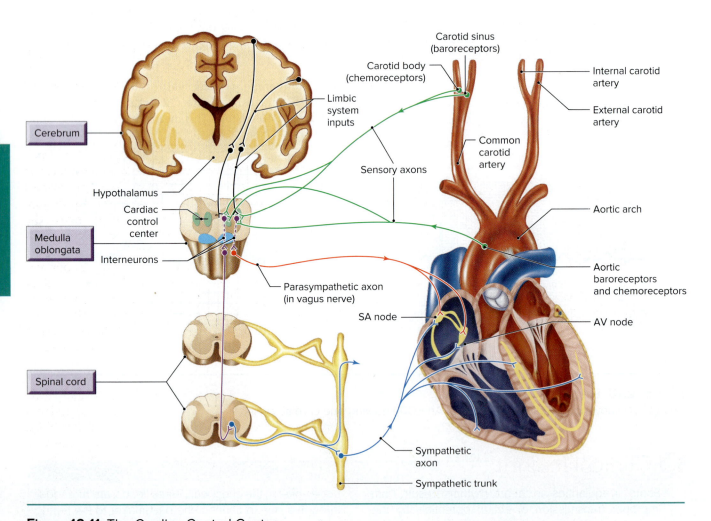

Figure 12.11 The Cardiac Control Center.
The rate and strength of heart contractions are regulated by the antagonistic actions of sympathetic (colored blue) and parasympathetic (colored red) parts of the autonomic division. Sensory axons are colored green.

Neurons of the sympathetic part extend axons from the cardiac control center down the spinal cord to the thoracic region. There the sympathetic axons exit the spinal cord to innervate the SA node, AV node, and portions of the myocardium. The transmission of action potentials causes the sympathetic axons to secrete *norepinephrine* at synapses in the heart. Norepinephrine increases the heart rate and strengthens the force of myocardial contraction. Physical and emotional stresses, such as exercise, excitement, anxiety, and fear, stimulate the sympathetic part.

Parasympathetic axons arise from the cardiac control center and exit in the vagus nerve (CN X) to innervate the SA and AV nodes. The transmission of action potentials causes the parasympathetic axons to secrete *acetylcholine* at the heart synapses, which decreases the heart rate. The greater the frequency of parasympathetic action potentials sent to the heart, the slower the heart rate. Excessive blood pressure and emotional factors, such as grief and depression, stimulate the parasympathetic part.

When the heart is at rest, more parasympathetic action potentials than sympathetic action potentials are sent to the heart. As cellular needs for blood increase, a decrease in the frequency of parasympathetic action potentials and an increase in sympathetic action potentials cause heart rate to increase.

Other Factors Affecting Heart Function

Age, sex, physical condition, temperature, epinephrine, thyroxine, and the blood levels of calcium and potassium ions also affect the heart rate and contraction strength.

The resting heart rate gradually declines with age, and it is slightly faster in females than in males. Average resting heart rates in females are 72 to 80 beats per minute, as opposed to 64 to 72 beats per minute in males. People who are in good physical condition have a slower resting heart rate than those in poor condition. Athletes may have a resting heart rate of only 40 to 60 beats per minute. An increase in body temperature, which occurs during exercise or when feverish, increases the heart rate.

Epinephrine, which is secreted by the adrenal glands during stress or excitement, affects the heart like norepinephrine–it increases the rate and strength of heart contractions. An excess of thyroxine produces a lesser, but longer-lasting, increase in heart rate.

A reduced level of blood Ca^{2+} decreases the rate and strength of heart contraction, while an increased blood Ca^{2+} level increases heart rate and contraction strength, and prolongs contraction. In extreme cases, an excessively prolonged contraction may result in death. An excessive blood K^+ level decreases both heart rate and contraction strength. A high dose of K^+ is often used in lethal injections, in which a dangerously high level of blood K^+ causes the heart to stop contracting. A low level of blood K^+ may cause potentially life-threatening heart rhythms.

 Check My Understanding

9. How are the heart rate and contraction strength regulated?
10. What other factors affect the heart rate and contraction strength?

12.5 Types of Blood Vessels

Learning Objectives
9. Describe the structure and function of arteries, arterioles, capillaries, venules, and veins.
10. Describe how materials are exchanged between capillary blood and interstitial fluid.

There are three basic types of blood vessels: arteries, capillaries, and veins. They form a closed system of tubes that carry blood from the heart to the tissue cells and back to the heart. Table 12.3 compares these three types.

Structure of Arteries and Veins

The walls of arteries and veins are composed of three distinct layers. The *tunica externa*, the outermost layer, is formed of dense irregular connective tissue that includes both collagen and elastic fibers. These fibers provide support and elasticity for the vessel. The *tunica media*, the middle layer, usually is the thickest layer. It consists of smooth muscle cells that encircle the blood vessel. The smooth muscle cells not only provide support but also produce changes in the diameter of the blood vessel by contraction or relaxation. The *tunica intima*, the innermost layer, forms the inner lining of blood vessels. It consists of a simple squamous epithelium, called the *endothelium*, supported by a thin layer of areolar connective tissue containing elastic and collagen fibers.

The walls of arteries and veins have the same basic structure. However, arterial walls are thicker because their tunica media contains more smooth muscle and elastic connective tissues as an adaptation to the higher blood pressure found in them. The tunica media of veins possesses very little smooth muscle, which leads to a much thinner wall. Veins possess larger lumens than arteries; as a result, they can hold a larger volume of blood. Another difference is that large veins, but not arteries, contain valves formed of endothelium. Venous valves prevent a backflow of blood. Compare the structure of arteries and veins in figure 12.12.

Table 12.3 Comparison of Arteries, Capillaries, and Veins

Type of Vessel	Function	Structure
Arteries	Carry blood from the heart to the capillaries Control blood flow and blood pressure	Composed of tunica intima, tunica media, and tunica externa Possess thick walls due to large amounts of smooth muscle and elastic connective tissue
Capillaries	Enable exchange of materials between blood and interstitial fluid	Microscopic vessels composed of endothelium supported by areolar connective tissue
Veins	Return blood from capillaries to the heart Serve as storage areas for blood	Composed of tunica intima, tunica media, and tunica externa Possess thin walls due to small amounts of smooth muscle and elastic connective tissue Large veins have venous valves.

Arteries

Arteries carry blood away from the heart. They branch repeatedly into smaller and smaller arteries and ultimately form microscopic arteries called **arterioles** (ar-tē′-rē-ōls). As arterioles branch and form smaller arterioles, the thickness of the tunica media decreases. The walls of the smallest arterioles consist of only the tunica intima and a few encircling smooth muscle cells. Arteries, especially the arterioles, play an important role in the control of blood flow and blood pressure.

Capillaries

Arterioles connect with **capillaries,** the most numerous and the smallest blood vessels. A capillary's diameter is so small that RBCs must pass through it in single file. The walls of capillaries consist of an endothelium supported by a layer of areolar connective tissue. These extremely thin walls facilitate the exchange of materials between blood in capillaries and tissue cells.

The continual exchange of materials between the blood and tissue cells is essential for life. Cells require oxygen and nutrients to perform their metabolic functions, and they produce carbon dioxide and other metabolic wastes that must be removed by the blood.

Tissue cells are enveloped in a thin film of extracellular fluid called **interstitial fluid,** or *tissue fluid,* that fills tissue spaces and lies between the tissue

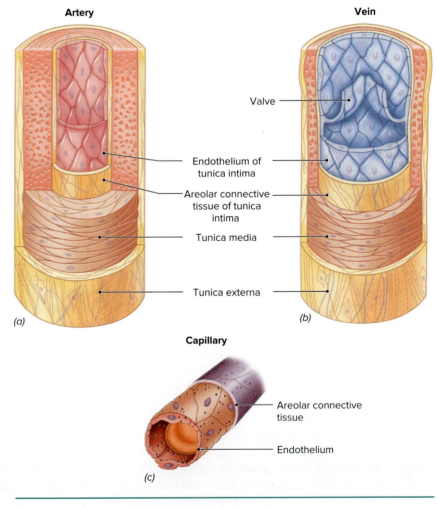

Figure 12.12 Comparison of Blood Vessels.
(a) The wall of an artery. *(b)* The wall of a vein. *(c)* The wall of a capillary.

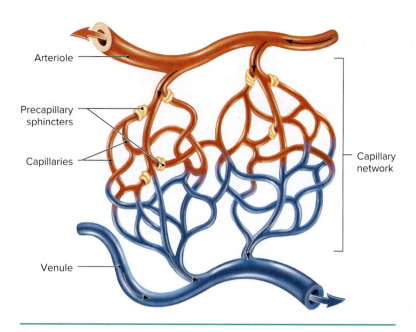

Figure 12.13 A Capillary Network.
Precapillary sphincters regulate the blood flow from an arteriole into a capillary. Oxygenated blood (red) enters a capillary network. Deoxygenated blood (blue) exits the capillaries and enters a venule.

Exchange of Materials

Recall that the capillary walls are so thin that materials can readily diffuse through them, and the junctions between these cells are not tight so fluid is able to move between the cells. Two opposing forces determine the movement of fluid between capillary blood and interstitial fluid: osmotic pressure and blood pressure. Osmotic pressure of the blood results from plasma proteins. Osmotic pressure tends to "pull" fluid from interstitial fluid into the capillaries by osmosis. The movement of interstitial fluid into the capillaries by osmosis is called **reabsorption.** Blood pressure against the capillary walls results from the force of ventricular contractions. It tends to push fluid out of the capillaries into the interstitial fluid. The process of forcing substances through a membrane due to greater hydrostatic pressure on one side of the membrane is known as **filtration.**

At the arteriolar end of a capillary, blood pressure exceeds osmotic pressure, so fluid moves out of the capillary into the interstitial fluid. In contrast, at the venular end of the capillary, osmotic pressure exceeds blood pressure, so fluid moves from the interstitial fluid into the capillary by osmosis (figure 12.14). About nine-tenths of the fluid that is filtered from the arteriolar end of a capillary into the interstitial fluid is reabsorbed into the venular end of the capillary. The remainder is picked up by the lymphoid system and ultimately is returned to the blood (see chapter 13).

cells and the capillaries. Therefore, all materials that pass between the blood and tissue cells must pass through the interstitial fluid. Dissolved substances such as oxygen and nutrients diffuse from blood in the capillary into the interstitial fluid and from the interstitial fluid into tissue cells. Carbon dioxide and metabolic wastes diffuse in the opposite direction.

The distribution of capillaries in body tissues varies with the metabolic activity of each tissue. Capillaries are especially abundant in active tissues, such as muscle and nervous tissues, where nearly every cell is near a capillary. Capillaries are less abundant in connective tissues and are absent in some tissues, such as cartilage tissue, epidermis, and the lens and cornea of the eye.

Blood flow in capillaries is controlled by *precapillary sphincters*, smooth muscle cells encircling the bases of capillaries at the arteriole–capillary junctions (figure 12.13). Contraction of a precapillary sphincter inhibits blood flow to its capillary network. Relaxation of the sphincter allows blood to flow into its capillary network to provide oxygen and nutrients for the tissue cells. The flow of blood in capillary networks occurs intermittently. When some capillary networks are filled with blood, others are not. Capillary networks receive blood according to the needs of the cells that they serve. For example, during physical exercise blood is diverted from capillary networks in the digestive tract to fill the capillary networks in skeletal muscles. This pattern of blood distribution is largely reversed after a meal.

Veins

After blood flows through the capillaries, it enters the **venules,** the smallest **veins.** Several capillaries merge to form a venule. The smallest venules consist only of endothelium and areolar connective tissue, but larger venules also contain smooth muscle tissue. Venules unite to form small veins. Small veins combine to form progressively larger veins as blood is returned to the heart. Larger veins, especially those in the upper and lower limbs, contain valves that prevent a backflow of blood and aid the return of blood to the heart.

Because nearly 60% of the blood volume is in veins at any instant, veins may be considered as storage areas for blood that can be carried to other parts of the body in times of need. Venous sinusoids in the skin, liver, and spleen are especially important reservoirs. If blood is lost by hemorrhage, both blood volume and pressure decline. In response, the sympathetic part sends action potentials to constrict the muscular walls of the veins, which reduces the venous volume while increasing blood volume and pressure in the heart, arteries, and capillaries. This effect compensates for the blood loss. A similar response occurs during strenuous muscular activity in order to increase the blood flow to skeletal muscles.

272 Chapter 12 Cardiovascular System

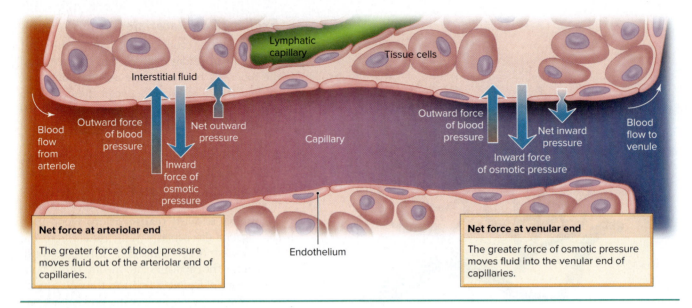

Figure 12.14 Fluid Exchange Across Capillary Walls.
Fluid moves out of or into capillaries according to the net difference between blood pressure and osmotic pressure. Solutes diffuse out of or into capillaries according to each solute's concentration gradient.

 Check My Understanding

11. Compare the structure and function of arteries, capillaries, and veins.
12. How does the exchange of materials occur between blood in capillaries and tissue cells?

12.6 Blood Flow

Learning Objective
11. Describe the mechanism of blood circulation.

Blood circulates because of differences in blood pressure. Blood flows from areas of higher pressure to areas of lower pressure. Blood pressure is greatest in the ventricles and lowest in the atria. Figure 12.15 shows the decline of blood pressure in the systemic circuit with increased distance from the left ventricle.

Contraction of the ventricles creates the blood pressure that propels the blood through the arteries. However, the pressure declines as the arteries branch into an increasing number of smaller and smaller arteries and finally connect with the capillaries. The decline in blood pressure occurs because of the increased distance from the ventricle. By the time blood has left the capillaries and entered the veins, there is very little blood pressure remaining to return the blood to the heart. Venous return is assisted by three additional forces: *skeletal muscle contractions, respiratory movements,* and *gravity.*

Contractions of skeletal muscles compress the veins, forcing blood from one valved segment to another and on toward the heart because the valves prevent a backflow of blood. This method of moving venous blood toward the

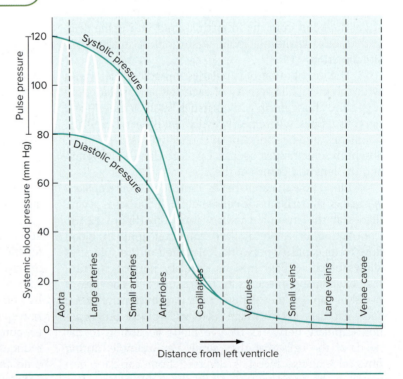

Figure 12.15 Blood Pressure in the Systemic Circuit.
Blood pressure decreases as distance from the left ventricle increases.

heart is especially important in the return of blood from the upper and lower limbs, and it is illustrated in figure 12.16.

Respiratory movements aid the movement of blood up toward the heart in the abdominopelvic and thoracic cavities. The downward movement of the diaphragm as it contracts during inspiration decreases the pressure within the thoracic cavity and increases the pressure within the abdominopelvic cavity. The higher pressure in the abdominopelvic cavity forces blood to move from the abdominopelvic veins up into thoracic veins, where the pressure is reduced. When the diaphragm relaxes and moves up, the thoracic and abdominopelvic pressures reverse.

Gravity aids the return of blood in veins above the heart.

Velocity of Blood Flow

The velocity of blood flow varies inversely with the overall cross-sectional area of the *combined* blood vessels. The changes in velocity are caused by changes in *peripheral resistance*, which is an opposition to blood flow created by friction between blood and blood vessel walls. The greater the cross-sectional area, the greater the resistance and the slower blood flows. Therefore, the velocity progressively decreases as blood flows through an increasing number of smaller and smaller arteries and into the capillaries. As the cross-sectional area decreases, there is less resistance, and blood flows faster. Then, the velocity progressively increases as the blood flows into a decreasing number of larger and larger veins on its way back to the heart.

Blood velocity is fastest in the aorta and slowest in the capillaries, an ideal situation providing for the rapid circulation of the blood and yet sufficient time for the exchange of materials between blood in the capillaries and the interstitial fluid surrounding tissue cells.

Check My Understanding

13. How does blood pressure change as blood moves farther from the heart?
14. What three mechanisms assist venous return?

12.7 Blood Pressure

Learning Objectives
12. Compare systolic and diastolic blood pressure.
13. Describe how blood pressure is regulated.

The term *blood pressure*, the force of blood against the wall of the blood vessels, usually refers to arterial blood pressure in the systemic circuit—in the aorta and its branches. Arterial blood pressure is greatest during ventricular contraction (systole) as blood is pumped into the aorta and its branches. This pressure is called the **systolic blood pressure,** and it optimally averages 110 millimeters of mercury (mm Hg) when measured in the brachial artery. The lowest arterial pressure occurs during ventricular relaxation (diastole). This pressure is called the **diastolic blood pressure,** and it optimally averages 70 mm Hg (figure 12.15).

The difference between the systolic and diastolic blood pressures is known as the *pulse pressure* (figure 12.15). The alternating increase and decrease in arterial blood pressure during ventricular systole and diastole causes a comparable expansion and contraction of the elastic arterial walls. This pulsating expansion of the arterial walls follows each ventricular contraction, and it may be detected as the *pulse* by placing the fingers on a superficial artery. Figure 12.17 identifies the name and location of superficial arteries where the pulse may be detected.

Factors Affecting Blood Pressure

Three major factors affect blood pressure: cardiac output, blood volume, and peripheral resistance. An increase in any of these factors causes an increase in blood pressure, while a decrease in any of these causes a decrease in blood pressure.

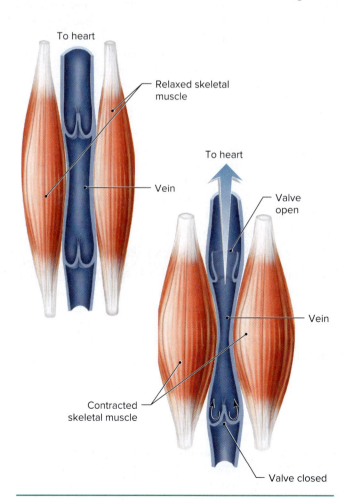

Figure 12.16 The Muscular Pump.
Contraction of skeletal muscles compresses veins and aids the movement of blood toward the heart.

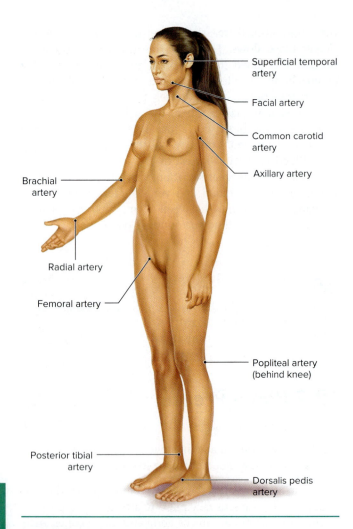

Figure 12.17 Pulse Points.
Locations and arteries where the pulse may be detected. See figures 12.19 and 12.20 for specific locations of these arteries.

 Clinical Insight

A blood pressure of 110/70 mm Hg is optimal. Each 20 mm Hg of systolic pressure over 115, and each 10 mm Hg of diastolic pressure over 75 doubles the risk of heart attack, stroke, and kidney disease.

Recall that cardiac output is determined by the heart rate and the stroke volume. An increase or decrease in cardiac output causes a comparable change in blood pressure.

Blood volume may be decreased by severe hemorrhage, vomiting, diarrhea, or reduced water intake. The decrease in blood volume causes a decrease in blood pressure. As soon as the lost fluid is replaced, blood pressure returns to a healthy pressure. Many drugs used to treat hypertension (abnormally high blood pressure) act as diuretics, meaning they increase urine volume, and as a result decrease blood volume. Conversely, if the body retains too much fluid, blood volume and blood pressure increase. A high-salt diet is a risk factor for hypertension because it causes the blood to retain more water as a result of osmosis, leading to an increase in blood volume.

Peripheral resistance is the opposition to blood flow created by friction of blood against the walls of blood vessels. Increasing peripheral resistance will increase blood pressure, while decreasing peripheral resistance decreases blood pressure. Peripheral resistance is determined by *vessel diameters, total vessel length,* and *blood viscosity.* Arterioles play a critical role in controlling blood pressure by changing their diameters. As arterioles constrict, peripheral resistance increases and blood pressure increases accordingly. As arterioles dilate, peripheral resistance and blood pressure decrease. Peripheral resistance is directly proportional to the total length of the blood vessels in the body: the longer the total length of the vessels, the greater their resistance to flow. Obese people tend to have hypertension partly because their bodies contain more blood vessels to serve the extra adipose tissue. *Viscosity* is the resistance of a liquid to flow. For example, water has a low viscosity, while honey has a high viscosity. Blood viscosity is determined by the ratio of plasma to formed elements and plasma proteins. Increasing viscosity, or shifting the ratio in favor of the formed elements and plasma proteins, increases peripheral resistance and blood pressure. Both dehydration (loss of water from plasma) and polycythemia (elevated RBC count) can increase viscosity. High levels of blood lipids and sugar is also a risk factor for hypertension because it increases blood viscosity, in addition to promoting the formation of plaque on the vessel walls. Decreasing viscosity through over-hydration or certain types of anemia (see chapter 11) will decrease peripheral resistance and blood pressure.

Control of Peripheral Resistance

The sympathetic part of the autonomic division controls peripheral resistance primarily by regulating the diameter of blood vessels, especially arterioles. The integration center is the **vasomotor center** in the medulla oblongata. An increase in the frequency of sympathetic action potentials to the smooth muscle of blood vessels produces **vasoconstriction,** which decreases vessel diameter and increases resistance. The increase in resistance increases blood pressure and blood velocity. This response accelerates the rate of oxygen transport to cells and the removal of carbon dioxide from blood by the lungs. A decrease in sympathetic action potential frequency results in **vasodilation,** which increases vessel diameter and decreases resistance. The decrease in resistance decreases blood pressure and blood velocity. The terms vasoconstriction and vasodilation are usually used to describe arteries. When describing changes in the diameter of veins, the terms *venoconstriction* and *venodilation* are used.

Like the cardiac control center, the activity of the vasomotor center is modified by action potentials from higher brain areas, and sensory action potentials from baroreceptors and chemoreceptors in the aortic arch and the carotid arteries. For example, a decrease in pressure, pH, or oxygen concentration of the blood stimulates vasoconstriction. Conversely, an increase in these values promotes vasodilation.

In addition, arterioles and precapillary sphincters are affected by local changes in the blood concentrations of oxygen, carbon dioxide, and pH. These local effects override the control by the vasomotor center, through a process called *autoregulation*. For example, if a particular muscle group is active for an extended period, a local decrease in oxygen concentration and an increase in carbon dioxide concentration result. These chemical changes stimulate the vasodilation of local arterioles and precapillary sphincters, which increases the flow of blood into capillary networks of the affected muscles to provide oxygen and to remove carbon dioxide more quickly.

Check My Understanding

15. How does blood pressure affect the flow of blood through blood vessels?
16. How are systolic and diastolic blood pressure different?
17. How do cardiac output, blood volume, and peripheral resistance affect blood pressure?

12.8 Circulation Pathways

Learning Objective
14. Compare the systemic and pulmonary circuits.

As noted earlier, the heart is a double pump that serves two distinct circulation pathways: the pulmonary and systemic circuits. These circuits were shown in figure 12.7.

Pulmonary Circuit

The **pulmonary circuit** carries deoxygenated blood to the lungs, where oxygen and carbon dioxide are exchanged between the blood and the air in the lungs. The right ventricle pumps deoxygenated blood into the pulmonary trunk, a short, thick artery that divides to form the left and right pulmonary arteries. Each pulmonary artery enters a lung and divides repeatedly to form arterioles, which in turn form alveolar capillaries that surround the air sacs (pulmonary alveoli) of the lungs (see chapter 14). Oxygen diffuses from the air in the alveoli into the capillary blood, and carbon dioxide diffuses from the blood into the air in the alveoli. Blood then flows from the capillaries into venules, which merge to form small veins, which, in turn, join to form progressively larger veins. Pulmonary veins emerge from each lung to carry oxygenated blood back to the left atrium of the heart.

Systemic Circuit

The systemic circuit carries oxygenated blood to the tissue cells of the body and returns deoxygenated blood to the heart. The left ventricle pumps the freshly oxygenated blood, received from the pulmonary circuit, into the aorta for circulation to all parts of the body except the lungs. The aorta branches to form many major arteries, which continually branch to form arterioles that branch to form systemic capillaries, where the exchange of materials between the blood and interstitial fluid takes place. Oxygen diffuses from the capillary blood into the interstitial fluid, while carbon dioxide diffuses from the interstitial fluid into the blood. From the capillaries, blood enters venules, which merge to form small veins, which join to form progressively larger veins. Ultimately, veins from the upper body (head, neck, shoulders, upper limbs, and thoracic and lumbar regions) join to form the superior vena cava, which returns blood from these regions back to the right atrium. Similarly, veins from the lower body (lower trunk and lower limbs) enter the inferior vena cava, which also returns blood into the right atrium. The **coronary sinus** drains the blood from the myocardium into the right atrium (see figure 12.5).

Check My Understanding

18. In what ways are the pulmonary and systemic circuits similar, and in what ways are they different?

12.9 Systemic Arteries

Learning Objective
15. Identify the major systemic arteries and the organs or body regions that they supply.

Major Branches of the Aorta

The aorta ascends from the heart, arches to the left and behind the heart, and descends through the thoracic and abdominal cavities just in front of the vertebral column. Because of its size, the aorta is divided into four regions: the ascending aorta, the aortic arch, the thoracic aorta, and the abdominal aorta. Figure 12.18 shows the major branches of the aorta and their relationships to the internal organs. Tables 12.4 and 12.5 list the major branches of the aorta and the organs and body regions that they supply.

The first arteries to branch from the aorta are the left and right coronary arteries, which supply blood to the heart. They branch from the aorta just above the aortic valve in the base of the *ascending aorta*.

276   Chapter 12   Cardiovascular System

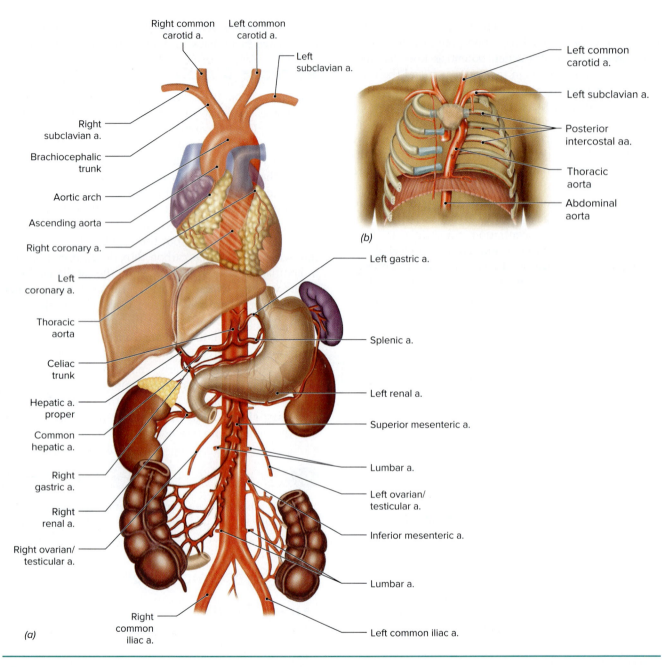

Figure 12.18 The Aorta.
(a) Major arteries that branch from the aorta. *(b)* Major arteries supplying the thoracic cage. (a. = artery; aa. = arteries)

Three major arteries branch from the *aortic arch*. In order of branching, they are the **brachiocephalic** (brāk-ē-ō-se-fal′-ik) **trunk,** the **left common carotid** (kah-rot′-id) **artery,** and the **left subclavian** (sub-klā′-vē-an) **artery.**

Pairs of **posterior intercostal** (in-ter-kos′-tal) **arteries** branch from the *thoracic aorta* to supply the intercostal muscles between the ribs and other organs of the thoracic wall. A number of other small arteries supply the organs of the thoracic cavity.

Once the aorta descends through the diaphragm, it is called the *abdominal aorta,* and it gives off several branch arteries to the abdominal wall and visceral organs. The **celiac** (sē′-lē-ak) **trunk** is a short artery that divides to form three branch arteries: (1) the **left gastric artery** supplies the stomach and esophagus, (2) the **splenic artery** supplies the spleen, stomach, and pancreas, and (3) the **common hepatic artery** supplies the liver, gallbladder, stomach, duodenum, and pancreas.

The **superior mesenteric** (mes-en-ter′-ik) **artery** supplies the pancreas, most of the small intestine, and the proximal half of the large intestine. The left and right **renal arteries** supply the kidneys. The left and right

Table 12.4 Major Arteries Branching from the Ascending Aorta, Aortic Arch, and Thoracic Aorta

Artery	Origin	Region Supplied
Coronary	Ascending aorta	Myocardium
Brachiocephalic trunk	Aortic arch	Branches as below
Right common carotid	Brachiocephalic trunk	Right side of head and neck
Right subclavian	Brachiocephalic trunk	Right shoulder and upper limb, thoracic wall
Left common carotid	Aortic arch	Left side of head and neck
External carotid	Common carotid	Scalp, face, and neck
Internal carotid	Common carotid	Brain
Left subclavian	Aortic arch	Left shoulder and upper limb, thoracic wall
Vertebral	Subclavian	Neck and brain
Axillary	Subclavian	Axilla and shoulder
Brachial	Axillary	Arm
Radial	Brachial	Forearm and hand
Ulnar	Brachial	Forearm and hand
Posterior intercostal	Thoracic aorta	Thoracic wall

Table 12.5 Major Arteries Branching from the Abdominal Aorta

Artery	Origin	Region Supplied
Celiac trunk	Abdominal aorta	Liver, stomach, spleen, gallbladder, esophagus, and pancreas
Common hepatic	Celiac trunk	Liver, gallbladder, stomach, duodenum, and pancreas
Left gastric	Celiac trunk	Stomach and esophagus
Splenic	Celiac trunk	Spleen, stomach, and pancreas
Renal	Abdominal aorta	Kidney
Superior mesenteric	Abdominal aorta	Pancreas, small intestine, and proximal half of the large intestine
Ovarian, testicular	Abdominal aorta	Ovaries or testes
Lumbar	Abdominal aorta	Lumbar region of back
Inferior mesenteric	Abdominal aorta	Distal half of the large intestine
Common iliac	Abdominal aorta	Pelvis and lower limb
Internal iliac	Common iliac	Pelvic wall, pelvic viscera, external genitalia, and medial thigh
External iliac	Common iliac	Pelvic wall and lower limb
Femoral	External iliac	Region surrounding the quadriceps and adductor muscles
Deep femoral	Femoral	Region surrounding the hamstring muscles and iliotibial tract
Popliteal	Femoral	Knee and some muscles of the thigh and leg
Anterior tibial	Popliteal	Anterior and lateral portions of the leg
Dorsalis pedis	Anterior tibial	Ankle and foot
Posterior tibial	Popliteal	Posterior leg
Fibular artery	Posterior tibial	Lateral leg muscles

ovarian arteries supply the ovaries in females. The left and right **testicular arteries** supply the testes in males.

Several pairs of **lumbar arteries** supply the walls of the abdomen and back. The **inferior mesenteric artery** supplies the distal half of the large intestine.

At the level of the iliac crests, the aorta divides to form two large arteries, the left and right **common iliac** (il′-ē-ak) **arteries,** which carry blood to the lower trunk and lower limbs.

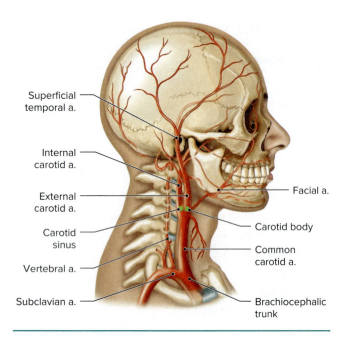

Figure 12.19 Major Arteries Supplying the Head and Neck.
(a. = artery)

Arteries Supplying the Head and Neck

The head and neck receive blood from several arteries that branch from the common carotid and subclavian arteries. Note in figures 12.18 and 12.19 that the brachiocephalic trunk branches to form the **right common carotid artery** and the **right subclavian artery.** The left common carotid and left subclavian arteries branch directly from the aortic arch.

Each common carotid artery divides in the neck to form an **external carotid artery** and an **internal carotid artery.** The external carotid arteries give rise to a number of smaller arteries that carry blood to the neck, face, and scalp. The internal carotid arteries enter the cranium and provide the major supply of blood to the brain.

The neck and brain are also supplied by the **vertebral arteries.** They branch from the subclavian arteries and travel up through the transverse foramina of cervical vertebrae to enter the cranium.

Arteries Supplying the Shoulders and Upper Limbs

The subclavian artery provides branches to the shoulder and passes under the clavicle to become the **axillary artery,** which supplies branches to the thoracic wall and axillary region. The axillary artery continues into the arm to become the **brachial artery,** which provides branches to serve the arm. At the elbow, the brachial artery divides to form a **radial artery** and an **ulnar artery,** which supply the forearm and wrist and merge to form a network of arteries supplying the hand (figure 12.20 and table 12.4).

Arteries Supplying the Pelvis and Lower Limbs

As noted earlier, the left and right common iliac arteries branch from the end of the aorta. Each common iliac branches within the pelvis to form internal and external iliac arteries. The **internal iliac artery** is the smaller branch that supplies the pelvic wall, pelvic organs, external genitalia, and medial thigh muscles. The **external iliac artery** is the larger branch, and it supplies the front of the pelvic wall and continues into the thigh, where it becomes the femoral artery (figure 12.20).

The **femoral artery** gives off branches that supply the regions surrounding the quadriceps femoris and adductor muscles. The largest branch is the **deep femoral artery,** which serves the regions surrounding the hamstring muscles and iliotibial tract. As the femoral artery descends, it passes behind the knee where it becomes the **popliteal** (pop-li-tē′-al) **artery,** which supplies certain muscles of the thigh and leg, as well as the knee. The popliteal artery branches just below the knee to form the anterior and posterior tibial arteries.

The **anterior tibial artery** descends between the tibia and fibula to supply the anterior and lateral portions of the leg, and it continues to become the **dorsalis pedis,** which supplies the ankle and foot. The **posterior tibial artery** lies posterior to the tibia and supplies the posterior portion of the leg, and it continues to supply the ankle and the plantar surface of the foot. Its largest branch is the **fibular artery,** which serves the lateral leg muscles (table 12.5).

⊕ Clinical Insight

The pulse may be taken at any surface artery, but the radial artery at the wrist and the common carotid artery in the neck are the most commonly used sites. Blood pressure is usually taken in the brachial artery of the arm. The radial artery at the wrist and the femoral artery at the groin are the common entry sites for angioplasty, a procedure in which a wire is fed into the arteries for widening narrowed or obstructed coronary or other systemic arteries.

✓ Check My Understanding

19. What is the arterial pathway of blood from the left ventricle to the right side of the brain?
20. What is the arterial pathway of blood from the left ventricle to the small intestine?
21. What is the arterial pathway of blood from the left ventricle to the top of the foot?

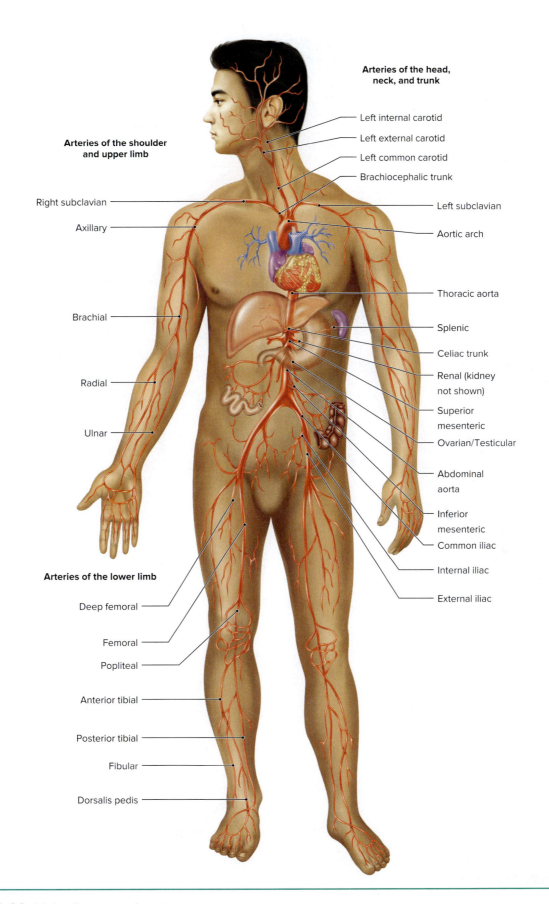

Figure 12.20 Major Systemic Arteries.

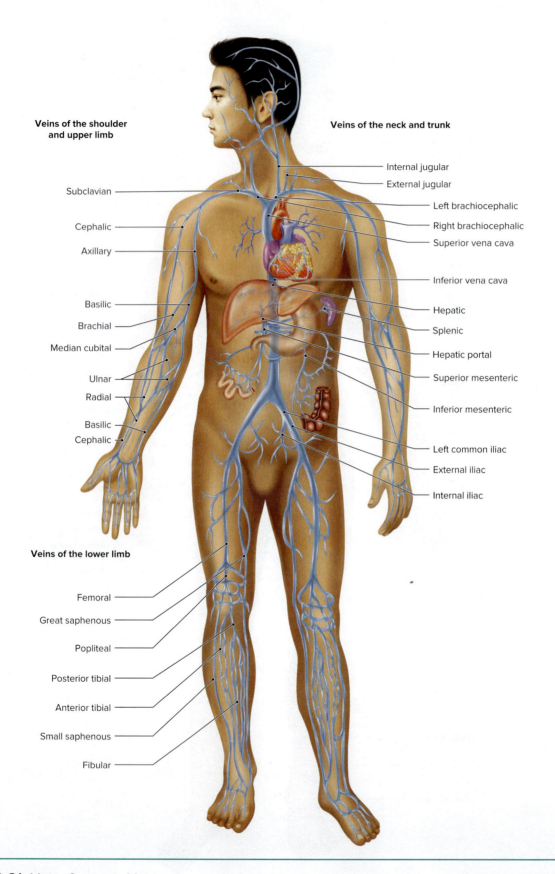

Figure 12.21 Major Systemic Veins.

12.10 Systemic Veins

Learning Objective

16. Identify the major systemic veins and the organs or body regions that they drain.

The systemic veins receive deoxygenated blood from capillaries and return the blood to the heart. Ultimately, all systemic veins merge to form two major veins, the superior and inferior venae cavae, that empty into the right atrium of the heart.

Veins Draining the Head and Neck

As shown in figure 12.22, superficial areas of the head and neck are drained by the left and right **external jugular** (jug′-ū-lar) **veins,** which lead into the left and right **subclavian veins,** respectively. The left and right **vertebral veins** carry blood from the cervical spinal cord and deep neck regions into the subclavian veins as well.

Most of the blood from the brain, face, and neck is carried by the left and right **internal jugular veins.** Each internal jugular vein merges with a subclavian vein to form a **brachiocephalic vein.** The left and right brachiocephalic veins join to form the **superior vena cava,** which returns blood to the right atrium of the heart (table 12.6).

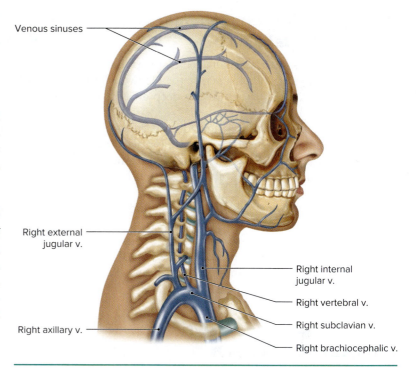

Figure 12.22 Major Veins Draining the Head and Neck. (v. = vein) APR

Table 12.6 Major Veins Draining to the Superior Vena Cava

Vein	Region Drained and Location	Receiving Vein
Head and Neck		
External jugular	Superficial areas of the head and neck	Subclavian
Vertebral	Cervical spinal cord and deep neck	Subclavian
Internal jugular	Brain, face, and neck	Brachiocephalic
Upper Limb and Shoulder		
Radial	Hand and forearm (deep)	Brachial
Ulnar	Hand and forearm (deep)	Brachial
Basilic	Medial upper limb (superficial)	Axillary
Cephalic	Lateral upper limb (superficial)	Axillary
Brachial	Arm (deep)	Axillary
Axillary	Axilla and shoulder	Subclavian
Subclavian	Shoulder and thoracic wall	Brachiocephalic
Trunk		
Brachiocephalic	Head, neck, shoulder, upper limb, and thoracic wall	Superior vena cava
Azygos	Thoracic and lumbar regions	Superior vena cava
Posterior intercostal	Thoracic wall	Azygos
Ascending lumbar	Lumbar region	Azygos

Veins Draining the Shoulders and Upper Limbs AP|R

Deep regions of the forearm are drained by the **radial** and **ulnar veins.** These two veins join at the elbow to form the **brachial vein,** which drains the deep areas of the arm (figure 12.21).

Superficial regions of the hand, forearm, and arm are drained by the laterally located **cephalic** (se-fal′-ik) **vein** and the medially located **basilic** (bah-sil′-ik) **vein.** Note the **median cubital** (kyū-bi-tal) **vein,** which connects the basilic and cephalic veins.

The basilic and brachial veins merge in the axilla to form the **axillary vein,** which then receives the cephalic vein. Once the axillary vein passes under the first rib, it becomes the **subclavian vein.** As noted earlier, the subclavian vein joins with the internal jugular vein to form the brachiocephalic vein (table 12.6).

Veins Draining the Pelvis and Lower Limbs AP|R

The **anterior** and **posterior tibial veins** drain the foot and deep regions of the leg. They join below the knee to form the **popliteal vein.** The **small saphenous** (sah-fē′-nus) **vein** drains the superficial posterior part of the leg and merges with the popliteal vein. The **fibular vein** drains the lateral portion of the leg and joins with the popliteal vein at the knee to form the **femoral vein,** which drains the deep regions of the thigh and hip.

The **great saphenous vein** originates from the venous arches in the foot, and it drains the medial and superficial portions of the foot, leg, and thigh. It merges with the femoral vein to form the **external iliac vein.** The external iliac vein and the **internal iliac vein** receive branches that drain the upper thigh and pelvic cavity, and they merge to form the **common iliac vein.** The left and right common iliac veins merge to form the **inferior vena cava,** which returns blood to the right atrium of the heart (see figure 12.21).

Clinical Insight

The median cubital vein is the vein of choice when drawing a sample of blood for clinical tests. It is easily located just under the skin on the front of the elbow joint. In coronary bypass surgery, a segment of the internal thoracic artery, saphenous vein, or radial artery is grafted to the afflicted coronary artery on each side of the blockage. The subclavian vein is a common site for implanting the central line, a long-term catheter for administering medications and taking blood samples.

Veins Draining the Abdominal and Thoracic Walls AP|R

The **azygos** (az′-i-gōs) **vein** drains most of the thoracic wall and lumbar region, and it empties into the superior vena cava near the right atrium. The azygos vein receives blood from a number of smaller veins, including the **posterior intercostal veins** and the **ascending lumbar vein,** which drains the lumbar region (figure 12.23).

Veins Draining the Abdominal Viscera AP|R

The **hepatic portal vein** carries blood from the stomach, intestines, spleen, and pancreas to the liver instead of the inferior vena cava. The hepatic portal vein is formed by the union of the **superior mesenteric vein,** which drains the small intestine and proximal half of the large intestine, and the **splenic vein,** which drains the spleen. The splenic vein receives blood from the **inferior mesenteric vein,** which drains the distal half of the large intestine, and the **pancreatic vein,** which drains the pancreas. The **gastric veins,** from the stomach, drain directly into the hepatic portal vein. All of these veins compose the **hepatic portal system.**

After entering the liver, the blood flows through the venous sinusoids, where materials are either removed or added before the blood enters the **hepatic veins,** which empty into the inferior vena cava (figure 12.24a). Note that 75% of the blood supply to the liver comes from the hepatic portal vein; the rest comes from the hepatic artery proper (see figure 12.18a). The hepatic portal system allows the liver to monitor and adjust the concentrations of substances in blood coming from the digestive tract before it enters the general circulation.

The left and right **renal veins** carry blood from the kidneys, and the left and right **ovarian** or **testicular veins** return blood from the ovaries in females or the testes in males, respectively. Both renal veins and the right ovarian or testicular vein drain into the inferior vena cava. The left ovarian or testicular vein empties into the left renal vein (figure 12.24b and table 12.7).

Check My Understanding

22. What is the venous pathway of blood from the left side of the head to the right atrium?
23. What is the venous pathway of blood from the small intestine to the right atrium?
24. What is the venous pathway of blood from the back of the ankle to the right atrium?

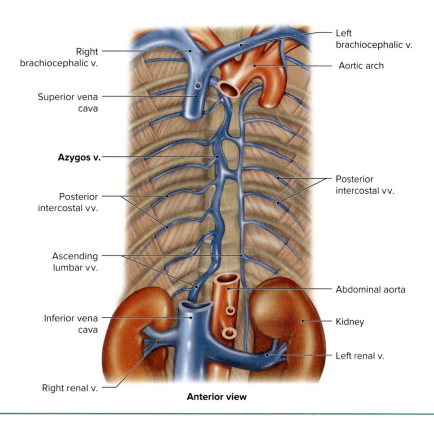

Figure 12.23 Major Veins of the Thoracic Cavity.
v. = vein; vv. = veins

Table 12.7 Major Veins Draining to the Inferior Vena Cava

Vein	Region Drained and Location	Receiving Vein
Lower Limb		
Anterior and posterior tibial	Foot and leg	Popliteal
Fibular	Lateral leg	Popliteal
Popliteal	Knee	Femoral
Small saphenous	Posterior leg (superficial)	Popliteal
Great saphenous	Lower limb (medial and superficial)	Femoral
Femoral	Deep regions of the hip and thigh	External iliac
Trunk		
Hepatic portal	Veins indented below	Liver sinusoids
Superior mesenteric	Small intestine and proximal large intestine	Hepatic portal
Splenic vein	Spleen	Hepatic portal
Inferior mesenteric	Distal large intestine	Splenic
Pancreatic	Pancreas	Splenic
Gastric	Stomach	Hepatic portal
Hepatic	Liver	Inferior vena cava
Renal	Kidneys	Inferior vena cava
Right ovarian or testicular	Right ovary or testis	Inferior vena cava
Left ovarian or testicular	Left ovary or testis	Left renal
External iliac	Pelvic wall and lower limb	Common iliac
Internal iliac	Viscera in pelvis, pelvic wall, and upper thigh	Common iliac
Common iliac	Pelvis and lower limb	Inferior vena cava

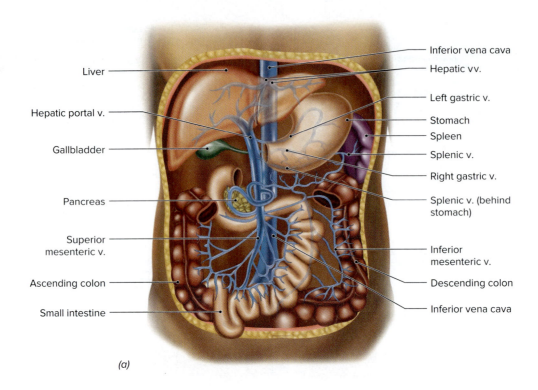

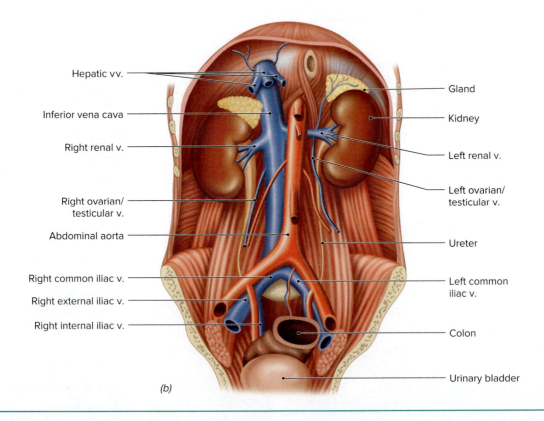

Figure 12.24 Veins of the Abdomen.
(a) An anterior view of the veins of the hepatic portal system. *(b)* An anterior view of the major veins draining into the inferior vena cava.

12.11 Disorders of the Heart and Blood Vessels

Learning Objective

17. Describe the common disorders of the heart and blood vessels.

These disorders are grouped according to whether they affect primarily the heart or the blood vessels. In some cases, the underlying cause of a heart ailment is a blood vessel disorder.

Heart Disorders

Arrhythmia (ah-rith'-mē-ah), or *dysrhythmia*, refers to an abnormal heartbeat. It may be caused by a number of factors, including damage to the conducting system of the heart, drugs, electrolyte imbalance, or a diminished supply of blood via the coronary arteries. In addition to irregular heartbeats, arrhythmia includes

- Bradycardia—a slow heart rate of less than 60 beats per minute. Note that the bradycardia in well-trained athletes is a healthy condition because it saves energy during resting heart contraction and has a greater potential to increase cardiac output.
- Tachycardia—a fast heart rate of over 100 beats per minute.
- Heart flutter—a very rapid heart rate of 200 to 300 beats per minute.
- Fibrillation—a very rapid heart rate in which the contractions are uncoordinated so that blood is not pumped from the ventricles. Ventricular fibrillation is usually fatal without prompt treatment.

Congestive heart failure (CHF) is the acute or chronic inability of the heart to pump out the blood returned to it by the veins. Symptoms include fatigue; edema (accumulation of fluid) of the lungs, feet, and legs; and excess accumulation of blood in internal organs. CHF may result from atherosclerosis of the coronary arteries, which deprives the myocardium of adequate blood.

Heart murmurs are unusual heart sounds. They are usually associated with defective heart valves, which allow a backflow of blood. Unless there are complications, heart murmurs have little clinical significance.

Myocardial infarction (mī-ō-kar'-dē-al in-fark'-shun) is the death of a portion of the myocardium due to an obstruction in a coronary artery. The obstruction is usually a blood clot that has formed as a result of atherosclerosis. This event is commonly called a "heart attack," and it may be fatal if a large portion of the myocardium is deprived of blood.

Pericarditis is the inflammation of the serous pericardium and is usually caused by a viral or bacterial infection. It may be quite painful as the inflamed membranes rub together during each heart cycle.

Blood Vessel Disorders

An **aneurysm** (an'-yū-rizm) is a weakened portion of a blood vessel that bulges out, forming a balloonlike sac filled with blood. Rupture of an aneurysm in a major artery may produce a fatal hemorrhage.

Arteriosclerosis (ar-te"-rēō-skle-rō'-sis) is hardening of the arteries. It results from calcium deposits that accumulate in the tunica media of arterial walls and is usually associated with atherosclerosis.

Atherosclerosis is the formation of fatty deposits (cholesterol and triglycerides) along the tunica intima of arterial walls. The atherosclerotic plaques reduce the lumen of the arteries and increase the probability of blood clots being formed. Such deposits in the coronary, carotid, or cerebral arteries may lead to serious cardiovascular problems (figure 12.25).

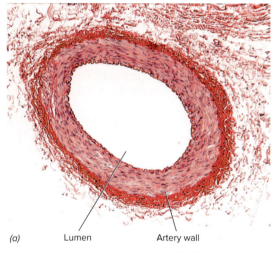
(a) Lumen · Artery wall

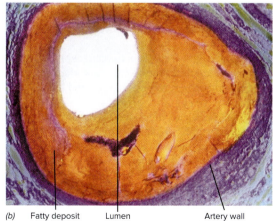
(b) Fatty deposit · Lumen · Artery wall

Figure 12.25 Comparison of Healthy and Diseased Arteries.
Cross sections of *(a)* a healthy artery and *(b)* an atherosclerotic artery whose lumen is diminished by fatty deposits. Atherosclerosis promotes the formation of a blood clot within the artery.

(a) ©Chuck Brown/Science Source; *(b)* ©CNRI/Science Source

Hypertension refers to chronic high blood pressure. It is the most common disease affecting the heart and blood vessels. Blood pressure that exceeds 140/90 mm Hg is indicative of hypertension. A systolic blood pressure of 120 to 139 mm Hg and a diastolic blood pressure of 80 to 89 mm Hg is considered to be *prehypertension*. Hypertension may be caused by a variety of factors, but persistent stress and smoking are commonly involved.

Phlebitis (flē-bī'-tis) is inflammation of a vein, and it most often occurs in a lower limb. If it is complicated by the formation of a blood clot, it is called *thrombophlebitis*.

Varicose veins are veins that have become dilated and swollen because their valves are not functioning properly. Heredity seems to play a role in their occurrence. Pregnancy, standing for prolonged periods, and lack of physical activity reduce venous return and promote varicose veins in the lower limbs. Chronic constipation promotes their occurrence in the anal canal, where they are called **hemorrhoids** (hem'-o-royds).

Chapter Summary

12.1 Anatomy of the Heart

- The heart wall consists primarily of the myocardium, a thick layer of cardiac muscle tissue. It is lined inside by the thin endocardium and outside by the thin epicardium.
- The pericardium is composed of the parietal layer of the serous pericardium and fibrous pericardium. The pericardial cavity is filled with pericardial fluid.
- The heart contains four chambers. The upper chambers are the left and right atria, which receive blood returning to the heart. The lower chambers are the left and right ventricles, which pump blood out of the heart. There are no openings between the atria or between the ventricles.
- Atrioventricular valves allow blood to flow between each atrium and its corresponding ventricle but prevent a backflow of blood. The mitral valve lies between the left atrium and left ventricle. The tricuspid valve lies between the right atrium and right ventricle.
- Semilunar valves allow blood to be pumped from the ventricles into their associated arteries but prevent a backflow of blood. The aortic valve is located in the base of the aorta. The pulmonary valve is located in the base of the pulmonary trunk.
- The right atrium receives deoxygenated blood from the superior and inferior venae cavae and the coronary sinus, and the left atrium receives oxygenated blood from the pulmonary veins.
- The right ventricle pumps deoxygenated blood into the pulmonary trunk, which divides into the left and right pulmonary arteries, which lead to the lungs. At the same time, the left ventricle pumps oxygenated blood into the aorta, which leads to all parts of the body except the lungs.
- The myocardium receives blood from the coronary arteries, which branch from the ascending aorta. Blood is returned from the myocardium by the cardiac veins, which open into the coronary sinus, which leads to the right atrium.

12.2 Cardiac Cycle

- The cardiac cycle includes both contraction (systole) and relaxation (diastole) phases.
- During atrial diastole, blood returns to the atria and flows on into the ventricles. Atrial systole forces more blood into the ventricles to fill them.
- During ventricular diastole, blood flows into the ventricles. Ventricular systole pumps blood from the ventricles into their associated arteries.
- The atrioventricular valves close and the semilunar valves open during ventricular systole due to increased blood pressure in the ventricles caused by contraction.
- The atrioventricular valves open and the semilunar valves close during ventricular diastole due to decreased blood pressure in the ventricles caused by relaxation.
- The lub-dup heart sound is caused by the closure of the heart valves. The first sound results from the closure of the atrioventricular valves. The second sound results from the closure of the semilunar valves.

12.3 Conducting System of the Heart

- The SA node is the pacemaker, which rhythmically initiates action potentials that cause the heart contractions.
- Action potentials pass through the atria, causing atrial systole, and simultaneously reach the AV node.
- Action potentials pass from the AV node along the AV bundle and bundle branches to the subendocardiac conducting network, which transmits the action potentials to the myocardium, causing ventricular systole.
- An electrocardiogram is a recording of the formation and transmission of action potentials through the conducting system of the heart.
- An electrocardiogram consists of a P wave, a QRS complex, and a T wave, and it is used in the diagnosis of heart ailments.

12.4 Regulation of Heart Function

- Cardiac output is a measure of heart function. It is determined by stroke volume and heart rate:

$$CO = SV \times HR$$

- Cardiac output is regulated by both intrinsic and extrinsic factors.
- Heart rate and stroke volume are controlled by the autonomic division. The cardiac control center is in the medulla oblongata. It receives sensory action potentials from baroreceptors and chemoreceptors, and is also affected by action potentials from the cerebrum and hypothalamus.

- Sympathetic axons release norepinephrine at heart synapses, which causes an increase in the heart rate and contraction strength. Parasympathetic axons release acetylcholine at heart synapses, which causes a decrease in heart rate.
- The dynamic balance in the frequency of sympathetic and parasympathetic action potentials reaching the heart adjusts the heart rate and stroke volume to meet body needs.
- Heart rate and stroke volume are also affected by age, sex, physical condition, temperature, epinephrine, thyroid hormones, and the blood concentration of Ca^{2+} and K^+.

12.5 Types of Blood Vessels

- The three basic types of blood vessels are arteries, capillaries, and veins. Large arteries and veins are formed of an outer tunica externa of dense irregular connective tissue, a middle tunica media of smooth muscle, and an inner tunica intima of endothelium supported by areolar connective tissue.
- Arteries have thick, muscular walls and carry blood from the heart. Large arteries divide repeatedly to form the smallest arteries, arterioles, until finally forming capillaries.
- Capillaries are the smallest and most numerous blood vessels. They are composed of an endothelium supported by a layer of areolar connective tissue. Their thin walls allow an exchange of materials between the blood and the interstitial fluid. Dissolved substances are exchanged by diffusion. Fluid exits the arteriolar end of a capillary because blood pressure is greater than osmotic pressure, and it reenters at the venular end of the capillary because osmotic pressure is greater than blood pressure.
- Veins have thinner walls than arteries and carry blood from capillaries toward the heart. The smallest veins are venules, which lead from capillaries and merge to form small veins. Large veins contain valves that prevent a backflow of blood.

12.6 Blood Flow

- Blood circulates from areas of higher pressure to areas of lower pressure. Blood pressure is highest in the ventricles and lowest in the atria.
- Systemic blood pressure declines as blood is carried from the arteries through the capillaries and through the veins. Skeletal muscle contractions and respiratory movements are important forces that aid the return of venous blood.
- Blood velocity varies inversely with the cross-sectional area of the combined blood vessels. Blood velocity is fastest in the aorta and slowest in the capillaries. The velocity progressively increases as the blood flows from capillaries to the larger veins.

12.7 Blood Pressure

- Optimal systolic blood pressure is 115 mm Hg. Optimal diastolic blood pressure is 75 mm Hg.
- The difference between systolic and diastolic pressures is the pulse pressure. The pulse may be detected by palpating superficial arteries.
- Blood pressure is determined by three factors: cardiac output, blood volume, and peripheral resistance.
- Peripheral resistance is determined by vessel diameters, total vessel length, and blood viscosity.
- The vasomotor center in the medulla oblongata provides the autonomic control of blood vessel diameter. In this way, the autonomic division controls peripheral resistance and blood pressure.
- Local autoregulation of arterioles overrides autonomic control and regulates blood flow in capillaries according to the needs of the local tissues.

12.8 Circulation Pathways

- The pulmonary circuit carries blood from the heart to the lungs and back again.
- The systemic circuit carries blood from the heart to all parts of the body, except the lungs, and back again.

12.9 Systemic Arteries

- The aorta is divided into the ascending aorta, aortic arch, thoracic aorta, and abdominal aorta.
- The major branch arteries of the aorta are the coronary, brachiocephalic trunk, left common carotid, left subclavian, posterior intercostals, celiac trunk, superior mesenteric, renal, ovarian/testicular, lumbar, inferior mesenteric, and common iliac arteries.
- The major arteries supplying the head and neck are paired arteries. Each common carotid artery branches to form the external and internal carotid arteries. The external carotid supplies the neck, face, and scalp. The internal carotid is the major artery supplying the brain. The vertebral arteries supply the neck and brain.
- Each shoulder and upper limb is supplied by a subclavian artery, which becomes the axillary artery, which becomes the brachial artery of the arm. The brachial artery branches to form the radial and ulnar arteries of the forearm.
- Each common iliac artery branches to form internal and external iliac arteries. The external iliac enters the thigh to become the femoral artery, which becomes the popliteal artery near the knee. The femoral artery has a branch called the deep femoral artery that goes to the hamstring area. The popliteal branches below the knee to form the anterior and posterior tibial arteries. The anterior tibial artery enters the foot to become the dorsalis pedis artery. The posterior tibial artery gives off a lateral branch called the fibular artery.

12.10 Systemic Veins

- Veins draining the head and neck are paired veins. On each side, the external jugular and vertebral veins empty into the subclavian vein. The internal jugular vein and subclavian merge to form the brachiocephalic vein. The left and right brachiocephalic veins join to form the superior vena cava.
- The ascending lumbar veins and the posterior intercostal veins enter the azygos vein, which opens into the superior vena cava.

- Radial and ulnar veins of the forearm join to form the brachial vein of the arm. The basilic vein joins the brachial vein to form the axillary vein, which, in turn, receives the cephalic vein. When the axillary vein passes under the first rib, it becomes the subclavian vein.
- Anterior and posterior tibial veins merge below the knee to form the popliteal vein. The popliteal vein receives the small saphenous and fibular veins to form the femoral vein. The great saphenous vein extends from the foot to join with the femoral vein near the hip, which forms the external iliac vein. The external iliac vein joins with the internal iliac vein to form the common iliac vein. The left and right common iliac veins merge to form the inferior vena cava.
- The splenic vein receives the inferior mesenteric vein and the pancreatic vein, and merges with the superior mesenteric vein to form the hepatic portal vein. The gastric veins drain into the hepatic portal vein. The hepatic portal vein empties into the liver sinusoids. The hepatic vein carries blood from the liver to the inferior vena cava.
- The right ovarian or testicular vein and the paired renal veins empty into the inferior vena cava. The left ovarian or testicular vein drains into the left renal vein.

12.11 Disorders of the Heart and Blood Vessels

- Disorders of the heart include arrhythmia, congestive heart failure, heart murmurs, myocardial infarction, and pericarditis.
- Disorders of blood vessels include aneurysm, arteriosclerosis, atherosclerosis, hypertension, phlebitis, and varicose veins.

Improve Your Grade

Connect Interactive Questions Reinforce your knowledge using multiple types of questions: interactive, animation, classification, labeling, sequencing, composition, and traditional multiple choice and true/false.

SmartBook Proven to help students improve grades and study more efficiently, SmartBook contains the same content within the print book but actively tailors that content to the needs of the individual.

Anatomy & Physiology REVEALED® Dive into the human body by peeling back layers of cadaver imaging. Utilize this world-class cadaver dissection tool for a closer look at the body anytime, from anywhere.

CHAPTER 13

Lymphoid System and Defenses Against Disease

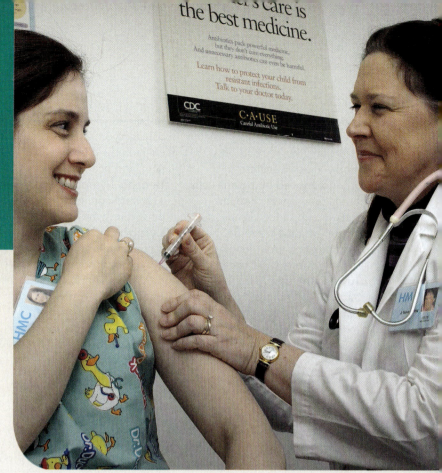

Source: James Ganthany/CDC

Adele, a 32-year-old receptionist, woke up this morning not feeling well and, while watching the morning news, heard another report about a new strain of influenza sweeping the country. Recently, several contagious diseases have begun to circulate through the people in her office. She and her 12 coworkers share printers, a water cooler, and a small kitchen space. It is easy to pass a viral or bacterial infection to another person if you are not careful. After suffering for almost two weeks with the flu last year, Adele was proactive regarding her health this year. She went to the local pharmacy a few months ago and received her flu and pneumonia vaccinations. Thanks to the vaccinations, her body now possesses defensive cells, called lymphocytes, to defend her against both diseases if she encounters them. However, these new lymphocytes will defend her against only the strains of flu and bacterial pneumonia in the vaccinations. As she begins to shiver slightly from a fever that has begun to develop, she now hopes that she has not been exposed to a new strain of flu or bacterial pneumonia, against which she will have no defense.

CHAPTER OUTLINE

13.1 Lymph and Lymphatic Vessels

13.2 Lymphoid Organs
- Red Bone Marrow
- Thymus
- Lymph Nodes
- Spleen

13.3 Lymphoid Tissues
- Tonsils
- Mucosa Associated Lymphoid Tissue

13.4 Nonspecific Resistance
- Mechanical Barriers
- Chemical Actions
- Phagocytosis
- Inflammation
- Fever
- Natural Killer Cells

13.5 Immunity
- Specialization of Lymphocytes
- Recognizing Pathogens
- Cell-Mediated Immunity
- Antibody-Mediated Immunity

13.6 Immune Responses
- Types of Immunity

13.7 Rejection of Organ Transplants

13.8 Disorders of the Lymphoid System
- Infectious Disorders
- Noninfectious Disorders

Chapter Summary

Module 10
Lymphatic System

Chapter 13 Lymphoid System and Defenses Against Disease

SELECTED KEY TERMS

Allergen An antigen that stimulates an allergic reaction.
Antibody (anti = against) A protein produced by plasma cells that binds to a specific antigen.
Antigen A substance capable of causing an immune response.
Complement A group of plasma proteins that destroy pathogens.
Immunity (immun = free) Resistance to specific antigens.
Immunocompetent (im-mu-no-kom′-pe-tent) Capable of recognizing and responding to a foreign antigen.
Inflammation (inflam = to set on fire) A localized response to damaged or infected tissues that is characterized by swelling, redness, pain, and heat.
Lymph (lymph = clear water) The fluid connective tissue transported in lymphatic vessels.
Lymph node A lymph-filtering secondary lymphoid organ.
Lymphatic vessel A vessel that transports lymph.
Pathogen A disease-causing organism or substance.
Primary lymphoid organ An organ where lymphocytes become immunocompetent.
Red bone marrow Primary lymphoid organ responsible for the production of all formed elements.
Secondary lymphoid organ An organ where immunocompetent lymphocytes proliferate and immune responses occur.
Spleen Secondary lymphoid organ located behind the stomach that stores formed elements and clears pathogens from the blood.
Thymus Primary lymphoid organ responsible for T cell maturation.

THE *LYMPHOID (LYMPHATIC) SYSTEM* is closely related to the cardiovascular system, both structurally and functionally (figure 13.1). A network of lymphatic vessels drains excess interstitial fluid (the approximate 10–15% that has not been returned directly to the blood capillaries) and returns it to the blood in a one-way flow that moves slowly toward the subclavian veins. Additionally, the lymphatic vessels in the small intestine function in lipid absorption and lymphocytes aid in the body's defense against disease-causing organisms or substances.

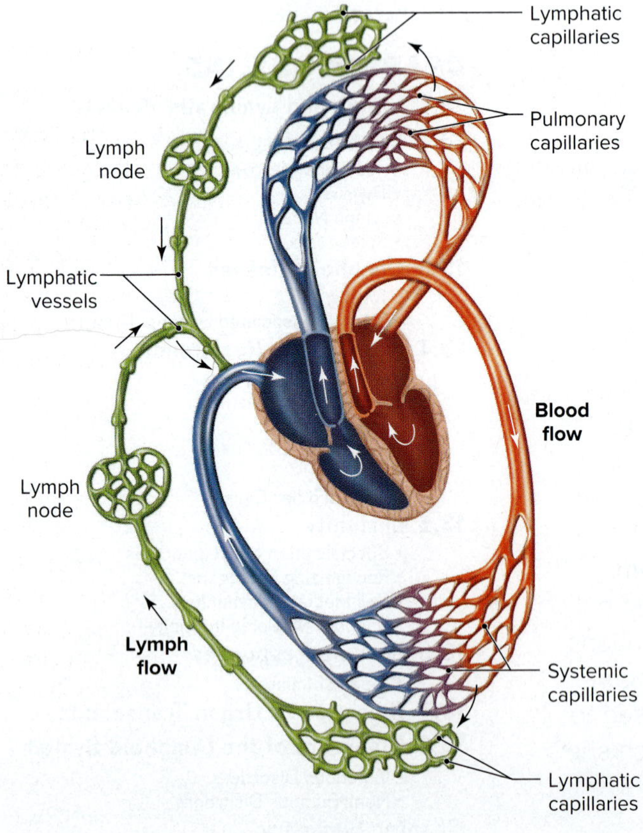

13.1 Lymph and Lymphatic Vessels

Learning Objectives

1. Describe the formation of lymph.
2. Describe the pathway of lymph in its return to the blood.
3. Describe how lymph is propelled through lymphatic vessels.

The lymphatic network of vessels begins with the microscopic **lymphatic capillaries.** Lymphatic capillaries are closed-ended tubes that form vast networks in the interstitial spaces within most vascular tissues (figure 13.2). Notably, these capillaries are not found in the CNS. Instead, the CNS relies on the flow of CSF to remove excess fluid from nerve tissue. Because the walls of lymphatic capillaries are composed of endothelial cells with unique junctions, interstitial fluid, proteins, and microorganisms can easily enter the vessels but cannot leave and reenter the interstitial space. Once fluid enters the lymphatic capillaries, it becomes a fluid connective tissue referred to as **lymph** (limf). Adequate lymphatic drainage is needed to prevent the accumulation of interstitial fluid, a condition called *edema* (ē-dē′-mă). Additionally, within the intestinal villi of the small intestine, lymphatic capillaries called *lacteals* (lak′-tē-alz) transport absorbed lipids and lipid-soluble vitamins away from the digestive tract.

Figure 13.1 Lymphatic Vessels. Lymphatic vessels transport fluid from interstitial spaces to the blood. AP|R

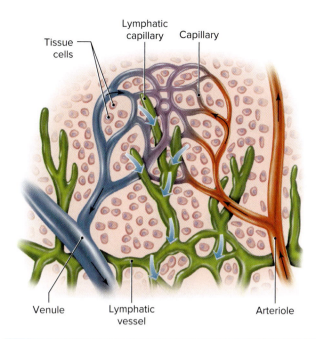

Figure 13.2 Lymphatic Capillaries.
Lymphatic capillaries (green) are microscopic, closed-ended tubes that begin in the interstitial spaces of most tissues.

From merging lymphatic capillaries, the lymph is carried into **lymphatic vessels.** These lymphatic vessels merge into even larger vessels called **lymphatic trunks** that are named after large body regions (figure 13.3). The walls of lymphatic vessels and trunks are much like those of veins. They have the same three layers and also contain valves to prevent backflow. The pressure that keeps the lymph moving comes from the massaging action produced by skeletal muscle contractions, intestinal movements, respiratory pressure changes (the same venous return mechanisms described in chapter 12), and from peristaltic contractions of some lymphatic vessels. Interconnecting lymphatic trunks eventually empty into one of the two principal vessels: the **thoracic duct** and the **right lymphatic duct** (figure 13.3). The larger thoracic duct drains lymph from the left thoracic region, left upper limb, left side of the head and neck, and all areas below the diaphragm. The thoracic duct begins in the abdominal cavity as a saclike enlargement called the **cisterna chyli** (sis-ter′-nă ki′-lē), which collects lymph from the lower limbs and the intestinal region. The thoracic duct then ascends along the vertebral column and drains into the left subclavian vein near the left internal jugular vein. The smaller right lymphatic duct receives lymph from the right upper limb, right thoracic region, and right side of the head and neck. The right lymphatic duct empties into the right subclavian vein near the right internal jugular vein (figure 13.3).

Clinical Insight

Edema is the swelling of localized tissues due to the accumulation of excess interstitial fluid. It results from either too much fluid exiting the blood in capillaries or insufficient removal of fluid by lymphatic vessels. There are a variety of causes for edema. For example, a sedentary lifestyle, which leads to the breakdown of valves in lower limb veins, can result in edema of the lower limbs. Also, removal of lymphatic vessels and lymph nodes during cancer surgery often leads to edema of the affected area.

Check My Understanding

1. What is lymph and how is it returned to the blood?
2. Why is the return of lymph to the blood important?

13.2 Lymphoid Organs

Learning Objective
4. Describe the locations and functions of the red bone marrow, thymus, lymph nodes, and spleen.

Lymphoid structures can be found throughout the body. While all lymphoid structures are capable of lymphocyte production, the red bone marrow and thymus are considered **primary lymphoid organs** because lymphocytes originate in these organs. After production in the red bone marrow, most lymphocytes and other immune cells go to **secondary lymphoid organs,** such as the lymph nodes and spleen that become the sites of proliferation of lymphocytes and immune responses.

Red Bone Marrow

Red bone marrow is a primary lymphoid organ found in the spongy bone of most of the axial skeleton and the proximal epiphyses of the humerus and femur. As described in chapter 11, red bone marrow is the site of origin of all formed elements. Not all lymphocytes formed in the red bone marrow are **immunocompetent,** capable of recognizing and attacking foreign substances, when they exit the marrow. To be immunocompetent, a lymphocyte must be able to elicit an immune response. **B cells,** or *B lymphocytes*, stay in the bone marrow until they are immunocompetent before moving on to secondary lymphoid organs. Other lymphocytes that will become **T cells,** or *T lymphocytes*, must first move to the thymus for maturation before moving to secondary lymphoid organs.

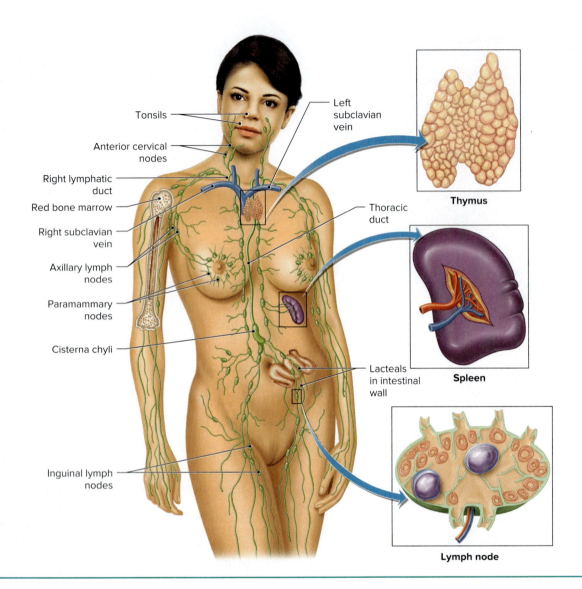

Figure 13.3 The Lymphoid System.
The lymphoid system is composed of several types of interconnected lymphoid organs, including the thymus, spleen, and lymph nodes. Lymph from the right upper limb, the right side of the head and neck, and the right thoracic region drains through the right lymphatic duct into the right subclavian vein. Lymph from the remainder of the body drains through the thoracic duct into the left subclavian vein. AP|R

Thymus AP|R

The **thymus** is a soft, bilobed gland located in the mediastinum above the heart (figure 13.3). This primary lymphoid organ is large (40 g) in infants and children, but after puberty it begins to atrophy and becomes quite small (12 g) in adults. The thymus plays a key role in the development of the lymphoid system before birth and during early childhood. Until the lymphoid system matures at about two years of age, an infant is more susceptible to disease than older children. The major function of the thymus is the differentiation of T cells into immunocompetent cells. The thymus produces hormones called *thymosins* that promote the differentiation and division of T cells, making them immunocompetent. After maturation, T cells are distributed by the blood to secondary lymphoid organs and lymphoid tissues throughout the body.

Lymph Nodes

Lymph nodes are secondary lymphoid organs that filter lymph. They usually occur in groups along the larger lymphatic vessels. They are widely distributed in the body, but they do occur in large collections in the inguinal, axillary, and cervical regions of the body, as well as within the thoracic and abdominopelvic cavities. There are no lymph nodes in the CNS.

Structure of Lymph Nodes

Lymph nodes are roughly bean-shaped and 1.0 to 2.5 cm in length. Figure 13.4 is a photograph of a lymph node

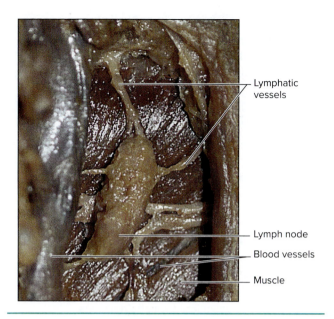

Function of Lymph Nodes

A major function of lymph nodes is the filtration and cleansing of the lymph as it passes through a node. They are the only lymphoid organs that filter the lymph. Damaged cells, cancerous cells, cellular debris, bacteria, and viruses become trapped in the reticular tissue of the lymph node and are destroyed by the action of lymphocytes and macrophages. Lymphocytes act against cancerous cells and **pathogens,** such as bacteria and viruses. Macrophages engulf cellular debris, immobilized or dead bacteria, and viruses.

Spleen

The **spleen** is a secondary lymphoid organ located behind the stomach just under the diaphragm in the left upper quadrant of the abdominopelvic cavity (see figure 13.3). The spleen is the largest lymphoid organ. The false ribs provide protection against physical injury. The spleen is a soft, purplish organ 5 to 7 cm (2-3 in) wide and 13 to 16 cm (5-6 in) long. It contains numerous centers for lymphocyte proliferation and large venous sinuses filled with blood.

Figure 13.4 A Lymph Node and Its Associated Vessels.
©Dr. Kent M. Van De Graaff RF

The spleen resembles a large lymph node. Like lymph nodes, it is enveloped by a thin capsule of dense irregular connective tissue and is subdivided by reticular tissue into many compartments. The compartments contain two basic types of tissues that are named for their appearance in fresh, unstained tissue. *White pulp* consists of large numbers of lymphocytes that cluster around tiny branches of the splenic artery. This tissue is mostly concerned with the immune functions of the spleen. *Red pulp* occupies the rest of a compartment, surrounding the white pulp and the venous sinuses. It is a storage area for RBCs and platelets and a site where worn-out red blood cells and pathogens are removed from the blood (figure 13.6). Before birth, the spleen and liver are the major blood-forming organs, but this function is later taken over by

in situ that shows lymphatic vessels leading to and from the node. Figure 13.5 shows the structure of a section of a lymph node. Note that the lymph node consists of a number of small subunits called **lymphoid nodules.** The lymphoid nodules are collections of lymphocytes and macrophages within reticular tissue and are the sites of activation and proliferation of lymphocytes.

Lymph enters a lymph node through several *afferent* (af′-er-ent) *lymphatic vessels* and flows through the *lymphatic sinuses,* which surround the lymphoid nodules. Lymph is collected from the sinuses and enters *efferent* (ef′-er-ent) *lymphatic vessels,* which carry lymph away from the lymph node. The indentation of the node where efferent lymphatic vessels emerge is called the *hilum.*

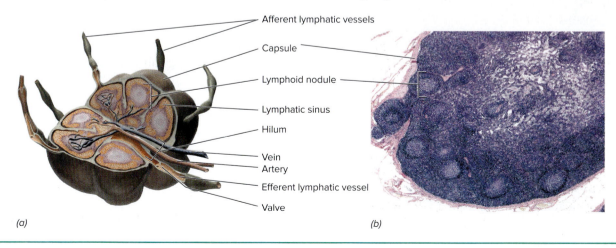

(a) (b)

Figure 13.5 Inner Structure of a Lymph Node.
(a) Cutaway showing inner structure. *(b)* Photomicrograph of a lymph node.
(b) ©McGraw-Hill Education/Al Telser, photographer

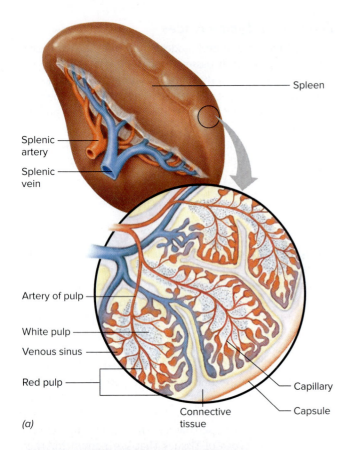

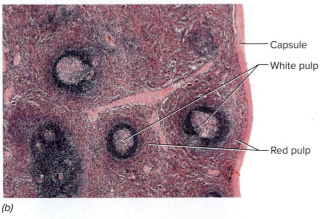

Figure 13.6 The Spleen.
(*a*) The spleen is subdivided into compartments by reticular tissue. (*b*) A photomicrograph of spleen tissue showing white pulp and red pulp. AP|R
(b) ©McGraw-Hill Education/Al Telser, photographer

red bone marrow. After birth, the spleen's role is related to both lymphoid and cardiovascular functions:

- It cleanses and filters the blood much like lymph nodes cleanse lymph. Lymphocytes and macrophages destroy pathogens, and macrophages clean up the debris.
- It stores a reserve supply of RBCs and platelets, which can be released into the blood in times of need, such as after a hemorrhage.
- It is a major site for RBC destruction and recycling, as described in chapter 11.
- It is a major site of lymphocyte activation and proliferation.

In spite of these important functions, the spleen is not essential for life. But following a splenectomy, a person may be more susceptible to potential pathogens and the effects of hemorrhage.

13.3 Lymphoid Tissues

Learning Objective
5. Describe the locations and functions of the tonsils and mucosa associated lymphoid tissues.

The tonsils and mucosa associated lymphoid tissues are not structurally organs; however, they function as secondary lymphoid organs because they are sites of immune responses.

Tonsils AP|R

Tonsils (ton'-sils) are clusters of lymphoid tissue located just beneath the mucous membrane in the pharynx (fayr'-inks), or throat, and oral cavity. Like all lymphoid tissue, they contain both lymphocytes and macrophages to fight pathogens. Their function is to intercept and destroy pathogens that enter through the nose and mouth before they can reach the blood.

There are three kinds of tonsils that are strategically located to carry out this function:

1. The *palatine tonsils* are located bilaterally at the junction of the oral cavity and the pharynx.
2. The *pharyngeal* (fah-rin'-jē-al) *tonsil* is located behind the nasal cavity in the upper portion of the pharynx. This tonsil is commonly called the *adenoid*.
3. The *lingual* (ling'-gwal) *tonsils* are located on the base of the tongue in the back of the oral cavity.

Tonsils are larger in young children and play an important role in their defense against pathogens. Sometimes the palatine tonsils and pharyngeal tonsil become so overloaded and swollen with pathogens, a condition called *tonsillitis*, that they must be surgically removed by a *tonsillectomy*.

Mucosa Associated Lymphoid Tissue

Individual lymphoid nodules, like those found within lymph nodes, are located throughout the body, especially in the areolar connective tissues of mucous membranes.

These collections of numerous macrophages and lymphocytes trapped in reticular tissue provide additional barriers to invasion by pathogens. Large clusters of many lymphoid nodules located in the mucous membranes of the respiratory, digestive, urinary, and reproductive tracts are referred to collectively as **MALT (mucosa associated lymphoid tissue).** The *appendix,* an extension of the large intestine located in the right lower quadrant of the abdominopelvic cavity, is part of the MALT that helps control bacterial growth in the large intestine. Table 13.1 outlines the components of the lymphoid system.

Check My Understanding

3. What are the functions of the red bone marrow, thymus, lymph nodes, spleen, tonsils, and MALT?

Clinical Insight

The spread of cancerous cells occurs when cells break away from the primary tumor and are carried to other sites via the lymphoid system or the blood. This process, called *metastasis* (me-tas′-ta-sis), generates secondary cancerous growths called *metastases* wherever the cancerous cells lodge. Because metastatic cells are frequently carried in the lymph, secondary cancerous growths often occur in lymph nodes on lymphatic vessels draining the region of the primary cancer. Knowledge of the location of lymphatic vessels and lymph nodes, and the direction of lymph flow, is important in the detection and treatment of metastases.

Table 13.1 Components of the Lymphoid System

Component	Characteristics	Function
Lymphatic Capillaries	Microscopic closed-ended tubes in interstitial spaces	Collect interstitial fluid from interstitial spaces; collect and transport dietary lipids and lipid-soluble vitamins; once in a lymphatic capillary, fluid is called lymph
Lymphatic Vessels	Formed by merging of lymphatic capillaries; structure similar to veins; contain valves; merge to form lymphatic trunks that drain into either the right lymphatic duct or the thoracic duct	Transport lymph and empty it into the subclavian veins
Lymphoid Organs		Sites of lymphocyte production or proliferation, and immune responses
Primary Lymphoid Organs		
Red bone marrow	Located mostly in spongy bone of the skeleton	Site of origination of all lymphocytes
Thymus	Bilobed gland located above the heart; size decreases with age	Site of T cell maturation; secretes hormones called thymosins, which stimulate maturation of T cells
Secondary Lymphoid Organs		
Lymph nodes	Small, bean-shaped organs arranged in groups along lymphatic vessels	Sites of lymphocyte proliferation; house T cells and B cells that are responsible for immunity; macrophages phagocytose pathogens and cellular debris from lymph
Spleen	Large lymphoid organ containing venous sinuses	RBC and platelet reservoir; macrophages phagocytose pathogens, cellular debris, and worn-out RBCs from the blood; houses lymphocytes
Lymphoid Tissues		
Tonsils	Masses of lymphoid tissue within the mucous membranes of pharynx and oral cavity	Protect against invasion of pathogens that are ingested or inhaled
MALT (mucosa associated lymphoid tissue)	Masses of lymphoid tissue within the mucous membranes of respiratory, digestive, urinary, and reproductive tracts	Guards against pathogens that penetrate the epithelium of a mucous membrane

13.4 Nonspecific Resistance

Learning Objective
6. Identify the components of nonspecific resistance.

Nonspecific resistance provides protection against all pathogens and foreign substances, but it is not directed against a specific pathogen. Nonspecific defense mechanisms include mechanical barriers, chemical actions, phagocytosis, inflammation, fever, and natural killer (NK) cells.

Mechanical Barriers

The most obvious *mechanical barriers* against pathogens are the skin and the mucous membranes. The closely packed epidermal cells of the skin make penetration by pathogens very difficult. Mucous membranes are less effective barriers than the skin. Mucus is continuously produced by the mucous membranes lining the respiratory and digestive tracts. The mucus entraps pathogens and airborne particles and usually prevents their contact with the underlying membranes. Pathogens entrapped upon entering the nose and mouth tend to be destroyed by the tonsils. The continuous flow of tears over the eyes, the production and swallowing of saliva, the movement of vaginal secretions, and the passage of urine through the urethra are examples of fluid mechanical barriers that help to flush away pathogens before they can attack body tissues.

Chemical Actions

Various body chemicals, including certain enzymes, provide a nonspecific defense against pathogens. A few examples will illustrate the effect of these chemicals.

Tears, saliva, nasal secretions, and perspiration contain the enzyme *lysozyme*, which destroys certain types of bacteria and helps to protect underlying tissues. In addition to being a mechanical barrier, the skin has a rather acidic pH that inhibits bacterial growth.

Pathogens the tonsils miss are swallowed at frequent intervals. Upon reaching the stomach, most pathogens are destroyed by gastric juice, either by its acidic pH or by the enzyme *pepsin*. Pepsin acts by digesting the proteins composing the pathogens.

Virus-infected cells produce *interferon*, a substance that stimulates uninfected cells to synthesize special proteins that inhibit the replication of viruses within them. In this way, the rapid growth of viruses may be inhibited.

The blood contains a group of plasma proteins known as **complement**, which are named because their actions complement the actions of antibodies. Complement proteins can bind to certain pathogens initiating a chain of events that leads to the destruction of the pathogen. The binding of complement is known as *complement fixation*. The fixed complement punches holes in the pathogen's plasma membrane, causing the cell to burst and the pathogen to be destroyed. Subsequently, the resulting debris is cleaned up by phagocytes (neutrophils and macrophages). Complement proteins also enhance phagocytosis and inflammation.

Phagocytosis

Phagocytosis (fag″-ō-sī-tō′-sis) is the engulfing and destruction (by digestion) of pathogens, damaged or cancerous cells, and cellular debris by neutrophils and macrophages. When an infection occurs, neutrophils and monocytes are quickly attracted to the infected tissues. Monocytes entering the tissues become transformed into macrophages, large cells that are especially active in phagocytosis.

Some macrophages wander among the tissues, searching out and phagocytizing pathogens and cellular debris. Others become fixed (stationary) in particular locations in the body, where they phagocytize pathogens that are passing by. Fixed macrophages are especially abundant along the inner walls of blood and lymphatic vessels and in the spleen, lymph nodes, liver, and red bone marrow. The wandering and fixed macrophages compose the **tissue macrophage system,** which plays a major role in the destruction of potential pathogens.

Inflammation

Inflammation is a localized response to infection or injury that promotes the destruction of pathogens and the healing process. It is characterized by redness, pain, heat, and swelling of the affected tissues.

Clinical Insight

The most common pathogens affecting humans are bacteria and viruses. Bacteria are very small, single-celled organisms that lack a true nucleus and other complex cellular organelles. Their DNA is concentrated, but it is not enclosed in a nuclear envelope as in higher organisms. Bacteria are simple organisms but they have the necessary metabolic machinery required for life and reproduction. Bacterial pathogens may cause disease by releasing toxins (poisons), releasing enzymes that damage cells, or entering and destroying cells. Antibiotics are effective in treating most bacterial infections.

In contrast, a virus is composed of nucleic acids, either DNA or RNA, enveloped by a protein coat. Viruses are thousands of times smaller than bacteria. A virus attaches to a cell's surface receptor and penetrates into the cell. Once inside, it takes over the cell's DNA and metabolic machinery, causing the cell to replicate hundreds or thousands of viruses, which burst forth as the cell is destroyed. The released viruses then move on to attack other cells. Antibiotics are not effective in treating viral infections.

When injury or infection occurs, several mechanisms produce chemicals, such as complement proteins and histamine, that cause dilation of the arterioles and increase the permeability of blood capillaries in the affected area. The increased blood flow to the local area produces redness and heat. The increased movement of fluids out of the blood capillaries produces swelling (edema) of the tissues. Pain results from irritation of nociceptors by pathogens, swelling, or chemicals released by infected cells.

Some of the chemicals of the inflammatory response attract WBCs to the affected area. In bacterial infections, neutrophils and macrophages actively phagocytize the pathogens and damaged cells. The accumulated mass of living and dead WBCs, tissue cells, and bacteria may form a thick, whitish fluid called **pus**.

Fluids from blood capillaries that enter the affected area contain both fibrinogen and fibroblasts. Fibrinogen may be converted into fibrin to form a clot that is subsequently penetrated and enveloped by fibers formed by the fibroblasts. This action tends to seal off the infected area and prevent the spread of pathogens to neighboring tissues.

The continued action of WBCs usually brings the infection under control. Then, the dead pathogens and cells are cleaned up by phagocytes, and new cells are produced by mitosis to repair any damage to the tissues.

Fever

Fever is a high body temperature that accompanies infections and is a necessary part of the immune response. It serves a useful purpose as long as the body temperature does not get too high. The increased body temperature inhibits growth of certain pathogens and increases the rate of body processes, including those that fight infection.

Natural Killer Cells

Natural killer (NK) cells, another type of lymphocyte, provide nonspecific defense rather than specific defense. NK cells provide "immune surveillance" by traveling through the body looking for unhealthy cells. Once these cells are located, NK cells kill virus-infected cells, bacteria, transplanted cells, and tumor cells. Unlike other lymphocytes, NK cells have no specificity for individual pathogens and do not form memory cells.

Table 13.2 summarizes the major components of nonspecific resistance.

Check My Understanding

4. What is the method of action of mechanical barriers, chemical actions, phagocytosis, inflammation, fever, and NK cells?

Table 13.2 Summary of Major Components of Nonspecific Resistance

Component	Function
Mechanical Barriers	
Intact skin	Closely packed cells and multiple cell layers prevent entrance of pathogens
Intact mucous membranes	Closely arranged cells retard entrance of pathogens; not as effective as intact skin
Mucus	Traps pathogens in digestive, respiratory, urinary, and reproductive tracts
Saliva	Washes pathogens from oral surfaces
Tears	Wash pathogens from surface of eye
Urine	Washes pathogens from urethra
Vaginal secretions	Wash pathogens from vaginal canal
Chemical Actions	
Acidic pH of skin	Retards growth of many bacteria
Gastric juice	Kills pathogens that are swallowed
Interferon	Helps to prevent viral infections
Lysozyme	Antimicrobial enzyme in nasal secretions, perspiration, saliva, and tears that kills some pathogens
Complement	Group of plasma proteins that enhance inflammation and phagocytosis. Also cause direct death of pathogens by puncturing plasma membranes.
Other Mechanisms	
Fever	Speeds up body processes and inhibits growth of pathogens
Inflammation	Promotes nonspecific resistance; confines infection; attracts WBCs
Natural killer (NK) cells	Destruction of unhealthy cells such as virus-infected cells, bacteria, transplanted cells, and tumor cells.
Phagocytosis	Phagocytes engulf and destroy pathogens

13.5 Immunity

Learning Objectives
7. Compare nonspecific resistance and specific resistance.
8. Explain the mechanism of cell-mediated immunity.
9. Explain the mechanism of antibody-mediated immunity.

In contrast to nonspecific resistance, **immunity** (i-mū'-ni-tē), or specific resistance, is directed at specific antigens. An **antigen** is any substance that can cause an immune response. An immune response involves the production of specific cells and substances to attack a specific antigen. Immunity has "memory"; that is, if the same pathogen should reenter the body at a later date, the immune response is quicker and stronger than during the first encounter. Lymphocytes play several important roles in immunity.

An immune response involves one or both of two distinct processes: **cell-mediated immunity** and **antibody-mediated immunity** (figure 13.7).

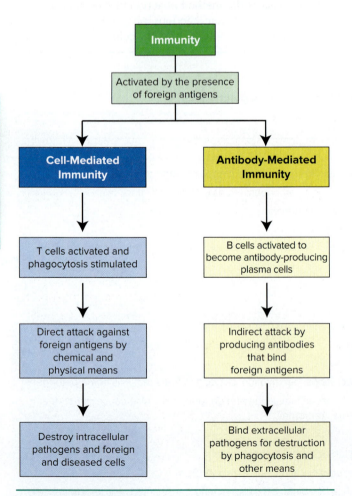

Figure 13.7 Overview of Immunity.

Specialization of Lymphocytes

During fetal development, all formed elements are initially produced by the yolk sac. Later in fetal development, the liver and spleen take over the production of formed elements. Around the time of birth, the red bone marrow takes over the production of most formed elements, including lymphocytes. However, lymphocytes must mature and become specialized within primary lymphoid organs before they can participate in immunity. About half of the unspecialized lymphocytes pass to the thymus, where they become immunocompetent T cells. The other half of the lymphocytes become specialized in the bone marrow to become immunocompetent B cells. T and B cells are carried by blood to secondary lymphoid organs, such as the lymph nodes, spleen, and tonsils, where they proliferate to form large populations of T and B cells, and some are released into the blood. About 75% of circulating lymphocytes are T cells, and about 25% are B cells (figure 13.8).

Recognizing Pathogens

All cells, including human cells, have recognition molecules that are antigens. (Recall that antigens are used in typing blood.) In fact, the cells of each person have a unique set of antigens. Antigens are usually large molecules, such as proteins and glycoproteins.

During the specialization process, lymphocytes "learn" to distinguish "self" antigens from foreign (nonself) antigens. Thereafter, lymphocytes can recognize an invading pathogen or body cell with abnormal antigens, such as cancerous cells and cells infected by a virus, and launch an attack. Unfortunately, lymphocytes also recognize a transplanted organ as foreign, as in graft rejection, and in some cases fail to recognize certain body tissues as "self" and attack the body's own tissues, as in autoimmune diseases.

The maturation process produces thousands (and perhaps millions) of different kinds of T and B cells, each with specific receptors capable of binding with (recognizing) a specific antigen. Therefore, when an antigen of a pathogen or foreign cell enters the body, only the T or B cells that have a receptor that can bind with the specific antigen are involved in the immune response. Table 13.3 summarizes the roles of B and T cells in immunity.

Cell-Mediated Immunity

In cell-mediated (cellular) immunity, T cells directly attack and destroy foreign cells or diseased body cells, such as transplanted cells or cancerous cells, and develop a memory of their antigens in case they should reappear in the future. Cell-mediated immunity also destroys *intracellular* pathogens, especially viruses.

A cell-mediated immune response begins when an **antigen-presenting cell (APC),** often a macrophage, engulfs a foreign antigen and displays the antigen, in

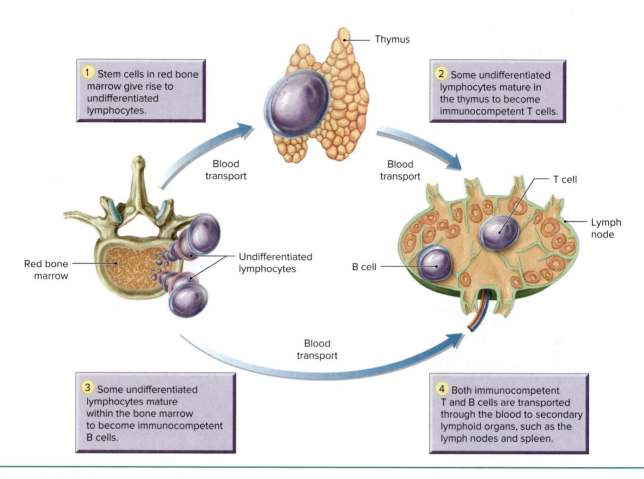

Figure 13.8 Differentiation of T Cells and B Cells.

Table 13.3 Roles of B Cells and T Cells in an Immune Response

Cell	Function
B cells	When activated, produce a clone of plasma cells and memory B cells that have the same antigen receptors as the activated parent B cell
Plasma cells	Produce specific antibodies against the antigen that activated the parent B cell
Memory B cells	Dormant cells that are activated if an antigen that fits their antigen receptors reenters the body; launch a secondary immune response
Helper T cells	When activated, start an immune response by releasing cytokines, which activate B cells, promote clone formation by B and T cells, and stimulate phagocytosis; form a clone of active helper T cells and memory T cells
Cytotoxic T cells	When activated, form a clone of cytotoxic T cells and memory T cells programmed to bind to the same antigen that activated the parent cytotoxic T cell; kill cells possessing the targeted antigen with toxic chemicals
Memory T cells	Dormant cells that are activated if an antigen that fits their antigen receptors reenters the body; launch a secondary immune response

combination with its own self proteins, on its plasma membrane. When a T cell, which has been programmed during maturation to recognize this particular antigen, binds to the antigen and the APC self proteins, it becomes activated. The activated T cell undergoes repeated mitotic divisions to form a **clone** of identical T cells capable of binding with the same antigen that activated the parent T cell (figure 13.9).

If the activated T cell is a **helper T cell** (T_H), it produces a clone consisting mostly of active T_H cells along with some **memory T cells** (T_M) capable of binding with the same antigen as the activated parent T cell. When active T_H cells bind with the antigen, they secrete *cytokines* (chemicals) that (1) attract neutrophils and macrophages to the site and stimulate their phagocytic activity, and

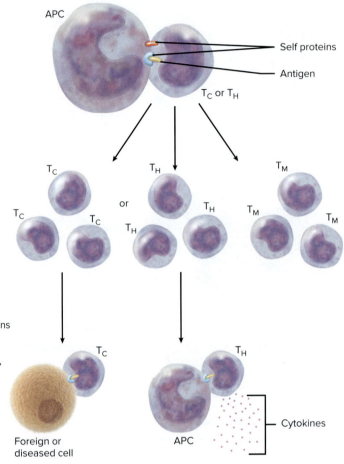

Figure 13.9 T Cell Activation and Cell-Mediated Immunity.
(APC = antigen presenting cell; T_H = helper T cell; T_C = cytotoxic T cell; T_M = memory T cell) **AP|R**

1. **Antigen recognition**
T cell binds to nonself antigen and self proteins of APC and becomes activated.

2. **Clone formation**
Activated T cell undergoes repeated mitotic division producing a clone of identical T cells.

3. **Action of clone cells**
T_C cells bind to cells displaying the targeted antigen and inject chemicals to destroy them.

T_H cells bind to antigen and self proteins on APC, become activated, and release cytokines, which promote activity of B and T_C cells, inflammation, and phagocytosis by neutrophils and macrophages.

T_M cells launch a secondary response if the targeted antigen reappears.

(2) stimulate mitotic division and immune responses of activated B and cytotoxic T cells.

If the activated T cell is a **cytotoxic T cell** (T_C), it produces a clone consisting mostly of active T_C cells along with some T_M. When an active T_C cell binds to any cell displaying the targeted antigen, it releases a lethal dose of chemicals to kill the cell. It then detaches and searches out and destroys other cells with the same antigen.

As more and more active T cells proliferate and the pathogen is eliminated, the immune response slows and stops. The long-lived T_M cells remain to launch a quicker and stronger attack if the targeted antigen should reappear.

Antibody-Mediated Immunity

Antibody-mediated (humoral) immunity involves both B cells and T_H cells. It provides a defense against *extracellular* pathogens. Rather than directly attacking foreign antigens, it involves the production of **antibodies** (proteins) that bind to foreign antigens tagging them for destruction by other means.

Clinical Insight

Cancer immunotherapy is an emerging science that is so groundbreaking, the reputable journal *Science* named it the "Breakthrough of the Year" in 2013. Cancer immunotherapy is unique because it targets the immune system, rather than the cancerous tumor cells, to treat cancer. The therapy uses drugs to program cytotoxic T cells to attack antigens it hasn't previously been able to recognize, such as antigens that are commonly found on cancerous cells.

An antibody-mediated immune response begins when foreign antigens bind to receptors of a B cell. The B cell engulfs the antigens and displays the antigen on its plasma membrane along with its own self antigens. When a T_H cell, which has been programmed to recognize this antigen, binds to this foreign antigen and self antigen complex, the T_H cell secretes cytokines that activate the B cell. The activated B cell produces a clone of identical

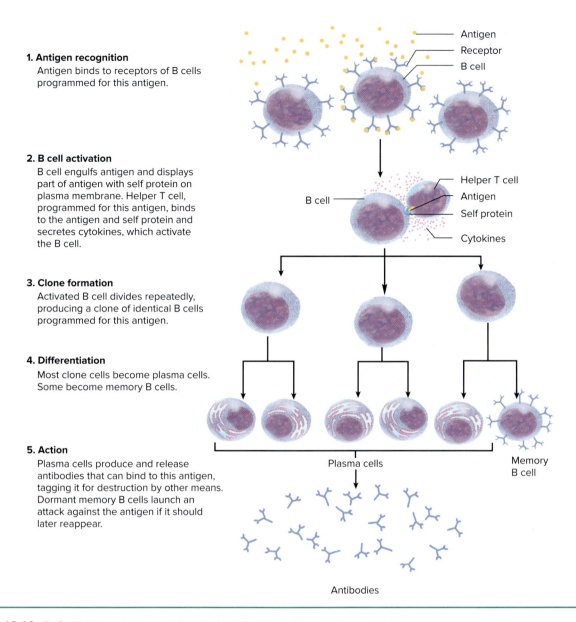

Figure 13.10 B Cell Activation and Antibody-Mediated Immunity. AP|R

1. **Antigen recognition**
Antigen binds to receptors of B cells programmed for this antigen.

2. **B cell activation**
B cell engulfs antigen and displays part of antigen with self protein on plasma membrane. Helper T cell, programmed for this antigen, binds to the antigen and self protein and secretes cytokines, which activate the B cell.

3. **Clone formation**
Activated B cell divides repeatedly, producing a clone of identical B cells programmed for this antigen.

4. **Differentiation**
Most clone cells become plasma cells. Some become memory B cells.

5. **Action**
Plasma cells produce and release antibodies that can bind to this antigen, tagging it for destruction by other means. Dormant memory B cells launch an attack against the antigen if it should later reappear.

B cells programmed against the same antigen. Most of the B cells in the clone become **plasma cells,** but some become **memory B cells** (figure 13.10).

Plasma cells rapidly produce and release antibodies capable of binding with the targeted antigen. Antibodies circulate throughout the body in blood and other body fluids binding to the nonself antigens so they are more easily destroyed by phagocytosis and other means. After the pathogen has been destroyed, antibodies remain to provide protection. If the targeted antigen reenters the body, memory B cells launch an even quicker and stronger attack.

Antibodies do not destroy pathogens directly. Instead, they bind to the antigens, forming *antigen–antibody complexes* that tag pathogens for destruction by other means. For example, an antigen–antibody complex involving bacteria or other cellular pathogens creates a site for binding complement, activating the same actions as described earlier in the nonspecific defenses section.

Antibodies neutralize bacterial toxins by binding to the antigens, which prevents the toxins from attaching to receptors of body cells. Body cells escape damage because the toxins cannot enter the cells. Subsequently, the antigen–antibody complexes are engulfed and destroyed by macrophages, eosinophils, and neutrophils.

Antibodies are proteins known as globulins, so another name for antibodies is **immunoglobulins,** which have a shorthand designation, *Ig*. The structure of an antibody determines its classification and each class plays a special role in antibody-mediated immunity. The amount of each type and their specific functions are listed in table 13.4.

Table 13.4 Classes of Antibodies

Class	Location	Functions
IgG (75%)	Plasma	Provides long-term immunity following vaccination or recovery from infection; crosses placenta to provide passive immunity for newborn infant; most important in fixing complement
IgA (15%)	Saliva, tears, mucus, breast milk, plasma	Protects mucous membranes from pathogens; provides passive immunity for breast-fed infants
IgM (10%)	B cells, plasma	Released from B cells and agglutinates antigens
IgD (0.2%)	B cells	Serves as a receptor on B cells
IgE (0.02%)	Binds to mast cells and basophils	Triggers allergic reactions by causing release of histamine when it binds with allergen

Check My Understanding

5. What is immunity?
6. How is an immune reaction started?

13.6 Immune Responses

Learning Objectives
10. Compare primary and secondary immune responses.
11. Explain the basis of vaccination.

When an antigen is encountered for the first time, it stimulates T cells and B cells to become activated and proliferate, producing clones that attack and destroy the invading antigen. This is the *primary immune response*, and it also produces memory cells that are able to recognize the same antigen if it should reenter the body. If another invasion of the same antigen occurs at a later date, the memory T cells and memory B cells recognize it and launch a *secondary immune response*, which is more rapid and intense than the primary immune response. A secondary immune response occurs each time the same antigen is detected by the memory cells. Figure 13.11 shows that the concentration of antibodies in the secondary immune response is much higher than in the primary response.

Types of Immunity

There is more than one way for a person to develop immunity to a particular pathogen, and these mechanisms may be grouped into two broad categories: active immunity and passive immunity. A person is directly involved in the development of **active immunity** but not in **passive immunity**. Active immunity is acquired through the use of a person's immune response, which leads to the development of memory cells. Passive immunity is acquired without the activation of a person's immune response, and therefore there is no memory. Further, immunity can be acquired naturally or, in some cases, artificially, through medical intervention.

Naturally acquired active immunity results after a person is exposed to a pathogen, gets sick, and recovers, leaving antibodies and memory B and T cells to fight the pathogen via a secondary immune response if it reenters the body.

Artificially acquired active immunity results after a person receives a vaccine (vak-sēn′) of weakened, dead, or inactivated pathogens or their antigenic parts, which trigger a primary immune response, leaving antibodies and memory B and T cells to fight the pathogen if it reenters the body. Booster shots may be used to trigger a secondary immune response to build up the concentration of antibodies even higher.

Naturally acquired passive immunity occurs in infants who have received maternal IgG via the placenta and IgA in breast milk. Antibodies from breast milk are an important aspect of defense in newborn infants.

Artificially acquired passive immunity results from receiving injections of antibodies produced in another person, an animal, or a synthetic source. This type of injection, called *antiserum*, is used in emergency situations

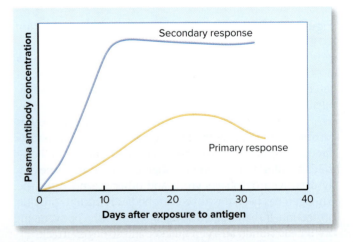

Figure 13.11 Immune Responses.
Antibodies are produced more rapidly and at a higher concentration during a secondary immune response.

Table 13.5 Types of Immunity

Type	Mechanism	Result
Naturally acquired active immunity	Infection by live pathogens	Person is ill with the disease; immune response destroys pathogens and leaves memory T and B cells to prevent later infection
Artificially acquired active immunity	Receives vaccine of weakened, dead, or inactivated pathogens or their antigenic parts	Immune response occurs without the person becoming ill; memory T and B cells remain to prevent later infection
Naturally acquired passive immunity	Antibodies passed from mother to fetus in utero or to newborn via breast milk	Enables short-term immunity for newborn infant without stimulating an immune response
Artificially acquired passive immunity	Receives injection of antibodies against a specific antigen	Enables short-term immunity without stimulating an immune response

when the pathogen (usually a toxin) acts too severely and quickly to wait for natural immunity to act. See table 13.5 for a summary of types of immunity.

Check My Understanding

7. Why are helper T cells so important to immunity?
8. How does a vaccination provide immunity against disease?

13.7 Rejection of Organ Transplants

Learning Objective

12. Explain how rejection of a transplanted organ occurs.

Organ transplants are viable treatment options for persons with a terminal disease of certain organs such as the heart, kidneys, and liver. Except for the surgery, the major problem encountered is that the patient's lymphoid system recognizes the transplanted organ as foreign and launches an attack against it. The problem is reduced by carefully determining compatibility of the tissues of both donor and recipient. This is usually done by comparing the antigens on the surfaces of leukocytes, called the *human leukocyte antigens (HLAs) group A*, of donor and recipient. A match does not have to be 100% to make a transplant, but the closer to 100%, the better the chance of avoiding rejection.

To overcome the typical immune response and organ rejection, immunosuppressive therapy is administered following transplant surgery. The lymphoid system must be suppressed sufficiently to prevent rejection of the organ but not enough to eliminate immunity against pathogens. Achieving this delicate balance has been aided by the use of *cyclosporine*, a selective immunosuppressive drug derived from fungi. Cyclosporine inhibits T cell functions but has minimal effects on B cells. Because T cells are primarily responsible for organ rejection, rejection is minimized, and B cells are able to provide antibody-mediated immunity against pathogens. In spite of advances in immunosuppressive therapy, bacterial and viral infections are the primary causes of death among organ transplant recipients. Immunosuppression also increases the risk of cancer.

13.8 Disorders of the Lymphoid System

Learning Objective

13. Describe the common disorders of the lymphoid system.

Infectious Disorders

Acquired immunodeficiency syndrome (AIDS) is a viral disease that is approaching epidemic proportions. It is caused by the *human immunodeficiency virus (HIV)*, which attacks and kills helper T cells and invades macrophages, which serve as a reservoir for the virus. In time, the immune defenses of the victim are greatly reduced, and the patient becomes susceptible to opportunistic diseases that ultimately lead to death. These secondary diseases include pneumonia caused by *Pneumocystis carinii* and a cancer known as *Kaposi sarcoma*, disorders that are rarely encountered except in AIDS patients. At present, there is no cure for AIDS, although major research efforts have provided drugs to slow its progress.

Although HIV has been found in most body fluids, it appears that sufficient HIV concentrations for transmission to other persons are not present in tears or saliva. Transmission does not occur through routine, nonintimate contact. Transmission occurs through exchanges of blood, most commonly by the use of contaminated hypodermic needles and by exposure of open wounds or mucous membranes to infected blood. Vaginal fluids and semen of infected persons are effective transmitting agents in sexual intercourse. Also, infected mothers may transmit HIV to infants during childbirth.

Elephantiasis (el-e-fan-tĭ′-ah-sis) is a chronic condition characterized by greatly swollen (edematous) lower limbs or other body parts, which become "elephantlike" in appearance. This occurs because the lymphatic vessels are blocked by masses of microscopic roundworms, which causes fluid to accumulate excessively in the tissues drained by the plugged lymphatic vessels. The microscopic worms are transmitted by the bites of certain species of mosquitoes found in tropical regions.

Lymphadenitis (lim-fad-en-ahy′-tes) is inflammation of the lymph nodes. It is a common complication of bacterial infections and is often described as "swollen glands." Doctors often feel the cervical region for swollen lymph nodes to see if they are inflamed due to fighting an infection in the head or neck, which drains into the cervical nodes.

A severely swollen lymph node is called a *bubo*. Buboes are named after the characteristic swelling of the inguinal and axillary lymph nodes in individuals infected with the bubonic plague, a bacterial infection of the lymphoid system that contributed to the Black Death. The Black Death is the largest known pandemic (large infectious outbreak), which killed about half of the population of Europe (estimated at up to 200 million people worldwide) from 1347 to 1351.

Noninfectious Disorders

Allergy (al′-er-jē), or *hypersensitivity*, refers to an abnormally intense immune response to an antigen that is harmless to most people. Such antigens are called **allergens** to distinguish them from antigens associated with disease. Sometimes allergies are triggered by a combination of environmental factors, rather than by an identifiable allergen. Such cases are difficult to treat because most drugs don't work against these types of reactions. Once a person is sensitized to an allergen, an allergic reaction results whenever subsequent exposure to that allergen occurs. Allergic reactions may be either immediate or delayed.

Immediate reactions result when allergens bind with IgE on the surface of mast cells. This interaction causes these cells to secrete substances—such as histamine—that stimulate an inflammatory response. Immediate allergic reactions may be either localized or systemic (whole body). Localized reactions, such as hay fever, hives, allergy-based asthma, and digestive disorders, are unpleasant but rarely life threatening. In contrast, systemic allergic reactions, also known as *anaphylaxis* (an-ah-fĭ-lak′-sis), are often life threatening. They quickly impair breathing and may cause circulatory failure due to a sudden drop in blood pressure as blood vessels dilate and fluid moves into the tissues. Allergic reactions to penicillin and bee stings can cause systemic allergic responses.

Delayed allergic reactions appear one to three days after exposure to the antigen. Delayed allergic reactions result from cytokines released by T cells. The dermatitis that occurs following contact with poison ivy and some cosmetic chemicals is a common delayed allergic reaction.

Autoimmune diseases result when T and B cells, for unknown reasons, recognize certain body tissues as foreign antigens and produce an immune response against them. This problem may result because certain body molecules have changed slightly and are no longer recognizable as self. Some of the most common autoimmune diseases are

- *rheumatoid arthritis (RA)*, which destroys joints;
- *type I diabetes*, which destroys beta cells in the pancreas;
- *multiple sclerosis (MS)*, which destroys the myelin sheath in the CNS;
- *Graves disease*, which stimulates the thyroid gland to produce excessive amounts of thyroid hormones;
- *myasthenia gravis*, which impairs synaptic transmission at neuromuscular junctions; and
- *Hashimoto disease*, which destroys the thyroid gland, creating hypothyroidism. It is the most common cause of hypothyroidism in the United States.

Lymphoma (lim-fō′-mah) is a general term referring to any tumor of lymphoid tissue. There are several types of lymphomas. One type of malignant lymphoma is *Hodgkin lymphoma*, which is cancer of lymphoid tissue involving the production of B cells. It is characterized by lymphadenitis, fatigue, and sometimes fever and night sweats. Early treatment with chemotherapy or radiation yields a high cure rate.

Severe combined immunodeficiency (SCID) is a group of disorders resulting from several different genetic defects. They are characterized by a marked deficit or absence of both T cells and B cells. Thus, the lymphoid system of affected individuals is basically nonfunctional. Infants have little or no protection against pathogens and usually die within a year without treatment. Transplants of healthy stem cells from red bone marrow or umbilical cord blood have proved to be successful in some cases, and gene therapy trials look promising for treating this devastating disease.

> ### ✓ Check My Understanding
> 9. What is so serious about the AIDS virus destroying helper T cells?
> 10. What is the cause of autoimmune diseases?

Chapter Summary

13.1 Lymph and Lymphatic Vessels

- As interstitial fluid accumulates in interstitial spaces, it is picked up by lymphatic capillaries. Once the fluid is within a lymphatic vessel, it is called lymph.
- Lymph is carried by lymphatic vessels and is ultimately returned to the blood. Lymphatic vessels contain valves that keep the lymph moving in one direction.
- Lymphatic vessels merge to form lymphatic trunks, which drain major regions of the body. Lymphatic trunks ultimately join one of two collecting ducts.
- The right lymphatic duct receives lymphatic trunks that drain the upper right portion of the body; it empties into the right subclavian vein.
- The thoracic duct receives lymphatic trunks from the rest of the body; it empties into the left subclavian vein.
- The propulsive forces that move lymph through the vessels are skeletal muscle contractions, intestinal movements, respiratory pressure changes (the same venous return mechanisms described in chapter 12), and peristaltic contractions of some lymphatic vessels.

13.2 Lymphoid Organs

- Lymphoid organs include red bone marrow, the thymus, lymph nodes, and the spleen.
- Primary lymphoid organs are sites of lymphocyte production.
- Secondary lymphoid organs are sites of lymphocyte proliferation and immune responses.
- Red bone marrow is a primary lymphoid organ that is the site of origination for all lymphocytes.
- B cells become immunocompetent in the red bone marrow, but T cells migrate to the thymus to become immunocompetent.
- Immunocompetence is the ability to recognize and respond to foreign antigens.
- The thymus is a primary lymphoid organ located above the heart within the mediastinum.
- Lymph nodes are secondary lymphoid organs occurring in groups along lymphatic vessels, where they filter lymph and serve as a site of lymphocyte activation and proliferation.
- The spleen is a secondary lymphoid organ that filters and cleanses the blood and contains a reservoir of RBCs and platelets that are squeezed into circulation if more blood is needed. It destroys old and damaged RBCs, as well as activates an immune response to foreign antigens in the blood.

13.3 Lymphoid Tissues

- Tonsils are groups of lymphatic tissues within the mucous membranes of the pharynx and oral cavity. They intercept pathogens entering the pharynx from the mouth and nose.
- Mucosa associated lymphoid tissue (MALT) consists of collections of WBCs within the mucous membranes of respiratory, digestive, urinary, and reproductive tracts.

13.4 Nonspecific Resistance

- Nonspecific resistance provides general protection against all pathogens, but it is not directed at any particular pathogen.
- Mechanical barriers include the skin, mucous membranes, mucus, tears, saliva, and urine.
- Protective chemicals include gastric juice, interferon, enzymes such as lysozyme and pepsin, and fluids with a low pH, including the surface of the skin.
- Complement is a group of plasma proteins that poke holes in the plasma membranes of pathogens. Complement also activates phagocytosis and inflammation.
- Phagocytosis is a major mechanism of nonspecific resistance. Neutrophils, monocytes, and wandering and fixed macrophages actively engulf and destroy pathogens. The wandering and fixed macrophages form the tissue macrophage system, which plays a major role in protection against pathogens.
- Inflammation helps control infection by attracting WBCs and macrophages and by increasing the blood supply to the affected area.
- Fever increases the rate of defense processes and inhibits the growth of certain pathogens.
- Natural killer (NK) cells provide immune surveillance by destroying unhealthy virus-infected cells, bacteria, transplanted cells, and tumor cells.

13.5 Immunity

- Immunity provides protective mechanisms against specific antigens. Lymphocytes and macrophages play key roles in immunity.
- An antigen is any substance that can cause an immune response.
- Maturing lymphocytes learn to distinguish molecules composing the body (self) from foreign (nonself) molecules. Large foreign antigens can stimulate an immune response.
- Undifferentiated lymphocytes are first formed in the fetal yolk sac with all formed elements. The liver and spleen take over formed element production in the fetal life as well before eventually moving to the red bone marrow in a newborn infant. About half of the lymphocytes, the B cells, stay in the red bone marrow to become immunocompetent. The other half are carried to the thymus, where they become immunocompetent T cells. Immunocompetent T cells and B cells are dispersed to secondary lymphoid organs and tissues throughout the body and also circulate in the blood.
- T cells provide cell-mediated immunity. T_H cells and T_C cells are activated when they bind to an antigen and self protein presented by an APC. Activated T_H cells form a clone of active T_H cells and T_M cells; activated T_C cells form a clone of active T_C cells and T_M cells.
- T_H cells release cytokines that stimulate mitosis and immune action of T and B cells and phagocytosis. T_C cells destroy foreign and diseased cells with chemicals. After the antigen is destroyed, T_M cells remain to attack the antigen if it reenters the body.

- B cells provide antibody-mediated immunity. When an antigen binds to receptors on a B cell, it is engulfed and displayed on the cell surface along with self antigens. A T_H cell binds to the antigen–self protein complex and secretes cytokines, which activate the B cell. The B cell forms a clone of identical B cells. Most of the cloned B cells become plasma cells, and a few become memory B cells. Plasma cells produce antibodies that bind to the antigen, enabling easier destruction of the antigen by phagocytosis and other means. Memory cells remain to attack the antigen if it reenters the body.
- Antibodies are specific proteins produced by plasma cells against specific antigens. There are five classes of antibodies: IgG, IgA, IgM, IgD, IgE. Each performs a special role.

13.6 Immune Responses

- The first contact with a specific antigen produces the primary immune response. Subsequent contacts with the same antigen produce secondary immune responses that are more rapid and intense.
- There are two basic types of immune responses: active and passive. Active immunity results when the body produces its own memory cells and antibodies. Passive immunity results from antibodies that have been produced by another person, an animal, or a synthetic source.

13.7 Rejection of Organ Transplants

- The healthy response of the lymphoid system is to attack and destroy organ transplants because they are foreign. Rejection is minimized by a good match between donor and recipient and by the administration of immunosuppressive drugs.
- The use of cyclosporine inhibits functions of T cells but leaves functions of B cells essentially intact.

13.8 Disorders of the Lymphoid System

- Infectious disorders include AIDS, elephantiasis, and lymphadenitis.
- Noninfectious disorders of the lymphoid system include allergy, autoimmune diseases, and lymphoma.

Improve Your Grade

Connect Interactive Questions   Reinforce your knowledge using multiple types of questions: interactive, animation, classification, labeling, sequencing, composition, and traditional multiple choice and true/false.

SmartBook   Proven to help students improve grades and study more efficiently, SmartBook contains the same content within the print book but actively tailors that content to the needs of the individual.

Anatomy & Physiology REVEALED®   Dive into the human body by peeling back layers of cadaver imaging. Utilize this world-class cadaver dissection tool for a closer look at the body anytime, from anywhere.

CHAPTER 14

Respiratory System

©Hero Images/Getty Images RF

One fall weekend, Jesse drives home from college to attend the homecoming football game at his old high school. When Jesse arrives at the game, it is as if the entire town has come out to watch their team battle the rival high school. For three hours, Jesse cheers loudly for every great play and yells at the referee for every bad call. In fact, the stadium is so loud that he has to shout to talk to some old friends who have sat down nearby. When the game ends, Jesse heads home feeling triumphant over his alma mater's victory. However, when he wakes up in the morning, his neck is sore and his voice is very raspy and barely audible. His mother, a registered nurse at the local hospital, diagnoses Jesse as having acute laryngitis. Jesse's loud cheering has inflamed his larynx, or "voice box," which is causing the soreness and making his speaking difficult. As he rests his voice and drinks some hot tea with honey, Jesse thinks to himself that supporting his school was completely worth a little discomfort.

CHAPTER OUTLINE

14.1 Introduction to the Respiratory System

14.2 Structures of the Respiratory System
- Nose
- Pharynx
- Larynx
- Trachea
- Bronchi, Bronchioles, and Pulmonary Alveoli
- Lungs

14.3 Breathing
- Inspiration
- Expiration

14.4 Respiratory Volumes and Capacities

14.5 Control of Breathing
- Respiratory Centers

14.6 Factors Influencing Breathing
- Chemicals
- Inflation Reflex
- Irritant Reflexes
- Higher Brain Centers
- Body Temperature

14.7 Gas Exchange
- Alveolar Gas Exchange
- Systemic Gas Exchange

14.8 Transport of Respiratory Gases
- Oxygen Transport
- Carbon Dioxide Transport

14.9 Disorders of the Respiratory System
- Inflammatory Disorders
- Noninflammatory Disorders

Chapter Summary

Module 11
Respiratory System

SELECTED KEY TERMS

Alveolar gas exchange The exchange of oxygen and carbon dioxide between the air in pulmonary alveoli and the blood in alveolar capillaries.
Breathing The movement of air into and out of the lungs.
Bronchial tree (bronch = windpipe) The branching bronchi.
Expiration (ex = from; spirat = breathe) Movement of air out of the lungs; exhalation.
Glottis The vocal folds and the opening between them within the larynx.

Inspiration Movement of air into the lungs: inhalation.
Larynx (laryn = gullet) Cartilaginous box providing a passageway for air between the pharynx and the trachea and containing the vocal folds.
Pharynx (pharyn = throat) Passageway behind the nasal and oral cavities that extends downward to the larynx and esophagus; the throat.
Pulmonary alveolus (alveol = small cavity) A microscopic air sac within a lung.

Surfactant A mixture of lipoproteins in pulmonary alveoli that reduces surface tension and prevents pulmonary alveolar collapse.
Systemic gas exchange The exchange of oxygen and carbon dioxide between the blood in systemic capillaries and the tissue cells.
Trachea (trache = windpipe) Airway extending between the larynx and the bronchi.

14.1 Introduction to the Respiratory System

Learning Objectives
1. Describe the five processes involved in respiration.
2. Identify processes involved in external respiration and internal respiration.
3. List the functions of the respiratory system.

The primary role of the respiratory system is to supply the blood with oxygen and to remove carbon dioxide from the blood. The entire process of respiration encompasses five unique and sequential processes:

1. **Breathing** (pulmonary ventilation)–the movement of air into and out of the lungs.
2. **Alveolar gas exchange**–the exchange of oxygen and carbon dioxide between the air in pulmonary alveoli and the blood in alveolar capillaries.
3. **Gas transport**–transport of oxygen and carbon dioxide between the lungs and tissues, accomplished by the cardiovascular system.
4. **Systemic gas exchange**–the exchange of oxygen and carbon dioxide between the blood in systemic capillaries and the tissue cells.
5. **Aerobic respiration**–the use of oxygen and production of carbon dioxide during ATP production.

The structures of the respiratory system are involved directly in only two of these processes: breathing and alveolar gas exchange, which are collectively referred to as **external respiration.** Systemic gas exchange and aerobic respiration together are referred to as **internal respiration.**

The respiratory system does more than just exchange respiratory gases. It also helps to detect odors, produce sounds, regulate blood pH, trap and defend the body from airborne pathogens, and assist in the movement of venous blood and lymph.

 Check My Understanding
1. What are the five processes involved in respiration?
2. What respiratory processes directly involve the respiratory system?
3. What are the functions of the respiratory system?

14.2 Structures of the Respiratory System

Learning Objective
4. Describe the structures and functions of the respiratory system.

The *respiratory system* is subdivided into upper and lower respiratory tracts. The *upper respiratory tract* includes the nose and pharynx. The *lower respiratory tract* includes the larynx, trachea, bronchi, and lungs (figure 14.1a).

Nose

The protruding portion of the *nose* is supported by bone and septal nasal cartilage (figure 14.1c). The nasal bones form a rigid support for the bridge of the nose, and septal nasal cartilage supports the remaining portions and is responsible for the flexibility of the nose. The *nostrils*, or *nares* (singular, *naris*), are the two external openings in the nose that allow air to enter and leave the nasal cavity. Stiff hairs around the nostrils tend to keep out large airborne particles and insects.

The **nasal cavity** is the inner chamber of the nose that is surrounded by skull bones. It is separated from the

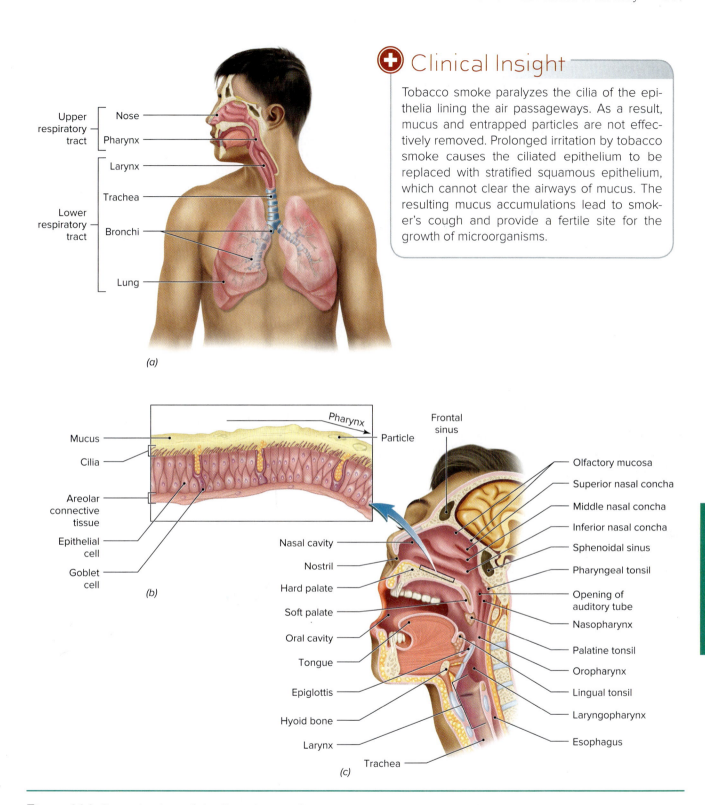

Figure 14.1 Organization of the Respiratory System.
(a) The general organization of the respiratory system. (b) Mucus and entrapped particles are moved by cilia of the mucosa from the nasal cavity to the pharynx. (c) Major structures of the upper respiratory tract. AP|R

oral cavity by the **palate** (roof of the mouth), which consists of two basic portions. In the front, the *hard palate* is formed by the palatine processes of the maxillae and the palatine bones. In the back, the soft palate is composed of skeletal muscle tissue. The nasal cavity is divided into left and right portions by the **nasal septum,** a vertical partition of bone and septal nasal cartilage that is located on the midline. Three *nasal conchae* (singular, concha) project from each lateral wall and serve to increase the surface area of and create air turbulence in the nasal cavity.

The superior nasal concha and superior nasal septum are lined with **olfactory mucosa,** a mucus membrane containing the olfactory epithelium (see chapter 9). The rest of the nasal cavity, larynx, trachea, and bronchi are covered by **respiratory mucosa,** a mucus membrane containing pseudostratified ciliated columnar epithelium. Goblet cells within the epithelium produce mucus to coat the epithelial surface. As air flows through the nasal cavity, it is warmed by the blood-rich respiratory mucosa and is moistened by the mucus. In addition, airborne particles, including microorganisms, are trapped in the mucus of the respiratory mucosa. Within the larynx, trachea, and bronchi, cilia of the epithelium slowly move the layer of mucus with its entrapped particles toward the pharynx (figure 14.1b), where it is swallowed. Upon reaching the stomach, most microorganisms in the mucus are destroyed by the gastric juice.

Several bones surrounding the nasal cavity contain **paranasal sinuses,** air-filled cavities. Sinuses are located in the frontal bone, ethmoid, maxillae, and sphenoid adjacent to the nasal cavity. The sinuses lighten the skull and serve as sound-resonating chambers during speech. The sinuses open into the nasal cavity, which increases nasal cavity surface area. They are also lined by respiratory mucosa, and the secreted mucus drains into the nasal cavity.

Pharynx

The **pharynx** (fayr'-inks), commonly called the throat, is a short passageway behind the nasal and oral cavities that extends downward to the larynx and esophagus. It has a muscular wall and is lined with a mucous membrane containing primarily stratified squamous epithelium. As shown in figure 14.1c, the pharynx consists of three parts: the *nasopharynx* behind the nose; the *oropharynx* behind the mouth; and the *laryngopharynx* behind the larynx.

The *auditory tubes* (figure 14.1c), which extend to the middle ear, open into the nasopharynx. Air moves in or out of the auditory tubes to equalize the air pressure on each side of the tympanic membrane.

The tonsils, clumps of lymphoid tissue, occur at the openings to the pharynx. The *palatine tonsils* are located bilaterally at the junction of the oropharynx and the oral cavity. The *pharyngeal tonsil* (adenoid) is located in the upper portion of the nasopharynx, and the *lingual tonsils* are found on the back of the tongue. Tonsils are sites of immune reactions and may become sore and swollen when infected. Enlargement of the palatine tonsils tends to make swallowing painful and difficult. A swollen pharyngeal tonsil tends to block the flow of air between the nasal cavity and the pharynx, which promotes mouth breathing and can create snoring. When a person breathes through the mouth, air is not adequately warmed, filtered, and moistened.

Larynx

The **larynx** (layr'-inks) is a boxlike structure, composed of several cartilages, that provides a passageway for air between the pharynx and the trachea. The three largest cartilages are the *thyroid cartilage,* which projects forward to form the Adam's apple; the *cricoid cartilage,* which forms the attachment to the trachea; and the *epiglottis,* a cartilaginous flap that helps to keep solids (e.g., food) and liquid from entering the larynx when swallowing. The larynx is supported by ligaments that extend from the hyoid bone (figure 14.2).

The **vocal folds** are two bands of elastic connective tissue covered by respiratory mucosa located within the larynx (figure 14.3). They are relaxed during resting breathing, but when contracted, they vibrate to produce vocal sounds when exhaled air passes over them. The pitch (high or low tone) of a sound is determined by the

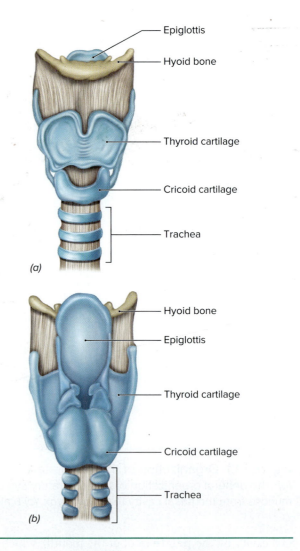

Figure 14.2 The Larynx.
(a) Anterior view. *(b)* Posterior view.

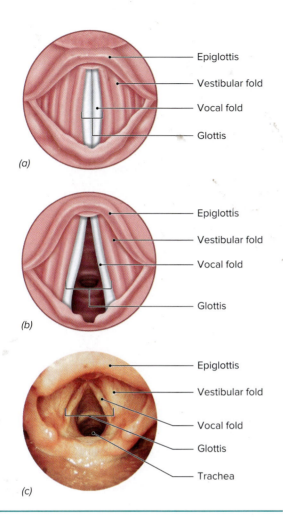

Figure 14.3 The Glottis.
The glottis viewed from above with the space between the vocal folds (a) closed and (b) open. (c) Photograph of the glottis with the space between the vocal folds open.
(c) ©CNRI/Science Source

vibration frequency of the vocal folds. High frequency vibrations lead to high pitch sound and vice versa. The loudness (volume) of a sound is related to the vibration amplitude of the vocal folds. The larger the amplitude, the louder the volume, and vice versa. The **glottis** consists of the vocal folds and the space between them and leads to the trachea. The **vestibular folds,** which lie above the vocal folds, are composed of a small amount of elastic connective tissue covered by respiratory mucosa. They prevent solids and liquid from entering the glottis and are not involved in sound production.

Because the oropharynx and laryngopharynx are also passageways for food, a mechanism exists to prevent food from entering the larynx and to direct food into the esophagus (ē-sof′-ah-gus), the flexible tube that carries food to the stomach. When swallowing, muscles lift the larynx upward, which causes the epiglottis to fold over and cover the opening into the larynx. This action directs food into the esophagus, whose opening is located just behind the larynx. Sometimes this mechanism does not work perfectly and a small amount of food or drink enters the larynx, stimulating a coughing reflex that usually expels the substance.

Trachea

The **trachea** (trā′-kē-ah), or windpipe, is the airway that extends from the larynx into the thoracic cavity, where it branches to form the right and left main bronchi. The walls of the trachea are supported by C-shaped *tracheal cartilages* that hold the passageway open in spite of the air pressure changes that occur during breathing (figure 14.4a). The open portion of the tracheal cartilages is against the esophagus, which is behind the trachea (see figure 14.2b). This orientation allows the esophagus to expand slightly as food passes down to the stomach.

The inner wall of the trachea is lined with respiratory mucosa. Mucus produced by the goblet cells coats the surface of the epithelium and traps airborne particles, including microorganisms. The beating cilia move the mucus and entrapped particles upward to the pharynx where they are coughed out or swallowed. Microorganisms are usually killed by gastric juice in the stomach.

Bronchi, Bronchioles, and Pulmonary Alveoli

The trachea branches at about midchest into the left and right *main bronchi* (brong′-kī; singular, bronchus). Each main bronchus enters its respective lung, where it branches to form smaller *lobar bronchi,* one for each lobe of the lung. Lobar bronchi branch to form segmental bronchi that lead to different segments within each lung lobe. The bronchi continue to branch into smaller and smaller bronchi. Because the bronchi resemble tree branches, they are collectively called the **bronchial tree** (figure 14.4a).

The walls of the bronchi contain cartilaginous rings similar to those of the trachea, but as the branches get progressively smaller, the amount of cartilage tissue gradually decreases and finally is absent in the very small tubes called the **bronchioles** (brong′-kē-ōls). As the amount of cartilage tissue decreases, the amount of smooth muscle increases. The smooth muscle plays an important role in regulating the airflow through the air passageways. Contraction of the smooth muscle causes *bronchoconstriction,* which decreases airflow. Relaxation of the smooth muscle results in *bronchodilation,* which increases airflow. Air passageways larger than bronchioles are lined with respiratory mucosa that continues to trap and remove airborne particles. Bronchioles are lined with a mucous membrane containing simple cuboidal epithelium, so foreign particles that reach them are not effectively removed. Bronchioles branch to form smaller and smaller bronchioles that lead to microscopic **alveolar ducts,** which terminate in

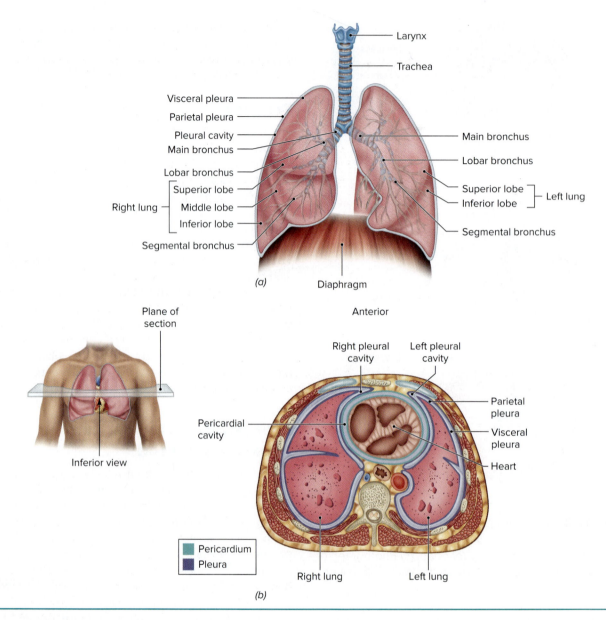

Figure 14.4 The Lower Respiratory Tract.
(a) Anterior view of the lower respiratory tract. (b) Transverse section of the thoracic cavity shows the pleurae and the relationship of the lungs to other thoracic organs. AP|R

tiny air sacs called **pulmonary alveoli** (al-vē'-ō-lī; singular, *alveolus*). Pulmonary alveoli resemble tiny grapes clustered about an alveolar duct and are composed of simple squamous epithelium (figure 14.5a).

The primary function of the bronchial tree and bronchioles is to carry air into and out of the pulmonary alveoli during breathing. The exchange of respiratory gases occurs between the air in the pulmonary alveoli and the blood in the capillary networks that surround the pulmonary alveoli (figure 14.5b). Oxygen and carbon dioxide diffuse readily through the thin **respiratory membrane**, which is composed of squamous cells of the alveolar wall and the capillary wall (figure 14.5c). Pulmonary alveoli are extremely numerous—about 300 million in each lung. They have a combined surface area of about 75 square meters and can hold about 5,800 ml of air.

Pulmonary alveoli contain very small spaces, which are coated with a watery fluid. The attraction (surface tension) between water molecules would cause the pulmonary alveoli to collapse if it were not for surfactant. **Surfactant** (ser-fak'-tant) is a mixture of lipoproteins secreted by special cells in pulmonary alveoli. It reduces the attraction between water molecules and keeps pulmonary alveoli open so they may fill with air more easily during inspiration. Without surfactant, pulmonary alveoli would collapse during expiration and become very difficult to reinflate.

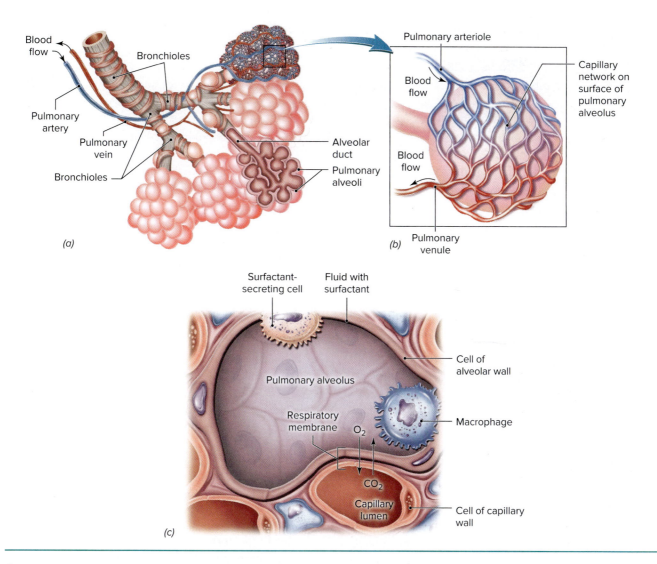

Figure 14.5 Microscopic Organization of the Lung.
(a) Bronchioles branch to form smaller and smaller bronchioles, which lead to microscopic alveolar ducts that terminate in a cluster of alveoli. (b) Each pulmonary alveolus is enveloped by a capillary network. Deoxygenated blood (blue) enters the capillary network. Oxygenated blood (red) exits the capillary network. (c) Cross-sectional view of a pulmonary alveolus shows the respiratory membrane consisting of alveolar and capillary walls. AP|R

Lungs

The paired **lungs** are large organs that occupy much of the thoracic cavity. The lungs consist primarily of air passageways, pulmonary alveoli, blood and lymphatic vessels, nerves, and connective tissues, giving the lungs a soft, spongy texture. They are roughly cone-shaped and are separated from each other by the mediastinum. Each lung is divided into lobes. The left lung has two lobes (superior and inferior) and is somewhat smaller than the right lung, which is composed of three lobes (superior, middle, and inferior) (see figure 14.4a). Each lobe is supplied by a lobar bronchus, blood and lymphatic vessels, and nerves.

Two layers of serous membranes, called **pleurae** (singular, *pleura*), enclose and protect each lung. The **visceral pleura** is firmly attached to the surface of each lung, and the **parietal pleura** lines the inner wall of the thoracic cage. The space between the visceral and parietal pleurae is known as the **pleural cavity.** A thin film of pleural fluid occupies the pleural cavity and reduces friction between the pleurae as the lungs inflate and deflate during breathing. Although lungs are elastic and tend to recoil, the attraction of water molecules in the pleural fluid within the pleural cavity keeps the visceral and parietal pleurae stuck together (figure 14.4b).

Table 14.1 summarizes the functions of the respiratory structures.

Table 14.1 Summary of Functions of the Respiratory Structures

Component	Function
Nose	Nostrils allow air to enter and exit the nasal cavity; the nasal cavity filters, warms, and moistens the inhaled air
Pharynx	Carries air between the nasal cavity and the larynx; filters, warms, and moistens the inhaled air; serves as passageway for food from the mouth to the esophagus; equalizes air pressure with middle ear via auditory tube
Larynx	Carries air between the pharynx and trachea; contains vocal folds for producing sounds in vocalization; prevents objects from entering the trachea
Trachea	Carries air between the larynx and the bronchi; filters, warms, and moistens the inhaled air
Bronchi	Carry air between the trachea and the bronchioles; filter, warm, and moisten the inhaled air
Bronchioles	Regulate the rate of airflow through bronchoconstriction and bronchodilation
Pulmonary alveoli	Allow the gas exchange between the air in the pulmonary alveoli and the blood in surrounding capillaries

Check My Understanding

4. What are the functions of the nose?
5. What are the three divisions of the pharynx, and where are they located?
6. What is the function of the epiglottis?
7. What are the functions of the vocal folds and vestibular folds?
8. What is the function of the cartilage tissue in the walls of the trachea and bronchi?
9. What are the functions of the bronchi and bronchioles?
10. Why is surfactant important?

14.3 Breathing

Learning Objective

5. Describe the mechanism of breathing.

Breathing, or pulmonary ventilation, is the process that exchanges air between the atmosphere and the pulmonary alveoli of the lungs. Air moves into and out of the lungs along an air pressure gradient–from regions of higher pressure to regions of lower pressure. There are three pressures that are important in breathing:

1. *Atmospheric pressure* is the pressure of the air that surrounds the earth. Atmospheric pressure at sea level is 760 mm Hg, but at higher elevations it decreases because there is less air at higher elevations.
2. *Intra-alveolar (intrapulmonary) pressure* is the air pressure within the lungs. As we breathe in and out, this pressure fluctuates between being lower than atmospheric pressure and higher than atmospheric pressure. The changes in pressure are so small that they are measured in cm H_2O instead of mm Hg. For example, if pressure reaches -1 cm H_2O, pressure has decreased 1 cm H_2O below atmospheric pressure. If pressure reaches $+1$ cm H_2O, pressure has increased 1 cm H_2O above atmospheric pressure.
3. *Intrapleural pressure* is the pressure within the pleural cavity. It is about -5 to -8 cm H_2O during various phases of breathing. This lower intrapleural pressure is often described as "negative pressure," and it keeps the lungs stuck to the inner walls of the thoracic cage and helps expand the lungs, even as the thoracic cage expands and contracts during breathing. If the intrapleural pressure were to equal atmospheric pressure, the lungs would collapse and be nonfunctional.

Inspiration

The process of moving air into the lungs is called **inspiration,** or inhalation. When the lungs are at rest, the air pressure in the lungs is the same as the atmospheric pressure. In order for air to flow into the lungs, the intra-alveolar pressure must be decreased to below atmospheric pressure. This change allows for air to flow from the higher air pressure in the atmosphere toward the lower air pressure within the lungs. The contraction of the diaphragm and the external intercostal muscles during inspiration causes an increase in lung volume, which results in a decrease in intra-alveolar pressure.

The dome-shaped **diaphragm** is a thin sheet of skeletal muscle separating the thoracic and abdominal cavities. When it contracts, the diaphragm moves downward and flattens, which increases the volume of the thoracic cavity. At the same time, contraction of the *external intercostal muscles* elevates and protracts the ribs and pushes

Clinical Insight

The presence of air in the pleural cavity is called a *pneumothorax* (nū-mō-thō'-raks). This may occur due to a thoracic injury or surgery that allows air to enter the pleural cavity. It also occurs in emphysema patients when air escapes from ruptured pulmonary alveoli into the pleural cavity. A pneumothorax causes the affected lung to collapse and become nonfunctional. Because each lung is in a separate pleural cavity, the collapse of one lung does not adversely affect the other lung. Treatment involves removing the intrapleural air to restore the healthy pressure so that the lung may inflate.

the sternum forward, which further increases the volume of the thoracic cavity (figures 14.6, 14.7a).

Because the negative intrapleural pressure and the surface tension of the pleural fluid keep the visceral pleura stuck to the parietal pleura, the lungs are pulled along when the thoracic cage expands. Therefore, the expansion of the thoracic cavity increases the volume of the lungs, which decreases the intra-alveolar pressure to −1 cm H_2O. Then, the higher atmospheric pressure forces air through the air passageways into the lungs until intra-alveolar and atmospheric pressures become equal once again. Resting inspiration requires the contraction of the diaphragm and the external intercostal muscles only. Forceful inspiration requires the involvement of additional muscles in the neck and chest, such as the sternocleidomastoid, scalenes, serratus anterior, and pectoralis minor (figure 14.6). The contraction of these muscles elevates and protracts the ribs to a greater extent, leading to a greater increase in the volume of the thoracic cavity. Through this further increase in thoracic volume, intra-alveolar pressure decreases to a greater extent, which results in greater airflow into the lungs.

Expiration

Expiration, or exhalation, is the movement of air out of the lungs. It occurs when the diaphragm and external intercostal muscles relax, allowing the thoracic cage and lungs to return to their original size. This results in a decrease in the volume of the thoracic cavity and lungs. The decrease in lung volume increases intra-alveolar pressure to +1 cm H_2O. The higher intra-alveolar pressure forces air out of the lungs until intra-alveolar and atmospheric pressures become equal once again.

Expiration during resting breathing is a rather passive process because the abundant elastic connective tissue in the lungs and thoracic wall causes them to return to their original size as soon as the muscles of inspiration relax. However, a forceful expiration is possible by contraction of the *internal intercostal muscles* (figure 14.6), which depresses and retracts the ribs, and by the muscles of the abdominal wall, which move the abdominal viscera and diaphragm upward. These contractions further decrease the volume of the thoracic cavity and lungs, which increases the intra-alveolar pressure, causing more air to flow out of the lungs.

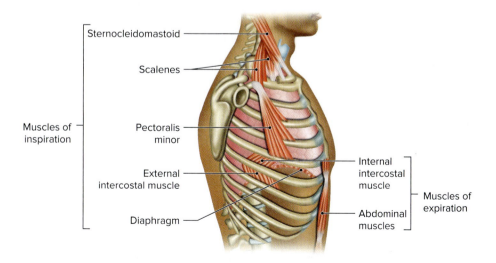

Figure 14.6 Respiratory Muscles.
Sternocleidomastoid, scalenes, and pectoralis minor are involved in forceful inspiration only. Internal intercostal muscles and abdominal muscles are involved in forceful expiration only. AP|R

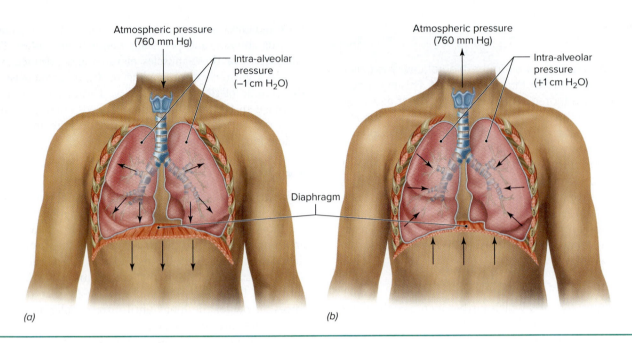

Figure 14.7 The Mechanisms of Breathing.
(a) During resting inspiration, the increasing volume of the thoracic cavity and lungs reduces the intra-alveolar pressure, leading to air flowing into the lungs. *(b)* During resting expiration, the decreasing volume of the thoracic cavity and lungs increases the intra-alveolar pressure, leading to air flowing out of the lungs. AP|R

Check My Understanding

11. How do intrapleural pressure and pleural fluid in the pleural cavity affect breathing?
12. Describe the mechanisms of inspiration and expiration.

14.4 Respiratory Volumes and Capacities

Learning Objective

6. Describe the various respiratory volumes and capacities and the significance of each.

Healthy adults average 12 to 15 resting breathing cycles per minute. A *breathing cycle* is one inspiration followed by one expiration. The volume of air inhaled and exhaled in a resting or forceful breathing cycle varies with size, sex, age, and physical condition. The average respiratory volumes have been determined by size, age, and sex in order to enable evaluation of pulmonary functions. Respiratory volumes that are 80% or less than the healthy range usually indicate some form of pulmonary disease.

An instrument called a *spirometer* is used to determine respiratory volumes. It produces a *spirogram*, a graphic record of the volume of air exchanged.

The volume of air inhaled or exhaled in a resting breathing cycle is about 500 ml, and it is known as the *tidal volume (TV)*. Forceful inspirations and expirations can exchange a much greater volume of air. The maximum volume of air that can be forcefully inhaled after a tidal inspiration is about 3,000 ml, and it is known as the *inspiratory reserve volume (IRV)*.

The maximum volume of air that can be forcefully exhaled after a tidal expiration is about 1,100 ml, and it is known as the *expiratory reserve volume (ERV)*. About 1,200 ml of air remains in the lungs after a maximum forced expiration. This residual air is known as the *residual volume (RV)*. Once an infant takes its first breath, there is always residual volume in the lungs. The surfactant in pulmonary alveoli keeps the pulmonary alveoli from collapsing. The intrapleural pressure and the surface tension of the pleural fluid keep the lungs partially inflated.

Respiratory capacities can be calculated by summation of two or more respiratory volumes. The maximum amount of air that can be forcefully exchanged is known as the *vital capacity (VC)*, and it is equal to the sum of the tidal volume, the inspiratory reserve volume, and the expiratory reserve volume—about 4,600 ml. The *total lung capacity (TLC)* is equal to the sum of the vital capacity and the residual volume—about 5,800 ml.

The respiratory volumes are summarized in table 14.2 and are graphically shown in figure 14.8.

Table 14.2 Summary of Respiratory Volumes and Capacities

Name	Definition	Average Volume
Tidal volume (TV)	Volume of air inhaled or exhaled during resting breathing	500 ml
Inspiratory reserve volume (IRV)	Volume of air that can be forcefully inhaled after a tidal volume inhalation	3,000 ml
Expiratory reserve volume (ERV)	Volume of air that can be forcefully exhaled after a tidal volume expiration	1,100 ml
Vital capacity (VC)	Maximum volume of air that can be forcefully exchanged VC = TV + IRV + ERV	4,600 ml
Residual volume (RV)	Volume of air remaining in the lungs after a maximum forceful exhalation	1,200 ml
Total lung capacity (TLC)	Maximum volume of air that the lungs can contain TLC = VC + RV	5,800 ml

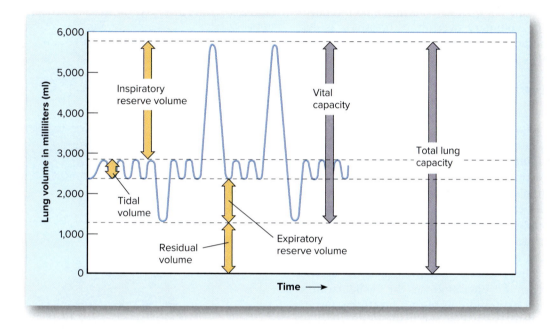

Figure 14.8 Respiratory Volumes and Capacities.
The wavy blue line is a spirometer tracing that indicates the volumes of air exchanged as the respiratory movements are performed.

Check My Understanding

13. What distinguishes inspiratory reserve volume and expiratory reserve volume?

14. What distinguishes tidal volume and vital capacity?

Clinical Insight

The determination of the lung volumes is useful in identifying the two basic categories of pulmonary disease: obstructive and restrictive disorders. Each category has a characteristic pattern of unhealthy test results. It is possible for a patient to exhibit both patterns simultaneously.

The *obstructive pattern* occurs where there is airway obstruction from any cause, such as in asthma, bronchitis, and emphysema. In this pattern, RV is increased and the RV/TLC ratio is increased.

The *restrictive pattern* occurs when there is a loss of lung tissue or when expansion of the lungs is limited. This pattern may result from lung tumors, weakness of respiratory muscles, pulmonary edema, or fibrosis of the lungs. In this pattern, the TLC is decreased, the RV/TLC ratio is healthy or increased, and the VC is decreased.

14.5 Control of Breathing

Learning Objective
7. Describe the neural control of breathing.

The resting rhythmic cycle of breathing is involuntary—we don't have to think about it. It continues when we are sleeping or even unconscious. However, we can voluntarily override the resting pattern and take deep breaths and breathe faster or slower if we wish. The centers for involuntary control of breathing lie in the brainstem, where two groups of neurons in the medulla oblongata and one group of neurons in the pons regulate breathing. The voluntary override of breathing is controlled by the primary motor area of the cerebral cortex (figure 14.9).

Respiratory Centers

Two bilateral groups of neurons compose the respiratory rhythmicity center in the medulla oblongata: the ventral respiratory group and the dorsal respiratory group. The *ventral respiratory group (VRG)* is responsible for the resting rhythmic cycle of breathing. It sends action potentials to the diaphragm and external intercostal muscles causing them to contract, which results in inspiration. Inspiration continues as long as the VRG sends out the action potentials; but when action potentials cease, the muscles of inspiration relax and expiration occurs. This pattern of alternating neural activity and inactivity of the VRG produces the cyclic nature of inspiration and expiration. In resting breathing, inspiration lasts about two seconds, and expiration lasts about three seconds.

However, breathing can be deeper or shallower and faster or slower as the needs of the body change. The VRG receives input from other sources that result in such changes in the breathing cycle. The *dorsal respiratory group (DRG)* serves as a center for receiving and integrating input from sensory sources. It sends action potentials to the VRG to make necessary changes in the breathing pattern in accordance with the sensory input. Sensory input is considered in the next section.

A third respiratory center, the *pontine respiratory group (PRG)*, is located in the pons. It receives input from higher brain centers and sends action potentials to the DRG and VRG that modify the breathing pattern. The PRG contains two types of neurons: those that stimulate and those that inhibit the DRG and VRG. Thus, the PRG can either speed up or slow down the transition from inspiration to expiration, which alters the rate and depth of breathing. The PRG plays a key role in adapting breathing to speaking, singing, exercise, sleep, and emotional respiratory responses such as crying or laughing.

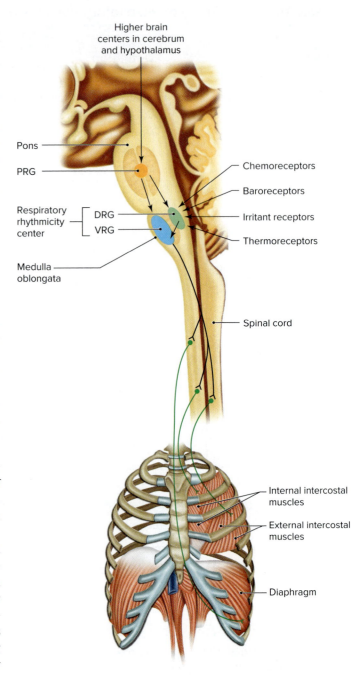

Figure 14.9 The Control of Breathing. Respiratory control centers are located in the medulla oblongata and the pons.

Check My Understanding
15. Where are the respiratory centers located?
16. How is breathing controlled by the nervous system?

14.6 Factors Influencing Breathing

Learning Objective

8. Describe the factors that influence breathing and how they produce their effect.

The respiratory areas of the medulla oblongata and pons are influenced by a number of factors that cause modifications in the rate and depth of breathing. Factors involved in involuntary control are detected by sensory receptors, which forward action potentials to the DRG. Higher brain centers involved in voluntary control send action potentials to the pontine respiratory group, which then transmits action potentials to the respiratory rhythmicity center (figure 14.9).

Chemicals

The most important chemical factors affecting respiration are the concentrations of CO_2, H^+, and O_2 in the blood or cerebrospinal fluid. Recall from chapters 9 and 12 that sensory receptors that are sensitive to these factors are called chemoreceptors. The chemoreceptors in the medulla oblongata are sensitive to increases in H^+ and CO_2 in the cerebrospinal fluid. The chemoreceptors in the carotid bodies and aortic bodies are sensitive to changes in CO_2, H^+, and O_2. The carotid bodies are located near the bifurcation of the common carotid arteries, while aortic bodies are located in the aortic arch. You can see that they are strategically located, especially to monitor blood going to the brain.

You may wonder why the concentration of H^+ is involved in respiratory control. The mechanism for transporting CO_2 in the blood releases H^+ as a by-product. Therefore, an increase in CO_2 concentration produces an increase in the H^+ concentration.

If the concentrations of CO_2 and H^+ in the blood or CSF increase, the DRG relays the information so that the VRG is stimulated to increase the rate and depth of breathing, which increases the rate of CO_2 and H^+ removal and returns their concentrations to homeostatic levels. Once homeostasis is restored, the rate and depth of breathing also return to resting levels.

If the CO_2 and H^+ concentrations in the blood or cerebrospinal fluid decrease, breathing is slow and shallow until their concentrations increase to homeostatic levels.

As mentioned previously, only the chemoreceptors in the carotid and aortic bodies are sensitive to changes in blood O_2 level, specifically to a decline in blood O_2 concentration. Usually, a drop in O_2 concentration is not a strong stimulus for increasing the rate and depth of breathing, and its main effect seems to be to increase the sensitivity of chemoreceptors to changes in the CO_2 concentration.

Inflation Reflex

Baroreceptors in the bronchi, bronchioles, and visceral pleurae are sensitive to inflation of the lungs. During inspiration, action potentials from the baroreceptors are sent to the DRG via the vagus nerves (CN X), where they inhibit the formation of action potentials causing inspiration. This promotes expiration and prevents excessively deep inspirations that may damage the lungs.

Irritant Reflexes

The respiratory tract contains irritant receptors that are sensitive to various chemical and physical irritants, such as smoke, dust, and excess amounts of mucus. When stimulated by irritants, these receptors send sensory action potentials to the DRG via the vagus nerves. The DRG then alters the function of the VRG, which triggers a reflex contraction of the respiratory muscles that leads to a sneeze or a cough in order to expel the irritants from the respiratory tract.

Higher Brain Centers

Action potentials from higher brain centers also alter the rhythmic cycle of breathing. These action potentials may be voluntarily generated in the cerebrum, as when a person chooses to alter the pattern of resting breathing. However, these voluntary controls are limited. For example, if a little child tries to "punish" his mother by holding his breath, the action potentials from higher brain centers are ignored and involuntary breathing resumes once the CO_2 level in his blood increases to a critical point.

Involuntary action potentials may be formed by higher brain centers in the cerebral cortex and the hypothalamus during emotional experiences, such as anxiety, fear, and excitement, which activate the autonomic division. At such times, the breathing rate is increased. Similarly, a sudden emotional experience or a sharp pain tends to momentarily stop breathing, a condition called *apnea* (ap′-nē-ah).

Body Temperature

An increase in body temperature, such as occurs during strenuous exercise or a fever, increases the breathing rate. Conversely, a decrease in body temperature decreases the breathing rate.

Check My Understanding

17. Which respiratory center receives input from higher brain centers?
18. What factors influence the control of breathing?

14.7 Gas Exchange

Learning Objective
9. Describe the mechanisms of gas exchange in the lungs and the body tissues.

Alveolar Gas Exchange

During **alveolar gas exchange,** respiratory gases are exchanged between the air in the pulmonary alveoli and the blood in the capillaries that surround them. Oxygen and carbon dioxide must diffuse through the respiratory membrane (see figure 14.5c).

Alveolar air has a higher concentration of oxygen and a lower concentration of carbon dioxide than does the capillary blood. Because molecules tend to move from an area of higher concentration to an area of lower concentration, oxygen diffuses from the alveolar air into the blood, and carbon dioxide diffuses from the blood into the alveolar air.

Blood entering a capillary network of a pulmonary alveolus is oxygen poor and carbon dioxide rich. Following the gas exchange, blood leaving the capillary network is oxygen rich and carbon dioxide poor (figure 14.10).

Systemic Gas Exchange

After blood has been oxygenated, it returns to the heart and is pumped throughout the body to supply the tissue cells through **systemic gas exchange.** Blood in the systemic capillaries supplying body tissues contains a higher concentration of oxygen and a lower concentration of carbon dioxide than the tissue cells. Therefore, oxygen diffuses from the blood into the interstitial fluid before entering the tissue cells, and carbon dioxide diffuses from the tissue cells into the interstitial fluid before entering the blood. In this way, cells are supplied with oxygen for their metabolic activities, and carbon dioxide, which is produced by cellular metabolism, is removed.

Blood entering a systemic capillary network at the tissue level is oxygen rich and carbon dioxide poor. Following gas exchange, blood leaving the systemic capillary network is oxygen poor and carbon dioxide rich. **APR**

Check My Understanding

19. What structure must oxygen and carbon dioxide diffuse through during alveolar gas exchange?
20. What is the difference between the alveolar capillary blood and the systemic capillary blood?

14.8 Transport of Respiratory Gases

Learning Objective
10. Describe how oxygen and carbon dioxide are transported by the blood.

The RBCs play a major role in the transport of both oxygen and carbon dioxide.

Oxygen Transport

In the lungs, oxygen diffuses from the air in pulmonary alveoli into the blood of surrounding capillaries. Most of the oxygen enters RBCs and combines with the heme

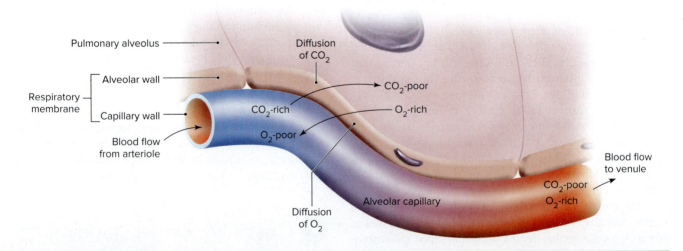

Figure 14.10 Alveolar Gas Exchange.
Exchange of oxygen and carbon dioxide between air in a pulmonary alveolus and blood in an alveolar capillary occurs by diffusion. **APR**

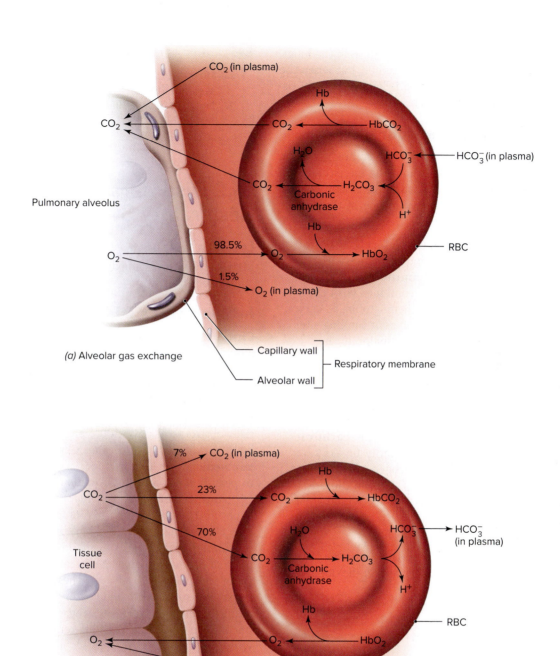

Figure 14.11 Exchange and Transport of Oxygen and Carbon Dioxide.

portions of hemoglobin (Hb) to form **oxyhemoglobin (HbO$_2$).** About 98.5% of the oxygen is transported as oxyhemoglobin. The remaining 1.5% is dissolved in the plasma (figure 14.11).

In body tissues, oxyhemoglobin releases oxygen and it diffuses from capillary blood into the interstitial fluid before entering the tissue cells. Actually, only about 25% of the oxygen is released in healthy individuals at rest, so oxyhemoglobin is present even in deoxygenated blood.

The reason that hemoglobin is such an effective carrier of oxygen is that the chemical bond between oxygen and hemoglobin is relatively unstable. When the surrounding oxygen concentration is high, as in the lungs, hemoglobin combines readily with oxygen; but when the surrounding oxygen concentration is low, as in body tissues, hemoglobin releases oxygen.

Carbon Dioxide Transport

The transport of carbon dioxide is more complex. Carbon dioxide diffuses from body cells into the interstitial fluid before entering the capillary blood. After carbon dioxide enters the blood, it is transported in one of these three ways (figure 14.11):

1. About 7% is dissolved in the plasma.
2. About 23% enters RBCs and combines with hemoglobin to form **carbaminohemoglobin (HbCO$_2$)**. Carbon dioxide combines with the globin (protein) portion of hemoglobin, so carbon dioxide and oxygen have different binding sites on hemoglobin. Therefore, hemoglobin can transport oxygen and carbon dioxide at the same time.
3. The remaining 70% of the carbon dioxide also enters RBCs, but it quickly combines with water to form **carbonic acid (H$_2$CO$_3$)**. This reaction is catalyzed by the enzyme **carbonic anhydrase**. Carbonic acid rapidly breaks down (dissociates) into hydrogen ions (H$^+$) and **bicarbonate ions (HCO$_3^-$)**. Carbon dioxide is now part of the bicarbonate ions, which then diffuse out of the RBCs and are transported to the lungs in plasma.

When the blood returns to the lungs, all of these reactions run in reverse, releasing carbon dioxide, which diffuses into the pulmonary alveoli.

 Clinical Insight

Carbon monoxide (CO) is an odorless, colorless gas that is produced by burning carbon fuels. It competes with oxygen for the same binding sites on hemoglobin molecules, and it combines with hemoglobin about 200 times more readily than oxygen. Further, CO binds so tightly with hemoglobin that it is hard to remove. Therefore, even low concentrations of CO can displace oxygen from hemoglobin molecules and deprive tissues of needed oxygen. CO poisoning is the leading cause of death from fires. It is especially treacherous because it kills quietly without attracting attention. Treatment includes the administration of 100% oxygen to flush out the CO.

 Check My Understanding

21. How does gas exchange occur in the lungs and in systemic tissues?
22. How are oxygen and carbon dioxide transported by the blood?

14.9 Disorders of the Respiratory System

Learning Objective

11. Describe the major disorders of the respiratory system.

Inflammatory Disorders

Chronic obstructive pulmonary disease (COPD) is a group of disorders in which there is a long-term obstruction that reduces airflow to and from the lungs. The two most important COPDs are chronic bronchitis and emphysema.

Bronchitis is inflammation of the bronchi, and it is characterized by excessive mucus production that partially obstructs air flow. Acute bronchitis is usually caused by viral or bacterial infections. Chronic bronchitis occurs in chronic asthmatics, and it is common in smokers due to persistent exposure to irritants in tobacco smoke.

Emphysema (em-fi-sē'-mah) results from long-term exposure to airborne irritants, especially tobacco smoke. It is characterized by a rupture of the pulmonary alveoli, forming larger spaces in the lungs, and excess mucus production, which plugs terminal bronchioles, trapping air in the pulmonary alveoli. These changes reduce the respiratory surface area and impair gas exchange. Exhaling requires voluntary effort by the patient. The disease is uncommon except among long-term smokers and people with long-term exposure to secondhand smoke. It usually can be prevented and progressive deterioration can be stopped by removing the airborne irritant—usually tobacco smoke. Otherwise, there is no cure.

Asthma (az'-mah) is another COPD but differs in that reduction in airflow is usually intermittent rather than constant. It is characterized by wheezing upon exhalation and **dyspnea** (labored breathing) that result from bronchoconstriction. It is often caused by an allergic reaction to airborne substances but also may result from hypersensitivity to bacteria or viruses infecting the bronchial tree.

The **common cold** is an infection of the upper respiratory tract. It may be caused by a number of viruses, and often involves rhinitis, laryngitis, and sinusitis. Excessive mucus production, sneezing, and congestion are common symptoms.

Rhinitis (rī-nī′-tis), **laryngitis,** and **sinusitis** are the inflammation of the respiratory mucosa lining the nasal cavity, the larynx, and the paranasal sinuses, respectively. They are characterized by an increased mucus secretion. Causes may be viral or bacterial infections or airborne allergens.

Influenza, or flu, is an infectious disease that may involve both the upper and the lower respiratory tracts. It is caused by one of several viruses. Symptoms are fever, chills, headache, and muscular aches, followed by cold-like symptoms. In comparison to the common cold, the effects of influenza are much more severe and may lead to the development of pneumonia.

Pneumonia (nū-mōn′-yah) is an acute inflammation of the pulmonary alveoli that may be caused by viral or bacterial infections. The pulmonary alveoli become filled with fluid, pathogens, and white blood cells, which reduce space for gas exchange. Blood oxygen level may be greatly reduced. Pneumonia is one of the common causes of death among older people. AP|R

Pleurisy (pler′-i-sē) is inflammation of the pleurae. It often results in a decrease in secretion of pleural fluid, which causes sharp pains with each breath. Pleurisy can also cause the opposite effect: an increase in pleural fluid secretion. This type exerts pressure on the lungs and impairs expansion of the lungs.

Tuberculosis (tū-ber″-kū-lō′-sis) is an inflammation caused by the bacterium *Mycobacterium tuberculosis,* which is transmitted by inhalation. When it infects the lungs, the destroyed lung tissue is replaced by dense irregular connective tissue that impairs gas exchange and reduces lung elasticity. Fortunately, modern drugs are effective in treating this disease.

Noninflammatory Disorders

Lung cancer is the second most common cancer and the leading cause of death from cancer in American males and females. It usually develops from long-term exposure to irritants, and the most common irritant producing this malignancy is tobacco smoke. The link between lung cancer and cigarette smoking has been firmly established. Lung cancer metastasizes rapidly and is not usually detected until it has spread to other parts of the body. Treatment includes surgical removal of the diseased lung, if detected prior to metastasis, and chemotherapy. More than 90% of lung cancers occur in smokers, so the most effective prevention is the elimination of cigarette smoking.

Pulmonary edema is the accumulation of fluid in the lungs. It results from excessive fluid passing from alveolar capillaries into the pulmonary alveoli, which may be due to congestive heart failure. Symptoms include labored breathing and a feeling of suffocation. Treatment includes administration of oxygen, diuretics, drugs that dilate the bronchioles, suctioning air passageways, and mechanical ventilation.

Pulmonary embolism refers to a blood clot or gas bubble that blocks a small artery in the lung and prevents blood from reaching a portion of a lung. Gas exchange cannot occur in the affected parts of the lung. A massive embolism affecting a large portion of a lung may cause cardiac arrest.

Infant respiratory distress syndrome (IRDS), or hyaline membrane disease, is a disease of newborn infants, especially premature infants. It results from an insufficient production of surfactant in the pulmonary alveoli, leading to alveolar collapse.

At birth, the respiratory system of an infant goes through a transition from a nonfunctional, fluid-filled system to a functional, air-filled system. Usually, an infant's first breath is the most difficult because it must open the collapsed pulmonary alveoli. Succeeding breaths are easier because surfactant keeps the pulmonary alveoli open after expiration. Without adequate surfactant, pulmonary alveoli tend to collapse at each expiration and the infant must expend a great amount of energy to force them open at each inspiration.

Chapter Summary

14.1 Introduction to the Respiratory System

- The primary role of the respiratory system is to supply the blood with oxygen and remove carbon dioxide from the blood. It also helps to detect odors, produce sounds, regulate blood pH, trap and defend the body from airborne pathogens, and assist in the movement of venous blood and lymph.
- Respiration includes five processes: breathing, alveolar gas exchange, gas transport, systemic gas exchange, and aerobic respiration.
- The respiratory system is only directly involved in breathing and alveolar gas exchange.

14.2 Structures of the Respiratory System

- The major structures of the respiratory system are the nose, pharynx, larynx, trachea, bronchi, and lungs.
- The outer portion of the nose is supported by bone and septal nasal cartilage, whereas the nasal cavity is surrounded by skull bones. The nasal conchae increase the surface area of the respiratory mucosa that lines the nasal cavity.

- Air enters and leaves the nasal cavity via the nostrils. Inhaled air is filtered, warmed, and moistened by the respiratory mucosa of the nasal cavity.
- The pharynx is a short passageway for both air moving between the nasal cavity and the larynx and food passing from the mouth to the esophagus. The pharyngeal, lingual, and palatine tonsils are clumps of lymphoid tissue associated with the pharynx.
- The larynx is a cartilaginous box that conducts air between the pharynx and the trachea, and it houses the vocal folds. During swallowing, the epiglottis prevents solids and liquid from entering the larynx and directs it into the esophagus.
- The trachea extends from the larynx into the thoracic cavity, where it branches to form the main bronchi, which enter the lungs.
- The bronchial tree consists of main, lobar, segmental, and smaller bronchi. They are supported by cartilaginous rings and are lined with respiratory mucosa. Bronchioles are composed of smooth muscle lined with a mucous membrane containing simple cuboidal epithelium. Bronchi and bronchioles carry air into and out of the pulmonary alveoli. They filter, warm, and moisten the inhaled air.
- Lungs fill most of the thoracic cavity. They consist of air passageways of the bronchial tree and bronchioles, pulmonary alveoli, blood and lymphatic vessels, nerves, and connective tissues. Gas exchange occurs between air in the pulmonary alveoli and the blood in the alveolar capillaries.
- Surfactant in the pulmonary alveoli prevents the collapse of the pulmonary alveoli.
- The visceral pleurae cover the outer surfaces of the lungs, and the parietal pleurae line the inner wall of the thoracic cage. The pleural cavity is the space between pleurae and is filled with pleural fluid.

14.3 Breathing

- Breathing involves inspiration and expiration. Air moves into and out of the lungs along a pressure gradient.
- Resting inspiration results from contraction of the diaphragm and external intercostal muscles, which increases the volume and decreases the pressure within the thoracic cavity and lungs. The higher atmospheric pressure causes air to flow into the lungs until the atmospheric and intra-alveolar pressures are equalized. A forceful inspiration also involves the contraction of neck and chest muscles.
- Resting expiration results from relaxation of these muscles, which decreases the volume and increases the pressure within the thoracic cavity and lungs. The higher intra-alveolar pressure causes air to flow out of the lungs until the intra-alveolar and atmospheric pressures are equalized. A forceful expiration involves the contraction of internal intercostal muscles and abdominal muscles.

14.4 Respiratory Volumes and Capacities

- Respiratory air volumes vary with size, sex, age, and physical condition. Variations from the norm usually indicate a pulmonary disorder.
- The average values for respiratory volumes and capacities are tidal volume–500 ml; inspiratory reserve volume–3,000 ml; expiratory reserve volume–1,100 ml; vital capacity–4,600 ml; residual volume–1,200 ml; and total lung capacity–5,800 ml.

14.5 Control of Breathing

- Breathing is controlled by the respiratory centers located in the pons and medulla oblongata.
- The two groups of neurons in the medulla oblongata are the ventral respiratory group (VRG) and the dorsal respiratory group (DRG).
- The VRG controls the resting rhythmic breathing cycle. The DRG integrates sensory input and stimulates the VRG to modify breathing to be faster or slower and deeper or shallower.
- The pontine respiratory group (PRG) in the pons relays action potentials, especially from higher brain centers, to the DRG and VRG to modify the breathing cycle.

14.6 Factors Influencing Breathing

- Chemoreceptors in the medulla oblongata are sensitive to changes in concentrations of carbon dioxide and hydrogen ions in the cerebrospinal fluid. An increase in their concentrations is the primary stimulus for inspiration. The breathing rate and depth vary directly with changes in blood carbon dioxide and hydrogen ion concentrations.
- Chemoreceptors in the carotid and aortic bodies are sensitive to the concentration of oxygen, carbon dioxide, and hydrogen ions in the blood. Blood oxygen concentration must be very low to produce a direct effect on breathing.
- The stretching of the bronchi, bronchioles, and visceral pleurae during inspiration triggers the inflation reflex, which inhibits excessive inspiration and promotes expiration.
- Higher brain centers can influence the respiratory centers either voluntarily or involuntarily. A sudden emotional experience or a sharp pain produces momentary apnea. Anxiety, fear, and excitement increase the breathing rate.
- Stimulation of irritant receptors by irritants triggers irritant reflexes.
- The breathing rate varies directly with changes in body temperature.

14.7 Gas Exchange

- Gas exchange between air in the pulmonary alveoli and the blood in alveolar capillaries occurs by diffusion, and it is called alveolar gas exchange.
- Oxygen diffuses from air in the pulmonary alveoli into the blood; carbon dioxide diffuses from the blood into air in the pulmonary alveoli.
- Gas exchange between tissue cells and blood in systemic capillaries occurs by diffusion, and it is called systemic gas exchange.
- Oxygen diffuses from the blood into the interstitial fluid before entering the tissue cells; carbon dioxide diffuses from the tissue cells into the interstitial fluid before entering the blood.

14.8 Transport of Respiratory Gases

- In the lungs, oxygen combines with hemoglobin to form oxyhemoglobin. In body tissues, oxyhemoglobin releases oxygen to tissue cells. About 98.5% of the oxygen is carried as oxyhemoglobin; only about 1.5% is carried dissolved in plasma.
- Carbon dioxide is mostly carried in bicarbonate ions in plasma. When carbon dioxide diffuses from tissue cells into the blood, 70% of it enters the RBCs. Carbonic anhydrase in RBCs catalyzes the combination of carbon dioxide and water to form carbonic acid, which ionizes to form hydrogen and bicarbonate ions. In the lungs, the reaction reverses to release carbon dioxide into the pulmonary alveoli.
- Twenty-three percent of the carbon dioxide is carried as carbaminohemoglobin, and 7% of the carbon dioxide is carried dissolved in the plasma.

14.9 Disorders of the Respiratory System

- Common inflammatory disorders include bronchitis, emphysema, asthma, common cold, rhinitis, laryngitis, sinusitis, influenza, pneumonia, tuberculosis, and pleurisy. Chronic bronchitis, emphysema, and asthma are COPDs.
- Common noninflammatory disorders include lung cancer, pulmonary edema, pulmonary embolism, and infant respiratory distress syndrome.

Improve Your Grade

Connect Interactive Questions Reinforce your knowledge using multiple types of questions: interactive, animation, classification, labeling, sequencing, composition, and traditional multiple choice and true/false.

SmartBook Proven to help students improve grades and study more efficiently, SmartBook contains the same content within the print book but actively tailors that content to the needs of the individual.

Anatomy & Physiology REVEALED® Dive into the human body by peeling back layers of cadaver imaging. Utilize this world-class cadaver dissection tool for a closer look at the body anytime, from anywhere.

CHAPTER 15

Digestive System

©Stockbyte/Getty Images RF

Janet, a young finance executive, has a high-stress job working in one of the top investment firms in New York City. On an average day, she rarely leaves her desk and works very long hours, which keep her from regularly exercising. Her days become so chaotic that she often forgets to eat and drink throughout the day. One day, Janet develops some abdominal discomfort but writes it off as nothing and continues with her day. After two more days of discomfort, Janet realizes she has not had a bowel movement in three days. Her 60-year-old mother frequently complains about being constipated, but Janet never dreamed it would happen to her at such a young age. At the recommendation of her mother, she increases her daily water intake and adds more fiber to her meals. For more exercise, Janet begins walking with a coworker at lunch. She even enrolls in a yoga class that meets twice per week to help manage her stress level. Within a couple of weeks, Janet's symptoms are completely relieved and she makes a mental promise to herself to never allow her life to abuse her colon again.

CHAPTER OUTLINE

15.1 Introduction to the Digestive System

15.2 Digestion: An Overview

15.3 Alimentary Canal: General Characteristics
- Structure of the Wall
- Movements

15.4 Mouth
- Cheeks
- Palate
- Tongue
- Teeth
- Salivary Glands
- Digestion and Absorption in the Mouth

15.5 Pharynx and Esophagus

15.6 Stomach
- Structure
- Gastric Juice
- Control of Gastric Secretion
- Digestion and Absorption

15.7 Pancreas
- Control of Pancreatic Secretion
- Digestion by Pancreatic Enzymes

15.8 Liver
- Structure
- Functions
- Bile
- Release of Bile

15.9 Small Intestine
- Structure
- Intestinal Juice
- Regulation of Intestinal Secretion
- Digestion and Absorption

15.10 Large Intestine
- Structure
- Functions
- Movements

15.11 Nutrients: Sources and Uses
- Energy Foods and Cellular Respiration
- Carbohydrates
- Lipids
- Proteins
- Vitamins
- Minerals
- MyPlate: A Visual Guide to Healthy Eating

15.12 Disorders of the Digestive System
- Inflammatory Disorders
- Noninflammatory Disorders

Chapter Summary

 Module 12 Digestive System

SELECTED KEY TERMS

Absorption The process by which nutrients pass from the alimentary canal into the blood.
Accessory organs Organs that have a digestive function but are not part of the alimentary canal.
Aerobic respiration The part of cellular respiration that does require oxygen and occurs within mitochondria.
Alimentary canal (aliment = food) The tube through which food passes from the esophagus to the anus.
Anaerobic respiration The part of cellular respiration that does not require oxygen and occurs within the cytosol of a cell.
Chyme (chym = juice) The acidic, semiliquid substance entering the small intestine from the stomach.
Dietary Reference Intake (DRI) Reference values used for planning and assessing nutrient intake.
Digestion (digest = to dissolve) Mechanical and chemical processes that convert nonabsorbable nutrient molecules into absorbable nutrient molecules.
Mastication Process of chewing food.
Nutrient (nutri = to nourish) A substance required for body cells to function.
Peristalsis (peri = around; stalsis = constriction) Wavelike contractions that move luminal contents through the alimentary canal.
Segmentation Localized contractions used to mix luminal contents with digestive secretions.
Sphincter (sphin = squeeze) A ring of smooth or skeletal muscle tissue that contracts to close an opening and relaxes to create an opening.

15.1 Introduction to the Digestive System

Learning Objectives
1. Identify the structures composing the alimentary canal.
2. Identify the accessory organs of the digestive system.

Body cells require a continuous supply of **nutrients** in order to carry out their vital functions. Nutrients are obtained through our diet and include carbohydrates, proteins, lipids, vitamins, minerals, and water. Most of the nutrient molecules in our food are too large to pass directly from the alimentary canal into the blood, meaning they must be broken down into smaller molecules. **Digestion** includes all of the mechanical and chemical processes that convert large, nonabsorbable nutrient molecules into small, absorbable nutrient molecules. **Absorption** is the process by which nutrients pass from the alimentary canal into the blood. The processes of digestion and absorption are the major functions of the digestive system.

The *digestive system* consists of the **alimentary canal** (or *digestive tract*), a long tube through which food passes, and **accessory organs,** organs with digestive functions that are not part of the alimentary canal. The major parts of the alimentary canal are the esophagus, stomach, small intestine, and large intestine, although the mouth and pharynx are usually included in most discussions because of their role in the digestive process. The major accessory organs are the teeth, tongue, salivary glands, liver, gallbladder, and pancreas (figure 15.1).

Check My Understanding
1. What substances are considered nutrients?
2. What are the major functions of the digestive system?
3. What structures compose the alimentary canal?

15.2 Digestion: An Overview

Learning Objectives
3. Compare mechanical and chemical digestion.
4. Describe the role of digestive enzymes.

Digestion involves both mechanical and chemical processes. *Mechanical digestion* is the physical breakdown of food into smaller pieces, which provides a greater surface area for contact with digestive secretions. *Chemical digestion* is the splitting of large, nonabsorbable nutrient molecules into small, absorbable nutrient molecules through the addition of water–a process known as *hydrolysis* (hī-drol′-i-sis). Because hydrolysis is usually very slow, it is the action of digestive enzymes that speeds up chemical digestion and enables the formation of small, absorbable nutrient molecules within the alimentary canal. Chemical digestion may be summarized as follows.

Nonabsorbable nutrient molecules + Water $\xrightarrow{\text{Digestive enzymes}}$ Absorbable nutrient molecules

A number of different types of enzymes are involved in chemical digestion. Each type of digestive enzyme acts on a specific type of nutrient molecule, or **substrate,** and

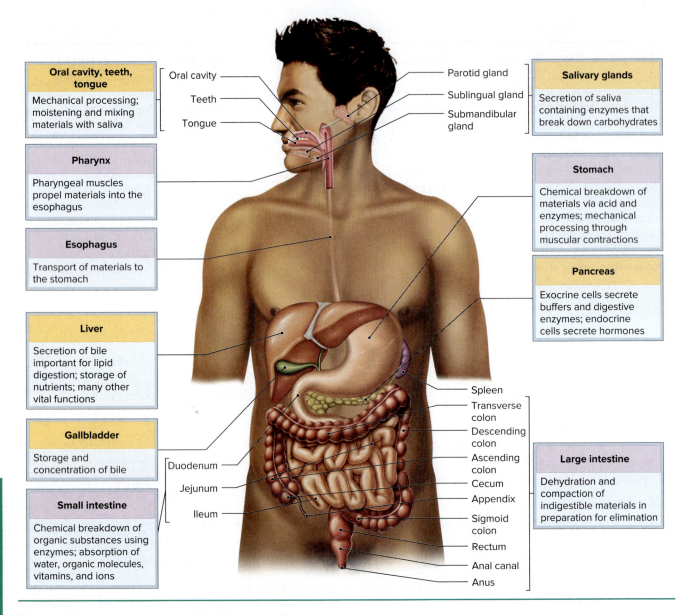

Figure 15.1 Major Organs of the Digestive System.
The structures of the alimentary canal are indicated by the purple boxes, while the accessory structures are indicated by yellow boxes. AP|R

speeds up its breakdown into smaller nutrient molecules. A series of digestive reactions involving several digestive enzymes from different parts of the digestive system are usually required to break down complex nutrient molecules into absorbable nutrient molecules.

 Check My Understanding

4. Why are mechanical digestion and chemical digestion important for healthy digestive system function?
5. Why are digestive enzymes required for effective chemical digestion to occur?

15.3 Alimentary Canal: General Characteristics

Learning Objective
5. Describe the four layers composing the wall of the alimentary canal and their functions.

The alimentary canal is a muscular tube about 6 m (20 ft) in length that extends from the esophagus to the anus. Various portions of the alimentary canal are specialized to perform different digestive functions. The hollow space within the alimentary canal through which food passes is called the *lumen*.

Structure of the Wall

The wall of the alimentary canal consists of four layers. From outermost to innermost, the layers are called the serosa, muscular layer, submucosa, and mucosa. These layers may be modified in the various regions but remain as distinct layers. Compare figure 15.2 to the discussion that follows.

The **serosa** is the outermost layer and consists of the *visceral peritoneum.* It is continuous with the *parietal peritoneum,* which lines the inner surface of the abdominal wall. Together, the visceral peritoneum and parietal peritoneum form the *peritoneum* (see chapter 1). These serous membranes secrete peritoneal fluid, which keeps the membrane surfaces moist and reduces friction as parts of the alimentary canal rub against each other and the abdominal wall.

The **muscular layer** lies just below the serosa. It consists of two layers of smooth muscle that differ in the orientation of their muscle cells. Muscle cells of the outer muscular layer are arranged longitudinally. Their contractions shorten the tube. Muscle cells of the inner muscular layer are arranged circularly around the tube. Their contractions constrict the tube. The functions of muscular layer are discussed next in the "Movements" section.

The **submucosa** lies between the muscular layer and the mucosa. It contains nerves, mucous glands, lymphatic vessels, and blood vessels embedded in areolar connective tissue.

The innermost layer is the **mucosa.** It consists of an epithelium that is in direct contact with the lumen and is supported by areolar connective tissue containing a few smooth muscle cells. The epithelium is often folded

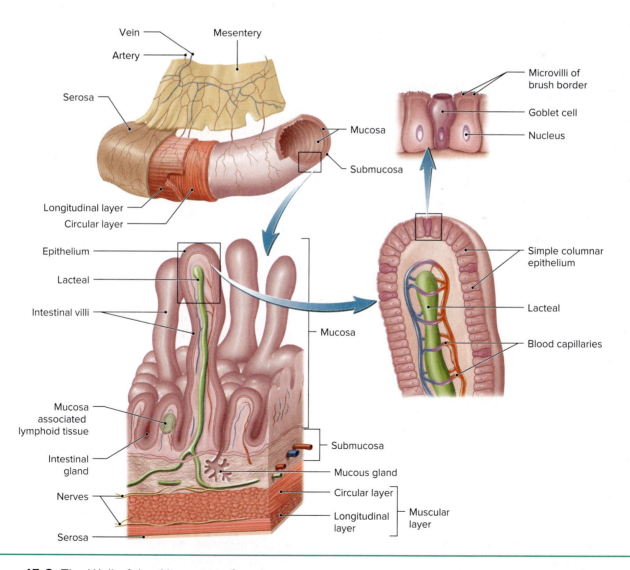

Figure 15.2 The Wall of the Alimentary Canal.
The wall of the alimentary canal consists of four layers as shown here in a section of the small intestine. From outermost to innermost they are the serosa, muscular layer, submucosa, and mucosa.

to increase the surface area that is in contact with food. The mucosa has different functions in different parts of the alimentary canal. In some regions, it secretes only mucus, which protects underlying cells. In others, it secretes mucus and digestive secretions containing enzymes and absorbs nutrients.

Movements

Contraction of the smooth muscle layers produces two different types of movement in the alimentary canal: mixing movements and propelling movements. **Segmentation** involves ringlike contractions followed by relaxation at multiple places along the alimentary canal. The alternating segmental contractions help to mix the luminal contents with the digestive secretions. These are especially frequent in the small intestine.

The movement that propels luminal contents through the alimentary canal is called **peristalsis** (per-i-stal′-sis). In peristalsis, coordinated contraction and relaxation of the circular and longitudinal muscular layers produce a wave of contraction along the alimentary canal that pushes the luminal contents in front of it. The movement resembles pushing toothpaste along a toothpaste tube toward its opening. In this way, peristaltic contractions move food from one portion of the alimentary canal to another.

 Check My Understanding

6. What organs compose the alimentary canal?
7. What is chemical digestion?
8. How are luminal contents moved through the alimentary canal, and what layer of the wall aids in this movement?

15.4 Mouth

Learning Objectives

6. Compare deciduous and permanent teeth.
7. Describe the basic structure of a tooth.
8. Describe digestion in the mouth.

The *mouth*, or *oral cavity*, is involved in the intake of food (ingestion), mechanically breaking it into small pieces, mixing it with saliva to begin chemical digestion, and swallowing it. The mouth is surrounded by the cheeks, palate, and tongue. Examine figures 15.3 and 15.4.

Cheeks

The *cheeks* form the lateral walls of the mouth. Skin covers their outer surfaces, and nonkeratinized stratified squamous epithelium lines their inner surfaces. The cheeks help to hold food within the mouth, while contractions of muscles within them produce facial expressions. The anterior portions of the cheeks form the *lips*, which surround the opening into the mouth. Lips are sensitive, highly mobile structures with important roles in producing speech and detecting touch and temperature stimuli. Their pinkish color results from numerous blood vessels near their surfaces.

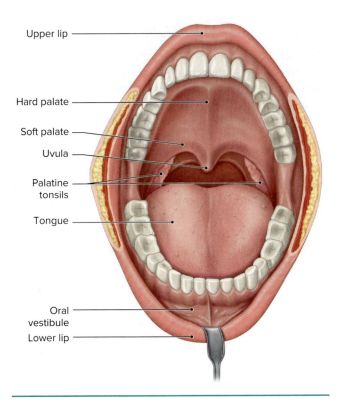

Figure 15.3 Major Structures Associated with the Mouth.

Palate

Recall from chapter 14 that the palate forms the roof of the mouth and separates the oral cavity from the nasal cavity. The hard palate provides protection to the nasal cavity while chewing, in addition to allowing simultaneous breathing and chewing. The soft palate ends posteriorly in a cone-shaped *uvula* that extends downward at the back of the oral cavity. The uvula is very sensitive to touch stimuli. This sensitivity causes the soft palate to move superiorly during swallowing, which closes off the nasal cavity and directs food downward into the pharynx. The uvula also plays a role in triggering the gag reflex.

Tongue

The *tongue* covers the floor of the oral cavity. It is composed primarily of skeletal muscle that is covered by a mucous membrane. An anterior fold of the mucous membrane on the bottom of the tongue attaches the tongue to the floor of the mouth. This membranous attachment, known as the *frenulum* (fren′-ū-lum) *of the tongue*, limits the backward movement of the tongue.

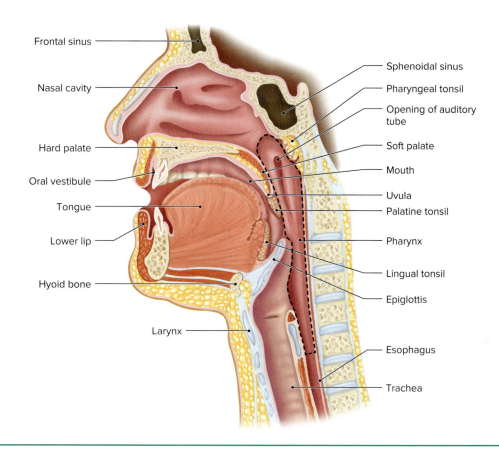

Figure 15.4 Digestive Structures Within the Head and Neck.
The structural relationships of the nasal cavity, mouth, pharynx, esophagus, and larynx are shown in sagittal section. The pharynx is outlined by the black dotted line. AP|R

Sometimes the frenulum of the tongue is too short and restricts tongue movements that are required for proper speech production. A person with this problem is said to be "tongue-tied." Cutting the frenulum of the tongue to allow freer movement usually solves the problem.

The top surface of the tongue contains numerous tiny projections called **lingual papillae** (pah-pil′-ē) that are the locations for taste buds, the sensory receptors for the sense of taste. Lingual papillae also give the tongue a rough texture, which aids in its manipulation of food during chewing and in mixing food with saliva. In swallowing, the tongue pushes food backward into the pharynx. The tongue possesses *lingual glands* that produce an enzyme called *lingual lipase*, which is a digestive enzyme that initiates lipid digestion in the stomach. The movement of the tongue is also important for the production of speech.

Teeth

Teeth are important accessory digestive structures that mechanically break food into smaller pieces during **mastication** (mas-ti-kā′-shun), or chewing. Humans develop two sets of teeth: deciduous and permanent teeth.

The **deciduous teeth,** the first set, start to erupt through the gums at about six months of age. Central incisor (in-sī-zer) teeth come in first, and second molar teeth erupt last. There are 20 deciduous teeth, 10 in each jaw, and all of them are in place by three years of age. Deciduous teeth are gradually shed starting at about six years of age, and they are usually lost in the same order in which they emerged.

The **permanent teeth** begin appearing at about six years of age when the first molar teeth (six-year molar teeth) erupt. All of the permanent teeth, except the third molar teeth, are in place by age 16. The third molar teeth (wisdom teeth) erupt between 17 and 21 years of age, or they may never emerge. In many persons, there is insufficient room for the third molar teeth, so they become impacted and often must be surgically removed. The 32 permanent teeth, 16 in each jaw, consist of four different types: incisor teeth, canine teeth, premolar teeth, and molar teeth (figure 15.5).

The chisel-shaped *incisor teeth* are adapted for biting off pieces of food. The *canine teeth* are used to grasp and tear tough food morsels. The broader, more textured

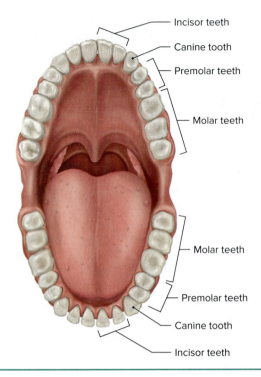

Figure 15.5 The Types of Human Teeth.
There are 32 permanent teeth, 16 in each jaw. Note the location and shape of the incisor teeth and molar teeth. AP|R

Table 15.1 Deciduous and Permanent Teeth

Type	Number Deciduous	Number Permanent
Incisor teeth		
Central	4	4
Lateral	4	4
Canine teeth	4	4
Premolar teeth		
First	0	4
Second	0	4
Molar teeth		
First	4	4
Second	4	4
Third	0	4
Totals	20	32

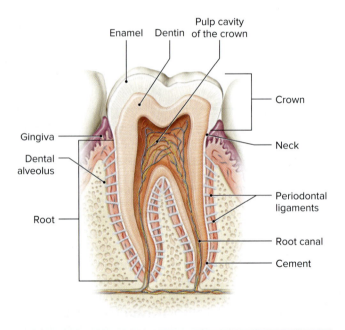

Figure 15.6 The Anatomy of a Tooth.
The structure of a tooth is shown in this section of a molar tooth. AP|R

surfaces of the *premolar teeth* and *molar teeth* are used to crush and grind food. Examine figure 15.5 and table 15.1.

Figure 15.6 shows the basic structure of a tooth. Each tooth consists of two major parts: a root and a crown. The *crown* is the portion of the tooth that is exposed and not covered by the *gingiva* (jin-ji-vah), or gum, covering the underlying bone. The *root* is embedded in a socket, called a *dental alveolus*. A hard substance called *cement* attaches the *root*, through tough *periodontal ligaments*, to the walls of the dental alveolus. The junction of the crown and root is known as the *neck* of the tooth.

Most of a tooth is composed of **dentin,** a hard, bonelike substance. The crown of the tooth has a layer of **enamel** overlying the dentin. Enamel is the hardest substance in the body, and it is appropriately located to resist the abrasion caused by chewing hard foods. Each tooth is supplied with blood vessels and a nerve, which enter the tip of a root and pass through a tubular *root canal* into the *pulp cavity of the crown,* a central space in the crown of a tooth. Together the pulp cavity of the crown and the root canal form the *pulp cavity* of the tooth. In a molar tooth, the pulp cavity of the crown is relatively large and roughly box-shaped. But in a canine tooth, the pulp cavity of the crown is an elongated enlargement of the root canal. *Dental pulp* is soft areolar connective tissue that fills the pulp cavity and supports the blood vessels and nerves.

Clinical Insight

Enamel and dentin are not replaced after they have been eroded by tooth decay. The repair of a decayed tooth requires removal of the decayed material, sterilization of the area, and the insertion of a filling. A filling seals the area and provides an abrasion-resistant surface.

Salivary Glands

The **salivary glands** secrete **saliva** into the mouth, where it is mixed with food during chewing. Various stimuli, such as the thought or smell of food and the presence of food in the mouth, activate a positive-feedback mechanism that leads to the production and secretion of saliva. Saliva consists mostly of water (99.5%), plus a small amount of other substances. The watery nature of saliva allows it to dissolve food substances and clean surfaces within the mouth. The sense of taste is dependent upon saliva because only dissolved chemicals can stimulate taste buds (see chapter 9). The mucus in saliva helps to hold food particles together during chewing and swallowing, in addition to providing lubrication. Saliva also contains many enzymes. One major enzyme in saliva is **salivary amylase,** which is a digestive enzyme that initiates carbohydrate digestion in the mouth. Another major enzyme is **lysozyme,** which kills certain types of bacteria. There are three pairs of major salivary glands: the parotid, submandibular, and sublingual glands (figure 15.7).

The largest salivary glands are the *parotid* (pah-rot′-id) *glands.* A gland is located in front of each ear on top of the masseter. Parotid glands secrete saliva that is rich in salivary amylase. Parotid secretions are emptied through a duct into the *oral vestibule* near the upper second molar teeth. The oral vestibule is the narrow space between the cheeks, lips, and teeth.

The *submandibular* (sub-man-dib′-ū-lar) *glands* are found in the floor of the mouth. They produce a watery saliva that contains relatively little mucus. Secretions of the submandibular glands are emptied through ducts into the front of the mouth at the base of the frenulum of the tongue.

The *sublingual* (sub-ling′-gwal) *glands* lie on the floor of the mouth below the tongue. They are the smallest of the major salivary glands. Their secretions consist mostly of mucus, and they are emptied by several ducts into the floor of the mouth below the tongue.

Digestion and Absorption in the Mouth

Both mechanical and chemical digestion take place in the mouth. Mechanical digestion in the mouth consists of breaking food into smaller pieces and mixing it with saliva during mastication. This improves chemical digestion because the smaller pieces have an increased surface area upon which digestive secretions may act. Chemical

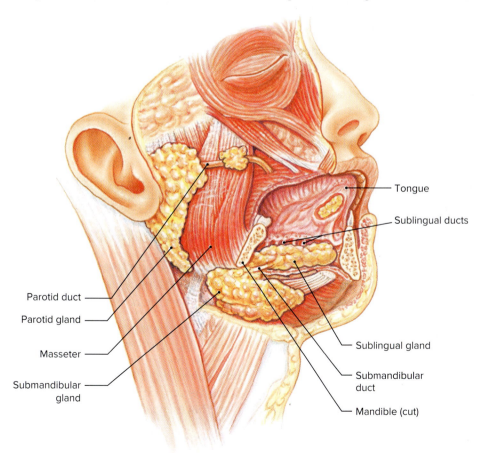

Figure 15.7 The Major Salivary Glands.

digestion starts in the mouth with the breakdown of certain polysaccharides. Salivary amylase acts on *starch* and *glycogen* and speeds up their breakdown into *maltose*, a disaccharide. The action of salivary amylase continues for a few minutes after food enters the stomach, but it soon stops when it is inactivated by the strong acidity of gastric juice. Lingual lipase acts on *triglycerides* and speeds up their breakdown into *monoglycerides* and *fatty acids*. Although lingual lipase is secreted by the tongue, it requires the strong acidity of gastric juice to be activated. Despite the small intestine being the primary site of nutrient absorption, the absorption of some vitamins, monosaccharides, alcohol, and certain types of drugs does occur in the mouth.

 Check My Understanding

9. What is the structure of a tooth?
10. What are the components and functions of saliva?
11. What digestive processes occur in the mouth?

15.5 Pharynx and Esophagus

Learning Objectives
9. Describe the location and function of the pharynx.
10. Describe the location and function of the esophagus.

The **pharynx** (fayr′-inks) is the passageway that connects the nasal and oral cavities with the larynx and esophagus. It is part of both the respiratory and the digestive systems. Its digestive function is the transport of food from the mouth to the esophagus during swallowing (see figure 15.4).

The swallowing reflex is activated when food is pushed into the pharynx by the tongue. The soft palate moves upward, preventing food from entering the nasal cavity, and directs food downward into the pharynx. At the same time, muscle contractions elevate the larynx, which causes the epiglottis to fold over and cover the opening into the larynx. This action prevents food from entering the larynx and directs it into the esophagus.

The **esophagus** (ē-sof′-ah-gus) is a muscular tube that extends from the pharynx downward through the mediastinum and the diaphragm to join with the stomach. When food is swallowed and enters the esophagus, peristaltic contractions propel the food toward the stomach. The esophageal mucosa produces mucus to lubricate the esophagus and aid the passage of food.

At the junction of the esophagus and stomach, the *lower esophageal*, or *cardiac*, *sphincter* (sfink′-ter) prevents regurgitation of stomach contents into the esophagus. But when the peristaltic wave that is propelling food toward the stomach reaches the sphincter, it relaxes and allows food to enter the stomach. A **sphincter** is a ring of smooth or skeletal muscle tissue that contracts to close an opening and relaxes to create an opening. However, the lower esophageal sphincter (LES) is a physiological sphincter rather than an anatomical one because it is not seen in cadavers. It is believed to be caused by muscle tone within the esophagus or surrounding diaphragm. **AP|R**

 Clinical Insight

Failure of the LES to close properly allows for gastric secretions to regurgitate into the lower portion of the esophagus. The *gastroesophageal reflux* (GER) irritates the esophageal lining and usually produces a burning sensation called "heartburn." Causes of GER include malfunction of the LES, alcohol consumption, smoking, and excess acid production in the stomach. Avoiding foods that stimulate acid production in the stomach, such as caffeine and tomatoes, can reduce GER symptoms. Medications that neutralize stomach acid or reduce its production can also alleviate GER symptoms. GER that occurs chronically may indicate a more serious condition called *gastroesophageal reflux disease* (GERD). The chronic esophageal irritation associated with GERD can lead to cellular changes that increase a person's risk of esophageal cancer or Barrett esophagus, which involves changes in the esophageal lining to resemble that of the intestine.

 Check My Understanding

12. How does the swallowing reflex push the food to the esophagus?
13. How does the esophagus carry food to the stomach?

15.6 Stomach

Learning Objectives
11. Describe the structure and functions of the stomach.
12. Explain the control of gastric secretions.

As shown in figure 15.8, the J-shaped **stomach** is a pouchlike portion of the alimentary canal. It lies just below the diaphragm in the left upper quadrant of the abdominopelvic cavity. The basic functions of the stomach are temporary storage of food, mixing food with gastric juice, and starting the chemical digestion of proteins.

Structure

The stomach may be subdivided into four regions: the cardia, fundus, body, and pyloric part. The *cardia* (closest to the heart) is a relatively small area that receives food from

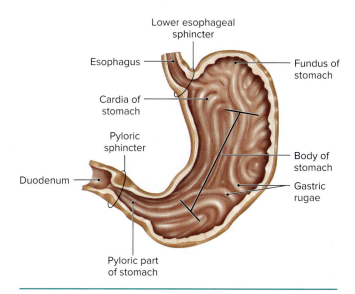

Figure 15.8 The Stomach.
The pouchlike stomach receives food from the esophagus and releases chyme into the duodenum. AP|R

the esophagus. The *fundus* expands above the level of the cardia and serves as a temporary storage area. The *body* is the largest region of the stomach, and it is located between the fundus and pyloric part. The *pyloric part* is the narrow portion located near the junction with the duodenum.

The *pyloric sphincter* is a thickened ring of circular smooth muscle cells that is located at the junction of the stomach and duodenum. This muscle usually is contracted, closing the stomach outlet, but it relaxes to let stomach contents pass into the small intestine.

The stomach possesses some specializations to accommodate its functions. The muscular layer contains a third layer of oblique muscle cells, which allows the stomach to better mix food with gastric secretions. The mucosa is quite thick, when compared to other organs of the alimentary canal. In an empty stomach, the mucosa and submucosa are organized into numerous folds called *gastric rugae* (rū-jē). These folds allow the lining to stretch as the stomach fills with food. The mucosa is dotted with numerous pores called *gastric pits*. Gastric pits receive secretions from **gastric glands** that extend deep into the mucosa.

Gastric Juice

The secretion of the gastric glands is known as **gastric juice.** *Mucous neck cells*, located near the opening to the gastric pit, secrete mucus to coat and protect the mucosa from the action of digestive secretions. *Chief cells*, located in the deepest portions of the gastric glands, secrete the digestive enzymes pepsinogen (inactive form of pepsin), gastric lipase, and rennin. *Parietal cells*, located in the midportion of the gastric glands, secrete hydrochloric acid (HCl) and intrinsic factor (figure 15.9).

As food is mixed with gastric juice and as chemical digestion occurs, it is converted into an acidic, semiliquid substance called **chyme** (kīm). Small amounts of chyme are released intermittently into the duodenum by the relaxing of the pyloric sphincter.

Control of Gastric Secretion

The rate of gastric secretion is controlled by both neural and hormonal means and is a good example of a positive-feedback mechanism. Gastric juice is produced continuously, but its secretion is greatly increased whenever food is on the way to, or already in, the stomach. The sight, smell, or thought of appetizing food, food in the mouth, or food in the stomach stimulates the transmission of parasympathetic action potentials that increase the secretion of gastric juice. These action potentials also, along with food in the stomach and stomach stretching, stimulate certain stomach cells to secrete a hormone called **gastrin.** Gastrin is absorbed into the blood and is carried to gastric glands, increasing their secretions (figure 15.10).

As stomach contents are gradually emptied into the small intestine, there is a decrease in the frequency of parasympathetic action potentials and an increase in the frequency of sympathetic action potentials to the stomach, which reduces the secretion of gastric juice. When chyme passes from the stomach into the small intestine, it stimulates the intestinal mucosa to release two hormones: **cholecystokinin** (kō-lē-sis-tō-kīn′-in) **(CCK)** and **secretin** (se′-krē′-tin), which reduce both the motility of the stomach and the secretion of gastric juice.

Digestion and Absorption

Food entering the stomach is thoroughly mixed with gastric juice by ripplelike, mixing contractions of the stomach wall. Gastric juice is very acidic (pH 2) due to an abundance of HCl. **Pepsin** is the most important digestive enzyme in gastric juice, and it is secreted in an inactive form that prevents digestion of the cells secreting it. Once it is released into the stomach, pepsin is activated by the strong acidity of gastric juice. Pepsin acts on *proteins* and breaks these complex molecules into shorter amino acid chains called *peptides*. However, peptides are still much too large to be absorbed and require further digestion in the small intestine. Recall that the lingual lipase is also activated by the strong acidity of gastric juice.

Gastric juice contains a substance known as **intrinsic factor** that is essential for the absorption of vitamin B12 by the small intestine.

The gastric juice of infants contains two unique enzymes that help to improve the digestion of milk proteins and lipids. **Rennin** (ren-in) curdles milk proteins, which keeps them in the stomach longer and makes them more easily digested by pepsin. **Gastric lipase** acts on *triglycerides* and breaks them into *fatty acids* and *monoglycerides*.

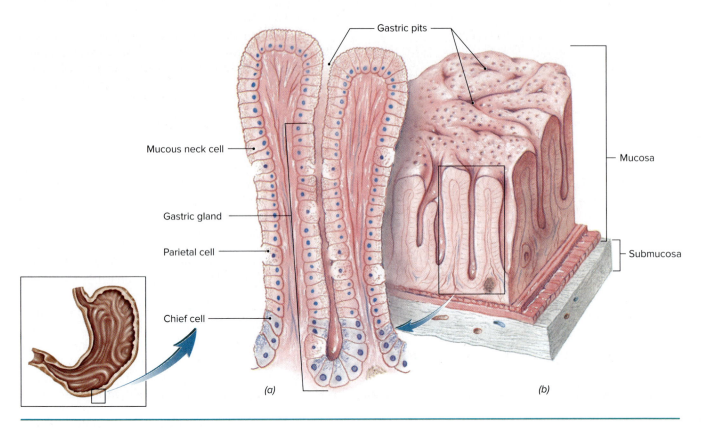

Figure 15.9 Gastric Glands.
(*a*) An enlargement of the gastric glands shows the locations of mucous neck cells, parietal cells, and chief cells. (*b*) The thick stomach mucosa is dotted with gastric pits, openings that receive the secretions from the gastric glands. AP|R

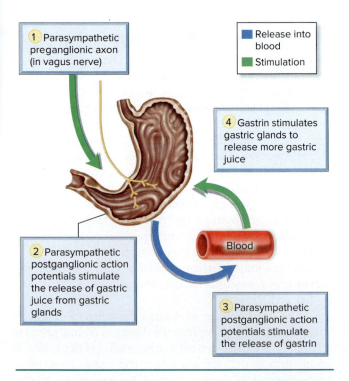

Figure 15.10 Neural and Hormonal Control of Gastric Secretions. AP|R

Clinical Insight

Gastric ulcers produce stomach pain one to three hours after eating. Without treatment, they can perforate the stomach wall, producing internal bleeding or peritonitis. Gastric ulcers result from persistent erosion of the alkaline mucus that coats the stomach lining. Most recurring gastric ulcers are caused by an acid-resistant bacterium, *Helicobacter pylori*, which erodes the protective mucosa, allowing gastric juice to attack cells within the other layers of the stomach wall. Other contributing factors that promote HCl secretion or reduce mucus production include stress, smoking, alcohol, coffee, aspirin, and nonsteroidal anti-inflammatory drugs. Treatment involves antibiotics to kill the bacteria and drugs to reduce gastric secretion.

Except for a few substances such as water, minerals, some drugs, and alcohol, little absorption occurs in the stomach.

When semiliquid chyme passes from the stomach into the duodenum, the first part of the small intestine, secretions from the pancreas and liver are emptied into the duodenum. The secretions from these accessory organs play important roles in digestion within the small intestine.

Check My Understanding

14. What digestive processes occur in the stomach?
15. How are gastric secretions controlled?

15.7 Pancreas

Learning Objective

13. Describe the control and functions of pancreatic secretions.

The **pancreas** is a small, pennant-shaped gland located behind the pyloric part of the stomach. It is connected by a duct to the duodenum, approximately 10 cm (4 in) from the pyloric sphincter. The pancreas has both endocrine and exocrine functions. The majority of the cells within the pancreas secrete **pancreatic juice,** which is the digestive (exocrine) function of the pancreas. Pancreatic juice is collected by tiny ducts that merge to form large ducts, which enter the **pancreatic duct.** The pancreatic duct extends the length of the pancreas and usually forms a smaller *accessory pancreatic duct*. The pancreatic duct joins with the bile duct where they both empty their secretions into the duodenum. Their common opening is controlled by a sphincter that dilates to allow pancreatic juice and bile to enter the duodenum. The accessory pancreatic duct allows pancreatic juice to enter the duodenum independently of bile (figure 15.11).

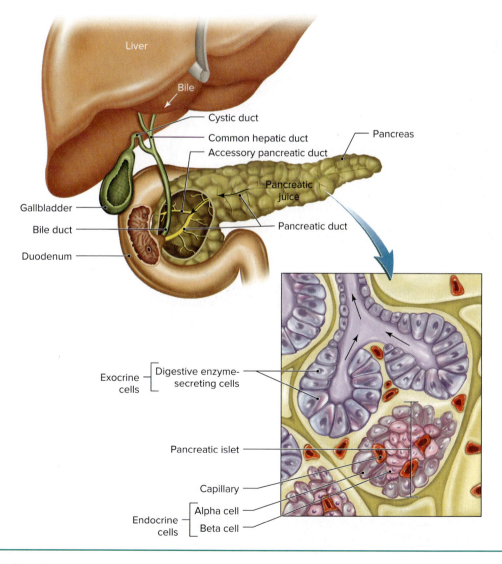

Figure 15.11 The Pancreas.
The pancreas is closely associated with the liver and duodenum. The inset shows secretory cells involved in the exocrine and endocrine functions of the pancreas. AP|R

Control of Pancreatic Secretion

Pancreatic secretion, like gastric secretion, is controlled by both neural and hormonal mechanisms. Neural control is via parasympathetic axons. When parasympathetic action potentials activate the stomach mucosa, they also stimulate the pancreas to secrete pancreatic juice.

Hormonal control of pancreatic secretion results from two hormones that stimulate different types of pancreatic cells. Acid chyme entering the duodenum stimulates the intestinal mucosa to release the hormone secretin. Secretin is carried by blood to the pancreas, where it stimulates secretion of pancreatic juice that is rich in bicarbonate ions (figure 15.12). Bicarbonate ions neutralize the acidity of the chyme entering the small intestine. Lipid-rich chyme stimulates production of cholecystokinin by the intestinal mucosa. CCK stimulates secretion of pancreatic juice that is rich in digestive enzymes.

Table 15.2 summarizes the major hormones that regulate digestive secretions.

Digestion by Pancreatic Enzymes

Pancreatic juice contains enzymes that act on each of the major classes of energy foods: carbohydrates, fats, and proteins. Their digestive actions occur within the small intestine.

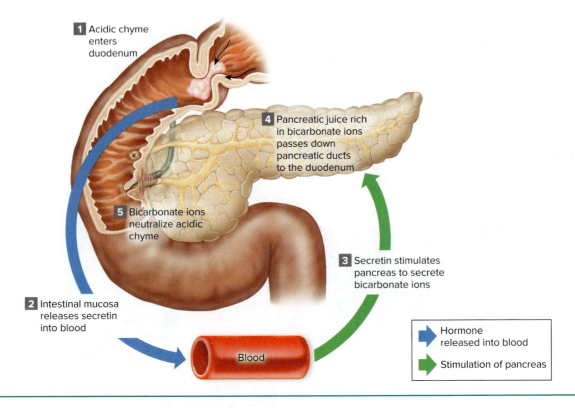

Figure 15.12 Hormonal Control of Pancreatic Secretion.

Table 15.2 Major Hormones Regulating Digestive Secretions

Hormone	Action	Source
Gastrin	Stimulates gastric juice secretion	Gastric mucosa; parasympathetic action potentials, stomach stretching, and food in the stomach stimulate release of gastrin
Cholecystokinin	Reduces gastric juice secretion; stimulates secretion of pancreatic juice that is rich in digestive enzymes; stimulates contraction of gallbladder and relaxes hepatopancreatic sphincter causing release of bile	Intestinal mucosa; lipid-rich chyme stimulates the release of cholecystokinin
Secretin	Stimulates secretion of pancreatic juice that is rich in bicarbonate ions; inhibits gastric secretion	Intestinal mucosa; acid chyme stimulates the release of secretin

Pancreatic amylase, like salivary amylase, acts on *starch* and *glycogen,* splitting these polysaccharides into *maltose,* a disaccharide. Maltose, however, is too big to be absorbed by the body.

Pancreatic lipase acts on *triglycerides* and splits them into *monoglycerides* and *fatty acids* that are absorbable.

Trypsin is the major pancreatic enzyme in pancreatic juice. Trypsin splits *proteins* into *peptides.* Like pepsin in the stomach, it is secreted in an inactive form and is activated only when mixed with intestinal secretions within the small intestine. This mode of secretion prevents the pancreatic cells from being digested by their own enzymatic secretions.

Check My Understanding

16. What are the functions of the enzymes in pancreatic juice?
17. How is pancreatic secretion controlled?

15.8 Liver

Learning Objectives
14. Describe the location and functions of the liver.
15. Explain how bile release is stimulated.

The **liver** is the largest gland in the body. It weighs about 1.4 kg (3 lb) and is dark reddish brown in color. The liver is located mostly in the right upper quadrant of the abdominopelvic cavity just below the diaphragm, where it is protected by the inferior ribs.

Structure

The liver is encased in a dense irregular connective tissue capsule that, in turn, is covered by the peritoneum for additional support. A ligament of dense regular connective tissue, called the *falciform ligament,* joins the liver to the diaphragm and the anterior abdominal wall and separates the two main lobes: a larger *right lobe* and a smaller *left lobe.* Several blood vessels and the common hepatic duct enter or exit the liver from a small area on the back of the liver (figure 15.13).

The liver receives blood from two sources. The *hepatic artery proper* brings oxygenated blood to the *hepatocytes* (liver cells). The *hepatic portal vein* brings deoxygenated, nutrient-rich blood from the digestive tract. As blood flows through the liver, hepatocytes remove, modify, or add substances to the blood before it leaves the liver via the *hepatic veins.*

Microscopically, the liver consists of multitudes of **hepatic lobules,** which serve as the structural and functional units. Each lobule is a short, roughly hexagonal cylinder with a central vein running through its core from which thin sheets of hepatocytes radiate. *Hepatic triads,* located at the corners where several lobules meet, are composed of three vessels: a branch of the hepatic artery proper, a branch of the hepatic portal vein, and a small *interlobular bile ductule* carrying bile. Between the sheets of hepatocytes are *hepatic sinusoids,* blood-filled spaces that carry blood from the hepatic artery proper and hepatic portal vein to the central vein of the lobule. As blood flows through the hepatic sinusoids, an exchange of materials occurs between the blood and the hepatocytes. Macrophages in the epithelium lining the sinusoids remove cellular debris and bacteria. The central veins of the lobules ultimately merge to form the hepatic veins.

As noted, the production of bile is the only digestive function of the liver. Bile is collected in tiny ducts that merge to form the interlobular ductules of the hepatic triads, which in turn unite to form the *right* and *left hepatic ducts* exiting the right and left lobes of the liver. The right and left hepatic ducts merge to form the *common hepatic duct,* which carries bile out of the liver. The common hepatic duct and the *cystic duct,* a short duct that extends from the gallbladder, merge to form the **bile duct,** which carries bile to the duodenum. The cystic duct carries bile to and from the **gallbladder,** a small, pear-shaped sac that stores bile temporarily between meals (figures 15.11 and 15.13).

Functions

The liver has many important and vital functions, though most are not associated with digestion. (1) The liver produces and secretes bile, a substance that aids in the digestion and absorption of lipids, and heparin, a blood anticoagulant. It also produces and secretes plasma proteins (see chapter 11). (2) The liver plays a critical role in carbohydrate metabolism. When blood glucose level is elevated, the liver can convert and store the excess as glycogen or triglycerides. When blood glucose level is low, the liver can convert glycogen, glycerol, fatty acids, and amino acids into glucose. (3) As part of lipid metabolism, the liver forms lipoproteins for the transport of fatty acids, triglycerides, and cholesterol. It also can synthesize cholesterol and use it to form bile salts. (4) During protein metabolism, the liver removes the amine groups from amino acids so that the remainder of the molecules can be used in aerobic respiration or in forming glucose and triglycerides. (5) The liver is used for the storage of triglycerides, glycogen, iron, and vitamins A, D, E, K, and B12. (6) It detoxifies the blood by modifying many drugs and toxic chemicals to form less toxic compounds. (7) Using phagocytosis, the liver removes worn-out blood cells and any bacteria present.

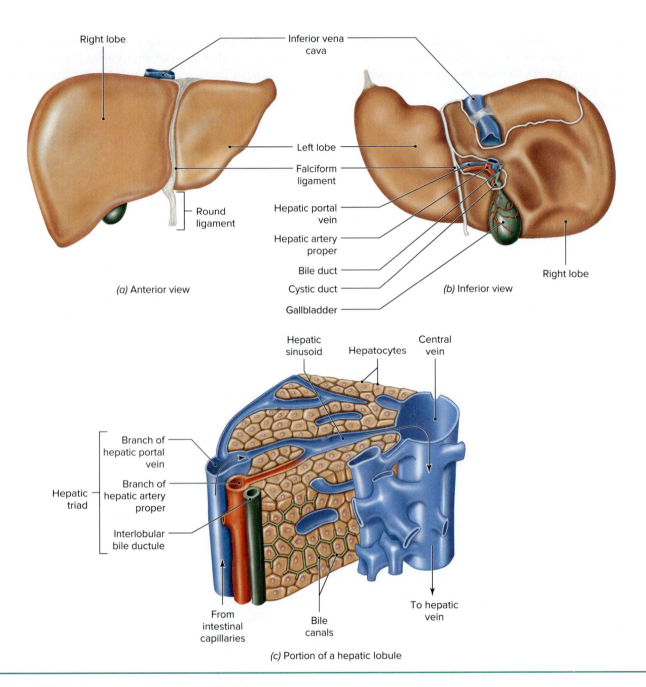

Figure 15.13 Anatomy of the Liver. AP|R

Bile

Hepatocytes continuously produce **bile,** a yellowish green liquid. Bile consists of water, bile salts, bile pigments, cholesterol, and minerals. Bile pigments, such as yellow-colored bilirubin, are waste products of hemoglobin breakdown that are excreted through bile (see chapter 11). *Jaundice* is a medical condition in which too much bilirubin is circulating in the blood due to liver or kidney malfunction or excessive red blood cell destruction. The excess bilirubin ends up being deposited within the skin, cornea, and mucous membranes, causing yellow discoloration.

Bile salts are the only bile components that play a digestive role. When in contact with fatty substances, they break up large fat globules into very small droplets, a process called *emulsification* (ē-mul-si-fi-kā´-shun). Emulsification greatly increases the surface area of the fats exposed to water and lipases. In this way, bile salts aid the digestion of fats. Bile salts also aid the absorption of fatty acids, cholesterol, and lipid-soluble vitamins by the small intestine.

Release of Bile

Bile usually enters the duodenum only when chyme is present. When the intestine is empty, the sphincter at the base of the bile duct constricts, which forces bile to enter the gallbladder for temporary storage.

When lipid-rich chyme enters the duodenum, it stimulates the release of cholecystokinin from the intestinal mucosa. CCK is carried by the blood to the gallbladder, where it stimulates contraction of muscles in the gallbladder wall. The contractions eject bile from the gallbladder into the bile duct. CCK also relaxes the hepatopancreatic sphincter at the base of the bile duct so bile is injected into the small intestine. Note that this hormonal control releases bile only when it is needed in the small intestine (figure 15.14).

Check My Understanding

18. Where is the liver located?
19. What is the digestive function of the liver?
20. How is bile release controlled?

15.9 Small Intestine

Learning Objectives

16. Describe digestion in the small intestine.
17. Explain how the end products of digestion are absorbed.

The small intestine is about 2.5 cm (1 in) in diameter and 4 to 5 m (12-15 ft) in length. It begins at the pyloric sphincter of the stomach, fills much of the abdominopelvic cavity, and empties into the large intestine. Most of the digestive processes and absorption of nutrients occur in the small intestine.

Structure

There are three sequential segments composing the small intestine. The **duodenum** (dū-o-dē′-num) is a very short section, about 25 to 30 cm (9-12 in) long, that receives chyme from the stomach. The middle section is the **jejunum** (je-jū′-num), and it is about 160 to 200 cm (5-6 ft) long. The last and longest segment is the **ileum** (il′-ē-um), which is about 170 to 215 cm (5-7 ft) long. The ileum joins with the large intestine at the *ileal* (il-ē-al) *orifice*.

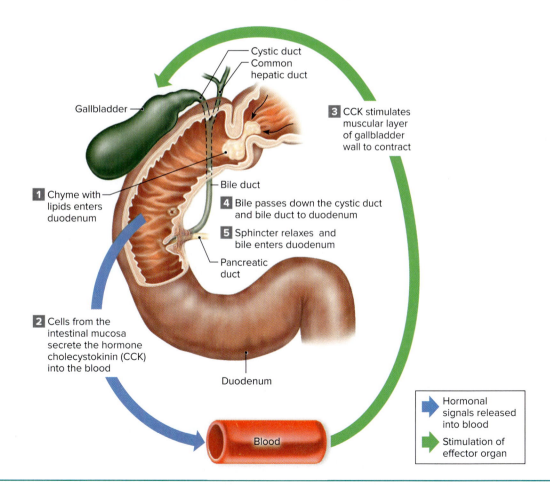

Figure 15.14 Hormonal Control of Bile Secretion.

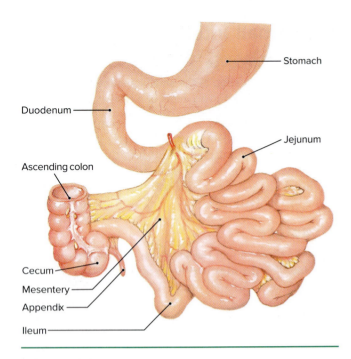

Figure 15.15 The Small Intestine.
The small intestine consists of the duodenum, jejunum, and ileum. Chyme from the stomach enters the duodenum. After digestion and absorption, chyme residues pass from the ileum into the large intestine. AP|R

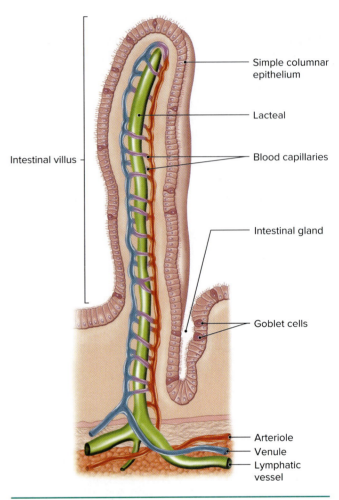

Figure 15.16 The Structure of an Intestinal Villus.

The small intestine is suspended from the posterior abdominal wall by the **mesentery** (mes´-en-ter-ē), double folds of the peritoneum that provide support but allow movement. Blood vessels, lymphatic vessels, and nerves serving the small intestine are also supported by the mesentery (figure 15.15).

The mucosa of the small intestine is modified to provide a very large surface area. The distinctive velvety appearance of the intestinal mucosa results from the presence of **intestinal villi,** tiny projections from the mucosa that are extremely abundant (see figure 15.2). Each intestinal villus is covered by simple columnar epithelium and contains a centrally located **lacteal,** a lymphatic capillary. A blood capillary network surrounds the lacteal. At the bases of the intestinal villi are tiny pits that open into **intestinal glands,** which secrete intestinal juice (figure 15.16).

The mucosal surface area in contact with chyme and digestive fluids is further increased by the presence of numerous microvilli forming a "brush border" (see chapter 4 and figures 15.2 and 15.17).

Intestinal Juice

The fluid secreted by the intestinal glands is known as **intestinal juice.** It is slightly alkaline and contains abundant water and mucus. Intestinal juice provides an appropriate environment for the action of bile salts and pancreatic digestive enzymes within the small intestine. Recall that trypsin in pancreatic juice is activated only after being mixed with intestinal secretions.

Regulation of Intestinal Secretion

The presence of chyme in the small intestine provides mechanical stimulation of the mucosa that activates the secretion of intestinal juice and enzymes. Chyme also causes an expansion of the intestinal wall, triggering a neural reflex that sends parasympathetic action potentials to the mucosa. The action potentials stimulate an increase in the rate of intestinal secretions.

Digestion and Absorption

Vigorous segmentation within the small intestine mixes chyme with bile, pancreatic juice, and intestinal juice. The emulsification of fats by bile and the continued digestion of carbohydrates, fats, and proteins by pancreatic and

Table 15.3 Summary of the Major Digestive Enzymes and Their Actions

Enzyme	Substrate	Product
Saliva		
Salivary amylase	Starch and glycogen	Maltose
Gastric Juice		
Pepsin	Proteins	Peptides
Pancreatic Juice		
Pancreatic amylase	Starch and glycogen	Maltose
Pancreatic lipase	Triglycerides	Monoglycerides* and fatty acids*
Trypsin	Proteins	Peptides
Brush Border Enzymes		
Maltase	Maltose	Glucose*
Sucrase	Sucrose	Glucose* and fructose*
Lactase	Lactose	Glucose* and galactose*
Peptidases	Peptides	Amino acids*

*End products of digestion.

brush border enzymes occur within the small intestine. *Brush border enzymes* are embedded within the brush border of the small intestine mucosa. These actions complete the digestive process within the small intestine.

There are three brush border enzymes that split disaccharides into monosaccharides. (1) **Maltase** converts *maltose* into *glucose;* (2) **sucrase** converts *sucrose* into *glucose* and *fructose;* and (3) **lactase** converts *lactose* into *glucose* and *galactose.*

Brush border enzymes acting on proteins are also present. Various **peptidases** split *peptides* into *amino acids.* Table 15.3 summarizes the enzymes involved in the digestion of carbohydrates, fats, and proteins.

Carbohydrate digestion begins in the mouth and concludes in the small intestine. The end products of carbohydrate digestion are monosaccharides, the simple sugars glucose, fructose, and galactose. These sugars are absorbed across the epithelium and into the capillaries of the intestinal villi through both facilitated diffusion and active transport.

Triglyceride digestion primarily occurs in the small intestine. The end products are monoglycerides and fatty acids. Very small, or short-chain, fatty acids are absorbed by simple diffusion across the epithelium and into the capillaries of the intestinal villi. All other lipids require an alternate means of absorption. Bile salts interact to form structures called *micelles,* small transportation spheres that are hydrophilic on their surface and hydrophobic in their core. Micelles absorb large fatty acids, monoglycerides, cholesterol, phospholipids, and lipid-soluble vitamins into their core and transport them to the intestinal brush border. The contents of the micelles move by simple diffusion into the epithelial cells. Once inside the epithelial cells, the fatty acids and monoglycerides recombine to form triglycerides.

The triglycerides combine in small clusters with phospholipids, steroids, and lipid-soluble vitamins. These clusters are coated with protein and form structures known as **chylomicrons** (kī-lō-mī´-krons). The protein coat makes chylomicrons water-soluble. Chylomicrons move out of the epithelial cells by exocytosis and enter the lacteals of the intestinal villi, as shown in figure 15.17. They are carried by lymphatic vessels to the left subclavian vein, where lymph from the intestine enters the blood.

Protein digestion begins in the stomach and concludes in the small intestine. The end products are amino acids, which are actively absorbed across the epithelium and into the capillaries of intestinal villi.

In addition to the end products of digestion, other needed substances are absorbed in the small intestine. For example, water, minerals, and water-soluble vitamins are absorbed into the capillaries of intestinal villi. Materials absorbed into the blood are carried from the intestines to the liver via the hepatic portal vein. After processing by the liver, appropriate concentrations of nutrients are

Clinical Insight

Lactose intolerance is caused by a deficiency or absence of lactase. The presence of undigested lactose in the intestines produces an osmotic gradient that prevents the reabsorption of water into the blood and, even worse, actually causes water to be drawn into the intestines from interstitial fluid. The result is diarrhea, flatulence, bloating, and intestinal cramps. Afflicted persons can avoid this problem if they take a tablet or liquid containing lactase before meals containing milk or milk products.

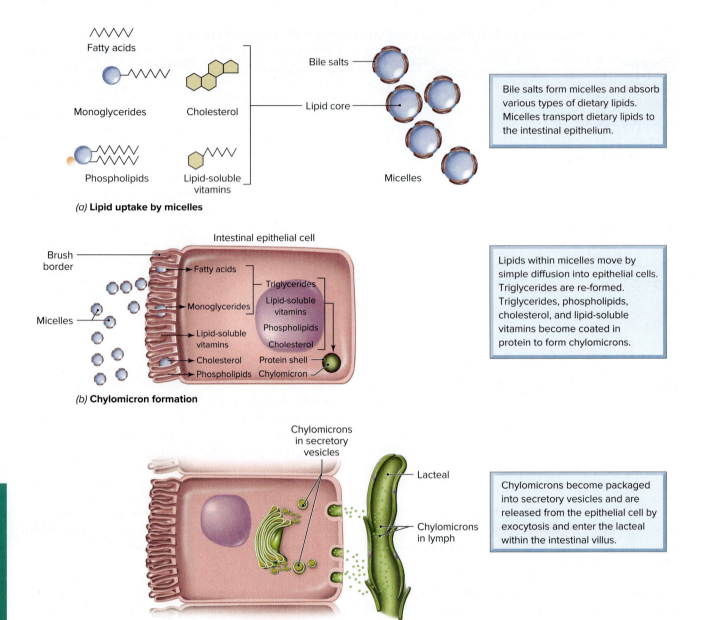

Figure 15.17 Absorption of Dietary Lipids.

released into the general circulation to serve the needs of tissue cells. In this way, the liver contributes to the overall homeostasis of the body.

 Check My Understanding

21. What are the three segments of the small intestine?
22. What digestive processes occur in the small intestine?
23. How are end products of digestion absorbed?

15.10 Large Intestine

Learning Objective

18. Describe the structure and functions of the large intestine.

The small intestine joins with the large intestine at the ileal orifice. This opening is closed most of the time but opens to allow chyme residue to enter the large intestine.

Structure

The *large intestine* gets its name because its diameter (6.5 cm; 2.5 in) is larger than that of the small intestine, although its length (1.5 m; 5 ft) is much shorter. The large intestine consists of four segments: cecum, colon, rectum, and anal canal.

The first segment of the large intestine is the pouch-like *cecum*, which bulges below the ileal orifice. The slender, wormlike **appendix** extends from the cecum and, although it has no digestive function, it contributes to the immune defense of the body.

The **colon** forms most of the large intestine and is subdivided into four regions. The *ascending colon* extends upward from the cecum along the right side of the abdominopelvic cavity. As it nears the liver, it turns left to become the *transverse colon*. Near the spleen, the transverse colon turns downward to become the *descending colon* along the left side of the abdominopelvic cavity. Near the pelvis, the descending colon becomes the *sigmoid colon,* which is characterized by an S-shaped curvature leading to the rectum (figure 15.18).

The **rectum** is the straight portion of the large intestine that continues downward from the sigmoid colon through the pelvic cavity and ends at the **anal canal.** The anal canal is the last 3 cm of the large intestine and its external opening is the **anus.** The mucosa of the anal canal is folded to form the *anal columns*, which contain networks of arteries and veins. The anus is kept closed except during defecation by the involuntarily controlled *internal anal sphincter* and the voluntarily controlled *external anal sphincter* (figure 15.19).

The colon has a puckered appearance when viewed externally. This results because the longitudinal muscles

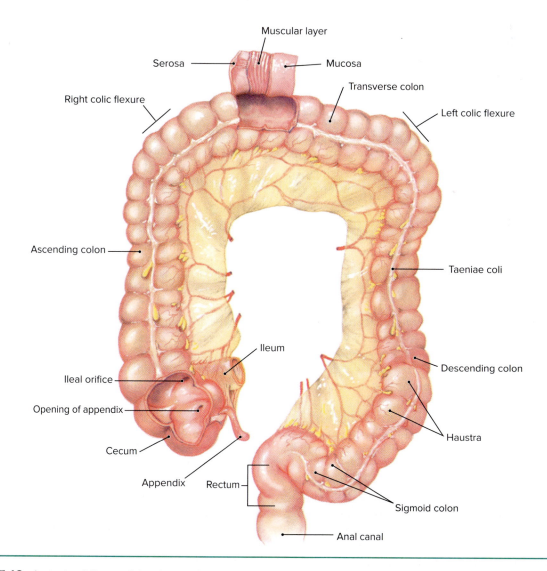

Figure 15.18 Anterior View of the Large Intestine.

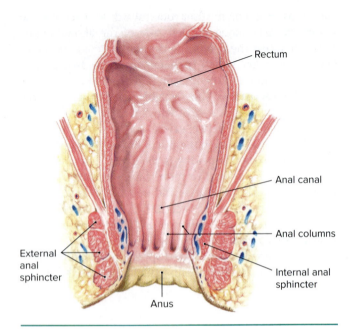

Figure 15.19 The Rectum and Anal Canal.
The rectum and the anal canal are located at the end of the alimentary canal.

are not uniformly layered but are reduced to three longitudinal bands, the *taeniae coli*, that run the length of the colon. Contraction of the taeniae coli gathers the colon into a series of pouches called *haustra* (singular, haustrum). Like the small intestine, the large intestine is supported by a mesentery.

The mucosa of the large intestine is also different from that of the small intestine. Villi are absent, and the simple columnar epithelium contains numerous mucus-producing goblet cells.

Functions

Chyme residue entering the large intestine contains water, minerals, bacteria, and other substances that were not digested or absorbed while in the small intestine. There are no digestive enzymes secreted by the large intestine. Instead, intestinal bacteria decompose the undigested food molecules. This action yields certain B vitamins and vitamin K, in addition to gas (flatus). The mucosa of the large intestine secretes large quantities of mucus that lubricate the intestinal lining and reduce abrasion as materials are moved along.

A major function of the large intestine is the absorption of water, some minerals, and vitamins as the contents slowly move through the colon. Much of this absorption occurs before the chyme residue reaches the descending colon, where it is congealed to form the **feces** (fē-sēz). Feces contain large amounts of bacteria, mucus, and water, as well as undigested food molecules.

Movements

Segmentation and peristalsis within the large intestine are more sluggish than those of the small intestine. Vigorous peristalsis occurs only two to four times a day, usually following a meal. These peristaltic contractions are called *mass movements* because they move the contents of the descending and sigmoid colons toward the rectum. The **defecation reflex** is activated when the rectum fills with feces and its wall is stretched. Parasympathetic action potentials stimulate muscular contractions that increase pressure within the rectum and relax the internal anal sphincter. Defecation, or expulsion of feces, occurs if the external anal sphincter is voluntarily relaxed. If its contraction is voluntarily maintained, defecation is postponed.

 Check My Understanding

24. What are the segments of the large intestine?
25. What are the functions of the large intestine?

15.11 Nutrients: Sources and Uses

Learning Objectives
19. Identify the sources and uses of carbohydrates, lipids, proteins, vitamins, and major minerals.
20. Explain the three major steps of cellular respiration.

Nutrients are chemicals in foods that provide energy for powering life processes; chemicals aiding or enabling life processes; or materials to construct molecules for the healthy development, growth, and maintenance of the body. There are six groups of nutrients: carbohydrates, lipids, proteins, vitamins, minerals, and water. All six groups provide raw materials for constructing new molecules, but only carbohydrates, lipids, and proteins provide energy to sustain life processes.

Essential nutrients are those nutrients that the body cannot synthesize and must obtain in food in order to construct other molecules necessary for life. The essential nutrients include certain amino acids, certain fatty acids, most vitamins, minerals, and water. Because the essential nutrients are not all present in any one food, a balanced diet is required. The conversion of raw materials into molecules for life processes is a major role of the liver.

The Institute of Medicine of the U.S. National Academy of Sciences has developed nutritional recommendations, called the **Dietary Reference Intake (DRI),** to assist in the planning and assessment of nutrient intake. DRI includes the **Recommended Daily Allowance (RDA)** for each nutrient. The RDA for a nutrient is the average daily intake level that is sufficient to meet the nutritional needs of a healthy person. Most nutritional labels provide RDA but not DRI recommendations. DRI and RDA recommendations can be found through the U.S. Department of Agriculture (USDA) website (fnic.nal.usda.gov).

Energy Foods and Cellular Respiration

Carbohydrates, fats, and proteins are called "energy foods" because they are used in cellular respiration to release the energy in their chemical bonds for ATP production. Recall from the discussion in chapters 3 and 7 that cellular respiration includes both anaerobic and aerobic components. **Anaerobic respiration,** or *glycolysis,* does not require oxygen and occurs within the cytosol of a cell, while **aerobic respiration** does require oxygen and occurs within mitochondria, where the enzymes catalyzing the reactions are located. There are two sequential, linked aerobic processes: the *citric acid cycle* and the *electron transport chain.* When oxygen is available, a nutrient, such as glucose, is completely degraded to carbon dioxide and water, in order to release energy. Approximately 40% of the energy is captured in ATP, while the remainder is lost as heat. ATP does not store energy, but it carries it to where it is needed to power life processes. The released heat energy is important in maintaining a homeostatic body temperature.

Follow the cellular respiration of glucose in figure 15.20. A molecule of glucose (containing 6 carbon atoms) is split during glycolysis to form 2 molecules of pyruvic acid (each containing 3 carbon atoms) and 2 ATP. Each pyruvic acid molecule is converted to acetyl-CoA (each molecule containing 2 carbon atoms), which releases CO_2. Each acetyl-CoA molecule enters the citric acid cycle. With each turn of the cycle, one acetyl-CoA molecule is broken down to release CO_2, H^+, and high-energy electrons. Substrate reactions associated with the citric acid cycle produce 2 ATP.

The higher-energy electrons pass along the molecular carriers of the electron transport chain from a higher-energy level to a lower-energy level—much like water flowing down a staircase. At each transfer (step) within the chain, energy is released to form ATP and the electrons move to the next lower energy level (next lower step in the staircase). Ultimately, all of the available energy is extracted, and the electrons, H^+, and O_2 combine to form water (H_2O). The electron transfer yields a total of 32 to 34 ATP, depending upon the cell in which aerobic respiration occurs. When added to the 2 ATP from glycolysis and the 2 ATP from the citric acid cycle, one molecule of glucose can yield a total of 36 to 38 ATP.

The end products of digestion for lipids and proteins also may be used in aerobic respiration, either directly or after conversion into compatible molecules. Figure 15.20 shows the major points of entry into aerobic respiration for these molecules. The number of ATP produced varies with the type of molecule broken down.

Carbohydrates

Nearly all **carbohydrates** in the diet come from plant foods. Glycogen is the only carbohydrate in animal foods. While there is very little of this polysaccharide in meat, animal liver is an abundant source. Monosaccharides are in honey and fruits; disaccharides are found in table sugar and dairy products; and starch, a polysaccharide, occurs in cereals, vegetables, and legumes (e.g., beans, peas, peanuts).

Cellulose is a polysaccharide that is abundant in plant foods, but it cannot be digested by humans because they lack the necessary digestive enzymes. However, it is an important dietary component, because it provides fiber (roughage) that increases the bulk of the intestinal contents, which aids the function of the large intestine. Evidence suggests that high-fiber diets reduce the risk of certain colon disorders, such as diverticulitis and colon cancer.

Carbohydrates are used mostly as an energy source, with glucose as the primary carbohydrate molecule used in cellular respiration. Recall from chapter 10 that the hormone insulin plays a crucial role in moving glucose into cells. Most cells can live by obtaining energy from fatty acids or amino acids via aerobic respiration, but some cells, notably neurons, are dependent upon a steady supply of glucose. For this reason, the functions of the nervous system decline if the concentration of blood glucose decreases.

The liver, along with the hormones insulin and glucagon, is involved in the regulation of glucose concentration in the blood. In response to insulin, excess glucose is converted into glycogen for storage primarily in the liver but also in skeletal muscles. If excess glucose still remains, it is converted into triglycerides and is stored in adipose tissue. When blood glucose level declines, glucagon signals the liver to convert glycogen into glucose. If still more glucose is needed, triglycerides are converted into glycerol and fatty acids. Then, glycerol may be converted into glucose.

Lipids

Lipids include triglycerides, phospholipids, steroids, and lipid-soluble vitamins, (A, D, E, and K), but triglycerides are the most common lipids in the diet. Triglycerides may be either saturated or unsaturated (see chapter 2). Fats and oils contain a mixture of saturated, monounsaturated (one double bond), and polyunsaturated (more than one double bond) fatty acids. Coconut and palm oils, dairy products, and beef fat contain mostly saturated fatty acids. Peanut, olive, and canola oils consist mostly of monounsaturated fatty acids. Safflower, sunflower, and corn oils

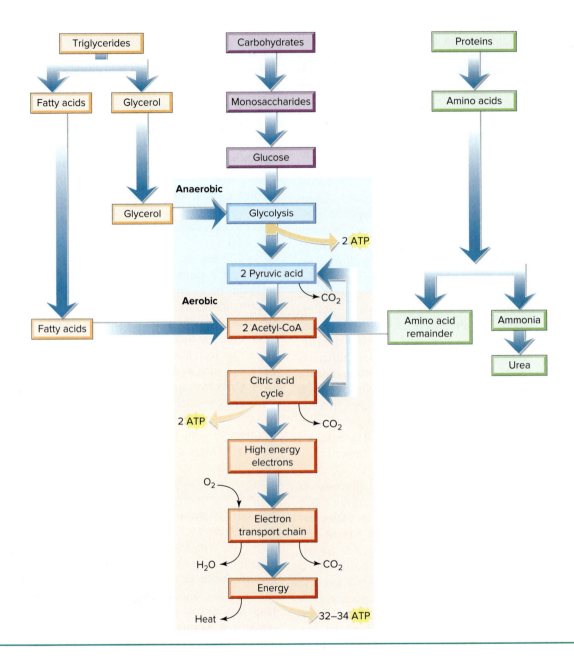

Figure 15.20 Energy Foods and Cellular Respiration.
Lipids, carbohydrates, and proteins may be broken down to release energy for ATP production. Note the three major steps of cellular respiration.

contain mostly polyunsaturated fatty acids. Cholesterol is present in dairy products, red meats, and egg yolks.

Lipids are essential components of the diet, although excessive amounts are not desirable. Phospholipids form the major portion of plasma membranes and the myelin sheaths of neurons. Triglycerides are important energy sources for many cells, including liver and skeletal muscle cells. Excess triglycerides are stored in adipose tissue, where they form the largest energy reserve in the body.

While cholesterol is not used as an energy source, it forms parts of plasma membranes and is used in the synthesis of bile salts and steroid hormones. The liver helps to regulate the concentration of triglycerides and cholesterol in the blood.

Lipids are hydrophobic and do not dissolve in the aqueous blood plasma. This poses a problem for their transport to and from body cells. To overcome this problem, lipids are combined with alpha and beta globulins to form complexes called *lipoproteins* (see chapter 11).

A lipoprotein has a core of triglycerides and cholesterol with a coating of protein, making the complex soluble in blood plasma. Lipoproteins are classified by their density. The greater the proportion of protein, the greater the density of a lipoprotein.

Low-density lipoproteins (LDLs), the "bad" cholesterol, are formed in the liver and transport cholesterol and triglycerides to body cells, including cells of arteries where dangerous plaques can form. *High-density lipoproteins (HDLs)*, the "good" cholesterol, are formed in the liver as almost empty shells of protein. They pick up cholesterol from cells throughout the body as they circulate and return the cholesterol to the liver. The cholesterol is then disposed of through bile, which helps decrease the risk of plaque formation in arteries. Thus LDLs transport cholesterol from the liver to body cells, and HDLs transport cholesterol from body cells to the liver.

Proteins

Prime sources of proteins include red meat, poultry, fish, milk products, eggs, nuts, cereals, and legumes. Of the 20 kinds of amino acids composing proteins, 8 cannot be synthesized from other amino acids by the liver. These eight amino acids are the **essential amino acids.** All essential amino acids must be present in the body in order for the body to synthesize proteins necessary for healthy growth and maintenance. Animal proteins contain all of the essential amino acids, but plant proteins lack one or more of them. However, if cereals and legumes (e.g., beans, peas) are eaten together, this combination provides all of the essential amino acids.

Amino acids are used by the body primarily to synthesize proteins: plasma proteins, certain hormones, enzymes, and proteins that form structural components of cells. However, they may be used as an energy source in aerobic respiration if there is a deficient supply of glucose or triglycerides. If there is an excess of amino acids, they may be converted into glucose or a triglyceride.

For amino acids to be used as an energy source or converted into glucose or a triglyceride, the liver first removes the amine groups ($-NH_2$) so that the remainders of the amino acid molecules are available for these alternative pathways. The amine groups react to form ammonia, a toxic substance. However, the liver converts ammonia into *urea*, a less toxic substance that is released into the blood and is excreted by the kidneys in urine (figure 15.20).

Vitamins

Vitamins are organic molecules that are required in minute amounts for the healthy functioning of the body. They are not energy sources, but they are essential for the utilization of energy foods. Most vitamins act with enzymes to speed up particular metabolic reactions.

Vitamins are classified according to their solubility: water-soluble or lipid-soluble. Water-soluble vitamins include vitamin C and the B vitamins. They are sensitive to heat and easily destroyed by cooking. Lipid-soluble vitamins are vitamins A, D, E, and K. They are more resistant to heat and are not easily destroyed by cooking.

The sources and functions of vitamins are shown in tables 15.4 and 15.5.

Minerals

Minerals are inorganic substances that plants absorb from the soil. They are present in both plant foods and animal foods because animals obtain them by eating plants. Humans need adequate amounts of the seven major minerals noted in table 15.6 but require only trace amounts of other minerals that occur in the body. About 4% of the body weight consists of minerals. Calcium and phosphorus, the most abundant minerals, account for about 75% of the total minerals in the body.

In the body, minerals may be incorporated into organic molecules, be deposited as mineral salts (as in bone), or occur as ions in body fluids. Table 15.6 identifies the sources and functions of the major minerals.

MyPlate: A Visual Guide to Healthy Eating

MyPlate, created by the USDA, is a unique guide to healthy eating when compared to the traditional food pyramids of the past. For a balanced daily diet, the following food group servings are recommended per day: 1½ to 2 cups of fruit, 1½ to 3 cups of vegetables, 4 to 8 ounces of grains, 2 to 6½ ounces of protein, 3 cups of dairy, and 5 to 7 teaspoons of fats and oils. The exact number of servings needed depends upon a person's age

Clinical Insight

Fad diets and TV ads indicate that people are concerned about maintaining or losing weight. Energy stored in foods and energy powering biological processes are measured in calories. A kilocalorie (kcal) or Calorie (with a capital C) is the amount of energy needed to raise one kilogram of water one degree Celsius. Carbohydrates and proteins contain about the same amount of energy, 4 kcal/g. Triglycerides, on the other hand, contain 9 kcal/g. This is why eating fatty foods so easily contributes to weight gain.

Table 15.4 Water-Soluble Vitamins

Vitamin	Sources	Importance
B1 (thiamine)	Meats, eggs, cereals, beans, peas, leafy green vegetables	Required for aerobic respiration and the synthesis of ribose
B2 (riboflavin)	Meats, cereals, leafy green vegetables; formed by colon bacteria	Required for aerobic respiration
B3 (niacin)	Meats, liver, beans, peas, peanuts	Required for aerobic respiration and synthesis of proteins, nucleic acids, and fats
B5 (pantothenic acid)	Meats, fish, cereals, beans, peas, fruits, vegetables, milk; formed by colon bacteria	Required for aerobic respiration, conversion of amino acids and lipids into glucose, synthesis of cholesterol and steroid hormones
B6 (pyridoxine)	Meats, poultry, fish, cereals, beans, peas, peanuts; formed by colon bacteria	Required for protein synthesis and formation of antibodies
B7 (biotin)	Liver, eggs, beans, peas, peanuts; formed by colon bacteria	Required for aerobic respiration and synthesis of fatty acids
B9 (folic acid)	Liver, cereals, leafy green vegetables; formed by colon bacteria	Required for synthesis of DNA, RNA, and healthy formed elements
B12 (cyanocobalamin)	Meats, poultry, fish, milk, eggs, cheese	Required for synthesis of nucleic acids, formation of red blood cells, and cellular respiration of amino acids
C (ascorbic acid)	Citrus fruits, cabbage, tomatoes, leafy green vegetables	Required for synthesis of steroid hormones, absorption of iron, and formation of connective tissues

Table 15.5 Lipid-Soluble Vitamins

Vitamin	Sources	Importance
A	Eggs, milk, butter, liver; green, yellow, and orange vegetables and fruits	Required for healthy skin and mucous membranes; for development of visual pigments in rods and cones; for healthy bones and teeth
D	Formed by skin when exposed to ultraviolet light; fish liver oils, milk, eggs	Required for healthy bones and teeth; promotes absorption of calcium and phosphorus
E	Cereals, vegetable oils, fruits, and vegetables	Prevents oxidation of fatty acids and vitamin A; helps keep plasma membranes intact
K	Synthesized by colon bacteria; leafy green vegetables, cabbage, pork, liver, soybean oil	Required for formation of prothrombin, an enzyme involved in the formation of blood clots

Table 15.6 Major Minerals in the Body

Mineral	Sources	Importance
Calcium (Ca)	Milk, milk products, leafy green vegetables	Forms bones and teeth; required for blood clotting, conduction of action potentials, and muscle contraction
Phosphorus (P)	Meats, cereals, nuts, milk, milk products, legumes	Forms bones and teeth; component of proteins, nucleic acids, ATP, and phosphate buffers
Potassium (K)	Meats, nuts, potatoes, bananas, cereals	Required for conduction of action potentials and muscle contractions
Sulfur (S)	Meats, milk, eggs, legumes	Component of vitamins biotin and thiamine, the hormone insulin, and some amino acids
Sodium (Na)	Table salt, cured meats, cheese	Helps maintain osmotic pressure of body fluids; required for action potential transmission
Chlorine (Cl)	Table salt, cured meats, cheese	Helps maintain osmotic pressure of body fluids; required for formation of HCl in gastric juice
Magnesium (Mg)	Milk, milk products, cereals, legumes, nuts, leafy green vegetables	Required for healthy nerve and muscle functions; involved in ATP–ADP conversions

Figure 15.21 MyPlate.
Source: US Department of Agriculture

and gender. However, many people have trouble maintaining appropriate portion sizes when actually placing food on their plates. Measuring food servings with cups and scales requires discipline and consistency, which many find difficult to maintain for long periods of time. MyPlate creates a visual guide, based on a 9-inch plate, for the portion sizes of each food group at every meal. The visual reference removes the need to measure food servings at meals, allowing individuals to more quickly place correct portion sizes on their plate. For example, one-half of your plate needs to be filled with fruits and vegetables at each meal (figure 15.21).

 Check My Understanding

26. How is energy produced from one glucose molecule through cellular respiration?
27. How are monosaccharides, cholesterol, and amino acids used in the body?

15.12 Disorders of the Digestive System

Learning Objective

21. Describe the major disorders of the digestive system.

Inflammatory Disorders

Appendicitis is an acute inflammation of the appendix. First symptoms include referred pain in the umbilical region and nausea. Later, pain is localized in the right lower quadrant of the abdominal wall. Surgical removal of the appendix is the standard treatment.

Inflammatory bowel disease (IBD) is a group of chronic disorders that cause painfully swollen, inflamed intestines, in addition to abdominal cramps and bloody diarrhea. IBDs result from an overactive immune response that continuously promotes inflammation, which causes permanent changes in the intestinal tissues and places sufferers at greater risk for developing intestinal cancer. There is a genetic predisposition for developing IBD. Treatment includes corticosteroids and immunosuppressant drugs. Surgical removal of the affected portions of the intestines is necessary when drug therapy fails to provide remission. The major IBDs are Crohn disease and ulcerative colitis. *Crohn disease* may affect any portion of the alimentary canal but most often affects the last part of the ileum. *Ulcerative colitis* affects only the large intestine and/or rectum.

Irritable bowel syndrome (IBS) is an intestinal disorder causing abdominal bloating, cramping pain, constipation, and diarrhea. Unlike IBDs, IBS does not permanently damage intestinal tissues, with symptoms usually controlled by dietary and lifestyle changes.

Diverticulitis (dī-ver-tik-ū-lī′-tis) is a disorder of the large intestine. Small, saclike outpockets of the colon often develop from a diet lacking sufficient fiber (bulk). Diverticulitis is the inflammation of these diverticula and may cause considerable pain, bloating, or diarrhea.

Hemorrhoids is a condition in which one or more veins in the anal canal become enlarged and inflamed. Chronic constipation and impaired venous return contribute to the development of hemorrhoids.

Hepatitis, an inflammation of the liver, may be caused by several factors, including viruses, drugs, or alcohol. It is characterized by jaundice, fever, and liver enlargement. Hepatitis is the number-one reason for liver transplants. There are three common viral types of hepatitis in the US.

Hepatitis A (infectious hepatitis) is caused by the hepatitis A virus, which is spread by sexual contact and fecal contamination of food and water. It is highly contagious, but an effective vaccine and treatment are available. Symptoms are usually mild. Recovery takes four to six weeks.

Hepatitis B (serum hepatitis) is a serious infection. It is spread through contaminated blood, saliva, vaginal secretions, and semen. Most people recover completely, but some retain live viruses for years and unknowingly serve as carriers of the disease. Chronic hepatitis may produce cirrhosis or lead to liver cancer. The hepatitis B vaccine is the best protection against hepatitis B infection. It consists of a series of three spaced injections and may be given to persons of all ages, including infants. Chronic hepatitis B patients are treated with antiviral medications.

Hepatitis C, the most common type of hepatitis in the US, is spread primarily by contaminated blood but also by contaminated food and water. Symptoms are usually so mild a person may not even know he or she has it until liver damage shows up, maybe decades later. It can lead to liver failure, cirrhosis, and cancer. In

Clinical Insight

Obesity has become a national health concern. Studies have shown that obesity is a significant health risk, perhaps second in importance only to smoking. Obesity results from consuming food containing more kilocalories than is necessary to meet the body's energy needs. Underlying causes of obesity include heredity, eating habits, inadequate exercise, and poor diet. People who are genetically predisposed toward obesity have a very difficult time controlling their weight. Excess kilocalories are stored as fats in adipose tissue. An excess of 3,500 kcal equals one pound of fat. The only way to get rid of stored fat is to consume fewer kilocalories than needed so the stored fat will be mobilized and broken down. Exercise increases the body's need for energy, which helps burn fat more rapidly.

An effective way to determine if a person is overweight or obese is to calculate the **body mass index (BMI)** and measure the waist. It is calculated by dividing a person's weight in kilograms by the square of his or her height in meters: $BMI = Wt/Ht^2$. To determine the weight in kilograms, divide the weight in pounds by 2.2. To determine the height in meters, divide the height in inches by 39.37. Someone with a BMI higher than 25 to 27 is considered overweight, and a person whose BMI has a value of 30 or more is considered obese. According to the National Institutes of Health, waist measurements over 35 inches in females and 40 inches in males indicate obesity.

---

some cases, the symptoms are mild and short-lived so medications are usually not needed. Most cases progress to the chronic form of the disease and require antiviral medications. There is no hepatitis C vaccine currently available.

Periodontal disease refers to a variety of conditions characterized by inflamed and bleeding gingivae, in addition to degeneration of the gingivae, cement, periodontal ligaments, and the dental alveoli which causes loosening of the teeth.

Peritonitis is the acute inflammation of the peritoneum that lines the abdominal cavity and covers abdominal organs. It may result from bacteria entering the peritoneal cavity due to contamination during accidents or surgery or by a ruptured intestine or appendix.

Noninflammatory Disorders

Cirrhosis of the liver is characterized by scarring, which results from dense irregular connective tissue replacing destroyed hepatocytes. It may be caused by hepatitis, alcoholism, nutritional deficiencies, or liver parasites.

Constipation is a condition in which defecation is difficult because feces are harder and drier than average. This results from feces remaining in the colon for a longer period of time, which allows more water to be absorbed.

Dental caries—tooth decay or "cavities"—result from the excess acid produced by certain microorganisms that live in the mouth and use food residues for their nutrients. Residues of carbohydrates, especially sugars, nurture these microorganisms and increase their acid production.

Diarrhea is the production of watery feces due to the unusually rapid movement of chyme residue through the colon, which decreases the amount of time available to absorb water. Increased peristalsis may result from a number of causes, including inflammation and chronic stress.

Eating disorders result from an obsessive concern about weight control, especially among young adult females. There are two major types of eating disorders: anorexia nervosa and bulimia.

Anorexia nervosa is self-imposed starvation that results in malnutrition and associated physiological changes. Patients with this disorder see themselves as overweight, although others see them as very thin. Death can occur due to the complications of prolonged starvation.

Bulimia is characterized by frequent overeating and purging by self-induced vomiting. Fears of being overweight, depression, and stress are associated factors. The exact cause is unknown. Bulimia may lead to such complications as an imbalance of electrolytes, erosion of tooth enamel by stomach acids, and constipation.

Gallstones result from crystallization of cholesterol in bile. They commonly occur in the gallbladder, but they may be carried into the bile duct where they block the flow of bile. Severe pain and often jaundice accompanies such blockage. Treatment may include drugs that dissolve the gallstones, shock-wave therapy to break up the stones, or surgical removal of the gallstones and gallbladder.

Chapter Summary

15.1 Introduction to the Digestive System
- Nutrients are needed by body cells to carry out vital functions. They include carbohydrates, proteins, lipids, vitamins, minerals, and water.
- Digestion includes all of the mechanical and chemical processes that convert large, nonabsorbable nutrient molecules into small, absorbable nutrient molecules.
- Absorption is the process by which nutrients pass from the alimentary canal into the blood.
- The structures of the alimentary canal are the esophagus, stomach, small intestine, and large intestine.
- The accessory organs of the digestive system include the teeth, tongue, salivary glands, liver, gallbladder, and pancreas.

15.2 Digestion: An Overview
- Digestion involves both mechanical and chemical processes. Mechanical digestion is the physical process of breaking food into smaller particles. Chemical digestion is an enzymatic process that converts nonabsorbable nutrient molecules into absorbable nutrient molecules.
- A number of different digestive enzymes are required for the chemical digestion of food molecules.

15.3 Alimentary Canal: General Characteristics
- The alimentary canal is the long tube through which food passes from the esophagus to the anus. Digestion and absorption of nutrients occur in some portions of the alimentary canal.
- The wall of the alimentary canal is composed of four layers. From outermost to innermost, they are the serosa, muscular layer, submucosa, and mucosa.
- Luminal contents are moved through the alimentary canal by peristaltic contractions. Segmentation mixes luminal contents with digestive secretions.

15.4 Mouth
- The mouth is surrounded by the cheeks, palate, and tongue. Teeth are embedded in the dental alveoli of the maxilla and mandible.
- The tongue is used to manipulate food during chewing and swallowing and to speak. It is the main location for taste buds. It produces lingual lipase, which aids in digesting triglycerides into fatty acids and glycerides within the stomach.
- Humans have two sets of teeth: deciduous and permanent. There are 32 permanent teeth divided into four types: incisor teeth, canine teeth, premolar teeth, and molar teeth.
- A tooth is composed of two major parts: a crown covered with enamel and a root embedded in a dental alveolus. Dentin forms most of the tooth. The centrally located pulp cavity contains blood vessels and nerves. The root is anchored to bone by cement and periodontal ligaments.
- Three pairs of salivary glands (parotid, submandibular, and sublingual) secrete saliva into the mouth. Salivary secretion is under reflexive neural control. Saliva cleanses and lubricates the mouth and binds food together.
- Saliva contains salivary amylase, which breaks down starch and glycogen into maltose, and lysozyme, which kills certain bacterial types.
- Lingual lipase, which is secreted by the tongue, acts on triglycerides and breaks them into monoglycerides and fatty acids within the stomach.

15.5 Pharynx and Esophagus
- The pharynx connects the oral and nasal cavities with the esophagus and larynx. Its digestive function is to transport food from the mouth to the esophagus.
- The swallowing reflex causes the epiglottis to cover the laryngeal opening, directing food into the esophagus.
- Pharyngeal, palatine, and lingual tonsils are located near the entrance to the pharynx.
- The esophagus carries food by peristalsis from the pharynx to the stomach.
- The lower esophageal sphincter relaxes to let food enter the stomach.

15.6 Stomach
- The stomach is a pouchlike enlargement of the alimentary canal. It is located in the left upper quadrant of the abdominopelvic cavity and consists of four regions: cardia, fundus, body, and pyloric part.
- The pyloric sphincter relaxes to allow chyme to enter the duodenum.
- The stomach mucosa contains gastric glands that secrete gastric juice, whose components include hydrochloric acid, pepsin, rennin, gastric lipase, and intrinsic factor. Gastric juice converts food into chyme.
- The secretion of gastric juice is regulated by neural reflexes and the hormones gastrin and cholecystokinin, which are produced by the gastric and intestinal mucosae, respectively.
- Pepsin acts on proteins and breaks them into peptides.
- Rennin is a gastric enzyme secreted by infants. It curdles milk, which aids digestion of milk proteins.
- In infants, gastric lipase breaks down triglycerides into fatty acids and monoglycerides.
- Gastric juice contains intrinsic factor, which is necessary for absorption of vitamin B12.
- Little absorption occurs in the stomach.

15.7 Pancreas
- The pancreas is located adjacent to the duodenum and pyloric part of the stomach. The pancreas secretes pancreatic juice, which is carried to the duodenum by the pancreatic duct. The hormones secretin and cholecystokinin from the intestinal mucosa stimulate the secretion of pancreatic juice.

- Pancreatic digestive enzymes act on each of the three types of energy food. Pancreatic amylase breaks down starch and glycogen into maltose. Pancreatic lipase breaks down triglycerides into monoglycerides and fatty acids. Trypsin breaks down proteins into peptides.

15.8 Liver

- The liver is located in the right upper quadrant of the abdominopelvic cavity.
- Hepatic lobules are the structural and functional units of the liver.
- The liver performs many functions. It plays a critical role in the metabolism of carbohydrates, lipids, and proteins. It stores fat, glycogen, iron, and lipid-soluble vitamins. It detoxifies drugs and toxins within the blood. It phagocytizes worn-out blood cells and bacteria. It produces and secretes plasma proteins, heparin, and bile.
- Bile is continuously secreted by the liver. Between meals, bile is stored in the gallbladder. It is released when the hormone cholecystokinin from the intestinal mucosa stimulates contraction of the gallbladder. Bile is carried to the duodenum by the bile duct.
- Bile emulsifies fats, which aids their digestion by lipases.

15.9 Small Intestine

- The small intestine occupies much of the abdominopelvic cavity and consists of three regions: duodenum, jejunum, and ileum. Most of the digestive processes and absorption of nutrients occur in the small intestine.
- The mucosa contains numerous intestinal glands that secrete intestinal juice and intestinal villi that absorb nutrients. Secretion of intestinal juice is activated by a neural reflex.
- The brush border enzymes maltase, sucrase, and lactase act on corresponding disaccharides to form the monosaccharides: glucose, fructose, and galactose. Brush border peptidases convert peptides into amino acids.
- End products of digestion are absorbed into the intestinal villi. Monosaccharides, amino acids, water-soluble vitamins, minerals, and very small fatty acids cross the epithelium and enter the blood capillaries of the intestinal villi. Large fatty acids, monoglycerides, phospholipids, cholesterol, and lipid-soluble vitamins are transported in micelles to the intestinal lining and are absorbed into the epithelial cells. Monoglycerides and fatty acids are then recombined to form triglycerides. Clusters of triglycerides, other lipids, and lipid-soluble vitamins are coated with protein to form chylomicrons, which enter the lacteals of the intestinal villi.
- Undigested and unabsorbed materials exit the small intestine and enter the large intestine through the ileal orifice.

15.10 Large Intestine

- The large intestine consists of the cecum, colon, rectum, and anal canal. The appendix is an appendage of the cecum.
- The large intestine is gathered into a series of pouches by the taeniae coli. Its mucosa lacks villi and secretes only mucus.
- The absorption of water and the formation and expulsion of feces are major functions of the large intestine. Bacteria decompose the undigested materials.
- Mass peristaltic movements propel the feces into the rectum, initiating the defecation reflex, which relaxes the internal anal sphincter. Voluntary relaxation of the external anal sphincter allows expulsion of the feces.

15.11 Nutrients: Sources and Uses

- Nutrients include carbohydrates, lipids, proteins, vitamins, minerals, and water. The liver is involved in the metabolism of many nutrients.
- Essential nutrients are nutrients that must be consumed because they cannot be synthesized by the body.
- Cellular respiration consists of glycolysis, the citric acid cycle, and an electron transport chain. Respiration of one molecule of glucose yields 36 to 38 ATP.
- End products of carbohydrate, lipid, and protein digestion may be used in aerobic respiration either directly or after modification.
- Dietary carbohydrates come primarily from plant foods. Cellulose is an indigestible polysaccharide that provides fiber in the diet.
- The liver regulates the concentration of glucose in the blood. Excess glucose is converted into glycogen or triglycerides for storage. These reactions may be reversed to release more glucose into the blood.
- Dietary lipids are mostly triglycerides that may be either saturated or unsaturated. Fats and oils contain a mixture of saturated, monounsaturated, and polyunsaturated fatty acids. Cholesterol occurs in egg yolks, milk, and meats.
- Phospholipids form important parts of plasma membranes and myelin sheaths of neurons. Triglycerides are an energy source for many cells. Excess triglycerides are stored in adipose tissue. The liver helps to regulate the concentration of triglycerides and cholesterol in the blood.
- Dietary proteins occur in meats, milk, eggs, cereals, nuts, and legumes. The eight essential amino acids cannot be synthesized by the body. Only animal proteins contain all of the essential amino acids.
- Amino acids are used primarily to synthesize protein in the body. These proteins form plasma proteins, enzymes, certain hormones, and structural parts of cells. Amino acids may have an amine removed by the liver and be used to form glucose or fat or be used as an energy source by cells.
- Vitamins are organic molecules required in minute amounts for healthy functioning of the body. Water-soluble vitamins include the B vitamins and vitamin C. Lipid-soluble vitamins include vitamins A, D, E, and K.
- Vitamins act with enzymes to speed up essential chemical reactions, such as cellular respiration, in cells.
- Minerals are inorganic substances that are necessary for the healthy functioning of the body. Minerals are obtained by plants from the soil and are passed on to animals, including humans, eating the plants.

- Many minerals are a part of organic molecules in the body. Other minerals are deposited as salts in bones and teeth, and some occur as ions in body fluids.
- Seven major minerals are required in moderate amounts in the diet: calcium, phosphorus, potassium, sulfur, sodium, chloride, and magnesium. Other minerals of the body are required in trace amounts.
- MyPlate is a visual guide to help people regulate the portion sizes for each food group at meals.

15.12 Disorders of the Digestive System
- Inflammatory disorders include appendicitis, inflammatory bowel disease (IBD), irritable bowel syndrome (IBS), diverticulitis, hemorrhoids, hepatitis, periodontal disease, and peritonitis.
- Other disorders include eating disorders (anorexia nervosa and bulimia), cirrhosis, constipation, dental caries, diarrhea, and gallstones.

Improve Your Grade

Connect Interactive Questions Reinforce your knowledge using multiple types of questions: interactive, animation, classification, labeling, sequencing, composition, and traditional multiple choice and true/false.

SmartBook Proven to help students improve grades and study more efficiently, SmartBook contains the same content within the print book but actively tailors that content to the needs of the individual.

Anatomy & Physiology REVEALED® Dive into the human body by peeling back layers of cadaver imaging. Utilize this world-class cadaver dissection tool for a closer look at the body anytime, from anywhere.

CHAPTER 16

Urinary System

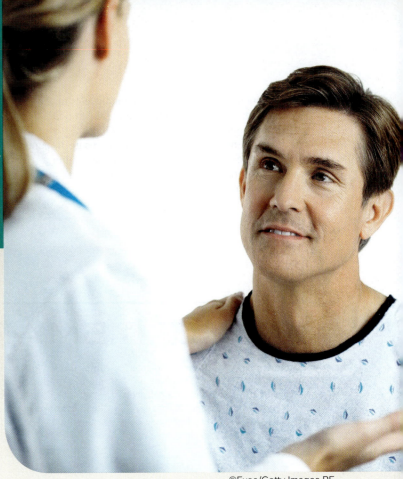
©Fuse/Getty Images RF

Because of his family history of hypertension (or high blood pressure), Peter lives an "anti-hypertensive" lifestyle to reduce his chances of developing the disorder. He engages in regular cardiovascular exercise to maintain a healthy body weight, in addition to avoiding canned and processed foods to reduce his dietary sodium intake. He even meditates regularly to help manage the stress effects of everyday life. However, Peter is surprised at his annual physical when he is diagnosed with hypertension at the age of 39. After discussing all of the available treatment options, Peter opts to try the diuretic Lasix in conjunction with his regular exercise regimen and diet. However, he does not understand why changing his kidney function will help to manage his blood pressure. The doctor explains that Lasix causes the kidneys to increase urine output, which will result in a decrease in Peter's overall blood volume. By decreasing his blood volume, his blood pressure will also decrease. If decreasing his blood pressure will decrease his risk of stroke or heart attack, Peter decides he is more than willing to use the restroom more often during the day.

CHAPTER OUTLINE

16.1 Overview of the Urinary System

16.2 Functions of the Urinary System

16.3 Anatomy of the Kidneys
- Gross Anatomy
- Microscopic Anatomy
- Types of Nephrons
- Renal Blood Supply
- Juxtaglomerular Complex

16.4 Urine Formation
- Glomerular Filtration
- Tubular Reabsorption and Tubular Secretion
- Water Conservation
- Characteristics of Urine

16.5 Excretion of Urine
- Ureters
- Urinary Bladder
- Urethra
- Micturition

16.6 Maintenance of Blood Plasma Composition
- Water and Electrolyte Balance
- Acid–Base Balance

16.7 Disorders of the Urinary System
- Inflammatory Disorders
- Noninflammatory Disorders

Chapter Summary

Module 13
Urinary System

SELECTED KEY TERMS

Acidosis Condition of arterial blood below pH 7.35.
Alkalosis Condition of arterial blood above pH 7.45.
Glomerular filtrate The fluid that enters the glomerular capsule during glomerular filtration.
Glomerular filtration The forcing of water and small solutes from the blood plasma in a glomerulus into a glomerular capsule.
Glomerulus (glomus = ball) The tuft of capillaries enveloped by the glomerular capsule.
Juxtaglomerular complex (juxta = next to) Specialized cells of the afferent and efferent glomerular arterioles and ascending limb of the nephron loop that are involved in controlling glomerular blood pressure.
Micturition (micture = to urinate) Urination.
Nephron (nephros = kidney) The structural and functional unit of the kidneys.
Peritubular capillaries (peri = around) Capillaries surrounding the cortical portion of a renal tubule.
Renal corpuscle (ren = kidney) The portion of a nephron composed of a glomerulus and its enveloping glomerular capsule.
Renal tubule The portion of a nephron composed of a proximal convoluted tubule, a nephron loop, a distal convoluted tubule, and the collecting duct.
Tubular fluid The fluid within the renal tubule.
Tubular reabsorption The movement of substances from the tubular fluid into the blood plasma.
Tubular secretion The movement of substances from the blood plasma into the tubular fluid.
Vasa recta (rectus = straight) Vessels surrounding the medullary portion of the nephron loops.
Water conservation Process of water reabsorption in the distal convoluted tubule and collecting duct, which prevents dehydration.

16.1 Overview of the Urinary System

Learning Objective

1. List the components of the urinary system.

The *urinary system* consists of the kidneys, ureters, urinary bladder, and urethra. The paired *kidneys* maintain the composition and volume of body fluids by removing wastes and excess substances in the formation of **urine,** the fluid waste produced by the kidneys. *Ureters* are slender tubes that carry urine from the kidneys to the *urinary bladder* for temporary storage. Urine is carried from the urinary bladder and is expelled from the body through the *urethra* (figure 16.1).

Check My Understanding

1. What organs compose the urinary system, and what are their functions?

16.2 Functions of the Urinary System

Learning Objectives

2. Describe the general functions of the urinary system.
3. Explain how nitrogenous wastes are kept within healthy limits in body fluids.

The metabolic activities of body cells produce a number of waste materials that tend to change the balance of water and dissolved substances in body fluids. The basic function of the urinary system is to maintain the volume and composition of body fluids within healthy limits.

1. **Maintenance of body fluid composition.** One major function of the kidneys is to keep the volume and composition of blood plasma at homeostasis. This is accomplished by balancing of the concentration of water and electrolytes, in addition to blood pH, through the formation of urine.
2. **Maintenance of blood pressure.** Whenever a kidney senses a decrease in blood pressure, it secretes **renin.** Renin is an enzyme that triggers the **renin-angiotensin mechanism,** which increases blood pressure.
3. **Secretion of erythropoietin.** When the blood oxygen falls below a healthy level, the kidneys release more erythropoietin, which stimulates RBC formation by red bone marrow. The increase in RBC number helps increase the blood oxygen level.
4. **Conversion of vitamin D.** In response to parathyroid hormone (PTH), the kidney converts inactive vitamin D to its active form. Active vitamin D is important in maintaining the blood Ca^{2+} level.
5. **Excretion of nitrogenous wastes.** The kidneys do not remove all nitrogenous wastes but keep their concentrations in the blood within tolerable

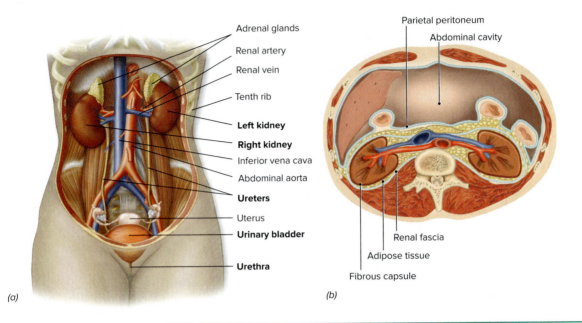

Figure 16.1 The Urinary System.
(a) The urinary system consists of two kidneys, two ureters, a urinary bladder, and a urethra. (b) Transverse section showing the retroperitoneal position of the kidneys and their surrounding support structures. **AP|R**

limits. The primary nitrogenous wastes produced by cellular metabolism are urea, uric acid, and creatinine (see table 16.1).

- **Urea** is a waste product of amino acid metabolism. In order for amino acids to be used as an energy source in aerobic respiration or converted into glucose or fat, the liver removes the amine ($-NH_2$) groups from them. The amine groups react to form ammonia, which is converted to the less toxic urea by the liver.
- **Uric acid** is a waste product of nucleic acid metabolism. An abnormally elevated concentration of uric acid in the blood and the deposition of uric acid crystals in joints are characteristic of a hereditary disorder called *gout*. Joints of the hands and feet are often the sites of uric acid deposition, which produces inflammation and severe pain.
- **Creatinine** is a waste product of muscle metabolism and, specifically, the breakdown of creatine phosphate.

 Check My Understanding

2. What are the general functions of the urinary system?

Table 16.1 Concentrations of Selected Chemicals in Blood, Glomerular Filtrate, and Urine*

Chemicals	Blood (g/l)	Glomerular Filtrate (g/l)	Urine (g/l)
Protein	44.4	0.0	0.0
Chloride (Cl^-)	3.5	3.5	6.3
Sodium (Na^+)	3.0	3.0	3.8
Bicarbonate (HCO_3^-)	1.7	1.7	0.4
Glucose	1.0	1.0	0.0
Urea	0.2	0.2	25.0
Potassium (K^+)	0.2	0.2	5.0
Uric acid	0.05	0.05	0.8
Creatinine	0.01	0.01	1.5

*Based on 180 liters of glomerular filtrate and 1.25 liters of urine produced in a 24-hour period.

16.3 Anatomy of the Kidneys

Learning Objectives

4. Describe the structure and blood supply of the kidney.
5. Describe the structure and functions of a nephron.

The *kidneys* are reddish brown, bean-shaped organs located on both sides of the vertebral column in the retroperitoneal space behind the abdominal cavity (figure 16.1b). The kidneys are located between the levels of the twelfth thoracic vertebra and the third lumbar vertebra and are partially protected by the floating ribs. Each kidney is protected by three layers of connective tissue. A thin *fibrous capsule* tightly envelops each kidney, supporting the soft inner tissues. A thick layer of adipose tissue serves as a cushioning shock absorber, and a fibrous *renal fascia* attaches each kidney to the abdominal wall.

Gross Anatomy

Each kidney is convex laterally and concave medially with a medial indentation called the *hilum*. Blood vessels, lymphatic vessels, nerves, and the ureter enter or exit at the hilum. An adult kidney is about 12 cm long, 7 cm wide, and 2.5 cm thick.

The inner macroscopic anatomy of a kidney is best observed in frontal section, as shown in figure 16.2a. Two functional regions of the kidney are evident: the renal cortex and the renal medulla. The **renal cortex** is the relatively thin, outer layer. Beneath the renal cortex is the **renal medulla,** which contains the cone-shaped *renal pyramids*. Each renal pyramid has a broad base adjacent to the renal cortex with an apex, or *renal papilla*, that extends inward toward the **renal pelvis,** the most central structure of the kidney. Narrow portions of the renal cortex, the *renal columns*, extend into the renal medulla between the renal pyramids.

The renal papilla of each renal pyramid fits into a funnel-shaped **minor calyx** (kā′-lix; plural, *calyces*), which receives urine from the renal papilla. Two or three minor calyces (kā′-li-sēz) converge to form a **major calyx,** and two or three major calyces merge to form the funnellike renal pelvis. The renal pelvis is contiguous with the ureter. Thus, the pathway of urine from nephrons to ureter is as follows: nephrons → papillary ducts → minor calyces → major calyces → renal pelvis → ureter. Urine is carried by the ureter to the urinary bladder.

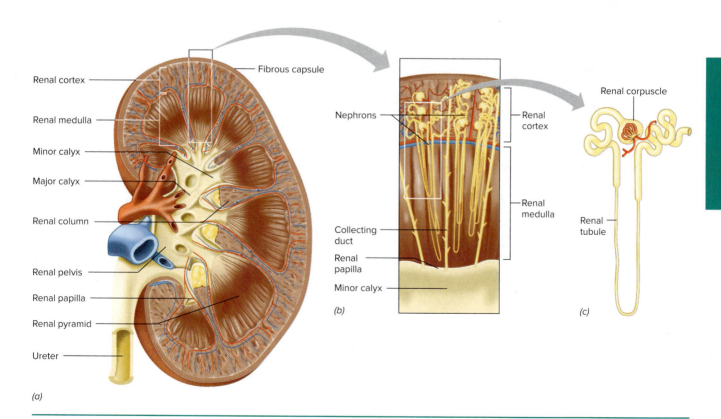

Figure 16.2 Internal Structure of the Kidney.
(a) Frontal section of a kidney. *(b)* A renal pyramid showing the orientation of nephrons and collecting ducts. *(c)* A single nephron. APR

Microscopic Anatomy

The work of the kidneys is performed by microscopic structures called **nephrons** (nef´-rons). Each kidney contains about 1 million nephrons, the structural and functional units of the kidneys. A nephron consists of two major parts: a renal corpuscle and a renal tubule. Figure 16.3 shows the structure of a nephron and its associated blood vessels.

Renal corpuscles are located in the renal cortex of the kidneys. Each renal corpuscle is composed of a **glomerulus** (glō-mer´-ū-lus; plural, *glomeruli*), a tuft of capillaries, which is enclosed in a double-walled **glomerular capsule.** The glomerular capsule is an expanded extension of a renal tubule.

A **renal tubule** leads away from the glomerular capsule and consists of sequential segments. The renal tubule begins with the *proximal convoluted tubule (PCT)*. It leads from the glomerular capsule to the *nephron loop*, the U-shaped second part of the tubule. The descending limb of the nephron loop descends into the renal medulla, and the ascending limb of the nephron loop ascends back into the renal cortex. The ascending limb of the nephron loop is continuous with the *distal convoluted tubule (DCT)*. Several DCTs unite with a single **collecting duct.** Collecting ducts begin in the renal cortex and extend the length of a renal pyramid to its papilla, where the collecting ducts merge to form **papillary ducts** before emptying into a minor calyx.

Types of Nephrons

There are two types of nephrons in the kidney: about 80% are cortical nephrons, and about 20% are juxtamedullary nephrons. The glomerular capsules of *cortical nephrons* are located near the surface of the renal cortex. The nephron loops of these nephrons are located almost entirely in the renal cortex of the kidney. Cortical nephrons are important in adjusting the composition of the urine. In contrast, the glomerular capsules of *juxtamedullary nephrons* are located deep in the renal cortex near the renal medulla. The nephron loops of these nephrons penetrate into the renal medulla. Juxtamedullary nephrons play an important role in regulating water content of the blood plasma.

Renal Blood Supply

The kidneys receive a large volume of blood–1,200 ml per minute, which is about one-fourth of the total cardiac output. Each kidney receives blood via a *renal artery*, which

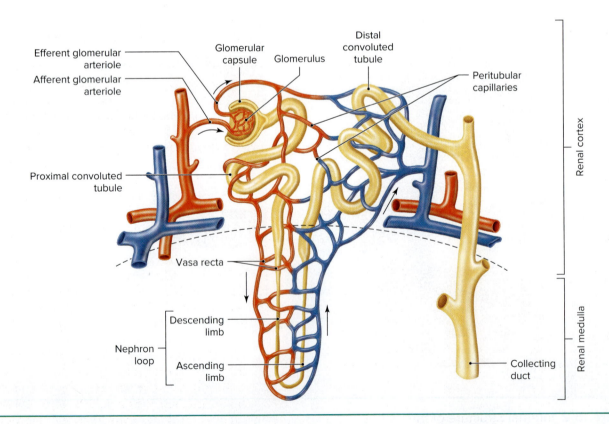

Figure 16.3 A Nephron and Its Associated Blood Vessels.
The nephron has been stretched out to show its parts more clearly. Arrows show the direction of blood flow.

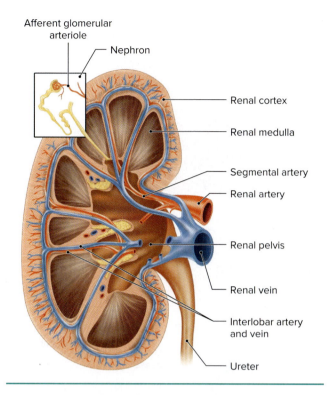

Figure 16.4 Main Branches of the Renal Artery and Renal Vein.

branches from the abdominal aorta. Within each kidney, the renal artery branches to form three or four *segmental arteries*, which branch further to form several *interlobar arteries* that run along the renal columns between the renal pyramids to the renal cortex. These arteries branch to form smaller and smaller arteries and finally form arterioles.

In the renal cortex, **afferent glomerular arterioles** branch from the smallest arteries, and each afferent glomerular arteriole carries blood to a glomerulus. Blood leaves a glomerulus through an **efferent glomerular arteriole**. Note that a glomerulus is a capillary ball between two arterioles. The efferent glomerular arteriole usually leads to **peritubular capillaries,** which surround the cortical portion of the renal tubule. Sometimes the efferent glomerular arteriole leads to the **vasa recta,** which are vessels surrounding the nephron loops and collecting ducts within the renal medulla. Blood from the peritubular capillaries and vasa recta enters venules. The venules merge to form the *interlobar veins*, which finally join to form the renal vein. A *renal vein* carries blood from each kidney to the inferior vena cava (figures 16.3 and 16.4; see figure 16.1).

Juxtaglomerular Complex

The **juxtaglomerular** (juks-tah-glo-mer´-u-lar) **complex** of each nephron is located where the ascending limb of the nephron loop contacts the afferent and efferent glomerular arterioles near the glomerulus (figure 16.5). It consists of two groups of specialized cells: granular cells and the macula densa. *Granular cells,* or juxtaglomerular epithelioid cells, are large smooth muscle cells in the walls of the afferent and efferent glomerular arterioles near their attachment to the glomerulus. The *macula densa* consists of rather narrow, tightly packed cells composing the ascending limb where it contacts the afferent and efferent glomerular arterioles. The juxtaglomerular complex helps to regulate blood pressure, as you will see shortly.

✓ Check My Understanding

3. What are the structural and functional differences between cortical and juxtamedullary nephrons?
4. How does blood flow to, through, and away from the nephron?

16.4 Urine Formation

Learning Objectives

6. Compare glomerular filtration, tubular reabsorption, and tubular secretion.
7. Explain how urine is formed.
8. Indicate the typical components of urine.

The formation of urine is a homeostatic mechanism that maintains the composition and volume of blood plasma within healthy limits. In the production of urine, nephrons perform three basic functions: (1) They regulate the concentration of solutes, such as nutrients and ions, in blood plasma, and this also regulates blood pH. (2) They regulate the concentration of water in blood plasma, which, in turn, helps regulate blood pressure. (3) They remove metabolic wastes and excess substances from the blood plasma.

Four processes are crucial to the formation of urine: (1) **Glomerular filtration** moves water and solutes, except plasma proteins, from blood plasma in the glomerulus into the glomerular capsule. The fluid that enters the glomerular capsule is called **glomerular filtrate.** Formed elements are not part of the glomerular filtrate. Once the glomerular filtrate passes from the glomerular capsule into the renal tubule, it is renamed **tubular fluid.** (2) **Tubular reabsorption** removes useful substances from the tubular fluid and returns them to the blood plasma. (3) **Tubular secretion** moves additional wastes and excess substances from the blood plasma into the tubular fluid. (4) **Water conservation** removes water from the tubular fluid of the distal convoluted tubule and collecting duct and returns it to the blood plasma

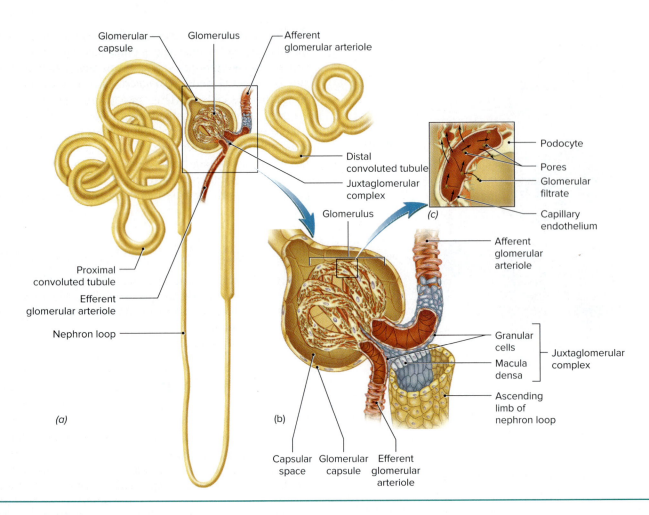

Figure 16.5 Juxtaglomerular Complex.
(a) The juxtaglomerular complex is located where the ascending limb of the nephron loop contacts the afferent and efferent glomerular arterioles near a glomerulus. (b) Enlargement showing the granular cells and macula densa, which compose the juxtaglomerular complex. (c) Magnification of the pores through which glomerular filtration occurs.

(figure 16.6). Once tubular fluid leaves the collecting duct and enters the papillary duct, it can no longer be modified and is thus called urine.

Glomerular Filtration

Urine formation starts with glomerular filtration, a process that forces some of the water and dissolved substances in blood plasma from the glomeruli into the glomerular capsules. Two major factors are responsible for glomerular filtration: (1) the increased permeability of glomerular capillary walls and (2) the elevated blood pressure within the glomeruli.

Glomerular capillaries are much more permeable to substances in the blood plasma than are other capillaries because their walls contain numerous pores, as shown in figure 16.5c. These pores allow water and most dissolved substances to easily pass through the capillary walls into the glomerular capsules. Unlike other capillaries, glomerular capillaries are enveloped by specialized cells called *podocytes*, which have numerous fingerlike cellular extensions that wrap around the capillaries. Podocytes help prevent plasma proteins and formed elements from entering glomerular capsules.

The elevated glomerular blood pressure is caused by the diameter of the efferent glomerular arteriole being smaller than that of the afferent glomerular arteriole. Because blood can enter a glomerulus at a faster rate than it can leave it, the greater blood volume within the glomerulus creates an increase in blood pressure. The glomerular blood pressure provides the force for glomerular filtration.

The result of glomerular filtration is the production of a glomerular filtrate consisting of the same substances that compose blood plasma, except for plasma proteins that are too large to pass through the pores of the glomerular capillaries. Because glomerular filtration is a nonselective process, the concentrations of these substances are the same in both blood plasma and glomerular filtrate (table 16.1).

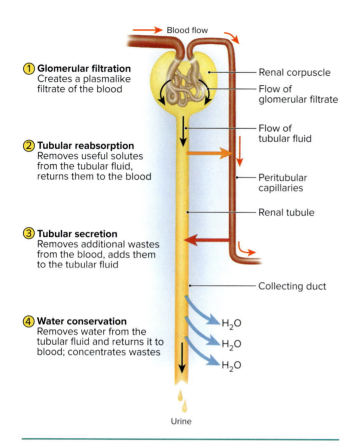

Figure 16.6 The Major Steps of Urine Formation.

Glomerular Filtration Rate

Glomerular filtration rate (GFR) is about 125 ml per minute, or 7.5 liters per hour. This means that the entire volume of blood is filtered every 40 minutes! In 24 hours, about 180 liters (nearly 45 gallons) of glomerular filtrate is produced. However, most of the glomerular filtrate is reabsorbed, as you will see shortly.

Maintenance of a relatively stable GFR is necessary for healthy kidney function. The GFR varies directly with glomerular blood pressure, which, in turn, is primarily determined by systemic blood pressure. The GFR is regulated by three homeostatic processes: renal autoregulation, sympathetic control, and the renin-angiotensin mechanism. These processes operate primarily by controlling the diameter of afferent glomerular arterioles to keep the GFR within healthy limits.

Renal autoregulation is the mechanism that keeps the GFR within healthy limits, without extrinsic neural or hormonal control, in response to moderate variations in systemic blood pressure. One way this occurs is by the response of smooth muscle in the afferent glomerular arteriole wall to blood pressure changes. If the afferent glomerular arteriole is stretched by increased blood pressure, it contracts; the smaller diameter of the afferent glomerular arteriole means less blood volume enters the glomerulus, thus decreasing blood pressure within the glomerulus. A decline in blood pressure causes the arteriole to dilate, allowing more blood volume to enter the glomerulus. The increased blood volume increases blood pressure, which keeps the GFR stable.

Another autoregulatory mechanism involves the juxtaglomerular complex. The juxtaglomerular complex monitors the GFR by using the macula densa to sense changes in the flow rate and chemical composition in the tubular fluid of the ascending limb. If the GFR increases, tubular fluid moves too quickly through the renal tubule and not enough ions are reabsorbed. The macula densa senses this because Na^+, K^+, and Cl^- concentrations are too high. As a result, the granular cells constrict the afferent glomerular arteriole, decreasing the GFR back to a healthy level. If the GFR declines, tubular fluid moves too slowly through the renal tubule and too many ions are reabsorbed. The macula densa senses this as meaning Na^+, K^+, and Cl^- concentrations are too low. As a result, the afferent glomerular arteriole relaxes to raise the GFR back to a healthy level.

Sympathetic control is a function of the sympathetic part of the autonomic division, which overrides renal autoregulation in times of large systemic blood pressure shifts or during the "fight or flight" response. If a drop in systemic blood pressure is detected, afferent glomerular arterioles are constricted, which decreases glomerular pressure and GFR. This decreases urine formation, which conserves water to maintain healthy blood pressure and volume. If an increase in blood pressure is detected, the afferent glomerular arteriole dilates as a result of the sympathetic control being removed, which increases glomerular pressure and the GFR. This results in an increase in urine production and water excretion to maintain healthy blood pressure and volume.

The *renin-angiotensin mechanism* is triggered when the juxtaglomerular complex detects a reduced GFR and releases the enzyme *renin*. Renin is secreted in response to (1) sympathetic stimulation; (2) a drop in blood pressure in the afferent glomerular arteriole; and (3) detection of a reduction of Na^+, K^+, and Cl^- levels in the tubular fluid in the ascending limb of the nephron loop by the macula densa (see figure 16.5). Renin converts a plasma protein (angiotensinogen), which is formed by the liver, into angiotensin I. Angiotensin I is rapidly converted into angiotensin II by the *angiotensin-converting enzyme (ACE)* released from endothelial cells of capillaries in the lungs and other organs. Angiotensin II constricts the efferent glomerular arterioles to maintain blood pressure in glomeruli, which maintains an adequate GFR in spite of a decline in systemic blood pressure. It also acts to restore blood volume and blood pressure by (1) constricting systemic arterioles; (2) stimulating aldosterone secretion by the adrenal cortex, which promotes the reabsorption of Na^+, which in turn promotes the reabsorption of water by osmosis;

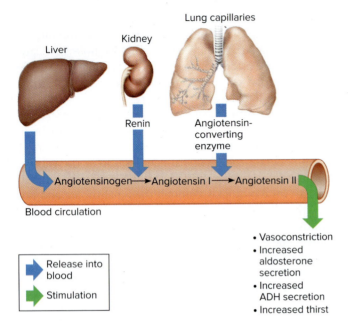

Figure 16.7 The Renin–Angiotensin Mechanism. The multiple actions of angiotensin II help to maintain the GFR and systemic blood pressure.

(3) stimulating secretion of antidiuretic hormone (ADH) by the posterior lobe of the pituitary gland, which promotes water reabsorption; and (4) stimulating thirst, which promotes water intake (figure 16.7).

Natriuretic peptides are secreted by the heart when its chambers are stretched by an excessive blood volume. Natriuretic peptides promote water excretion by increasing GFR, and inhibiting Na^+ reabsorption in the DCT and collecting duct, which results in a decrease in blood volume and, in turn, a decrease in blood pressure.

Tubular Reabsorption and Tubular Secretion

Tubular reabsorption and tubular secretion involve both active and passive transport mechanisms. The effectiveness of passive transport of any substance depends upon (1) the permeability of the renal tubule, peritubular capillaries, and vasa recta to the substance, and (2) the concentration gradient of the substance.

Events in the Proximal Convoluted Tubule

About 65% of the tubular fluid is reabsorbed in the PCT. All nutrients, such as glucose and amino acids, are actively reabsorbed here. Cations, such as those of Na^+, K^+, and Ca^{2+}, are also actively reabsorbed. The active reabsorption of cations causes anions, such as Cl^- and HCO_3^-, to be passively reabsorbed by electrochemical attraction. The reabsorption of these substances increases the osmotic pressure of the blood plasma in the peritubular capillaries and decreases the osmotic pressure of the tubular fluid. This causes water to be passively reabsorbed from the tubular fluid by osmosis (figure 16.8).

Tubular secretion is the process that extracts substances from blood plasma in the peritubular capillaries and secretes them into the tubular fluid in the renal tubule. Metabolic wastes, such as urea and uric acid, and drugs, are removed from the blood in this way. It not only removes unwanted wastes but also helps regulate the pH of body fluids by selectively removing H^+ and HCO_3^-.

Events in the Nephron Loop

Water is passively reabsorbed by osmosis from the descending limb of the nephron loop into the capillaries supplied by the vasa recta. The ascending limb is impermeable to water, but solutes are reabsorbed passively by diffusion from the proximal ascending limb (figure 16.9).

The ascending limb actively pumps Na^+ out of the tubular fluid, and K^+ and Cl^- follow passively into the interstitial fluid. Some K^+ reenter the tubule, but Na^+ and Cl^- remain in the interstitial fluid. The accumulation of ions in the interstitial fluid of the renal medulla establishes a strong osmotic gradient for the reabsorption of water from the descending limb and the collecting duct. Establishing this osmotic gradient is a major function of the nephron loop.

Tubular secretion does not occur in the nephron loop. At the end of the nephron loop, the tubular fluid is quite dilute due to the removal of many solutes. It still contains about 20% of the water and 10% of the salts that were present in the glomerular filtrate.

Events in the Distal Convoluted Tubule

Reabsorption in the DCT and collecting duct is under hormonal control. The active reabsorption of Na^+ from the tubular fluid into peritubular capillaries is promoted by the hormone aldosterone. Passive reabsorption of negative ions, such as Cl^- and HCO_3^-, then occurs by electrochemical attraction. The reabsorption of ions decreases the osmotic pressure of the tubular fluid, which causes water to be reabsorbed into the blood plasma by osmosis. However, the amount of water reabsorbed from the tubular fluid

 Clinical Insight

ACE inhibitors are a group of drugs commonly used to treat hypertension. These drugs help to reduce the blood pressure by decreasing the activity of ACE, resulting in decreased production of angiotensin II. Decreased production of angiotensin II leads to decreased blood pressure.

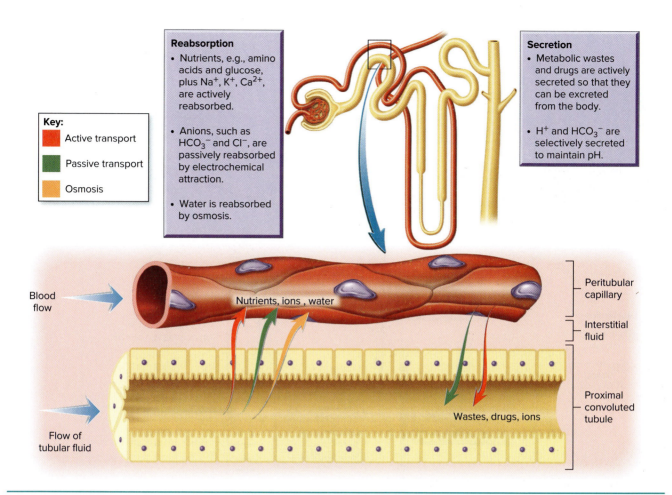

Figure 16.8 Reabsorption and Secretion in the Proximal Convoluted Tubule.

in the DCT is controlled by ADH (figure 16.9). Activating the renin-angiotensin system causes these two hormones to increase sodium and water reabsorption, which increases blood volume and blood pressure.

Parathyroid hormone (PTH) and active vitamin D also affect the DCT. They control the amount of Ca^{2+} reabsorption that takes place in the DCT and are very important in maintaining blood Ca^{2+} homeostasis, as will be discussed further in the "Electrolyte Balance" section of this chapter.

Aldosterone release can also be triggered by an increased blood K^+ level (hyperkalemia). Aldosterone causes the active secretion of K^+ from blood plasma in the peritubular capillaries into the distal convoluted tubule and collecting duct. Hydrogen ions are also actively secreted, as necessary, to maintain the blood pH. When necessary, the DCT is the site of active and passive drug secretion and elimination.

Water Conservation

The tubular fluid entering the collecting duct is approximately isotonic with the blood plasma, so it still contains a lot of water. The concentration of ions in the interstitial fluid surrounding the collecting duct increases as it extends from the DCT in the renal cortex to the renal papilla in the renal medulla. Recall that Na^+, K^+, and Cl^- have been moved into the interstitial fluid by the ascending limb of the nephron loop. This creates a strong osmotic gradient that causes the passive reabsorption of water from the tubular fluid in the DCT and collecting duct by osmosis. The reabsorption of water causes the tubular fluid to become concentrated and hypertonic to blood plasma by the time it exits the collecting duct. The reabsorption of water from the DCT and collecting duct is regulated by ADH, which increases the DCT and collecting duct's permeability to water (figure 16.9). Only 1% of the original glomerular filtrate volume remains as urine.

Table 16.2 summarizes the functions of each segment of a nephron in the formation of urine. AP|R

Characteristics of Urine

The volume of urine produced in a 24-hour period usually is 1.5 to 2.0 liters. In healthy persons, fresh urine is usually clear with a pale yellow to light amber color and

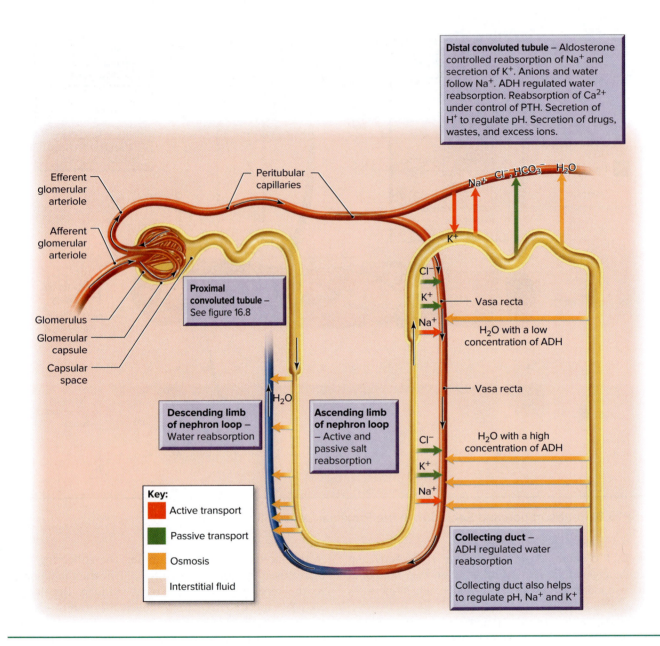

Figure 16.9 Reabsorption and Secretion in the Nephron Loop, Distal Convoluted Tubule, and Collecting Duct.

a characteristic odor. The color is due to the presence of *urochrome*, a substance produced by the breakdown of bile pigments in the intestine. The more concentrated the urine, the darker is its color.

Urine is usually slightly acidic (about pH 6), although the healthy pH range extends from 4.8 to 7.5. Variations in pH usually result from the diet. High-protein diets increase the acidity. Conversely, vegetarian diets tend to make the urine more alkaline.

Urine is more dense than water because of the many solutes it contains. The term *specific gravity* is used to compare how much denser urine is than water. Pure water has a specific gravity of 1.000, and the specific gravity of healthy urine ranges from 1.002 (dilute urine) to 1.030 (concentrated urine). Specific gravity is a measure of the concentration of a urine sample.

The usual solutes of urine have been discussed earlier in this chapter (table 16.1). Substances not typically found in urine are glucose, proteins, formed elements, hemoglobin, and bile pigments. The presence of any of these substances suggests possible pathological conditions. Healthy values of urine components and some indications of abnormal values are listed in Appendix D.

Table 16.2 Summary of the Functions of Nephrons and Papillary Ducts

Structure	Function
Renal Corpuscle	
Glomerulus	Glomerular filtration
	Glomerular blood pressure forces some of the water and dissolved substances (except proteins) from the blood plasma through the pores in the glomerular capillary walls
Glomerular capsule	Receives glomerular filtrate from glomerulus
Renal Tubule	
Proximal convoluted tubule	Active reabsorption of all nutrients, including glucose and amino acids
	Active reabsorption of cations like sodium, potassium, calcium, and magnesium
	Passive reabsorption by electrochemical attraction of anions such as bicarbonate
	Passive reabsorption of water by osmosis
	Active secretion of hydrogen and bicarbonate ions
Nephron loop	
Descending limb	Passive reabsorption of water by osmosis
Ascending limb	Active reabsorption of sodium ions into the interstitial fluid of the renal medulla
	Passive reabsorption of chloride and potassium ions
Distal convoluted tubule	Active reabsorption of calcium ions under the influence of PTH and active vitamin D
	Active reabsorption of sodium ions under the influence of aldosterone
	Passive reabsorption by electrochemical attraction of anions such as chloride and bicarbonate
	Passive reabsorption of water by osmosis under the influence of ADH
	Active secretion of hydrogen ions
	Active secretion of potassium ions under the influence of aldosterone
Collecting duct	Passive reabsorption of water by osmosis under the influence of ADH
	Active reabsorption of sodium ions under the influence of aldosterone
	Active secretion of potassium ions under the influence of aldosterone
Papillary duct	Tubular fluid enters it from the collecting ducts, is renamed urine, and is transported to the minor calyx

 Check My Understanding

5. What are the mechanisms of glomerular filtration, tubular reabsorption, tubular secretion, and water conservation?
6. For each section of a nephron, what substances are reabsorbed and secreted during the formation of the urine entering the minor calyx?

16.5 Excretion of Urine

Learning Objectives

9. Describe the structure and function of the ureters, urinary bladder, and urethra.
10. Describe the control of micturition.

The term *urinary tract* refers collectively to the renal pelvis, the ureters, the urinary bladder, and the urethra. These structures function to carry urine from the kidneys to the external environment. Urine passes from the renal pelvis into the ureter and is carried to the urinary bladder. Urine is voided from the urinary bladder through the urethra.

Ureters

Each **ureter** is a slender tube about 25 cm (10 in) long that extends from a kidney to the urinary bladder. It begins at the kidney with the funnel-shaped renal pelvis and enters the bottom sides of the urinary bladder (figure 16.10; see figure 16.1).

The wall of a ureter is formed of three layers. The outer fibrous layer is composed of dense irregular connective tissue. The middle layer consists of smooth muscle cells that produce peristaltic waves for urine transport. Urine transport is also due in part to gravitational pull and hydrostatic pressure gradients. The inner layer is a mucous membrane that is continuous with that of the renal pelvis and the urinary bladder. A flaplike fold of mucosa in the urinary bladder covers the opening of the ureter, and it functions as a valve that prevents backflow of urine into the ureter.

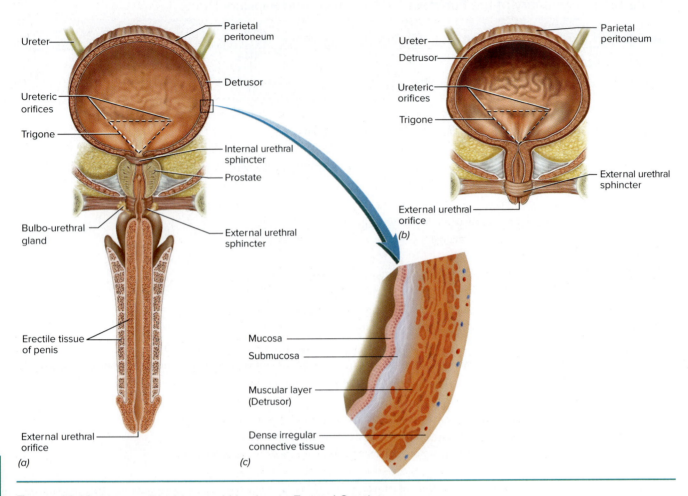

Figure 16.10 Urinary Bladder and Urethra in Frontal Section.
(a) Male. (b) Female. (c) Magnification of the wall of the urinary bladder. AP|R

Clinical Insight

A routine urinalysis is a common clinical test that provides information about kidney function and also about general health of the body. Kidney function is also assessed by two blood tests. The *blood urea nitrogen (BUN)* test evaluates how effectively the kidney removes urea from the blood. The healthy BUN value is 6 to 20 mg per 100 ml of blood. In acute kidney failure and in later stages of chronic kidney failure, BUN may range from 50 to 200 mg per 100 ml of blood. *Blood (serum) creatinine* is another test that assesses kidney effectiveness. The creatinine level in the blood is usually stable (0.6–1.5 mg/100 ml), so an increase indicates a decrease in kidney function.

Urinary Bladder

The **urinary bladder** is a hollow, muscular organ located behind the pubic symphysis within the pelvic cavity. It lies below the parietal peritoneum. The urinary bladder provides temporary storage of urine, and its size and shape vary with the volume of urine it contains. When filled with urine, it is almost spherical as its upper surface expands. When it is empty, its upper surface collapses, giving a deflated appearance.

The inner floor of the urinary bladder contains the *trigone* (trī'-gōn), a smooth, triangular area that contains an opening at each of its angles. The openings of the ureters are located at the two side rear angles, and the opening of the urethra is located in the front angle (figure 16.10).

Four layers compose the wall of the urinary bladder. The innermost layer is the *mucosa*, which is composed of transitional epithelium that is adapted to the repeated stretching of the urinary bladder wall. The epithelium stretches, and its thickness decreases as the urinary bladder fills with urine.

The mucosa is supported by the underlying *submucosa* formed of areolar connective tissue containing an abundance of elastic fibers. Blood vessels and nerves supplying the urinary bladder are present in the submucosa.

Smooth muscle cells compose the third, and thickest, layer. These cells form a muscle called the *detrusor* (dē-trū′-sor). The detrusor is relaxed as the urinary bladder fills with urine, and it contracts as urine is expelled. Cells of the detrusor form an *internal urethral sphincter* at the junction of the urinary bladder and the urethra in males.

The outer layer consists of the parietal peritoneum, but it covers only the upper portion of the urinary bladder. The remainder of the urinary bladder surface is coated with dense irregular connective tissue.

Urethra

The **urethra** is a thin-walled tube that carries urine from the urinary bladder to the external environment. The urethral wall contains smooth muscle cells and is supported by connective tissue. The inner lining is a mucosa that is continuous with the mucosa of the urinary bladder. An *external urethral sphincter*, which is composed of skeletal muscle fibers, is located where the urethra penetrates the pelvic floor.

The female urethra is quite short, about 3 to 4 cm (1.5 in) in length. The external urethral orifice, its outer opening, lies in front of the vaginal orifice. The male urethra is much longer, about 16 to 20 cm (6–8 in) in length, because the urethra runs the length of the penis. The external urethral orifice is at the tip of the penis.

Micturition

Micturition (mik-tū-rish′-un), or urination, is the act of expelling urine from the urinary bladder. Although the urinary bladder may hold up to 1,000 ml of urine, micturition usually occurs long before that volume is attained. When 200 to 400 ml of urine have accumulated in the urinary bladder, baroreceptors in the urinary bladder wall are stimulated and they trigger the *micturition reflex*. This reflex sends parasympathetic action potentials to the detrusor, causing rhythmic contractions. As this reflex continues, it causes the involuntarily controlled internal urethral sphincter to open in males and the person becomes aware of the desire to urinate. Females become aware of the desire to urinate when urine enters into the beginning of the urethra and causes it to stretch. The act of urinating then becomes a consciously controlled process. If the voluntarily controlled external urethral sphincter is relaxed, micturition occurs; if it is not relaxed, micturition is postponed.

Micturition may be postponed by keeping the external urethral sphincter voluntarily closed, and in a few moments the urge to urinate subsides. After more urine enters the urinary bladder, the micturition reflex is activated again, and the urge to urinate returns. Micturition cannot be postponed for long periods of time. After a while, the reflex overwhelms voluntary control and micturition occurs, ready or not.

An infant is not able to be toilet trained until neural development allows control of the external urethral sphincter muscle. Voluntary control is usually possible shortly after two years of age.

> ### Check My Understanding
> 7. How is micturition controlled?

16.6 Maintenance of Blood Plasma Composition

Learning Objectives

11. Explain how water balance is maintained in body fluids.
12. Explain how electrolyte balance is maintained in body fluids.
13. Explain how pH balance is maintained in body fluids.

The composition and volume of blood plasma are affected by diet, cellular metabolism, and urine production. The intake of food and liquids provides the body with water and a variety of nutrients, including minerals, that are absorbed into the blood. Cellular metabolism uses nutrients and produces waste products, including nitrogenous wastes. Urine production retains essential nutrients and minerals in the blood plasma but removes some water along with excess substances and nitrogenous wastes. In healthy people, the kidneys are able to keep the composition and volume of the blood plasma relatively constant in spite of variations in diet and cellular activity.

Water and Electrolyte Balance

Two important components of blood plasma and other body fluids are water and electrolytes, and their concentrations in body fluids must be maintained within healthy limits. Recall that water is the solvent of body fluids in which the chemical reactions of life occur. Recall that electrolytes are substances that form ions when dissolved in water, and they are so named because they can conduct an electric current when dissolved in water. For example, sodium chloride is an electrolyte that forms sodium and chloride ions when dissolved in water.

The concentrations of water and electrolytes in body fluids are interrelated because the concentration of one affects the concentration of the other. For example,

Clinical Insight

Substances that increase the production of urine are known as *diuretics*. Physicians often prescribe a diuretic to reduce the volume of body fluids in patients with edema or hypertension.

the concentration of electrolytes establishes the osmotic pressure that enables water to be reabsorbed by osmosis.

Water Balance

The intake of water is largely regulated by the *thirst center* located in the hypothalamus of the brain. The thirst center is activated when it detects an increase in solute concentration in the blood. It is also activated by angiotensin II when blood pressure declines significantly. An awareness of thirst stimulates water intake to replace water lost from body fluids. Water intake must balance water loss, and this averages about 2,500 ml per day.

The body loses water in several ways, but about 60% of the total water loss occurs in urine. By regulating the volume of water lost in urine, the kidneys have the ability to regulate the concentration of water in blood plasma. In addition, water is lost in the humidified air exhaled from the lungs, in feces, and in perspiration (figure 16.11).

The volume of water lost in urine varies with both the volume of water lost by other means and the volume of water intake. These factors affect the action of the kidneys simultaneously, but we consider them separately to better understand how they influence kidney function.

In general, the more water that is lost through other means, the less water that is lost in urine. For example, if excessive water loss occurs through perspiration or diarrhea, more water is reabsorbed from the renal tubule. The result is a smaller volume of more concentrated urine. Conversely, if water loss through other means is minimal, water reabsorption is reduced, and a larger volume of more dilute urine is produced.

Similarly, the greater the intake of water, the less water is reabsorbed and a larger volume of more dilute urine is produced. Conversely, a lower water intake means more water is reabsorbed and a smaller volume of more concentrated urine is produced.

You can see that regulating water balance is a dynamic process and that water balance is largely controlled by the amount of water reabsorbed from renal tubules into the blood plasma. Whether more or less water is reabsorbed is dependent upon ADH secreted by the posterior lobe of the pituitary gland. ADH promotes water reabsorption by increasing the permeability of the DCT and collecting ducts to water.

When the water concentration of blood is excessive, natriuretic peptides are secreted and ADH secretion declines. The combined effect is that less water is reabsorbed and a greater volume of urine (and water) is excreted. The result is a decrease in water concentration in the blood. Conversely, when the water concentration of blood decreases, natriuretic peptides are not secreted, ADH secretion is increased, more water is reabsorbed, and a smaller volume of urine is produced. ADH minimizes water loss in urine, but it cannot prevent it. Thus, water must be replenished daily by fluid intake.

Electrolyte Balance

Important electrolytes in body fluids include ions of sodium, potassium, calcium, chloride, phosphate, sulfate, and bicarbonate. Electrolytes are obtained from the intake of food and fluids. A craving for salt results when electrolytes are in low concentration in body fluids.

Electrolyte balance is regulated largely by active reabsorption of cations, which, in turn, secondarily controls the passive reabsorption of anions by electrochemical attraction. Sodium ions are the most important ions to be regulated because they compose about 90% of the cations in extracellular fluids. Certain hormones play important roles in maintaining electrolyte balance.

Aldosterone is a hormone that regulates the balance of sodium and potassium ions in the blood plasma by stimulating the active reabsorption of sodium ions and the active secretion of potassium ions by the DCT and collecting duct. Thus, aldosterone causes an exchange of sodium and potassium ions between the tubular fluid and

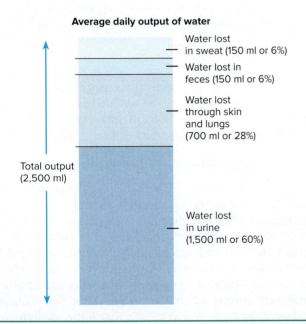

Figure 16.11 Pathways of Water Loss.
Urine formation is the most important process that regulates water loss.

Clinical Insight

If water loss significantly exceeds water intake for several days, extracellular fluids may become more concentrated, causing water to move out of the cells by osmosis. This condition, known as *cellular dehydration*, may lead to serious complications unless the water loss is quickly restored. In serious cases, dehydration may result in fever, mental confusion, or coma.

the blood plasma until the blood concentrations of these two ions return to healthy levels. The adrenal cortex is stimulated to secrete aldosterone by (1) an increase of K^+ in the blood, (2) a decrease of Na^+ in the blood, and (3) angiotensin II. As long as blood concentrations of sodium and potassium ions are within healthy limits, aldosterone is not secreted. In contrast to aldosterone, natriuretic peptides promote the excretion of sodium ions and water by inhibiting sodium reabsorption, and thus osmosis, in the DCT and collecting duct when excess blood volume is detected by the heart.

The blood concentration of Ca^{2+} is regulated mainly by the actions of PTH and active vitamin D. When the blood Ca^{2+} concentration declines, the parathyroid glands are stimulated to secrete PTH. PTH promotes an increase in blood Ca^{2+} by stimulating three different processes: (1) the reabsorption of Ca^{2+} ions from the DCT, (2) the movement of Ca^{2+} from bone tissue into the blood, and (3) the activation of vitamin D. Active vitamin D assists PTH; in addition, it increases the absorption of Ca^{2+} from foods by the small intestine. When the blood Ca^{2+} level returns to a healthy level, PTH secretion is decreased. The lack of PTH is usually sufficient to decrease the blood Ca^{2+} level. Table 16.3 summarizes the effect of hormones that act on the kidneys.

During times of rapid bone remodeling, such as childhood or pregnancy, *calcitonin* is secreted by the thyroid gland. Calcitonin plays an antagonistic role to PTH by promoting the deposition of calcium in bones, which reduces the level of blood calcium.

Acid–Base Balance

The arterial blood pH must be maintained within rather narrow limits–pH 7.35 to pH 7.45–for body cells to function properly. Arterial blood pH below 7.35 is called **acidosis,** and arterial blood pH above 7.45 is called **alkalosis.** Cellular metabolism produces products that tend to upset the acid-base balance. These products, such as lactate, phosphoric acid, and carbonic acid, tend to make the blood more acidic, as shown in figure 16.12.

Acids are substances that release hydrogen ions (H^+) when they are in water, which decreases the pH and increases the acidity of the liquid. Strong acids release more H^+ than weak acids. **Bases** are substances that, when placed in water, release ions that can combine with hydrogen ions, such as OH^- or HCO_3^-. Body fluids contain both acids and bases, and the balance between them determines pH. The balance of acids and bases in the body is regulated by three processes: (1) **buffers,** which act directly in the body fluids; (2) the **respiratory mechanism,** which controls the carbonic acid level; and (3) the **renal mechanism,** which regulates H^+ and HCO_3^- levels.

Buffers

The blood and other body fluids contain chemicals known as buffers (see chapter 2) that prevent significant changes in pH. Buffers are able to combine with or release H^+ ions as needed to stabilize the pH. If the H^+ concentration is excessive, buffers combine with some H^+ to reduce their concentration. Conversely, if too few H^+ are present, buffers release some H^+ to increase their concentration to within healthy limits. In this way, buffers help to keep the blood pH relatively constant.

The **bicarbonate buffer system** relies on a mixture of carbonic acid and HCO_3^-. Carbonic acid (H_2CO_3) forms by the hydration of carbon dioxide and then

Table 16.3 Hormones Acting on the Kidneys

Hormone	Source	Action
Aldosterone	Adrenal cortex	Stimulates reabsorption of Na^+ from the tubular fluid into the blood plasma; stimulates the secretion of K^+ from blood plasma into the tubular fluid
Antidiuretic hormone	Posterior lobe of pituitary gland	Stimulates the reabsorption of water from the tubular fluid into the blood plasma by making the DCT and collecting ducts more permeable to water; decreases the volume of urine produced
Natriuretic peptides	Heart	Inhibit reabsorption of Na^+ from the DCT and collecting ducts
Parathyroid hormone and active vitamin D	Parathyroid glands Liver and kidneys	Stimulates the reabsorption of Ca^{2+} from the tubular fluid into the blood plasma

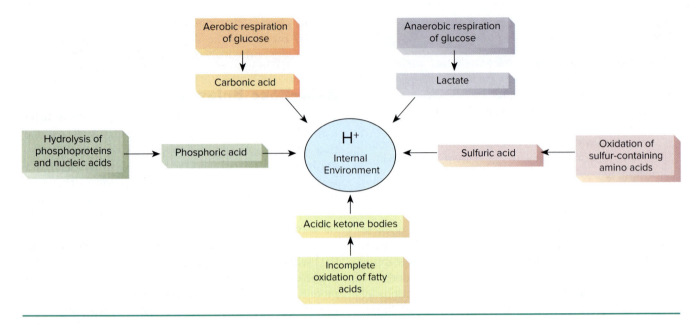

Figure 16.12 Examples of Metabolic Processes That Increase the Hydrogen Ion Concentration of Body Fluids.

Clinical Insight

The kidneys have a tremendous functional reserve. Renal insufficiency becomes evident only after about 75% of the renal functions have been lost. As the development of renal failure progresses, patients must rely on *hemodialysis* as a means of removing wastes and excessive substances from the blood. In hemodialysis, the patient's blood is pumped through selectively permeable tubes that are immersed in a dialyzing solution within a "kidney machine." Nitrogenous wastes and excessive electrolytes diffuse from the blood into the dialyzing solution, while certain needed substances, such as buffers, diffuse from the dialyzing solution into the blood. In this way, the concentrations of wastes and electrolytes in the patient's blood are temporarily restored within healthy limits. Hemodialysis may be required two to three times per week for patients with chronic kidney failure.

An alternative method is called *continuous ambulatory peritoneal dialysis (CAPD)*. In this technique, 1 to 3 liters of dialyzing fluid are introduced into the peritoneal cavity through an opening made in the abdominal wall. Waste products and excessive substances diffuse from blood vessels in the peritoneum into the dialyzing solution, which is drained after two to three hours. This technique is less costly, may be done at home, and allows the patient to move about during the procedure. However, it must be done more frequently than dialysis using a kidney machine, and there is a greater chance of serious infection.

---

dissociates into HCO_3^- and H^+. The bicarbonate buffer system is particularly important in regulating the acid–base balance of extracellular fluids, such as blood. Additionally, as you will soon see, the respiratory mechanism and the renal mechanism directly influence the bicarbonate buffer system.

$$CO_2 + H_2O \leftrightarrow H_2CO_3 \leftrightarrow HCO_3^- + H^+$$

The **phosphate buffer system** relies on a mixture of HPO_4^{2-} and $H_2PO_4^-$. The phosphate buffer system is most important in regulating the acid–base balance of intracellular fluid. The following reaction can proceed to the right to release H^+ and decrease pH, or it can proceed to the left to bind H^+ and increase pH.

$$H_2PO_4^- \leftrightarrow HPO_4^{2-} + H^+$$

The most abundant and powerful buffer system in the body is the **protein buffer system.** Proteins are able to act as buffers because amino acids have acidic (–COOH) side groups, which release H^+ when pH is elevated and decrease the pH, and amino acids also have amine (–NH₂) side groups that bind H^+ when pH is decreased and elevate the pH. Below are the reactions of the protein buffer system.

$$-COOH \rightarrow -COO^- + H^+ \quad \text{and} \quad -NH_2 + H^+ \rightarrow -NH_3$$

Respiratory Mechanism

The respiratory system also plays a significant role in regulating H^+ concentration of body fluids. The respiratory mechanism alters the bicarbonate buffer system by changing the level of CO_2 in the body. Recall that the production of CO_2 results in the formation of carbonic acid, which dissociates to release H^+. When CO_2 production increases and blood pH decreases, the respiratory rhythmicity center in the medulla oblongata stimulates an increase in the rate and depth of breathing to remove the excess CO_2. Conversely, when the H^+ concentration of the blood is decreased, the rate and depth of breathing are decreased until the blood H^+ concentration increases to a healthy level.

Renal Mechanism

The renal mechanism is able to control the bicarbonate buffer system. By selectively excreting excess H^+ or HCO_3^- in urine, kidneys help to maintain a healthy pH of body fluids. The kidneys also have the ability to produce HCO_3^- and H^+ from carbon dioxide in times of shortage.

Check My Understanding

8. How do ADH, aldosterone, PTH, and natriuretic peptides regulate water and electrolyte concentrations in blood?
9. What are the three buffer systems?
10. How does the kidney adjust blood pH?

16.7 Disorders of the Urinary System

Learning Objective

14. Describe the common disorders of the urinary system.

Inflammatory Disorders

Cystitis (sis-tī′-tis) is the inflammation of the urinary bladder. It is often caused by bacterial infection. Females are more prone to cystitis because their shorter urethra makes it easier for bacteria to reach the urinary bladder.

Glomerulonephritis (glō-mer-ū-lō-ne-frī′-tis) is the inflammation of a kidney involving the glomeruli. It may be caused by bacteria or bacterial toxins. The inflamed glomeruli become more permeable, allowing formed elements and proteins to leak into the glomerular filtrate and remain in the urine.

Pyelonephritis (pī-e-lō-ne-frī′-tis) is the inflammation of the renal pelvis and nephrons. If only the renal pelvis is involved, the condition is called *pyelitis*. These infections result from bacteria carried by blood from other places in the body or by migration of bacteria from distal portions of the urinary tract.

Urethritis is the inflammation of the urethra. It may be caused by several types of bacteria, but the bacterium *Escherichia coli* is the most common. Urethritis is more common in females.

Noninflammatory Disorders

Diuresis, or polyuria, is the excessive production of urine. It results from inadequate tubular reabsorption of water and is characteristic of diabetes insipidus and diabetes mellitus.

Renal calculi (kal′-kū-lī), or kidney stones, result from crystallization of uric acid or of calcium or magnesium salts in the renal pelvis. They can cause extreme pain, especially when moving through a ureter by peristalsis. Ultrasound waves can be used to break up the stones, as an alternative to surgery.

Renal failure is characterized by a reduction in urine production and a failure to maintain a healthy volume and composition of body fluids. It may occur suddenly (acute) or gradually (chronic). Renal failure leads to *uremia*, a toxic condition caused by excessive nitrogenous wastes in the blood, and ultimately to *anuria*, a cessation of urine production. Hemodialysis and/or a kidney transplant may be necessary.

Chapter Summary

16.1 Overview of the Urinary System

- The urinary system consists of the kidneys, ureters, urinary bladder, and urethra.
- The kidneys produce urine, and the rest of the urinary system collects and excretes the urine.

16.2 Functions of the Urinary System

- The functions of the urinary system are the maintenance of blood plasma composition, the secretion of renin and erythropoietin, and the excretion of nitrogenous wastes.

- A major function of the kidneys is the removal of excess nitrogenous wastes—urea, uric acid, and creatinine—in order to keep their concentrations in the blood within healthy limits.

16.3 Anatomy of the Kidneys

- The paired kidneys are located on both sides of the vertebral column in the retroperitoneal space behind the abdominal cavity.
- Internal structure of a kidney consists of two recognizable functional parts: an outer renal cortex, and an inner renal medulla composed of renal pyramids.
- Nephrons are the structural and functional units of the kidneys. Each nephron consists of a renal corpuscle and a renal tubule. A renal corpuscle is composed of a glomerulus and a glomerular capsule. A renal tubule is composed of a proximal convoluted tubule, the nephron loop, a distal convoluted tubule, and the collecting duct.
- Collecting ducts merge to form papillary ducts in renal papilla, which in turn empty into a minor calyx of the renal pelvis.
- The blood supply for each kidney is provided by a renal artery, and drainage is through the renal vein.
- An afferent glomerular arteriole brings blood to a glomerulus. Blood exits the glomerulus via an efferent glomerular arteriole and flows through either the peritubular capillaries, which surround the cortical portion of the renal tubule, or the vasa recta, which surround the medullary portion of the nephron loop and collecting duct.
- The juxtaglomerular complex consists of modified cells of the afferent and efferent glomerular arterioles and the ascending limb of the nephron loop at their point of contact.

16.4 Urine Formation

- The process of urine formation regulates the composition and volume of blood plasma by removing excess nitrogenous wastes and surplus substances from the blood plasma.
- Urine is formed by four sequential processes: glomerular filtration, tubular reabsorption, tubular secretion, and water conservation.
- In glomerular filtration, water and dissolved substances (except plasma proteins and formed elements) in blood plasma are filtered from the glomerulus into the glomerular capsule. Glomerular filtration results from the increased permeability of glomerular capillaries and the elevated blood pressure within the glomerulus.
- About 180 liters of glomerular filtrate are formed in a 24-hour period.
- Glomerular filtration rate is proportional to the glomerular blood pressure. Glomerular blood pressure is maintained by mechanisms that control the diameters of the afferent and efferent glomerular arterioles.
- Glomerular blood pressure generally varies directly with systemic blood pressure.
- Glomerular filtration rate is regulated by renal autoregulation, sympathetic control, and the renin-angiotensin mechanism.
- In tubular reabsorption, needed substances are reabsorbed back into the blood plasma of the peritubular capillaries and vasa recta by either active or passive transport.
- Cations are actively reabsorbed. Anions are passively reabsorbed by electrochemical attraction to the cations. Water is passively reabsorbed by osmosis.
- Most tubular reabsorption occurs in the PCT, especially of nutrients such as amino acids and glucose, but other portions of the renal tubule are also involved.
- In tubular secretion, certain substances are actively or passively secreted into the tubular fluid from the blood plasma. Uric acid and hydrogen ions are actively secreted. Potassium ions are secreted both actively and passively. Most tubular secretion occurs in the DCT.
- The nephron loop selectively reabsorbs water in the descending limb and Na^+ and Cl^- in the ascending limb, creating an osmotic gradient in the renal medulla.
- Tubular fluid is concentrated by water reabsorption from the DCT and collecting duct (water conservation), so it is hypertonic to blood plasma.
- The daily production of urine is 1.5 to 2.0 liters. Healthy urine is a clear, pale yellow to amber fluid with a characteristic odor. The color is due to the presence of urochrome.
- Urine is usually slightly acidic but the pH may range from 4.8 to 7.5.
- Urine is more dense than water due to the dissolved substances it contains.
- Abnormal substances that may be in urine are glucose, proteins, formed elements, hemoglobin, and bile pigments.

16.5 Excretion of Urine

- A ureter is a slender tube that carries urine by peristalsis, gravitational pull, and hydrostatic pressure gradient into the urinary bladder.
- Urine is temporarily held in the urinary bladder. The urinary bladder is located behind the pubic symphysis in the pelvic cavity.
- The wall of the urinary bladder consists of the mucosa, submucosa, muscular layer, and dense irregular connective tissue. The parietal peritoneum covers only its upper surface. The muscular layer consists of smooth muscle cells forming the detrusor.
- A thickening of the detrusor at the urinary bladder-urethra junction forms the internal urethral sphincter in males. The external urethral sphincter in both genders is formed of skeletal muscle fibers in the floor of the pelvis.
- Micturition is the process of voiding urine from the urinary bladder. Urine is expelled from the urinary bladder through the urethra.
- When the urinary bladder contains 200 to 400 ml of urine, the micturition reflex is triggered, causing the detrusor to contract rhythmically. Continued contractions open the involuntarily controlled internal urethral sphincter in males. If the voluntarily controlled external urethral sphincter is relaxed, micturition occurs. If not, micturition is postponed.

16.6 Maintenance of Blood Plasma Composition

- The prime function of the kidneys is to maintain the volume and composition of the blood plasma in spite of variations in diet and metabolic processes.

- Water intake must equal water loss. Most water is lost in urine, but other avenues include exhaled air, perspiration, and feces.
- The volume of water lost in urine is decreased when water loss via other means is increased, and vice versa.
- Antidiuretic hormone, which is released from the posterior lobe of the pituitary gland, increases the permeability of DCT and collecting ducts to water and thereby promotes water reabsorption by osmosis.
- Electrolyte intake must replace electrolyte loss. Electrolytes are conserved largely by the active reabsorption of cations that passively pull along anions by electrochemical attraction.
- Aldosterone regulates the blood concentration of sodium and potassium ions by stimulating the active reabsorption of sodium ions from, and the active secretion of potassium ions into, the DCT and collecting duct.
- Natriuretic peptides promote the excretion of sodium ions from the DCT and collecting duct.
- The concentration of calcium ions in the blood is regulated by the actions of two hormones acting in the DCT.
- Parathyroid hormone stimulates the reabsorption of calcium from bones into the blood and the reabsorption of calcium ions by the kidneys. Active vitamin D assists PTH and also increases the absorption of calcium by the small intestine. In contrast, calcitonin promotes the deposition of calcium in bones.
- The maintenance of the blood pH between 7.35 and 7.45 includes three major mechanisms: buffers in the blood either combine with or release hydrogen ions as needed; carbon dioxide is removed by the lungs; and renal tubules regulate the rate of hydrogen and bicarbonate ions secreted into the tubular fluid.
- Arterial blood pH less than 7.35 is called acidosis. Arterial blood pH greater than 7.45 is called alkalosis.

16.7 Disorders of the Urinary System

- Inflammatory disorders include cystitis, glomerulonephritis, pyelonephritis, and urethritis.
- Noninflammatory disorders include diuresis, renal calculi, and renal failure.

Improve Your Grade

Connect Interactive Questions Reinforce your knowledge using multiple types of questions: interactive, animation, classification, labeling, sequencing, composition, and traditional multiple choice and true/false.

SmartBook Proven to help students improve grades and study more efficiently, SmartBook contains the same content within the print book but actively tailors that content to the needs of the individual.

Anatomy & Physiology REVEALED® Dive into the human body by peeling back layers of cadaver imaging. Utilize this world-class cadaver dissection tool for a closer look at the body anytime, from anywhere.

CHAPTER 17

Reproductive System

©Lisette Le Bon/Purestock/SuperStock RF

Out of all of the organ systems, the reproductive system is the only one that is not essential for the survival of an individual. However, it is not insignificant in its ability to influence homeostasis. Consider Debbie and Patrick, fraternal twins born in Pennsylvania. From birth through childhood, their physical attributes were quite similar aside from their external genitalia. The general shape of their torsos and limbs showed no significant differences, indicating similar skeletal and muscular development. The voices of both children exhibited the higher pitch common to young children. However, at puberty, these similarities ended. Debbie's pelvis widened and her mammary glands developed. Patrick's skeletal mass increased greatly to match the enlargement of his musculature, resulting in a broadening of his shoulders and chest. The similarities in their voices ended as Patrick's larynx enlarged, resulting in a decrease in vocal pitch. These visible and audible changes matched the internal changes designed to prepare the children for reproduction as adults. All of these changes in form and function are the product of reproductive system activity. Though not essential for personal survival, this system clearly is needed for the survival of the human race.

CHAPTER OUTLINE

- **17.1 Introduction to the Reproductive System**
- **17.2 Male Reproductive System**
  - Testes
  - Accessory Ducts
  - Accessory Glands
  - Semen
  - Male External Genitalia
- **17.3 Male Sexual Response**
- **17.4 Hormonal Control of Reproduction in Males**
  - Action of Testosterone
  - Regulation of Male Sex Hormone Secretion
- **17.5 Female Reproductive System**
  - Ovaries
  - Uterine Tubes
  - Uterus
  - Vagina
  - Female External Genitalia
- **17.6 Female Sexual Response**
- **17.7 Hormonal Control of Reproduction in Females**
  - Female Sex Hormones
  - Female Reproductive Cycles
- **17.8 Mammary Glands**
- **17.9 Birth Control**
  - Contraception
  - Anti-Implantation Devices
  - Sterilization
  - Induced Abortion
- **17.10 Disorders of the Reproductive System**
  - Male Disorders
  - Female Disorders
  - Sexually Transmitted Diseases (STDs)

Chapter Summary

Module 14
Reproductive System

SELECTED KEY TERMS

Estrogens A group of female sex hormones produced primarily by the ovaries.
Gamete A sex cell, either a sperm or secondary oocyte.
Gonads (gone = seed) The primary sex glands–the ovaries and testes.
Meiosis (mei = less) A form of cell division in which the daughter cells contain one-half the number of chromosomes as the parent cell.
Menopause (men = month; paus = stop) The cessation of regular menstrual cycles.

Menstrual cycle The series of changes in the endometrium that occur each month unless pregnancy occurs.
Oogenesis (oo = ovum, egg; genesis = origin) The process of ova formation.
Ovarian cycle The monthly formation and release of a secondary oocyte and the ovarian events that take place in preparation for pregnancy.
Ovulation The release of a secondary oocyte from an ovary.

Progesterone The female sex hormone produced primarily by the corpus luteum.
Puberty (puber = grown up) The age at which reproductive organs mature.
Semen (semin = seed) Fluid composed of sperm and secretions of male accessory glands.
Spermatogenesis The process of sperm formation in the testes.
Testosterone The primary male sex hormone produced by the testes.

17.1 Introduction to the Reproductive System

Learning Objective
1. Define the key terms of reproduction.

Reproduction is the process by which life is sustained from one generation to the next. The human reproductive system is specially adapted for its role in reproduction. The **gonads** (gō′-nads)–the ovaries and testes–form the **gametes** (sex cells). Other reproductive organs nurture gametes or transport them to sites where they may unite. After fertilization occurs, the new life develops within the female reproductive system and culminates in the birth of an infant. Sexual maturation and the development of gametes and pregnancy in females are regulated by hormones secreted by the pituitary gland and the gonads.

Check My Understanding
1. What are the cells that unite during fertilization called?

17.2 Male Reproductive System

Learning Objectives
2. Describe the locations and functions of the male reproductive organs.
3. Describe spermatogenesis.
4. Describe the sources, contents, chemical characteristics, and functions of semen.

The primary functions of the male reproductive system are the production of male sex hormones, the formation of **sperm** (male gamete), and the placement of sperm in the female reproductive tract, where one sperm can unite with a **secondary oocyte** (female gamete). The organs of the male reproductive system include (1) paired testes, which produce sperm and male sex hormones; (2) accessory ducts that store and transport sperm; (3) accessory glands, whose secretions form part of the semen; and (4) external genitalia, including the scrotum and penis (figure 17.1).

Testes

The paired **testes** (tes′-tēz; singular, *testis*) are the male gonads, or sex glands. Each testis is protected and supported by a capsule of dense irregular connective tissue. Septa (partitions) of connective tissue radiate into the testis from its rear surface, dividing the testis into inner subdivisions called *lobules*. Each lobule contains several highly coiled **seminiferous** (se-mi-nif′-er-us) **tubules.** Seminiferous tubules are lined with **spermatogenic epithelium,** which is formed of spermatogenic cells and supporting cells. *Spermatogenic cells* divide to produce sperm, while *supporting cells* support and nourish the spermatogenic cells and help regulate sperm formation. The cells that fill the spaces between the seminiferous tubules are known as **interstitial** (in-ter-stish′-al) **cells,** and they produce **testosterone** (tes-tos′-te-rōn), the primary male sex hormone (figure 17.2a, b).

Spermatogenesis

Spermatogenesis (sper-mah-tō-jen′-e-sis) is the process that produces sperm by the division of the spermatogenic cells in the spermatogenic epithelium. Spermatogenesis begins at **puberty** (pū′-ber-tē), the age at which

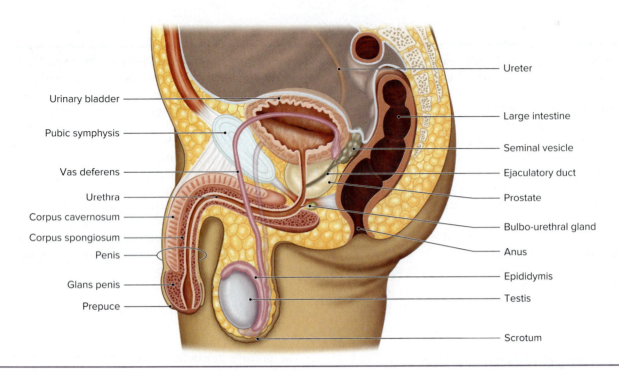

Figure 17.1 Male Reproductive System in a Median Section. AP|R

reproductive organs mature, and continues throughout the life of a male. Sexual maturity and sperm production are controlled by follicle-stimulating hormone (FSH) and luteinizing hormone (LH) from the anterior lobe of the pituitary gland and by testosterone from interstitial cells of the testes. LH is often called interstitial cell-stimulating hormone (ICSH) in males. Hormonal relationships are discussed later in this chapter.

The large cells near the basement membrane of a seminiferous tubule are known as **spermatogonia** (sper-mah-to-gō´-nē-ah; singular, *spermatogonium*). Each spermatogonium contains 46 chromosomes (23 pairs), the usual number of chromosomes for human body cells. Each spermatogonium divides by mitosis to produce two spermatogonia, referred to as type A and type B spermatogonia, each with 46 chromosomes. The type A spermatogonium remains next to the basement membrane of the tubule. It will serve as the "stem" spermatogonium and will divide repeatedly by mitosis. The type B spermatogonium that is pushed toward the lumen of the tubule undergoes changes to become a **primary spermatocyte** (figure 17.2c).

Primary spermatocytes divide by **meiosis** (mī-ō´-sis), a special type of cell division. Meiosis requires two successive divisions and reduces the number of chromosomes in the daughter cells by one-half.

Meiosis in spermatogenesis is summarized in figures 17.2c and 17.3. Each primary spermatocyte, containing 46 chromosomes, divides in meiosis I to form two **secondary spermatocytes,** each containing 23 chromosomes.

Prior to meiosis I, the chromosomes replicate. Each replicated chromosome is composed of two *chromatids* joined together at a region called a *centromere*. During metaphase of meiosis I, the replicated chromosomes are arranged as homologous pairs. During cytokinesis, the members of each chromosome pair are separated into different daughter cells. Thus, each secondary spermatocyte contains 23 replicated chromosomes (figure 17.3).

The genetic diversity of the sperm is created by the following two mechanisms. First, *random alignment* of the paired homologous chromosomes on the cellular equator occurs during meiosis I, so that the daughter cells contain different combinations of maternal and paternal chromosomes. Second, *crossover*, the exchange of some DNA between the paired homologous chromosomes, occurs during meiosis I, so that some chromosomes contain genes from both parents.

In meiosis II, the chromatids separate into different daughter cells so each secondary spermatocyte divides to form two **spermatids,** each containing 23 chromosomes. Each spermatid attaches to a supporting cell, gradually loses much of its cytoplasm, and develops a flagellum to form a sperm containing 23 chromosomes.

Examine figure 17.2c and note how cells in stages of spermatogenesis are arranged in sequence with sperm located in the lumen of the tubule. Note that the spermatids are connected by cytoplasmic bridges that allow for their development to be synchronous. Once the sperm are completely formed, they are carried into the epididymis, where they are temporarily stored while they mature.

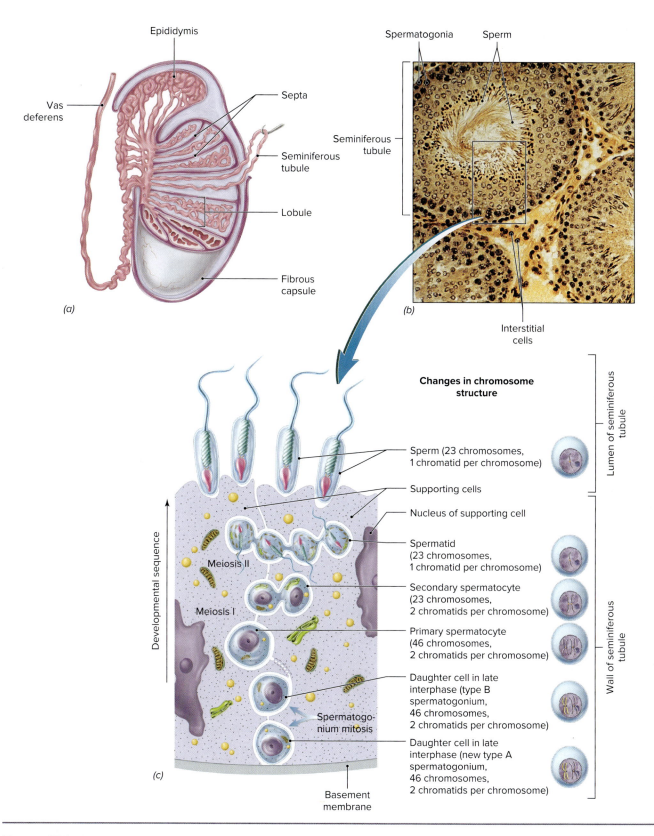

Figure 17.2 Testis and Seminiferous Tubule.
(a) Testis in sagittal section; *(b)* A photomicrograph of seminiferous tubules in cross section (200×); *(c)* A diagram showing the sequence of stages in spermatogenesis. Note that cells in developmental stages are embedded in supporting cells until sperm are released into the lumen of a seminiferous tubule. APR

(b) ©Biophoto Associates/Science Source

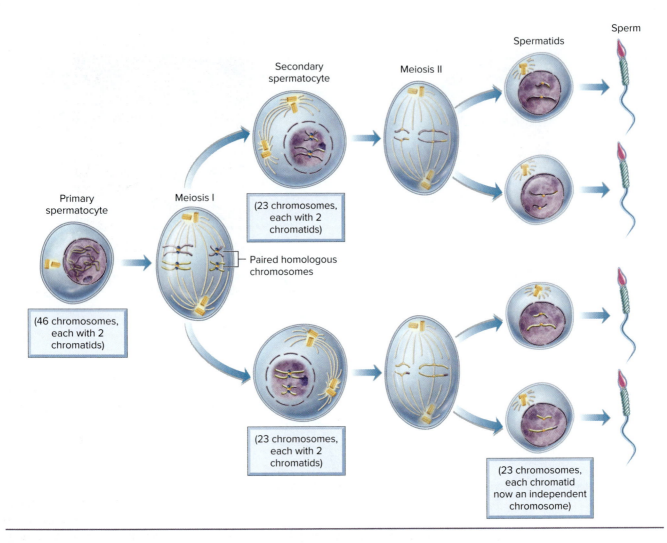

Figure 17.3 Spermatogenesis.
Spermatogenesis requires meiosis plus the development of spermatids into sperm. AP|R

A mature sperm consists of a head, neck, and flagellum. The flattened *head* is composed of a compact nucleus containing 23 chromosomes and a caplike structure, the *acrosome*, which covers the nucleus. The acrosome contains enzymes that help the sperm penetrate a secondary oocyte. The *neck* connects the head to the flagellum. The neck contains the centriole donated by the sperm to the new offspring. The *flagellum* has a middle piece, a principal piece, and an end piece. The middle piece contains mitochondria, where ATP is formed to power movement of the principal and end pieces, which enables movement of the sperm (figure 17.4).

Accessory Ducts

Sperm pass through a series of accessory ducts as they are carried from the testes to the external environment. These accessory ducts include the epididymis, vas deferens, ejaculatory duct, and urethra. These structures are collectively referred to as the *male reproductive tract*.

Epididymis

The seminiferous tubules of a testis lead to a number of small ducts that open into the epididymis. The **epididymis** (ep-i-did′-i-mis; plural, *epididymides*) appears as a comma-shaped organ that lies along the top and back of a testis (figure 17.5; see figure 17.1). Upon close examination, the epididymis is shown to be a long (6 m [20 ft]), tightly coiled, slender tube that is continuous with the vas deferens (see figure 17.2a).

Sperm mature as they are slowly moved (10–14 days) through the epididymis by weak peristaltic contractions. The mature sperm are stored in the epididymis until they are ejaculated. Sperm stored for more than two months are destroyed and absorbed by the epididymis.

Vas Deferens

As shown in figures 17.1 and 17.5a, a **vas deferens** (vas def′-er-enz; plural, *vasa deferentia*) extends from the epididymis upward in the scrotum, passes through the inguinal canal, and enters the pelvic cavity. It runs along the side of

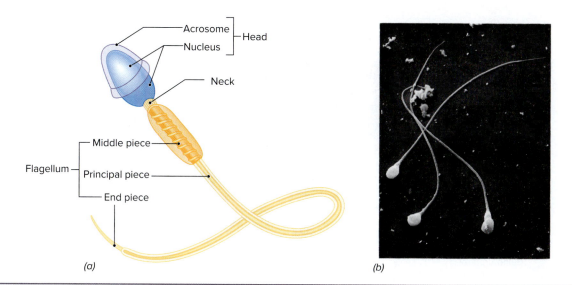

Figure 17.4 Sperm Structure.
(a) Parts of a mature sperm. *(b)* Scanning electron photomicrograph of human sperm (1,400×).
(b) ©David M. Phillips/Science Source

the urinary bladder and merges with the duct from a seminal vesicle under the urinary bladder. The duct formed by this merger is an ejaculatory duct. The vasa deferentia have rather thick, muscular walls that move the sperm by peristalsis.

Ejaculatory Duct

Each short **ejaculatory duct** is formed by the merger of a vas deferens and a duct from a seminal vesicle. The ejaculatory ducts enter the prostate and merge with the urethra within the prostate (figure 17.5a; see figure 17.1). During ejaculation, muscular contractions of the ejaculatory ducts mix seminal vesicle secretions with sperm and propel them into the urethra.

Urethra

The **urethra** is a thin-walled tube that extends from the urinary bladder through the penis to the external environment (figure 17.5; see figure 17.1). The urethra serves a dual role in the male. It transports urine from the urinary bladder during micturition, as noted in chapter 16, and it also carries semen, which includes sperm, during ejaculation. Control mechanisms prevent urine and semen passing at the same time.

Accessory Glands

Three different types of exocrine glands produce secretions involved in the reproductive process. These glands are the seminal vesicles, prostate, and bulbo-urethral glands (figure 17.5a; see figure 17.1).

Seminal Vesicles

The **seminal vesicles** are paired glands located on the back of the urinary bladder. The duct of each seminal vesicle merges with the vas deferens on the same side to form an ejaculatory duct near the back of the prostate. The alkaline secretions of the seminal vesicles help to keep semen alkaline and contain fructose and prostaglandins. Secretions by the seminal vesicles compose about 60% of semen.

Prostate

The **prostate** is a pear-shaped gland that encircles the urethra where it exits the urinary bladder. The ejaculatory ducts pass through the rear half of the prostate to join with the urethra within the prostate. Prostatic fluid is forced through 20 to 30 tiny ducts into the urethra during ejaculation. The secretion is an alkaline, milky fluid containing substances that activate the swimming movements of sperm. It forms about 30% of semen.

Bulbo-Urethral Glands

The **bulbo-urethral** (bul-bō-ū-rē′-thral) **glands** are two small, spherical glands that are located below the prostate near the base of the penis (figure 17.5a; see figure 17.1). These glands secrete an alkaline, mucuslike fluid into the urethra in response to sexual stimulation. This secretion

 Clinical Insight

Male infertility may be caused by a number of factors, such as hormone imbalances, duct blockage, a low sperm count, abnormal sperm, or a low fructose concentration in the semen. The minimum sperm count for male fertility is considered to be at least 20 million sperm per milliliter of semen.

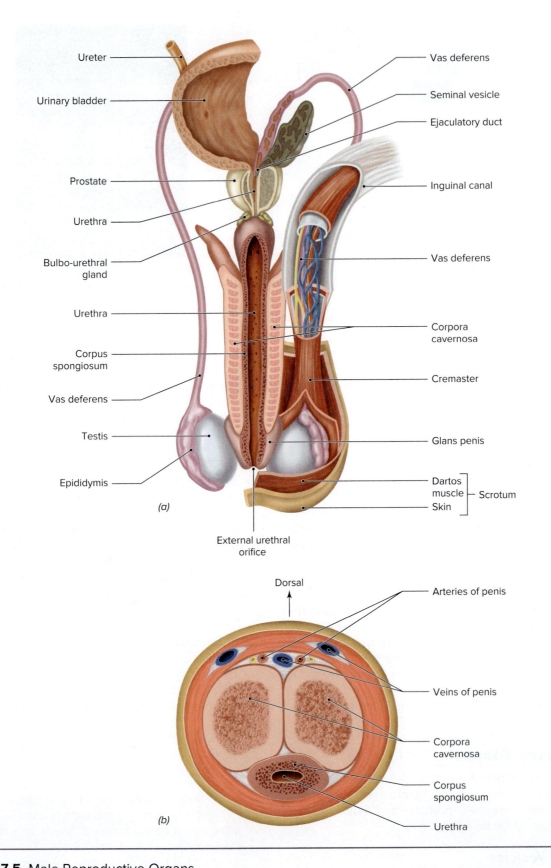

Figure 17.5 Male Reproductive Organs.
(a) Testes, accessory ducts, accessory glands, and a longitudinal section of penis. *(b)* A cross section of penis.

neutralizes the acidity of the urethra and lubricates the end of the penis in preparation for sexual intercourse.

Semen

The **semen** (sē′-men) is the fluid passed from the urethra during ejaculation. It consists of the fluids secreted by the bulbo-urethral glands, seminal vesicles, and prostate along with sperm and fluid from the testes. The alkalinity (pH 7.5) of semen protects the sperm by neutralizing the acidity of the male's urethra and the female's vagina. Fructose from seminal vesicles provides the nutrient energy for sperm, and prostatic fluid activates their swimming movements. After semen is deposited in the vagina during sexual intercourse, prostaglandins from the seminal vesicles stimulate reverse peristalsis of the uterus and uterine tubes, which accelerates the movement of sperm through the female reproductive tract. The volume of semen in a single ejaculation may vary from 2 to 5 ml, with 50 to 150 million sperm per milliliter. Although only one sperm participates in fertilization, many sperm are necessary for fertilization to occur.

Male External Genitalia

Male external genitalia are the visible parts of the male reproductive system, including the scrotum and penis.

Scrotum

The **scrotum** (skrō′-tum) is the sac of skin and subcutaneous tissue that contains the testes (see figure 17.1). It hangs from the trunk midline behind the penis. A medial partition keeps each testis in a separate chamber within the scrotum. Testes develop within the pelvic cavity but descend into the scrotum through the inguinal canals near the end of the seventh month of fetal development. This migration occurs under the stimulation of testosterone. The descent of the testes into the scrotum keeps their temperature 2–3°F below average body temperature. This lower temperature is necessary for the production of viable sperm.

The subcutaneous tissue of the scrotum contains a layer of smooth muscle called the *dartos muscle* (figure 17.5a). In addition, there are two thin, ribbonlike skeletal muscles, named the *cremaster*, that are attached to the testes. Through the actions of these muscles, the testes are elevated closer to the body in cold temperatures and depressed in warm temperatures.

Penis

The **penis** is the male copulatory organ that deposits semen in the female vagina during sexual intercourse. It contains specialized erectile tissues that enable it to become enlarged and rigid during sexual excitement.

Three columns of erectile tissue compose the body of the penis. The *corpora cavernosa* are two columns located on the top of an erect penis. The single corpus spongiosum, through which the urethra extends, is located on the bottom of an erect penis (figure 17.5; see figure 17.1). The corpus spongiosum expands at the tip to form the *glans penis*, which contains numerous sensory receptors and the external urethral orifice. A loose sheath of skin, the *prepuce*, extends distally to cover the glans.

Table 17.1 summarizes the functions of male reproductive organs.

Clinical Insight

Failure of the testes to descend into the scrotum, called cryptorchidism, produces sterility and increases the chance of testicular cancer. Undescended testes may be stimulated to descend by testosterone administration, or surgical means may be used to move the testes into the scrotum. Both treatments are done in early childhood.

Table 17.1 Summary of Functions of Male Reproductive Organs

Organ	Function
Testis	Seminiferous tubules produce sperm and secrete inhibin; interstitial cells secrete testosterone
Epididymis	Site of sperm maturation and temporary storage; carries sperm to vas deferens
Vas deferens	Carries sperm to ejaculatory duct
Ejaculatory duct	Carries sperm and secretions from the seminal vesicle to the urethra
Urethra	Carries semen to the external environment
Bulbo-urethral gland	Secretes mucuslike fluid that neutralizes acidity of the urethra and lubricates the glans penis
Seminal vesicle	Secretes alkaline fluid containing nutrients for sperm and prostaglandins for stimulating reverse peristalsis of the uterus and uterine tubes; helps keep semen slightly alkaline
Prostate	Secretes alkaline fluid that helps keep semen slightly alkaline and activates motility of sperm
Scrotum	Contains and protects testes; regulates temperature of testes
Penis	Inserted into vagina during sexual intercourse; deposits semen in vagina; contains sensory receptors associated with feelings of sexual pleasure

Clinical Insight

For many American male babies, *circumcision*, the surgical removal of the prepuce, is performed in the delivery room or within three days after birth. In Jewish culture, it may be performed on the eighth day as a religious rite. It is believed that circumcision improves male hygiene and decreases the risk of penile infections. It may also reduce the risk of cervical cancer in female sexual partners.

Check My Understanding

2. How are sperm formed?
3. What is the path of sperm from a testis to outside the body?
4. What are the components of semen?
5. What is the function of the scrotum?

17.3 Male Sexual Response

Learning Objective
5. Describe the male sexual response.

In the absence of sexual stimulation, the vascular sinusoids in the erectile tissue of the penis contain a small amount of blood and the penis is flaccid (flak´-sid), or soft. Sexual stimulation initiates parasympathetic action potentials that cause the dilation of the arterioles and constriction of the venules supplying the erectile tissue. These vascular changes cause the erectile tissue to become engorged with blood, which produces **erection,** a condition in which the penis swells, lengthens, and becomes erect. The parasympathetic action potentials also stimulate the secretion from the bulbo-urethral glands.

Continued sexual stimulation of the penis, as in sexual intercourse, culminates in **orgasm,** which is characterized by **ejaculation** (ē-jak-ū-lā´-shun) of semen and a feeling of intense pleasure. Just prior to ejaculation, sympathetic action potentials stimulate peristaltic contractions of the epididymides, vasa deferentia, and ejaculatory ducts along with contractions of the seminal vesicles and prostate. These contractions force semen into the urethra. Then, ejaculation occurs as certain skeletal muscles contract rhythmically from the proximal penis distally, forcing semen through the urethra to the external environment. A feeling of general relaxation follows.

Immediately after ejaculation, the vascular changes that produced erection are reversed. Sympathetic action potentials cause the constriction of the arterioles and dilation of the venules supplying the erectile tissue, allowing the accumulated blood to leave the penis. The penis becomes flaccid as blood is carried away. After orgasm, erection is not possible for a time period that varies from less than an hour to several hours.

Check My Understanding

6. What part of the autonomic division stimulates peristalsis in the male reproductive tract?

17.4 Hormonal Control of Reproduction in Males

Learning Objectives
6. Describe the actions of testosterone.
7. Describe the hormonal control of testosterone secretion.
8. Describe the hormonal control of spermatogenesis.

The onset of male sexual development begins around the age of 11 or 12 and is completed by ages 15 to 17. The mechanisms initiating the onset of puberty are not well understood, but the sequence of events is known. Hormones of the hypothalamus, anterior lobe of the pituitary gland, and testes are involved. Melatonin from the pineal gland may also be involved.

Puberty begins when unknown stimuli trigger the hypothalamus to secrete **gonadotropin-releasing hormone (GnRH),** which is carried to the anterior lobe of the pituitary gland by the blood. GnRH activates the release of **luteinizing hormone (LH)** and **follicle-stimulating hormone (FSH)** from the anterior lobe of the pituitary gland. LH promotes growth of the interstitial cells of the testes and stimulates their secretion of testosterone. FSH and testosterone in combination act on the seminiferous tubules, stimulating spermatogenesis (figure 17.6).

Action of Testosterone

Male sex hormones are collectively called **androgens** (an´-dro-jens), and testosterone is the most important one. They are produced primarily by the testes. Testosterone secretion starts in fetal development, continues for a brief time after birth, and then nearly ceases until puberty.

At puberty, testosterone stimulates the maturation of the male reproductive organs, the continuation of spermatogenesis, and the development of the male secondary sex characteristics. **Secondary sex characteristics** are the physical features that distinguish sexes from each other. Male secondary sex characteristics include (1) growth of body hair, especially on the face, axillary, and pubic regions; (2) increased muscular development;

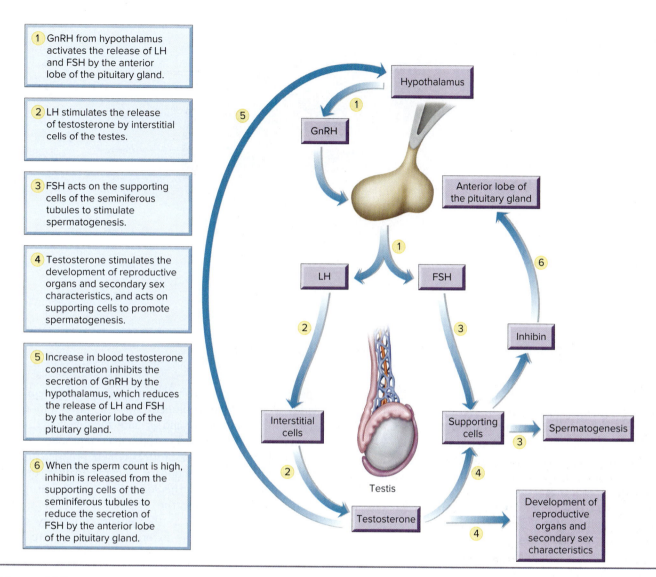

Figure 17.6 Hormonal Control of Spermatogenesis and Testosterone Secretion.

(3) development of heavy bones, broad shoulders, and narrow pelvis; and (4) deepening of the voice due to enlargement of the larynx and thickening of the vocal folds. Less obvious effects are increases in the rate of cellular metabolism and RBC production. Both metabolic rate and the concentration of RBCs in the blood are greater in males than in females. Testosterone production starts to decline at about 40 years of age, resulting in a gradual decline in the functions of reproductive organs and secondary sex characteristics.

Regulation of Male Sex Hormone Secretion

Like many other hormones, testosterone production is controlled by a negative-feedback mechanism, as shown in figure 17.6. Note that the secretion of GnRH starts the feedback mechanism. After puberty, this mechanism maintains the testosterone level in the blood within healthy limits.

As testosterone concentration increases in the blood, it inhibits the production of GnRH by the hypothalamus, which, in turn, reduces the release of LH and FSH from the anterior lobe of the pituitary gland. The decrease in LH production causes a decrease in testosterone secretion by the interstitial cells of the testes, which reduces testosterone concentration in the blood. The reduction in FSH decreases sperm production.

Conversely, as testosterone concentration in the blood declines, the hypothalamus is stimulated to secrete GnRH, which, in turn, promotes the release of LH and FSH by the anterior lobe of the pituitary gland. The increase in LH production causes an increase in testosterone secretion by the testes, which increases the testosterone concentration in the blood. The increase in FSH increases sperm production.

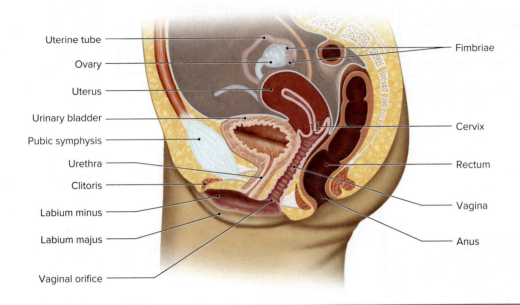

Figure 17.7 Female Reproductive System in a Median Section. AP|R

Testosterone maintains the male reproductive organs and spermatogenesis. However, sperm production is fine-tuned by the hormone **inhibin,** which is secreted by supporting cells within the seminiferous tubules. When the sperm count is high, secretion of inhibin is increased, which decreases sperm production by decreasing FSH secretion by the anterior lobe of the pituitary gland. When the sperm count is low, inhibin secretion decreases, and sperm production is increased by an increase in FSH secretion. Thus, inhibin works to keep sperm concentration in semen within healthy limits without altering the secretion of LH and testosterone.

 Check My Understanding

7. How is sperm production regulated?
8. What are the functions of testosterone?
9. How is testosterone secretion regulated?

17.5 Female Reproductive System

Learning Objectives

9. Describe the location and functions of the organs of the female reproductive system.
10. Describe oogenesis and the development of the corresponding ovarian follicles.

The female reproductive system produces female sex hormones and secondary oocytes and transports the secondary oocytes to a site where they may unite with sperm. In addition, the female reproductive system provides a suitable environment for the development of the embryo and fetus and is actively involved in the birth process. The organs of the female reproductive system include the paired ovaries, which produce female sex hormones and secondary oocytes; paired uterine tubes, which transport the secondary oocytes; a uterus, where development of the embryo and fetus occurs; a vagina, which serves as the female copulatory organ and birth canal; and the external genitalia (figure 17.7; figure 17.10). The uterine tubes, the uterus, and the vagina are collectively referred to as the *female reproductive tract*.

Ovaries

The **ovaries** are located near the upper side walls of the pelvic cavity, one on each side of the uterus. They are about the same size and shape as large almonds, and they are supported by several ligaments. The outer surface of an ovary is covered by the **ovarian mesothelium** (figure 17.8a). Under the ovarian mesothelium is a more dense region, which contains ovarian follicles of various sizes. Each **ovarian follicle** is a spherical structure composed of an oocyte enveloped by supporting cells. An **oocyte** is an immature ovum. The inner region of an ovary consists of areolar connective tissue that supports nerves and blood vessels.

Oogenesis

Oogenesis (ō-uh-jen´-e-sis) is the process of producing **ova** (singular, *ovum*). Oogenesis is similar to spermatogenesis with a few notable exceptions. The process of oogenesis and the development of the corresponding ovarian follicles are shown in figures 17.8b and 17.9. Refer to

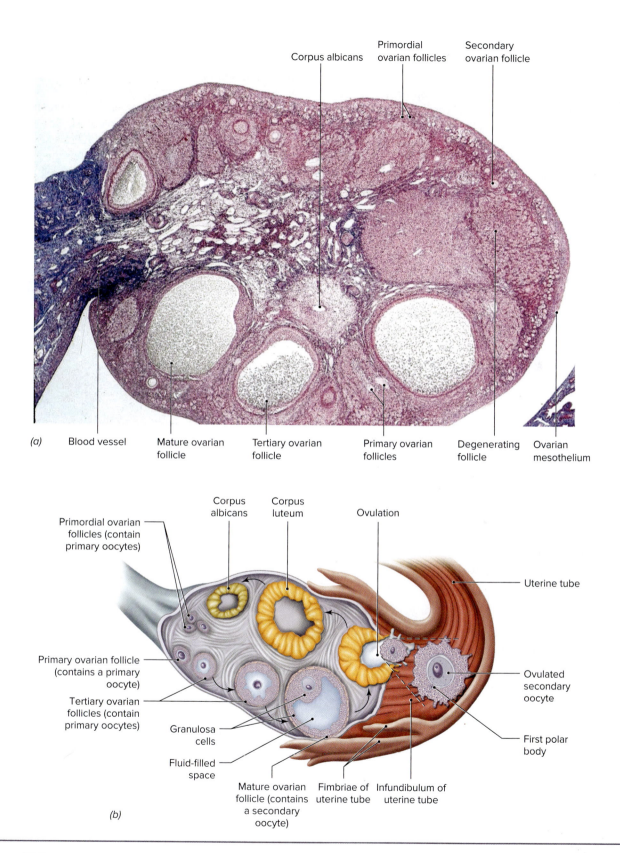

Figure 17.8 Histology of the Ovary.
(a) Photomicrograph of an ovary (30×). (b) Stages of ovarian follicular development in an ovary, release of a secondary oocyte at ovulation, and formation of a corpus luteum. APR
(a) ©Victor P. Eroschenko

these figures as you study the following process of oogenesis. Oogenesis is closely tied to the ovarian cycle, which is discussed later in this chapter.

By the fifth month of fetal development of a human female, the ovaries contain several million **oogonia** (ō-uh-gō′-nē-ah), the stem cells of the ovaries. Gradually, most of the oogonia mature into **primary oocytes** (ō′-uh-sitz) surrounded by a single layer of squamous *follicular epithelial cells* forming the *primordial ovarian follicles*. When a female is born, the infant's ovaries lack oogonia and contain about 2 million primary oocytes within primordial ovarian follicles. Primary oocytes contain 46 chromosomes, so they are capable of meiosis. Starting at puberty, a few primordial ovarian follicles are activated each month. The development of the oocyte and ovarian follicle takes nearly a year to complete. However, only one of the primary oocytes will complete meiosis I each month.

As shown in figure 17.9, chromosomes are replicated prior to meiosis I and line up as homologous pairs at metaphase. Each chromosome consists of two chromatids

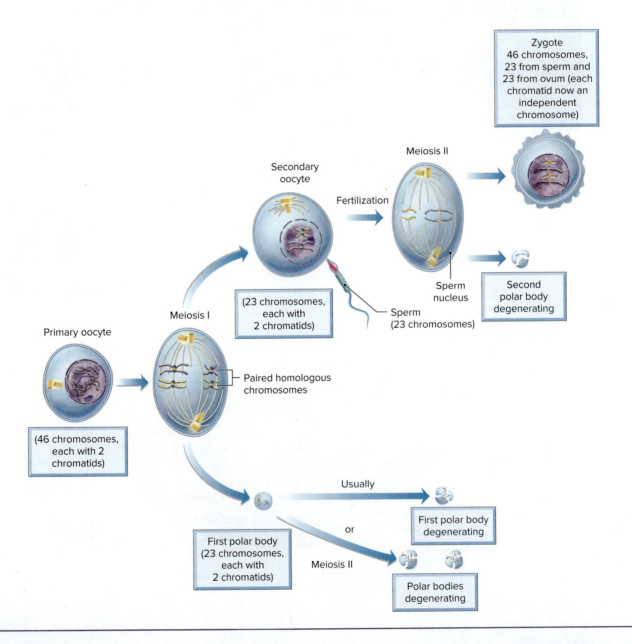

Figure 17.9 Oogenesis.
In meiosis I, the primary oocyte divides to form a secondary oocyte and a polar body. If the secondary oocyte is penetrated by a sperm, it divides in meiosis II to form an ovum and a second polar body. Union of ovum nucleus and sperm nucleus forms a zygote. Polar bodies degenerate.

joined by the centromere. In meiosis I, the primary oocyte divides to form two dissimilar cells, each with only 23 chromosomes. Only one member of each chromosome pair is distributed to each daughter cell. The smaller cell, the first **polar body**, contains almost no cytoplasm and performs no further function. It is formed as a convenient way to get rid of 23 chromosomes. The larger cell is the secondary oocyte, and it contains nearly all of the cytoplasm that was present in the primary oocyte. It is the secondary oocyte that is released from an ovary each month during a female's reproductive life. **Ovulation** is the process of an ovary releasing a secondary oocyte. As in spermatogenesis, the genetic diversity of the secondary oocytes results from the random alignment of the paired homologous chromosomes and the crossover between the paired homologous chromosomes during meiosis I.

Meiosis II is completed only if the secondary oocyte is penetrated by a sperm. Meiosis II forms another polar body and an ovum, which receives nearly all of the cytoplasm. The chromatids of each chromosome separate, and they are distributed into different daughter cells. Therefore, both the ovum and the second polar body contain 23 chromosomes. The fusion of the ovum nucleus and sperm nucleus forms a **zygote,** the first cell of a preembryo. The polar bodies simply degenerate.

Uterine Tubes

The paired **uterine tubes** receive and transport the secondary oocyte, are the sites of fertilization, and transport the preembryo if fertilization occurs. Each uterine tube

 Clinical Insight

Infections of the female reproductive tract can easily migrate through the uterine tubes into the pelvic cavity, where the infections become far more serious.

extends laterally from the upper side of the uterus to an ovary. The lateral end of each tube forms a funnel-shaped expansion, the *infundibulum* (in-fun-dib'-ū-lum), that partially envelops an ovary but is not connected to it. Each infundibulum is subdivided into a number of fingerlike processes called *fimbriae* (fim'-brē-ē) that may touch the ovary (figure 17.10).

The inner lining of a uterine tube consists of simple ciliated columnar epithelium and secretory cells. The beating of the cilia creates a current that helps draw the ovulated secondary oocyte into the infundibulum. Then, the beating cilia and the peristaltic contractions of the uterine tube move the oocyte slowly toward the uterus.

Uterus

The **uterus** (ū'-ter-us) is located in the pelvic cavity behind the urinary bladder. As shown in figure 17.7, it is located above the vagina and is bent forward over the urinary bladder. The uterus is a hollow organ with thick, muscular walls. Its primary function is to provide an appropriate internal environment for a developing embryo and fetus.

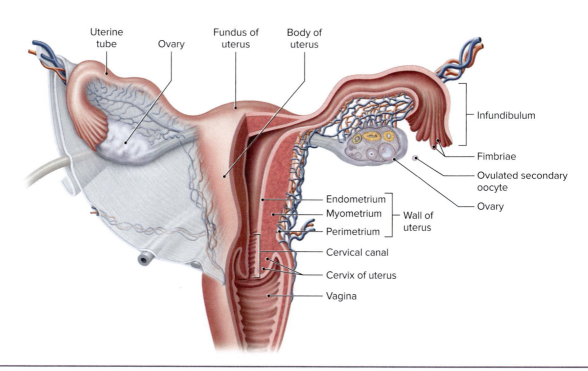

Figure 17.10 Posterior View of the Female Reproductive Organs.

The uterus has three major regions. The *fundus*, the upper rounded region, joins with the uterine tubes laterally. The *body*, the middle portion, is enlarged and rounded. The *cervix*, the lower tubular portion, is inserted into the upper end of the vagina.

The walls of the uterus are composed of three layers. The *endometrium* (en-dō-mē′-trē-um), the mucosal layer, is the inner layer of the uterus. The *myometrium* is the thick layer of smooth muscle that forms most of the wall thickness. The *perimetrium*, the serous layer, is the outer layer of the uterus (figure 17.10).

Vagina

The **vagina** (vah-ji′-nah) is the collapsible tube that extends from the uterus to the external environment. It serves as the female copulatory organ and the birth canal. The vagina is located behind the urethra and in front of the rectum (figure 17.10; see figure 17.7).

The vaginal wall consists of three layers: an inner mucosa of stratified squamous epithelium and areolar connective tissue, a thin layer of smooth muscle, and an outer layer of dense irregular connective tissue.

Female External Genitalia

The orifices of the urethra and vagina are surrounded by the **female external genitalia,** or *vulva*. Figure 17.11 illustrates the female external genitalia and **perineum** (per-i-nē′-um), the area between the mons pubis and the anus. The portion of the perineum between the vaginal orifice and the anus is often called the *obstetrical perineum*. It is so named because this tissue is often torn during childbirth.

The paired **labia majora** (lā′-bē-ah ma-jō′-rah) (singular, *labium majus*) are rounded longitudinal folds of adipose tissue and a thin layer of smooth muscle covered by skin. The labia majora enclose the other external genitalia, and their medial margins are separated by a narrow cleft. They join in front of the pubic symphysis at the *mons pubis*, a rounded cushion of adipose tissue. The labia majora are formed of the same embryonic tissues that form the scrotum in males.

The **labia minora** (singular, *labium minus*) are paired, thinner, longitudinal folds that lie medial to the labia majora. They join in the front to form a hoodlike covering over the clitoris. In the back, they fuse with the labia majora.

The narrow space between the labia minora is known as the **vestibule.** The urethra opens into the front portion of the vestibule and the vagina opens behind the urethra. The female external genitalia contain the same erectile tissues as the male penis. The **bulbs of the vestibule** are composed of corpus spongiosum and are located beneath the labia minora surrounding the beginning of the vagina. They become engorged with blood and are involved in the female sexual response. The *vestibular glands* lie behind the bulbs on each side of the vaginal orifice and release their secretions into the vestibule. The **clitoris** (klit′-er-is) is formed of two columns of corpora cavernosa that extend bilaterally near the pubis. The *glans of the clitoris* is formed by the union of the two columns of the clitoris. The glans is the only portion of the female erectile tissues that is visible in the perineum and is located just behind the mons pubis where the labia minora meet. The glans contains abundant sensory receptors that are involved in the female sexual response.

Table 17.2 summarizes the functions of the female reproductive organs.

Check My Understanding

10. How is a secondary oocyte formed?
11. What are the parts of the vulva?

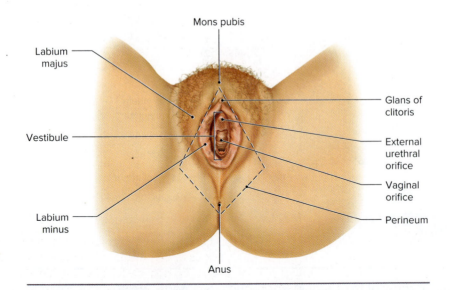

Figure 17.11 Female External Genitalia. AP|R

Table 17.2 Summary of Functions of Female Reproductive Organs

Organ	Function
Ovary	The site of oocyte production and maturation; secretes estrogens, progesterone, and inhibin
Uterine tube	Receives and transports secondary oocyte, site of fertilization; carries the preembryo if fertilization occurs
Uterus	Provides an appropriate internal environment for the development of embryo and fetus during pregnancy
Vagina	Serves as the copulatory organ and birth canal
Labia majora	Protect other female external genitalia
Labia minora	Enclose vestibule; protect vaginal and urethral orifices
Clitoris	Contains sensory receptors associated with feelings of sexual pleasure
Bulbs of the vestibule	Erectile tissue of the vestibule that fills with blood during sexual stimulation
Vestibular glands	Secrete fluid that lubricates vaginal orifice and vestibule during sexual stimulation

17.6 Female Sexual Response

Learning Objective

11. Describe the female sexual response.

In the absence of sexual stimulation, the erectile tissues of the bulbs of the vestibule and the clitoris contain a small amount of blood. When a woman is sexually stimulated, the parasympathetic action potentials cause dilation of the arterioles and constriction of the venules supplying the erectile tissues. These vascular changes cause the erectile tissues to become engorged with blood and produce erection.

Simultaneously, parasympathetic action potentials cause enlargement of the vaginal mucosa and breasts and erection of the nipples due to increased blood flow to these areas. Secretion of the vestibular glands is increased, lubricating the vestibule and aiding entry of the penis.

As in the male, sexual response in the female culminates in orgasm. Sympathetic action potentials and the prostaglandins in semen cause the muscles of the pelvic floor and walls of the uterus and uterine tubes to contract rhythmically. This reverse peristalsis aids the movement of sperm through the uterus and toward the upper ends of the uterine tubes. Orgasm produces a sensation of intense pleasure followed by general relaxation and a feeling of warmth throughout the body.

 Check My Understanding

12. What part of the autonomic division triggers erection?

17.7 Hormonal Control of Reproduction in Females

Learning Objective

12. Describe the ovarian and menstrual cycles and their regulation by hormones.

Reproduction in females is controlled by hormones produced by the hypothalamus, anterior lobe of the pituitary gland, and ovaries.

Female Sex Hormones

The ovaries produce the two major groups of female sex hormones–estrogens and progesterone–plus inhibin, which aids estrogens in exerting an inhibitory effect on the anterior lobe of the pituitary gland via a negative-feedback mechanism. Ovarian follicles under the stimulation of FSH produce **estrogens,** a group of female sex hormones produced primarily by the ovaries. Estrogens stimulate the maturation of the female reproductive organs and the development and maintenance of the female secondary sex characteristics. The female secondary sex characteristics include development of the mammary glands and breasts, a broad pelvis, increased deposition of subcutaneous tissue (especially in the breasts, buttocks, and thighs), and increased blood supply to the skin. The development of axillary and pubic hair is stimulated by the small amount of androgens produced by the adrenal glands.

After ovulation, the portion of the ovarian follicle remaining in the ovary becomes the **corpus luteum** (kor′-pus lū′-tē-um) (see figure 17.8b). Under stimulation by LH, the corpus luteum secretes the other female sex hormone, **progesterone** (prō-jes′-te-rōn), as well as estrogens. The major role of progesterone is the development and maintenance of the endometrium in pregnancy, but it also inhibits uterine contractions and dilation of the cervix during pregnancy. Both estrogens and progesterone play major roles in the regulation of the female reproductive cycles.

Female Reproductive Cycles

The two female reproductive cycles are hormonally controlled and occur simultaneously starting at puberty: the ovarian cycle and the menstrual cycle. The **ovarian cycle** involves the monthly formation and release of a secondary oocyte and the ovarian events that take place in preparation for pregnancy. The **menstrual cycle** involves repetitive changes in the endometrium that lead to monthly menstruation if pregnancy does not occur. The lengths of these cycles range from 24 to 35 days in different women, but 28 days is about average. Figure 17.12 illustrates the stages of the menstrual and ovarian cycles along with the blood concentrations of the hormones that control them in a 28-day cycle.

Except for periods of pregnancy and nursing, women experience monthly reproductive cycles from puberty, at about 11 years of age, until menopause.

Ovarian Cycle

Puberty in females begins at about 11 years of age, with the first menstruation (menarche) occurring at about 13 years

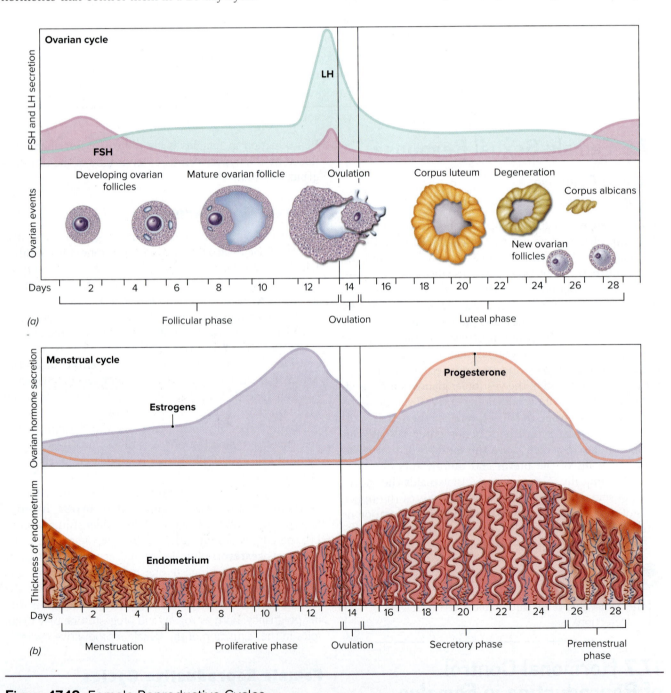

Figure 17.12 Female Reproductive Cycles.
The ovary and endometrium undergo characteristic changes as a result of hormonal fluctuations during the ovarian and menstrual cycles. *(a)* In the ovarian cycle, the ovary responds to FSH and LH released from the anterior lobe of the pituitary gland. *(b)* In the menstrual cycle, the endometrium responds to the estrogens and progesterone that are subsequently released from the ovaries. **AP|R**

of age. The female reproductive cycles begin when the hypothalamus secretes GnRH, which activates the anterior lobe of the pituitary gland to release FSH and a small amount of LH. Refer to figures 17.8, 17.9, 17.12, and 17.13 as you study the following process in the ovarian cycle.

Recall that the primary oocytes are formed within the primordial ovarian follicles of the ovary prior to birth. The primordial ovarian follicles do not continue to mature at birth; instead, they arrest development for several years. Beginning at puberty, during the *follicular phase* of each

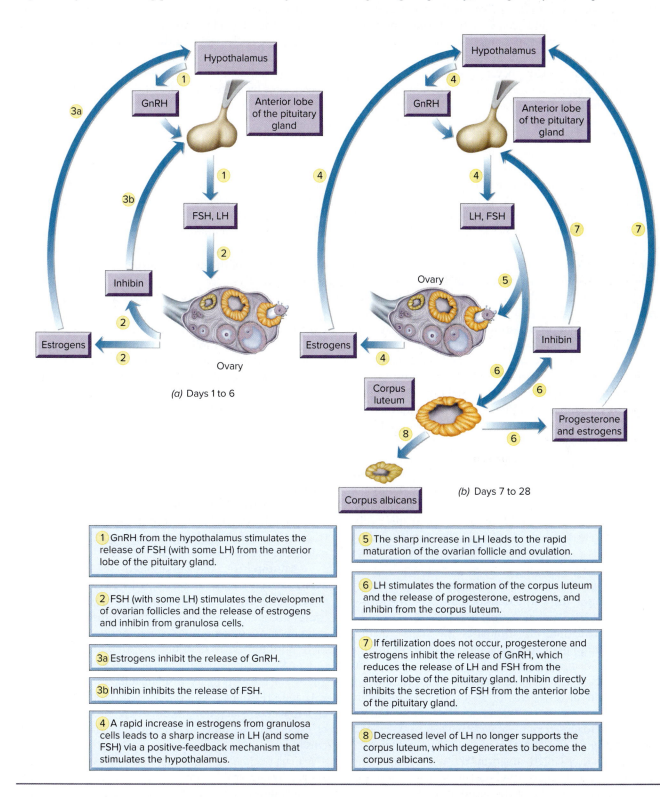

1. GnRH from the hypothalamus stimulates the release of FSH (with some LH) from the anterior lobe of the pituitary gland.

2. FSH (with some LH) stimulates the development of ovarian follicles and the release of estrogens and inhibin from granulosa cells.

3a. Estrogens inhibit the release of GnRH.

3b. Inhibin inhibits the release of FSH.

4. A rapid increase in estrogens from granulosa cells leads to a sharp increase in LH (and some FSH) via a positive-feedback mechanism that stimulates the hypothalamus.

5. The sharp increase in LH leads to the rapid maturation of the ovarian follicle and ovulation.

6. LH stimulates the formation of the corpus luteum and the release of progesterone, estrogens, and inhibin from the corpus luteum.

7. If fertilization does not occur, progesterone and estrogens inhibit the release of GnRH, which reduces the release of LH and FSH from the anterior lobe of the pituitary gland. Inhibin directly inhibits the secretion of FSH from the anterior lobe of the pituitary gland.

8. Decreased level of LH no longer supports the corpus luteum, which degenerates to become the corpus albicans.

Figure 17.13 Hormonal Regulation of the Ovarian Cycle.

ovarian cycle, FSH promotes the development of about 20 of the primordial ovarian follicles into *primary ovarian follicles*, each containing a primary oocyte whose follicular epithelial cells have grown from squamous to cuboidal cells. Some of these primary ovarian follicles then develop further into *secondary ovarian follicles* as their follicular epithelial cells multiply and become stratified. The appearance of small fluid-filled spaces within the secondary ovarian follicles marks the formation of *tertiary ovarian follicles*, each still containing a primary oocyte. The follicular epithelial cells surrounding the primary oocyte have now grown to become **granulosa cells.** The granulosa cells are responsible for secreting estrogens—as well as passing nutrients to the developing oocyte.

During each ovarian cycle, one dominant tertiary ovarian follicle will continue development. The remaining ovarian follicles undergo atresia, or death. The developing tertiary ovarian follicle continues to secrete low levels of estrogens, in addition to *inhibin*, into the blood. The low blood levels of estrogens, aided by inhibin, initiate a negative-feedback mechanism on GnRH and FSH, respectively. The inhibition of FSH secretion prevents development of additional ovarian follicles.

Even though the blood level of FSH is low, increased sensitivity of the granulosa cells to FSH results in a rapid rise in the production of estrogens starting on day 7 of the ovarian cycle, and reaching a peak on day 12. At high blood levels, estrogens stimulate the hypothalamus to secrete GnRH via a positive-feedback mechanism, leading to the release of LH (and some FSH) from the anterior lobe of the pituitary gland. This results in a sharp increase in LH level and a lesser increase in FSH level in the blood, which peak on day 13.

The abrupt increase in LH stimulates the rapid maturation of the dominant tertiary ovarian follicle into a *mature ovarian follicle*. LH also stimulates the completion of meiosis I to form a secondary oocyte and the first polar body. Additionally, the high blood level of LH stimulates ovulation, which releases the secondary oocyte surrounded by granulosa cells. Ovulation occurs 14 days before the onset of the next menstruation, regardless of the length of the cycle. In addition to causing the release of the secondary oocyte, ovulation marks the transition between the phases of both the ovarian and menstrual cycles.

After ovulation, LH stimulates the transformation of the remaining granulosa cells into the corpus luteum, marking the beginning of the *luteal phase*. The corpus luteum secretes increasing amounts of progesterone and lesser amounts of estrogens for about 10 days following ovulation. The increasing blood level of progesterone, aided by estrogens, begins a negative-feedback mechanism that inhibits the secretion of GnRH by the hypothalamus. The lack of GnRH, aided by the release of inhibin from the corpus luteum, inhibits the release of LH and FSH by the anterior lobe of the pituitary gland.

If the secondary oocyte is not fertilized, the corpus luteum degenerates into the nonfunctional corpus albicans when blood LH level declines. Loss of the corpus luteum causes a rapid decline in the blood levels of estrogens and progesterone during the last few days of the cycle. Once the low blood level of progesterone no longer inhibits the hypothalamus and allows the secretion of GnRH, the next reproductive cycle begins. If the secondary oocyte is fertilized, the corpus luteum continues to enlarge and produce increasing amounts of progesterone and estrogens, and degenerates by 16 to 20 weeks of development (see chapter 18 for details).

Menstrual Cycle

The menstrual, or uterine, cycle refers to the series of changes in the endometrium that occur each month unless pregnancy occurs. These changes are responses to fluctuating blood levels of estrogens and progesterone. The menstrual cycle has four phases: menstruation, proliferative phase, secretory phase, and premenstrual phase (see figure 17.12b).

Menstruation starts a new menstrual cycle. It begins on the first day of the next menstrual cycle and lasts from three to five days.

The *proliferative phase* is characterized by a buildup of the endometrium under stimulation by estrogens, whose concentration in the blood increases as the dominant tertiary ovarian follicle develops. This phase begins at the end of menstruation and ends at ovulation.

Following ovulation, both progesterone and estrogens are produced by the corpus luteum, and they continue the development of the endometrium in the *secretory phase*. Estrogens promote the continued thickening of the endometrium. Progesterone stimulates the formation of blood vessels and glands in the endometrium, preparing it to receive a preembryo. If fertilization of a secondary oocyte does not occur, the *premenstrual phase* begins. The blood levels of estrogens and progesterone drop rapidly, triggering the breakdown of the endometrium. This eventually leads to menstruation on the first day of the next menstrual cycle.

Menopause, the cessation of regular menstrual cycles, usually begins around age 45–55 and can last up to ten years. The onset is usually gradual with menstrual cycles becoming irregular. During this time, a woman can still conceive, so menopause is not considered to be complete until the cycles have not occurred for one year.

Aging of the ovaries is the cause of menopause. There are fewer primary ovarian follicles to respond to FSH and LH from the anterior lobe of the pituitary gland, and ovulation does not occur. Therefore, the secretion of estrogens and progesterone by the ovaries is greatly curtailed.

The decline in female sex hormones often is accompanied by physical symptoms such as headaches, insomnia, and depression. Hot flashes, caused by sudden and temporary dilation of dermal blood vessels, are perhaps the most common symptom. Hormone replacement therapy (HRT), the administration of estrogens and progesterone, was previously used to treat the unpleasant symptoms of menopause. Its use has been greatly curtailed because studies have shown that it increases the risk of breast cancer, strokes, and heart disease.

 Check My Understanding

13. How are FSH, LH, inhibin, estrogens, and progesterone involved in ovarian and menstrual cycles?

17.8 Mammary Glands

Learning Objective

13. Describe the structure of mammary glands and breasts.

Both males and females possess **mammary glands.** The mammary glands of males and immature females are similar. At puberty, estrogens and progesterone stimulate the development of female mammary glands. Estrogens start breast and mammary gland development, and progesterone stimulates the maturation of the mammary glands so that they are capable of secreting milk. Another hormone from the anterior lobe of the pituitary gland, *prolactin,* is required for milk production.

Mature mammary glands are female accessory reproductive structures that are specialized for milk production (figure 17.14). They are located in the subcutaneous tissue of female breasts, and are considered integumentary glands.

The breasts are formed on top of the pectoralis major muscles. Breasts contain large amounts of areolar and adipose tissues that surround and cushion the mammary glands. Dense irregular connective tissue within each breast is attached to the dermis and to the fascia of the pectoralis major to provide support for the breast and mammary gland. A circle of pigmented skin, the *areola* (ah-rē′-ō-lah), is located near the apex of each breast. A *nipple* containing erectile tissue is located in the center of each areola.

Each mammary gland consists of 15 to 25 lobes containing lobules. The lobules contain glandular alveoli that produce milk under stimulation by prolactin after the birth of an infant. Milk is carried from the lobules by lactiferous ducts, which open into *lactiferous sinuses* that lead to the nipple and the external environment. Regulation of lactation is discussed in chapter 18.

 Check My Understanding

14. What hormone is required for the production of milk?

17.9 Birth Control

Learning Objective

14. Describe how various methods of birth control work.

Birth control methods may be categorized into several groups based on their mode of action: hormonal, chemical, and behavioral contraceptive methods; antiimplantation devices; sterilization; and induced abortion.

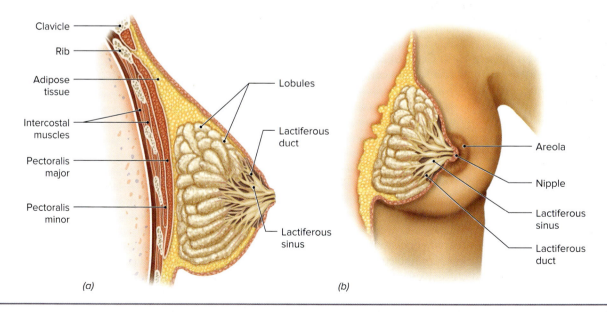

Figure 17.14 Anatomy of the Female Breast and Mammary Glands. *(a)* Sagittal section. *(b)* Anterior view.

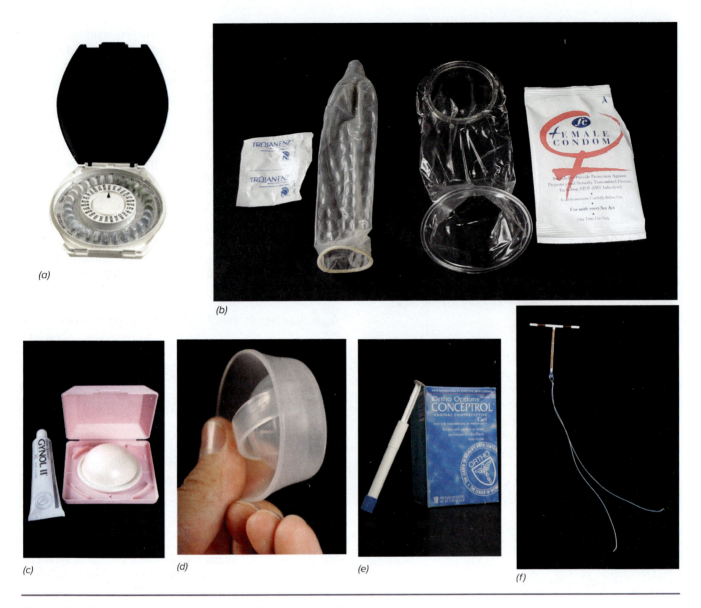

Figure 17.15 Examples of Birth Control Devices and Chemicals.
(a) Oral contraceptive, (b) male and female condoms, (c) diaphragm and spermicide, (d) cervical cap, (e) spermicide gel, and (f) intrauterine device.

(a) ©Brand X Pictures/Getty Images RF; (b–d, f) ©McGraw-Hill Education/Jill Braaten, photographer; (e) ©McGraw-Hill Education/Christopher Kerrigan, photographer

Figure 17.15 shows a few birth control devices, and Table 17.3 summarizes the effectiveness of the various birth control methods.

Contraception

Contraception is the prevention of conception, which is the union of sperm nucleus and ovum nucleus. Contraceptive methods are designed to prevent sperm from reaching and penetrating a secondary oocyte, which leads to the formation of an ovum. There are several types of contraceptives, including the use of hormones, barriers to sperm, spermicides (chemicals that kill sperm), and behavioral methods.

Hormonal Methods

Hormonal birth control methods use a higher concentration of synthetic progesterone and a lower concentration of synthetic estrogens to prevent ovulation. These hormones provide a negative-feedback control that inhibits the secretion of GnRH by the hypothalamus, which reduces the secretion of FSH and LH by the anterior lobe of the pituitary gland. This prevents maturation of ovarian follicles so ovulation does not occur. Use of hormonal birth control methods is timed to a woman's menstrual cycle.

Hormonal birth control methods are very effective, but some women may experience unpleasant side

Table 17.3 Effectiveness of Birth Control Methods

Effectiveness values are approximate because they vary with how the birth control methods are used.

Method	Effectiveness*
Vasectomy	Nearly 100%
Tubal ligation	Nearly 100%
Depo-Provera	99%
Norplant	99%
Oral contraception	97%
Skin patch	97%
Vaginal ring	97%
Intrauterine device (IUD)	95%
Male condom	85%
Female condom	85%
Diaphragm with spermicide	80%
Cervical cap with spermicide	80%
Withdrawal	75%
Rhythm method	70%

*Effectiveness is the percentage of women who do not become pregnant in one year while using this method of birth control.

effects, such as headache, irregular menstruation, nausea, and bloating. Women who smoke or have a history of blood clots, stroke, heart disease, liver disease, or uncontrolled diabetes have an elevated health risk when using hormonal birth control methods.

There are several ways to use hormonal birth control.

The **oral contraceptive,** "the pill," was the first available method of delivery, and it is still widely used. Several types are available, but the combination pill, which contains both synthetic progesterone and estrogens, is most commonly used. One pill is taken each day for three weeks of the menstrual cycle.

Some combination birth control pills may be used in higher dosages as *morning after pills (MAPs),* for postcoital emergency contraception. When they are used within 72 hours of unprotected intercourse, the hormone combination disrupts hormonal controls and tends to prevent either fertilization or implantation.

A **birth control patch** is a small adhesive skin patch containing the synthetic hormones. It is applied to the skin, and the hormones are absorbed through the skin. A new patch is used each week for three weeks.

A **vaginal ring** is a soft plastic ring that is wedged into the upper part of the vagina, where it remains and releases synthetic hormones over a three-week period. It is removed at the start of the fourth week.

Norplant is a set of tiny silicone rods containing synthetic progesterone that are surgically implanted under the skin of the arm or over the scapula. The released progesterone acts similarly to an oral contraceptive by inhibiting ovulation, and the implant is effective for up to five years. The implant may be removed to terminate contraception.

Depo-Provera is a synthetic progesterone that is administered by muscular injection at three-month intervals. It protects against pregnancy by preventing ovulation and altering the endometrium to inhibit implantation of a preembryo. Its use often causes weight gain as a side effect, and it can be a considerable health risk, so it must be used under a physician's care. Women with a personal or strong family history of high blood pressure, asthma, kidney disease, migraine headaches, or breast cancer are at increased risk if using Depo-Provera.

Barriers

Three types of **barriers** are designed to prevent sperm from entering the uterus: the condom, diaphragm, and cervical cap.

Condoms act by preventing sperm from being deposited in the vagina. A *male condom* is a thin sheath of latex rubber that is slipped over the penis prior to sexual intercourse. A *female condom* is a thin polyurethane bag with a flexible ring at each end. The woman inserts the bag into the vagina prior to sexual intercourse. Condoms reduce but do not eliminate the chance of infection by sexually transmitted diseases (STDs).

A **diaphragm** is a dome-shaped sheet of rubber supported by a firm but somewhat flexible ring. It is placed in the upper vagina over the cervix prior to intercourse. It prevents sperm from entering the uterus. Spermicidal substances are used along with a diaphragm.

A **cervical cap** is a thimble-shaped piece of latex rubber that fits snugly over the cervix. Spermicidal substances are used with a cervical cap. Females with abnormal Pap smears or cervical infections should not use a cervical cap.

Spermicides

A variety of **spermicides** are available, including creams, jellies, and suppositories. These chemicals kill sperm by destroying their plasma membranes.

Behavioral Methods

The **rhythm method** requires abstinence from sexual intercourse from three days before the day of ovulation to three days after ovulation, for a total of seven days. It is based on the fact that the secondary oocyte may be penetrated by a sperm for only about 24 hours after ovulation. One difficulty with this method is determining the time of ovulation, because few women have perfectly regular cycles. The failure rate is higher in women who have irregular cycles.

Withdrawal, or **coitus interruptus,** is the removal of the penis from the vagina just prior to ejaculation. Failures of this method result from preejaculatory emission of semen or failure to withdraw before ejaculation.

Anti-Implantation Devices

An *intrauterine device (IUD)* is a small plastic and copper object that is inserted into the uterus by a physician. The IUD causes inflammation of the endometrium, preventing implantation of a preembryo. An IUD may remain in place for long periods of time, but it should be checked at regular intervals by a physician because an IUD may be spontaneously expelled from the uterus. Undesirable side effects include excessive menstrual bleeding, painful cramps, and an increased risk of pelvic inflammatory disease and infertility.

Sterilization

Sterilization surgeries may be performed on both males and females. In males, a **vasectomy** is performed by cutting and blocking the vasa deferentia within the scrotum so that sperm do not form part of the semen. The sperm disintegrate and are reabsorbed. In females, a **tubal ligation** is performed through a small abdominal incision. The uterine tubes are cut and blocked to prevent the transport of a secondary oocyte toward the uterus. The ovulated secondary oocytes disintegrate and are reabsorbed. Vasectomies and tubal ligations do not affect the production of sex hormones or the sexual response.

Induced Abortion

An **abortion** is the premature expulsion of an embryo or fetus from the uterus. Spontaneous abortions are called *miscarriages*. They usually result from hormonal disorders or serious abnormalities of the developing embryo. An *induced abortion* may be used to terminate an unwanted pregnancy. Induced abortion involves dilation of the cervix and removal of the embryo or fetus by suction or surgical means. Possible side effects of induced abortions include prolonged bleeding and perforation of the uterus.

The so-called abortion pill, *mifepristone* (RU 486), is a progesterone antagonist. It causes the endometrium to break down, thereby detaching the embryo or fetus, which is passed from the uterus after administration of prostaglandins to promote uterine contractions. Its use is limited to the first five weeks of pregnancy and requires the supervision and care of a physician.

Check My Understanding
15. How do hormonal birth control methods work?
16. How does an IUD prevent pregnancy?

17.10 Disorders of the Reproductive System

Learning Objective
15. Describe the common disorders of the reproductive system.

Male Disorders

Prostatitis is acute or chronic inflammation of the prostate and is often associated with tenderness and enlargement of the prostate. It is usually caused by bacteria in connection with urinary tract infections or sexually transmitted diseases. It also occurs without a known cause.

Benign prostatic hyperplasia (BPH) is the enlargement of the prostate without inflammation, resulting from an increase in the number of glandular cells. It occurs in about 33% of males over 60 years of age and is usually detected by a rectal examination. In some cases, it may restrict the flow of urine and prevent the control of micturition. It could lead to urinary retention and urinary infections, but does not lead to prostate cancer. The cause of BPH is unknown, although it may be related to the changes in the ratio of androgens to estrogens associated with aging. Surgical correction usually is by *transurethral resection*, a procedure in which an instrument is inserted into the urethra to remove portions of the prostate compressing the urethra.

Prostate cancer is the most common cancer and the second leading cause of death from cancer in American males. The cause is unknown, although it is probably related to genetic and hormonal factors. It usually grows slowly and shows no symptoms at its early stage. Males over 40 years of age should have annual prostate examinations.

Testicular cancer is the most common male cancer between 15 and 35 years of age, although it is not generally a common form of cancer. The cause is unknown, although it may be related to cryptorchidism, physical damage, or environmental pollutants. It is one of the most easily detected cancers and has one of the highest cure rates of all the cancers. Monthly testicular self-exams are recommended for men over age 14.

Erectile dysfunction (ED) is the inability to attain and maintain an erection long enough for sexual intercourse. It may result from organic or psychological factors. Treatment is available in most cases.

Infertility is the inability of a male to produce and deposit in the vagina sufficient numbers of viable sperm to bring about fertilization of a secondary oocyte. A low sperm count is a common cause.

Female Disorders

Amenorrhea is the absence of menstruation. *Primary amenorrhea* is the failure of a woman to begin menstruation, and it is caused by endocrine disorders or abnormal reproductive development. *Secondary amenorrhea* is the absence of one or more menstrual periods without pregnancy. This may result from excessive physical exertion or excessive weight loss.

Dysmenorrhea refers to painful menstruation that prevents a woman from doing her usual activities for one or more days during menstruation. Uterine contractions are thought to be responsible for the pain.

Premenstrual syndrome is characterized by severe physical or emotional distress after ovulation and prior to menstruation. The cause is unknown, although it is probably related to ovarian hormone production.

Toxic shock syndrome is characterized by high fever, fatigue, headache, sore throat, vaginal irritation, vomiting, and diarrhea. It results from a toxin produced by a strain of bacteria (*Staphylococcus aureus*) whose growth is apparently enhanced by the use of highly absorbent tampons.

Endometriosis is the growth of endometrial tissue outside of the uterus. The tissue migrates through the uterine tubes into the pelvic cavity. It may cause premenstrual or menstrual pain due to its breakdown during menstruation. Infertility may result from tubal obstruction.

Infertility in females is the inability to conceive. It may be caused by tubal obstruction; hypothalamus, pituitary gland, or ovarian diseases; or a lack of maintenance of the endometrium.

Pelvic inflammatory disease (PID) is a collective term referring to any infection of the female reproductive organs and/or other pelvic tissues. The most common pathogens are those of sexually transmitted diseases that migrate into the pelvic cavity via the uterine tubes.

Cancer of the female reproductive system most often occurs as breast cancer or cervical cancer. *Breast cancer* is the most common cancer and the second leading cause of death from cancer in American females. Although the precise cause is unknown, it is strongly related to genetic factors and estrogens. It rarely occurs before age 30 and is more prevalent after menopause. Monthly breast self-exams over age 20 and yearly mammography screens over age 40 are recommended. Any lump in a breast should be brought to a physician's attention immediately because breast cancer has a high fatality rate unless it is treated early. *Cervical cancer* is a rather slow-growing cancer, and most of the cases are caused by human papilloma viruses (HPV). Two newly developed vaccines, Cervarix and Gardasil, are available for preventing the types of HPV that cause most cases of cervical cancer. These vaccines are recommended for females and males at ages 11 to 26 and ages 11 to 21, respectively. Annual Pap smear exams, the common test for detecting early cervical cancer, are recommended for females over age 21.

Sexually Transmitted Diseases (STDs)

Acquired immunodeficiency syndrome (AIDS) results from infection with the *human immunodeficiency virus (HIV)*, which attacks a group of lymphocytes known as helper T (T_H) cells. HIV is transmitted via sexual intercourse and by blood transfer, including through infected needles shared by drug users. Also, it may be passed from an infected mother to her developing fetus. It is not transmitted by casual contact. AIDS is discussed in more detail in the disorders section of chapter 13.

Gonorrhea is caused by the bacterium *Neisseria gonorrhoeae*. In males, it infects the urethra, causing painful urethritis. In females, it infects the vagina and may spread to the urethra, uterus, uterine tubes, and pelvic cavity. Females may not experience symptoms until advanced stages. Gonorrhea is a major cause of female sterility due to damage to the uterine tubes, and it can cause blindness in newborn babies. Antibiotics usually provide effective treatment. Most newborn infants born in hospitals receive eye drops of antibiotics at birth to protect against possible gonorrhea infection.

Syphilis is caused by the bacterium *Treponema pallidum*. Without treatment, syphilis progresses through several recognizable stages. The *first stage* is characterized by an open sore, a *chancre* (shang′-ker), at the site of entrance by the bacterium. It may not be readily noticed, especially in females. The chancre heals within one to five weeks. Several weeks later, the *second stage* appears as muscle and joint pain, fever, and skin rash; it persists from four to eight weeks. Then, the disease enters a latent period of variable length. When the bacterium begins to destroy organs, such as the brain and liver, the *third stage* is recognized. Treatment with antibiotics is effective prior to the third stage.

Chlamydia (klah-mid′-ē-ah) is caused by the bacterium *Chlamydia trachomatis*, and it infects nearly 5 million people each year. In males, it causes painful urethritis. In females, it may spread throughout the reproductive tract, causing damage to uterine tubes that can lead to sterility. Chlamydia may also be transmitted from an infected mother to her infant during childbirth when the bacterium enters the infant's eyes. Antibiotics are an effective treatment.

Genital herpes is caused by the herpes simplex virus type 2. This disorder is characterized by painful blisters on the reproductive organs and may be accompanied by fever or flu-like symptoms. It may be transmitted from an infected mother to her infant during delivery. In the newborn, it may cause mild discomfort, serious neural damage, or death. Treatment with the drug acyclovir inhibits viral replication but does not eliminate the virus. The virus remains in the body and intermittently produces the genital blisters.

Genital warts are caused by HPV. About 1 million new cases occur each year. Patients with genital warts may have an increased risk of certain cancers of the reproductive organs. Although there is no treatment that eliminates the virus, the warts can be removed by electrocautery (burning), cryosurgery (freezing), or laser surgery. Two newly developed vaccines, Cervarix and Gardasil, are available for preventing the types of HPV that cause most of the genital warts.

Chapter Summary

17.1 Introduction to the Reproductive System

- The gonads are the male testes and female ovaries.
- The gonads produce sex cells called gametes, which join during fertilization to produce an offspring.
- The female reproductive tract is where the offspring matures until it is born.

17.2 Male Reproductive System

- The male reproductive system consists of the testes, accessory ducts, accessory glands, and external genitalia.
- Testes are the male gonads. Each testis is divided into lobules containing seminiferous tubules that produce sperm. Interstitial cells between the tubules produce testosterone.
- Spermatogenesis is the process of producing sperm from spermatogonia. Meiosis of a primary spermatocyte produces four spermatids that mature into sperm, each with 23 chromosomes.
- Sperm pass from the seminiferous tubules into the highly coiled epididymis, where they mature and are stored. From the epididymis, sperm pass in sequence through the vas deferens, ejaculatory duct, and urethra to the external environment.
- Accessory glands secrete alkaline fluids. Seminal vesicles release their secretions into the ejaculatory ducts. The prostate and bulbo-urethral glands release their fluids into the urethra.
- Semen consists of sperm and fluids from the testes and accessory glands. About 2 to 5 ml of semen is produced in each ejaculation, with 50 to 150 million sperm in each milliliter. Semen neutralizes the acidity of the male's urethra and the female's vagina, and contains substances that protect and nourish the sperm and activate the swimming movement of sperm.
- Testes are contained in the saclike scrotum away from the body. The cremaster and dartos muscle control the distance the testes are from the body to keep the temperature of the testes at about 2–3°F below average body temperature.
- The penis contains three columns of erectile tissue that become engorged with blood during sexual excitement. The erect penis is inserted into the vagina in sexual intercourse.

17.3 Male Sexual Response

- Sexual stimulation causes erection of the penis. In sexual intercourse, continued stimulation results in orgasm, which is characterized by ejaculation and intense pleasure.
- After orgasm, blood leaves the erectile tissues and the penis returns to its flaccid state.

17.4 Hormonal Control of Reproduction in Males

- At puberty, the hypothalamus secretes GnRH, which stimulates the anterior lobe of the pituitary gland to release FSH and LH. LH stimulates the production of testosterone. FSH and testosterone stimulate the production of sperm.
- Testosterone stimulates the maturation of the male reproductive organs, and the continuation of sperm production. It also develops and maintains the male secondary sex characteristics.
- Testosterone secretion is regulated by a negative-feedback mechanism.
- Sperm production is kept within healthy limits by inhibin exerting a negative-feedback control on FSH secretion.

17.5 Female Reproductive System

- The female reproductive system consists of the ovaries, uterine tubes, uterus, vagina, and external genitalia.
- Ovaries are the female gonads, which produce secondary oocytes and female sex hormones. They are located in the pelvic cavity on each side of the uterus.
- Oogenesis is the process of producing ova from oogonia. Millions of oogonia are formed during fetal development, but most degenerate. Those that remain become primary oocytes containing 46 chromosomes and are enveloped by a single layer of squamous follicular epithelial cells, forming primordial ovarian follicles.
- Starting at puberty, a few primary ovarian follicles are stimulated to grow each month. The dominant tertiary ovarian follicle enlarges, and its primary oocyte undergoes meiosis I, forming a secondary oocyte and the first polar body.
- The mature ovarian follicle grows until it ruptures, releasing the secondary oocyte in ovulation.
- The infundibulum of a uterine tube receives the secondary oocyte, and the uterine tube carries it toward the uterus. If the secondary oocyte is penetrated by a sperm, meiosis II occurs, forming the ovum and a second polar body. Then, the sperm nucleus and ovum nucleus unite at fertilization.
- The uterus is located in the pelvic cavity behind the urinary bladder. Its major function is to provide a suitable environment for the developing offspring. The layers composing the uterine wall are the endometrium, myometrium, and perimetrium.
- The vagina serves as the copulatory organ and the birth canal.
- The female external genitalia are collectively called the vulva, which includes the lateral labia majora, the medial labia minora enclosing the vestibule, and the clitoris, two columns of corpora cavernosa that join in front of the vestibule.

17.6 Female Sexual Response

- Sexual stimulation causes erection of the clitoris, bulbs of the vestibule, and nipples and secretion by the vestibular glands.
- Sexual response culminates in orgasm, which produces rhythmic contractions of muscles in the walls of the uterus and uterine tubes and a feeling of intense pleasure.

17.7 Hormonal Control of Reproduction in Females

- The ovaries secrete estrogens and progesterone, the female sex hormones. Estrogens stimulate the maturation of the female reproductive organs and the development and maintenance of female secondary sex characteristics. Progesterone and estrogens maintain the endometrium during pregnancy.
- Female reproductive cycles involve simultaneous ovarian and menstrual cycles, which are controlled by interactions of FSH, LH, estrogens, and progesterone.
- The start of an ovarian cycle is initiated by the secretion of GnRH by the hypothalamus, which stimulates the release of FSH and a small amount of LH from the anterior lobe of the pituitary gland. FSH promotes the development of ovarian follicles. The dominant ovarian follicle enlarges and secretes estrogens and some progesterone. Estrogens, aided by inhibin, suppress FSH secretion to prevent development of additional ovarian follicles. Levels of estrogens continue to increase under FSH stimulation. The high levels of estrogens trigger the hypothalamus to produce LH and a small amount of FSH. Ovulation results from follicle stimulation by a sharp increase in LH.
- After ovulation, the ovarian follicle remnants become a corpus luteum under stimulation by LH. The secretion of GnRH is suppressed by a large amount of progesterone and a lesser amount of estrogens. If fertilization does not occur, secretion of progesterone and estrogens declines rapidly, which allows the hypothalamus to secrete GnRH, starting another ovarian cycle.
- A menstrual cycle begins with the first day of menstruation, which results from the decline in the blood levels of progesterone and estrogens. Increasing levels of estrogens cause thickening of the endometrium during the proliferative phase. After ovulation, estrogens stimulate continued development of the endometrium, and progesterone promotes the formation of blood vessels and glands in the endometrium in the secretory phase.
- If fertilization does not occur, the drop in blood levels of progesterone and estrogens begins the premenstrual phase. The premenstrual phase ends with the onset of menstruation.
- Menopause is the cessation of regular menstrual cycles. It usually occurs between ages 45 and 55.
- Menopause results when primary ovarian follicles no longer respond to FSH and LH, resulting in a marked decline in estrogen and progesterone secretion.

17.8 Mammary Glands

- Mammary glands are adapted to secrete milk after childbirth. They are located within the female breasts.
- Estrogens and progesterone stimulate development of the mammary glands and breasts. Prolactin stimulates the production of milk.

17.9 Birth Control

- Contraceptive methods include hormonal methods, barriers, spermicides, and behavioral methods.
- Anti-implantation methods use intrauterine devices that change the endometrium, preventing implantation.
- Sterilization methods involve a vasectomy in males or a tubal ligation in females.
- Induced abortion involves dilation of the cervix and evacuation of the uterine contents.

17.10 Disorders of the Reproductive System

- Common male disorders include benign prostatic hyperplasia, prostate cancer, testicular cancer, impotence, and infertility.
- Common female disorders include amenorrhea, dysmenorrhea, premenstrual syndrome, toxic shock syndrome, endometriosis, infertility, pelvic inflammatory disease, and cancer.
- Common sexually transmitted diseases include AIDS, gonorrhea, syphilis, chlamydia, genital herpes, and genital warts.

Improve Your Grade

Connect Interactive Questions Reinforce your knowledge using multiple types of questions: interactive, animation, classification, labeling, sequencing, composition, and traditional multiple choice and true/false.

SmartBook Proven to help students improve grades and study more efficiently, SmartBook contains the same content within the print book but actively tailors that content to the needs of the individual.

Anatomy & Physiology REVEALED® Dive into the human body by peeling back layers of cadaver imaging. Utilize this world-class cadaver dissection tool for a closer look at the body anytime, from anywhere.

CHAPTER 18

Development, Pregnancy, and Genetics

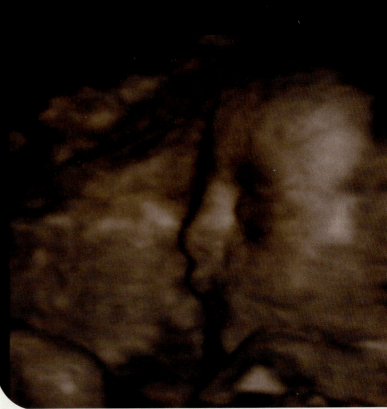

Courtesy of Jason LaPres

Sandra and Colin, a married couple in their early 40s, discover that Sandra is pregnant with their first child and are very excited. However, they are aware that later-in-life pregnancies involve higher risk. Decades of research show that there is a higher risk of chromosomal abnormalities, most notably Down syndrome, in fetuses conceived by men and women over 40. The likelihood of physical malformations, such as clubbed foot, or organ malfunction is also greatly increased. In light of all of these potential problems, Sandra and Colin elect to have an amniocentesis test performed at 18 weeks of development. During the test, a small amount of amniotic fluid is withdrawn from around the fetus. The fetal tissues within the fluid show no evidence of chromosomal anomalies, nor do they indicate other fetal issues such as neural tube defects. Their 3D ultrasound at 24 weeks of development also shows healthy development of the major organs and limbs. With confirmation that their fetus is healthy and developing as expected, Sandra and Colin breathe a little easier and move into the third trimester of the pregnancy with happy anticipation.

CHAPTER OUTLINE

18.1 Fertilization and Early Development
- Fertilization
- Preembryonic Development

18.2 Embryonic Development
- Germ Layers
- Extraembryonic Membranes
- Placenta
- External Appearance

18.3 Fetal Development

18.4 Hormonal Control of Pregnancy

18.5 Birth
- First Breath

18.6 Cardiovascular Adaptations
- Fetal Cardiovascular Adaptations
- Postnatal Cardiovascular Changes

18.7 Lactation

18.8 Disorders of Pregnancy, Prenatal Development, and Postnatal Development
- Pregnancy Disorders
- Prenatal and Postnatal Disorders

18.9 Genetics
- Sex Determination
- Genes
- Gene Expression
- X-Linked Traits
- Predicting Inheritance

18.10 Inherited Diseases
- Chromosomal Abnormalities
- Single-Gene Disorders
- Genetic Counseling

Chapter Summary

apre vealed.com

Module 14
Reproductive System

SELECTED KEY TERMS

Allele An alternate form of a gene.
Embryo Name of the developing offspring from the beginning of the third week through the end of the eighth week of development.
Fertilization The union of sperm nucleus and ovum nucleus.
Fetus Name of the developing offspring from the beginning of the ninth week of development to birth.
Gene A unit of inheritance; part of a DNA molecule in a chromosome.
Genotype The genetic composition of an individual.

Germ layer One of three embryonic tissues from which all subsequent tissues develop.
Heterozygous (hetero = different) A condition in which alleles of a gene pair are different.
Homozygous (homo = the same) A condition in which alleles of a gene pair are identical.
Implantation The attachment of a blastocyst to the endometrium.
Parturition The process of giving birth.
Phenotype (phen = visible) The observable traits determined by genes.

Placenta (plac = flat) Structure formed from embryonic and maternal tissues that allows for the exchange of substances between maternal blood and embryonic/fetal blood.
Preembryo Name of the developing offspring from fertilization through the end of the second week of development.
Zygote (zygo = yolk) Cell formed by the union of the sperm nucleus and ovum nucleus that leads to the development of a human being.

EACH NEW LIFE BEGINS AS A SINGLE CELL formed by the fusion of a sperm and an ovum. At the instant of its formation, this first cell contains all of the inherited information necessary for growth and development, plus all of the inherited characteristics that provide an individual's unique identity. After about 38 weeks of *prenatal* (fertilization to birth) growth and development or 40 weeks from the beginning of the last menstruation (pregnancy), an infant is born.

18.1 Fertilization and Early Development

Learning Objective

1. Describe the processes of fertilization, preembryonic development, and implantation.

Recall from chapter 17 that each **primary oocyte** undergoes the first meiotic division while still in the ovarian follicle. This division forms a **secondary oocyte** and the first **polar body,** each containing 23 chromosomes. At ovulation, the secondary oocyte and first polar body, still enclosed within a sphere of granulosa cells, are released into a uterine tube. They are slowly moved toward the uterus through peristalsis and the beating cilia of epithelial cells lining the uterine tube.

Fertilization

After semen is deposited in the vagina, sperm begin their long journey into the uterus and on into the uterine tubes. Sperm inherently swim against the slight current of fluid that flows from the uterine tubes through the uterus and into the vagina, which helps to guide sperm toward the uterine tubes. Of the millions of sperm in the semen, only a few hundred sperm actually enter the uterine tubes. Prostaglandins in semen stimulate reverse peristalsis of the uterus and uterine tubes that greatly aids the migration of sperm. Sperm reach the upper portions of the uterine tubes within one hour after sexual intercourse. Usually, only one uterine tube contains a secondary oocyte because only one secondary oocyte is usually released at ovulation. Sperm entering an empty uterine tube have no chance for fertilization.

Within the uterine tube, sperm are chemically attracted to the secondary oocyte. Many sperm cluster around the oocyte and attempt to penetrate the granulosa cells (figure 18.1). The acrosomes of the sperm release enzymes that dissolve the "glue" holding granulosa cells together

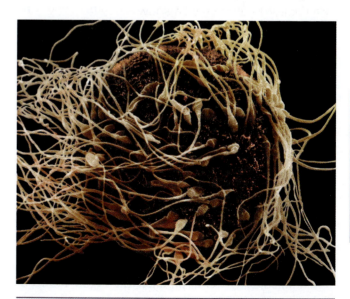

Figure 18.1 Secondary Oocyte Surrounded by Sperm. Scanning electron photomicrograph (1,200×). Only one sperm may enter the oocyte.
©David M. Phillips/Science Source

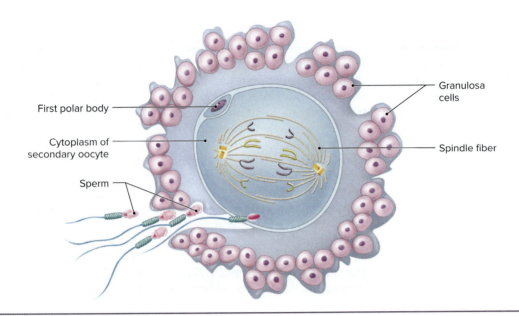

Figure 18.2 Fertilization of a Secondary Oocyte.
Enzymes from many sperm are required to separate the granulosa cells, enabling one sperm to penetrate the plasma membrane of the secondary oocyte. Sperm penetration triggers the second meiotic division within the secondary oocyte in order to form the ovum.

so they can reach the oocyte. It takes many sperm to disperse the granulosa cells, so that one sperm can eventually wriggle between them to contact the oocyte. The acrosome then releases a different enzyme that enables the sperm to penetrate the oocyte membrane and enter the oocyte. Once this happens, changes in the oocyte plasma membrane prevent other sperm from entering (figure 18.2).

When a sperm enters the secondary oocyte, it triggers the second meiotic division, which forms the **ovum** and a second polar body. Then, the sperm nucleus and ovum nucleus unite in **fertilization** to form a **zygote**. The zygote contains 46 chromosomes, 23 from the sperm and 23 from the ovum, and leads to the development of a human being.

A secondary oocyte remains viable for about 24 hours after ovulation. Most sperm remain viable in the female reproductive tract for about 72 hours, although some may be viable for up to five days. Therefore, fertilization is most likely to occur when sexual intercourse occurs from three days before ovulation to one day after ovulation.

Preembryonic Development

Preembryonic development occurs between fertilization and the end of the second week of development. During this time, the developing offspring is called a **preembryo.**

Immediately after fertilization, the zygote begins to divide by mitosis. The early cell divisions are collectively known as *cleavage*. These divisions occur so rapidly that maximum cellular growth between divisions is not possible, which results in increasingly smaller cells. During this time, the preembryo is carried along the uterine tube by peristalsis and the beating cilia of epithelial cells lining it. By the time the preembryo reaches the uterus, it consists of a solid ball of cells called a *morula* (mor'-u-lah) that is not much larger than the zygote.

A continued but slightly slower rate of mitosis forms a larger hollow ball of cells called a **blastocyst** (blas'-to-sist). Located within the blastocyst is the *embryoblast* or *inner cell mass*, a specialized group of cells from which the embryo later develops. The outer wall of the blastocyst is called the *trophoblast*, which later will form the embryonic portion of the placenta.

About the seventh day of development, the blastocyst attaches to the endometrium. Digestive enzymes, released

 Clinical Insight

Identical, or *monozygotic, twins* develop from a single zygote; this means that the twins possess identical genetic characteristics. The embryoblast of the blastocyst separates completely, usually by the end of the first week of embryonic development, and results in two embryos within separate amnion sacs yet sharing a common chorion and placenta. *Fraternal*, or *dizygotic, twins* develop from two zygotes: two different secondary oocytes are fertilized by different sperm. These twins do not possess identical genetic characteristics and develop within separate amnions and chorions. Each embryo develops its own placenta, though the placentas may fuse if they are located near each other within the uterus.

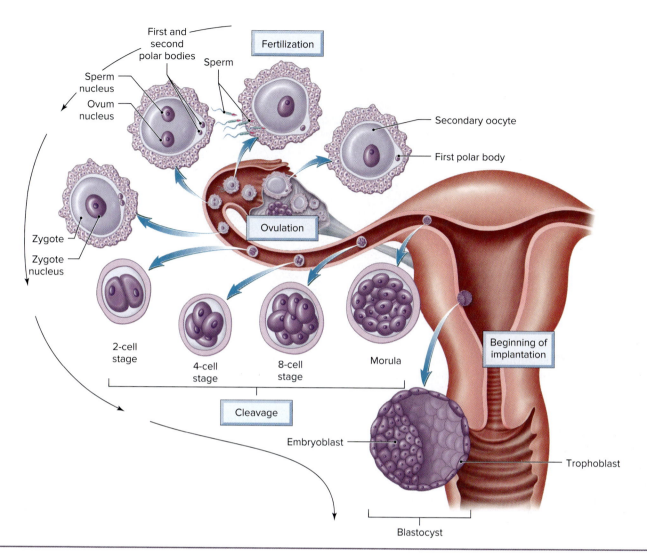

Figure 18.3 The Stages of Fertilization, Cleavage, and Implantation. Implantation begins on about the seventh day of development. AP|R

from the trophoblast, enable the blastocyst to penetrate into the endometrium, where it is soon covered by the outer endometrium. This entire process is called **implantation** and is completed by the fourteenth day (figure 18.3).

 Check My Understanding

1. How do sperm reach the secondary oocyte?
2. What are the major events that occur from sperm penetration of a secondary oocyte to implantation?

18.2 Embryonic Development

Learning Objectives
2. Describe the major events of embryonic development.
3. Explain the importance of the three germ layers.

4. Identify the extraembryonic membranes and their functions.
5. Describe the structure and function of the placenta.

The embryonic stage of development begins at the start of the third week of development and is completed at the end of the eighth week. The developing offspring during this time is called an **embryo.** The embryo undergoes rapid development, forming the rudiments of all body organs, extraembryonic membranes, and the placenta. By the end of the eighth week, it has a distinct human appearance.

Germ Layers

After implantation, the embryoblast grows to become the *embryonic disc*, which is supported by a short stalk extending from the wall of the blastocyst. The embryonic disc consists of three embryonic tissues: ectoderm,

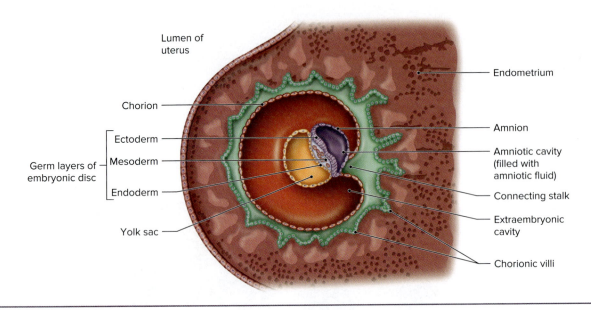

Figure 18.4 The Germ Layers.
Development of the germ layers of the embryonic disc and three extraembryonic membranes—chorion, amnion, and yolk sac—in an embryo. Note the chorionic villi.

mesoderm, and endoderm. The ectoderm forms the back of the developing embryo, while the endoderm forms the front. The mesoderm is the middle tissue layer. These embryonic tissues are called **germ layers** because all body tissues and organs are formed from them. Figure 18.4 illustrates the formation of the germ layers.

Briefly, **ectoderm** forms all of the nervous system and the epidermis of the skin. **Mesoderm** forms muscles, bone tissue, blood, and other forms of connective tissues. **Endoderm** forms the epithelial lining of the digestive, respiratory, and urinary tracts. Table 18.1 provides a more detailed listing of the major structures formed by each primary germ layer.

Extraembryonic Membranes

While the embryonic disc is forming, slender extensions from the trophoblast grow into the surrounding endometrium, firmly anchoring the blastocyst. The trophoblast of the blastocyst is now called the **chorion** (kō'-rē-on), the outermost extraembryonic membrane, and the slender extensions are known as **chorionic** (kō-rē-on'-ik) **villi**.

Table 18.1 Structures Formed by the Primary Germ Layers

Ectoderm	Mesoderm	Endoderm
All nerve tissue	All connective tissues including bone tissue and cartilage tissue	Epithelial lining of alimentary canal (except oral cavity and anal canal)
Sensory epithelium of sense organs		
Pituitary gland, pineal gland, and adrenal medulla	Skeletal, cardiac, and most smooth muscle tissue	Liver and pancreas
Epidermis of the skin including nails, hair follicles, sebaceous glands, sweat glands, and mammary glands	Red bone marrow and lymphoid tissues	Epithelial lining of respiratory system (except nasal cavity and paranasal sinuses)
	Blood and lymphatic vessels	Epithelial lining of urethra and bladder
Cornea, lens, and muscles within the eye	Dermis of the skin	Thymus, thyroid and parathyroid glands
Epithelial lining of oral and nasal cavities, paranasal sinuses, and anal canal	Kidneys, ureters, ovaries, testes, and reproductive ducts	Epithelial lining of accessory reproductive glands
Salivary glands	Adrenal cortex	Epithelial lining of tonsils, auditory tubes, and tympanic cavity
Tooth enamel	Synovial and serous membranes	
Cranial and spinal meninges	Spleen	

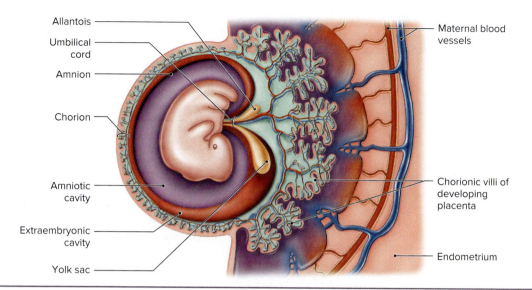

Figure 18.5 Extraembryonic Membranes.
Amniotic fluid surrounds the embryo as the amnion continues to grow. The yolk sac and allantois become part of the developing umbilical cord. The chorion surrounding the embryo becomes thinner, the chorionic villi grow into the endometrium, and embryonic blood vessels extend into the chorionic villi as the placenta develops.

At about the same time, two other extraembryonic membranes separate from the embryonic disc. The **amnion** (am′-nē-on) forms posterior to the embryo and the **yolk sac** forms anterior to the embryo (figure 18.4). *Amniotic fluid* fills the *amniotic cavity*, the space between the embryonic disc and the amnion. As the embryo develops, the amnion margins move toward the front of the embryo. In a short time, the embryo is enveloped by the amnion (figure 18.5).

Amniotic fluid serves as a shock absorber for the developing embryo. It also prevents adhesions from developing between various parts of the embryo. Later in development, the fetus swallows and inhales amniotic fluid and discharges dilute urine into it.

The yolk sac forms an outpocketing that becomes the **allantois** (al-lan′-to-is), the last of the extraembryonic membranes. Both the allantois and the yolk sac subsequently become part of the **umbilical** (um-bil′-i-kal) **cord,** which attaches the embryo to the placenta (figure 18.5). The yolk sac forms the early formed elements and germ cells for the embryo. It also serves as a shock absorber for the embryo, in addition to forming the primitive gut. The allantois also produces formed elements and brings umbilical blood vessels to the placenta.

Placenta

As the embryo continues to grow and the chorion enlarges, the layer of the endometrium covering the blastocyst becomes increasingly thinner. The chorionic villi in this region disintegrate, and only those in contact with the thick, spongy endometrium persist. This leads to the formation of the **placenta** (plah-sen′-tah), a disc-shaped structure formed of both embryonic and maternal tissues. The embryonic portion is formed of the chorion and chorionic villi, and the maternal portion is formed of the associated endometrium (figure 18.5).

The developing embryo is attached to the placenta by the umbilical cord. Two umbilical arteries bring embryonic blood to the placenta, and a single umbilical vein returns the blood to the embryo. There are no nerves in the umbilical cord.

The placenta provides an interface between the embryonic blood and maternal blood for the exchange of water, respiratory gases, nutrients, wastes, hormones,

⊕ Clinical Insight

Many substances are harmful to a developing embryo and fetus because they can pass across the placenta from the maternal blood into the blood of the embryo and fetus. Such harmful substances include alcohol, cocaine, heroin, nicotine, caffeine, and many therapeutic drugs. Physicians recommend that no drugs or other chemical substances be taken during pregnancy except those crucial for the mother's health.

In addition, pathogens of certain maternal infections can pass across the placenta and infect the fetus. These infections include AIDS, German measles, syphilis, and toxoplasmosis (transmitted from cat feces). All of these diseases can cause serious fetal disorders.

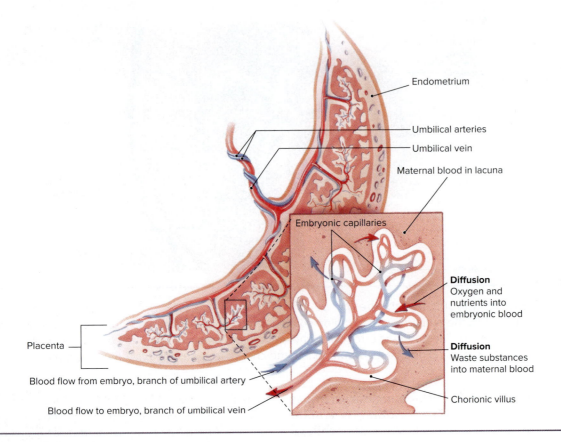

Figure 18.6 The Placenta.
The placenta is formed of both embryonic and maternal tissues. The chorion and chorionic villi form the embryonic portion, while the endometrium of the uterine wall forms the maternal portion. Materials are exchanged between embryonic blood and maternal blood by diffusion.

and antibodies. The embryonic blood must receive all required substances from the mother's blood and pass metabolic wastes into the mother's blood. The placenta is usually fully functional by the end of the eighth week.

Embryonic blood and maternal blood do not mix in the placenta. The maternal blood vessels open into blood-filled spaces called *lacunae* into which the embryonic blood vessels extend and branch repeatedly. This arrangement provides a large surface area for the exchange of substances between embryonic blood and maternal blood. Observe this relationship in figure 18.6. It is important to note that the placenta continues to serve in this capacity during fetal development as well.

External Appearance

By the fourth week of development, the head and limb buds of the embryo are recognizable. By the seventh week, the rudiments of all organs are present, and the eyes and ears are visible. Figure 18.7 illustrates the visible changes in the embryo from the fourth to the seventh week.

 Check My Understanding

3. What are the three primary germ layers and what does each form?
4. What are the extraembryonic membranes, and what are their roles?
5. What composes the placenta and what is its function?

18.3 Fetal Development

Learning Objective

6. Describe the major changes that occur during fetal development.

At the beginning of the ninth week of development, the developing offspring has a distinctively human appearance and is now referred to as a **fetus** (figure 18.8). Table 18.2 highlights the changes in physical appearance that take place during fetal development.

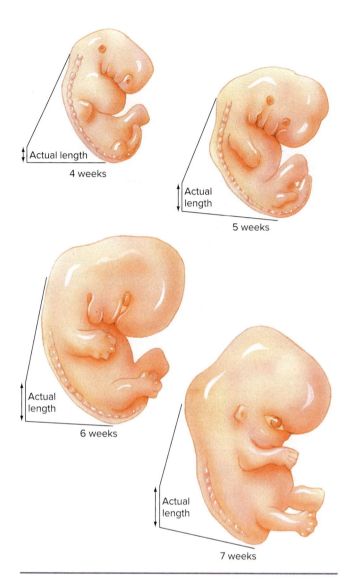

Figure 18.7 Embryonic Development.
Outer appearance of the embryo from the fourth to the seventh week of development.

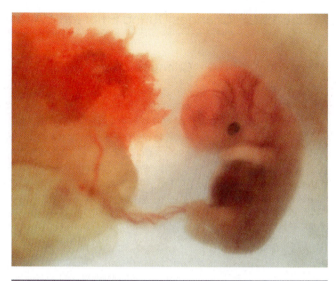

Figure 18.8 Photograph of an Embryo.
This photograph of an eight-week embryo shows the umbilical cord and the enveloping amnion.
©Scott Camazine/Phototake

20th week, the fetus weighs about 460 grams, or approximately 1 pound.

Continued growth of organ systems during the 21st through 29th weeks enables the head, body, and limbs to attain infant proportions. However, the fetus is still lean, with wrinkled, rather translucent skin. By the 29th week, the fetus weighs about 1,300 grams, or approximately 3 pounds.

Deposition of adipose tissue in the subcutaneous tissue plus continued growth and development of organ systems, as well as the descent of the testes in males, occur in the eighth and ninth months. By the time the fetus is full term, the lanugo has been shed, the skin is pinkish with ample subcutaneous tissue, hair covers the scalp, and all organ systems are ready for birth. The fetus weighs about 3,400 grams, or about 7.5 pounds.

6. When does fetal development begin?

18.4 Hormonal Control of Pregnancy

Learning Objective
7. Explain the hormonal control of pregnancy.

Without the formation of a preembryo, the corpus luteum degenerates about two weeks after ovulation as a result of the decline in luteinizing hormone from the anterior lobe

At the start of the fetal stage, the head is as large as the body and all rudimentary organs are present. The ossification of bones begins during weeks 9 through 12 and organs continue to develop. By the 12th week, the fetus weighs about 45 grams (1.6 ounces) and is about the size of a candy bar.

During the 13th through 16th weeks, the eyes and ears reach their final positions and a heartbeat may be detected with a stethoscope. During the 17th through 20th weeks, fine hair (*lanugo*) covers the body and hair appears on the scalp. Sebum from sebaceous glands and dead epidermal cells form the *vernix caseosa*, which protects the skin from the digestive and urinary wastes in the surrounding amniotic fluid. Movements of the fetus may now be detectable by the mother. By the end of the

Table 18.2 Changes During Embryonic And Fetal Development

Weeks of Development	Changes
5–8 Weeks	Recognizable human shape; head as large as the body; eyes far apart; upper and lower limbs present with digits; heart possesses four chambers; tail disappears; formed elements produced by liver; organ systems present in rudimentary form
9–12 Weeks	Head is half the length of the body; nails form on digits; brain enlarges; eyes nearly fully developed but still far apart and eyelids still fused; ears formed but low set; nose bridge forms; upper limbs almost fully lengthened; heartbeat detectable; gender distinguishable by external genitalia; heartbeat detectable but not with stethoscope; ossification begins; fetus moves but not detectable by mother
13–16 Weeks	Body is larger than head; eyes and ears reach characteristic positions; facial features well developed; lips exhibit sucking movements; bones distinct; lower limbs lengthen; kidneys well formed; meconium (fetal feces) begins to form in intestines; heartbeat detectable by stethoscope
17–20 Weeks	Head more proportional to body; lanugo and vernix caseosa cover body; eyebrows, eyelashes, and head hair present; lower limbs fully lengthened; brown fat forms for future heat production; fetus assumes fetal position due to limited space; fetal movements felt by mother
21–25 Weeks	Rapid increase in weight gain; skin pink and wrinkled; surfactant formed by lungs
26–29 Weeks	Eyes open; body is lean; head and body proportional; skin still wrinkled and pinkish; subcutaneous tissue begins to form; testes begin to descend toward scrotum in males; red bone marrow begins formed element production; organ systems continue to develop
30–34 Weeks	Skin pinkish and smoother due to subcutaneous tissue deposition; testes continue to descend in males; bones still continuing to ossify; fetus assumes "head down" position
35–38 Weeks	More subcutaneous tissue is deposited; skin smoother and pinkish because melanin is not produced until skin is exposed to light; testes located within scrotum in males; lanugo is shed; vernix caseosa still present; skull bones largely ossified except at fontanelles; body larger than head

of the pituitary gland. The resulting decline in blood levels of estrogens and progesterone causes the endometrium to break down and to be shed with menstruation. Pregnancy causes hormone changes that maintain the endometrium.

The trophoblast of the blastocyst (see figure 18.3) secretes **human chorionic gonadotropin (hCG)** (kō-rē-on′-ik gōn-ah-dō-trō′-pin), an LH-like hormone that maintains the corpus luteum for the first 10 to 12 weeks of development. Recall that the trophoblast forms the chorion, which forms the embryonic portion of the placenta and is also a source of hCG. Under stimulation of hCG, the corpus luteum continues to secrete progesterone and estrogens to maintain the endometrium. Pregnancy tests are designed to detect the presence of hCG in a woman's blood or urine, usually within eight to ten days after fertilization.

 Clinical Insight

If the corpus luteum stops secreting progesterone and estrogens too quickly or if the placenta is too slow in starting to secrete these hormones, a decline in their concentrations results. Such a decline may detach the placenta from the uterine wall, causing a miscarriage.

The concentration of hCG in the blood rises sharply and peaks after approximately 10 weeks of development. Blood hCG level then declines and levels off between 16 and 20 weeks of development, leading to the degeneration of the corpus luteum. However, by approximately 12 weeks, the placenta takes over the role of producing estrogens and progesterone, which prevents loss of the endometrium when the corpus luteum degenerates. The ovaries remain inactive during the remainder of the pregnancy because the high level of progesterone in the blood suppresses secretion of gonadotropin-releasing hormone (GnRH) by the hypothalamus, which in turn prevents the release of follicle-stimulating hormone (FSH) and luteinizing hormone (LH) from the anterior lobe of the pituitary.

The blood concentrations of estrogens and progesterone continue to increase throughout pregnancy (figure 18.9). Note that the levels of estrogens increase faster than the level of progesterone. As pregnancy continues, placental estrogens and progesterone stimulate development of the mammary glands in preparation for milk secretion.

 Check My Understanding

7. What hormonal changes are triggered by the presence of an implanted embryo?

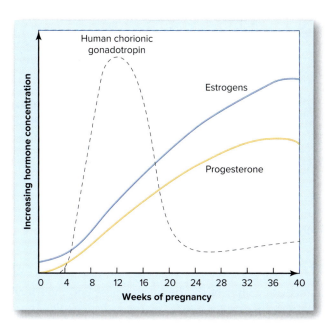

Figure 18.9 Hormone Levels During Pregnancy. This graph shows the relative concentrations of human chorionic gonadotropin (hCG), estrogens, and progesterone during pregnancy. Weeks of pregnancy are measured from the first day of the last menstruation. However, weeks of development are measured from the time of fertilization. Recall that fertilization usually occurs approximately two weeks after the beginning of the last menstruation, so that four weeks of pregnancy equals two weeks of development, eight weeks of pregnancy equals six weeks of development, and so on.

18.5 Birth

Learning Objective

8. Describe the neuroendocrine positive-feedback mechanism controlling labor.

During the latter stages of pregnancy, the blood concentrations of estrogens become increasingly greater than the blood concentration of progesterone, as noted earlier. Whereas progesterone inhibits uterine contractions, estrogens promote them. Therefore, there is an increasing tendency toward the onset of uterine contractions as the pregnancy approaches full term. These uterine contractions are often referred to as *Braxton Hicks contractions* or "false labor." Throughout pregnancy, the hormone **relaxin** is secreted first by the corpus luteum and then by the placenta. It helps in the development of blood vessels within the placenta, in addition to other cardiovascular changes that occur in the mother.

Birth usually occurs within two weeks of the calculated due date, which is 280 days from the beginning of the last menstruation. The birth process is called **parturition** (par-tū-rish′-un), and the events associated with parturition are collectively called **labor.** The fetus is usually in a "head down" position at this time (figure 18.10). As the fetus reaches full term, the high blood levels of estrogens override progesterone's inhibition of uterine contractions, allowing uterine contractions to occur. Pressure of the fetus on the cervix stretches the cervix, stimulating a neuroendocrine positive-feedback mechanism that promotes uterine contractions. In fact, physicians sometimes initiate labor by breaking the amnion so that increased pressure is placed on the cervix. The process of labor can be divided into three stages.

The *first stage of labor* is the dilation of the cervix. The stretching of the cervix triggers the formation of action potentials that are sent to the hypothalamus. When these reach a critical frequency, the hypothalamus activates the posterior lobe of the pituitary gland to release *oxytocin*, increasing its concentration in the blood. Oxytocin stimulates the characteristic rhythmic contractions that begin at the upper end of the uterus and move toward the cervix, pushing the fetus toward the *vagina*, or birth canal.

The continued contractions of the uterus force the fetus's head against the cervix, which results in greater stretching of the cervix. The increase in cervical stretching causes more action potentials to be sent to the hypothalamus. The hypothalamus then stimulates the posterior lobe of the pituitary to release more oxytocin into the blood. The higher blood level of oxytocin triggers more intense and frequent uterine contractions to occur, which in turn produce greater cervical stretching. This

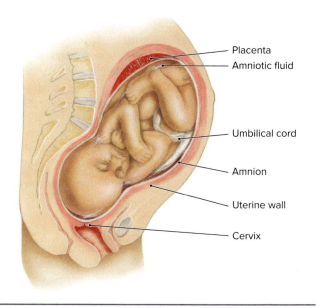

Figure 18.10 Full-Term Fetal Position. A full-term fetus positioned with the head against the cervix.

positive-feedback mechanism will produce increasingly stronger uterine contractions until birth occurs.

Dilation of the cervix is the longest stage of labor. It may last from 6 to 12 hours, depending on the size of the fetus and whether the mother has given birth previously. During this time, the amnion ruptures (or "water breaks") and the cervix dilates to the size of the fetus's head (figure 18.11).

The *second stage of labor* is the delivery (expulsion) of the fetus. It usually lasts less than an hour, with contractions occurring every two to three minutes and lasting about one minute. Once the head is expelled, the rest of the body exits rather quickly.

The *third stage of labor* is the delivery of the placenta. Within 15 minutes after birth of the infant, the placenta detaches. Continued contractions expel the placenta (the afterbirth). The placenta is checked carefully to see that all of it has been removed from the uterus because any residue may cause a serious uterine infection. Detachment of the placenta produces some bleeding because endometrial blood vessels at the placental site are ruptured. However,

 Clinical Insight

About 5% of births are *breech births* in which the fetus is presented buttocks first. This complicates the delivery and may require delivery by *cesarean section*. In a cesarean section, a transverse incision is made just above the pubic symphysis through the walls of the abdomen and uterus, through which the infant is extracted.

uterine contractions compress the broken blood vessels so that serious bleeding is usually avoided. Subsequently, the uterus decreases in size rather quickly.

First Breath

Immediately after birth, the infant's nose and mouth are aspirated to remove mucus or fluid that would impair breathing. The umbilical cord is clamped and cut, separating the infant from the placenta, which has served as its

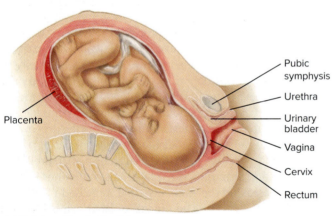
(a) **Fetal position before labor**

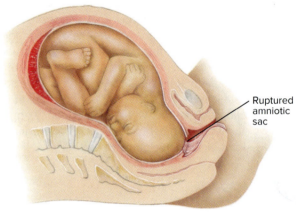
(b) **Dilation of the cervix**

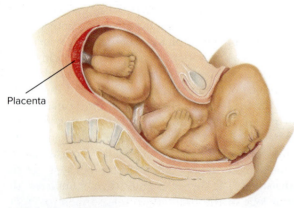
(c) **Expulsion of the fetus**

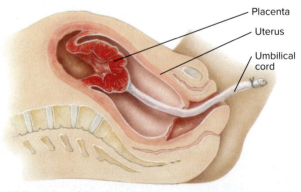
(d) **Expulsion of the placenta**

Figure 18.11 Stages of Labor.

Clinical Insight

Premature infants born before 24 weeks of development seldom survive. The respiratory system of such an infant is not sufficiently developed, even with the use of synthetic surfactant, to enable rapid gas exchange with blood in the lungs.

prenatal respiratory organ. As carbon dioxide increases in the infant's blood, the respiratory rhythmicity center in the medulla oblongata is activated and the infant's first inspiration is stimulated.

The first breath is difficult because the lungs are collapsed. In an infant, surfactant in pulmonary alveoli reduces surface tension, making the first breath and subsequent breathing easier.

Check My Understanding

8. How do increased levels of estrogens in the blood contribute to the onset of labor?
9. How does the neuroendocrine positive-feedback mechanism control uterine contractions?

18.6 Cardiovascular Adaptations

Learning Objectives

9. Explain the fetal cardiovascular adaptations and their value to the fetus.
10. Describe the cardiovascular changes that occur in postnatal development.

Fetal circulation is quite different from adult circulation because the digestive tract, lungs, and kidneys are not functioning. Oxygen and nutrients are obtained from the maternal blood in the placenta, while carbon dioxide and other metabolic wastes are removed via the maternal blood. The pattern of fetal circulation is an adaptation to these conditions. Birth immediately separates the infant from its supply of nutrients and oxygen and stimulates cardiovascular changes to accommodate independent living as an air-breathing human.

Fetal Cardiovascular Adaptations

Figure 18.12 illustrates the pattern of fetal circulation. Oxygenated and nutrient-rich blood is carried from the placenta to the fetus by the *umbilical vein*, which enters the fetus at the umbilicus (navel). Within the fetus, the umbilical vein passes toward the liver, where it divides into two branches. About half of the blood carried by the vessel enters the liver, while the other half bypasses the liver by flowing through the **ductus venosus** (duk′-tus ven-ō′-sus) and into the inferior vena cava. Full blood flow through the fetal liver is not necessary because the fetal intestines are nonfunctional and the mother's liver removes potentially hazardous substances before her blood enters the placenta. The oxygenated blood from the umbilical vein is mixed with deoxygenated blood in the inferior vena cava. The addition of blood from the ductus venosus increases the blood pressure within the inferior vena cava and the right atrium, which keeps the foramen ovale open.

Most of the blood entering the right atrium of the fetal heart passes directly through the **foramen ovale** (ō-vah′-lē), an opening in the interatrial septum, into the left atrium. The blood in the left atrium flows into the left ventricle and is pumped into the aorta for transport throughout the body. The blood that does enter the right ventricle is pumped through the pulmonary trunk. However, most of it bypasses the lungs by flowing through the **ductus arteriosus** (duk′-tus ar-te-rē-ō′-sus) into the aortic arch. These two lung bypasses work together to provide better oxygen and nutrient delivery to fetal tissues by providing additional blood for transport to the body. However, sufficient blood flows through the pulmonary circuit to maintain the nonfunctional lungs.

Blood is returned to the placenta by two *umbilical arteries* that branch from the internal iliac arteries. Trace the flow of blood through the fetal cardiovascular system shown in figure 18.12. Table 18.3 summarizes these fetal cardiovascular adaptations.

Postnatal Cardiovascular Changes

After the infant is breathing, changes are made to convert the pattern of fetal circulation into that of an air-breathing infant. This involves closure of the foramen ovale and the constriction of all of the vessels used to get blood quickly from the umbilical vein into the aorta.

The distal portions of umbilical arteries constrict, inhibiting the flow of blood to the placenta. Subsequently, the proximal portions of the umbilical arteries

Clinical Insight

Umbilical cord blood is rich in the stem cells required for the production of formed elements. After delivery and after the umbilical cord is cut, cord blood may be collected, saved, and kept frozen and available in case it is needed by the child later in life. If the child develops leukemia later in life and needs a transfusion of stem cells, the cord blood is available for transfusion. Because it is essentially a self-transfusion, rejection is not a problem.

Table 18.3 Fetal Cardiovascular Adaptations

Structure	Function
Umbilical vein	Carries oxygenated and nutrient-rich blood from the placenta to the fetus
Ductus venosus	Carries about half of the blood in the umbilical vein into the inferior vena cava, bypassing the liver and mixing oxygenated and deoxygenated blood
Foramen ovale	Allows a large portion of the blood entering the right atrium to pass through the interatrial septum directly into the left atrium, bypassing the pulmonary circuit and providing as much oxygen and as many nutrients as possible for body cells via the systemic circuit
Ductus arteriosus	Carries most blood from the pulmonary trunk directly into the aorta, bypassing the nonfunctional lungs and providing more blood with oxygen and nutrients for the systemic circuit
Umbilical arteries	Carry deoxygenated blood from the internal iliac arteries back to the placenta

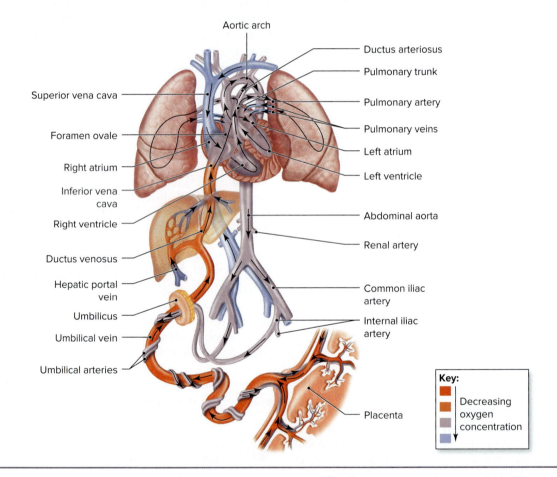

Figure 18.12 Cardiovascular Adaptations in a Fetus.

will persist as *superior vesical arteries* and the distal portions will become *medial umbilical ligaments*. Blood continues to flow through the umbilical vein from the placenta to the newborn for about one minute, and then it constricts. It will become the *round ligament*. At the same time, the ductus venosus constricts. It will subsequently become the *ligamentum venosum* in the wall of the liver.

As the umbilical vein constricts and the pulmonary circulation becomes functional, blood pressure in the right atrium decreases, whereas blood pressure in the left atrium increases due to increased blood flow to and from the lungs. The higher blood pressure in the left atrium closes a tissue flap in the left atrium over the foramen ovale, separating the pulmonary and systemic

glands in preparation for milk secretion, or **lactation** (lak-tā′-shun). Blood prolactin level also increases during pregnancy due to the action of various prolactin-releasing factors on the anterior lobe of the pituitary gland. Prolactin stimulates milk secretion by the mammary glands. However, elevated levels of estrogens and progesterone interfere with the activity of prolactin during pregnancy so that milk secretion does not occur.

After birth, the blood levels of estrogens and progesterone drop dramatically, which allows prolactin to stimulate milk secretion. However, its effects are not evident for two to three days. In the meantime, the mammary glands produce *colostrum* (ko-los′-trum), which differs from true milk by containing higher concentrations of protein and essentially no fat. The high protein content provides an added boost of essential nutrients for protein synthesis, which is needed for the continued development and growth of the infant.

The continued secretion of both prolactin and milk are maintained by mechanical stimulation of the nipples by the suckling infant. Suckling triggers the formation of action potentials that stimulate the hypothalamus to secrete prolactin-releasing factors, which in turn cause continued prolactin secretion by the anterior lobe of the pituitary. Thus, lactation is maintained by a *positive-feedback* mechanism.

Milk does not simply flow from the breasts. Instead, it is ejected after about 30 seconds of suckling by the infant. Stimulation of the nipple by suckling sends action potentials to the hypothalamus, which triggers the release of oxytocin by the posterior lobe of the pituitary. *Oxytocin* stimulates contraction of specialized epithelial

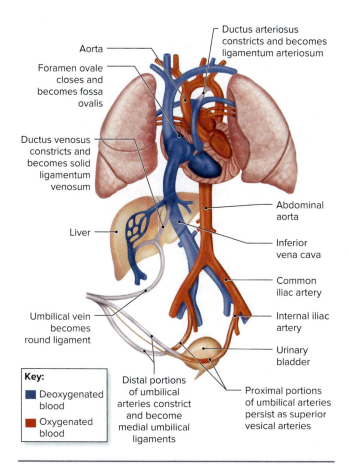

Figure 18.13 Cardiovascular Changes in a Newborn Infant.

circuits. Fusion of the tissue flap leaves a slight depression in the interatrial septum, called the *fossa ovalis*, at the site of closure. About the same time, the ductus arteriosus constricts and ultimately becomes the *ligamentum arteriosum*. These cardiovascular changes are functionally complete within 30 minutes after birth, but it takes about one year for tissue growth to make them permanent (figure 18.13).

 Check My Understanding

10. What are the cardiovascular adaptations in a fetus?
11. What cardiovascular changes occur in a newborn infant?

18.7 Lactation

Learning Objective

11. Describe the control of lactation and milk ejection.

High blood levels of estrogens and progesterone during pregnancy stimulate the development of the mammary

 Clinical Insight

Breast-feeding seems to provide advantages for the infant, including (1) better nutrition because nutrients are easier to absorb, (2) rapid bonding due to prolonged contact with the mother, (3) antibodies that prevent digestive inflammation and provide defense against pathogens, and (4) enhanced cognitive development.

 Clinical Insight

Pitocin, the brand name for oxytocin, can be administered intravenously to induce labor, reinforce ongoing labor contractions, and control postpartum bleeding and hemorrhage. The oxytocin released during breast-feeding promotes the uterine contractions needed to return the uterus to near its nonpregnant size.

cells surrounding the ducts and alveolar glands within the mammary glands, resulting in milk ejection or "letdown."

Milk production may continue for as long as the nipple is suckled, but the volume of milk produced gradually declines. If suckling is stopped, milk accumulates, the secretion of prolactin is inhibited, and milk production ceases in about a week.

Check My Understanding

12. How do hormones control lactation?

18.8 Disorders of Pregnancy, Prenatal Development, and Postnatal Development

Learning Objective
12. Identify the major disorders of pregnancy, prenatal development, and postnatal development.

Pregnancy Disorders

Eclampsia (ē-klamp'-sē-ah), or *toxemia of pregnancy*, is a disorder that occurs in two forms. *Preeclampsia* of late pregnancy is characterized by increased blood pressure, edema, and proteinuria (protein in the urine). The cause is unknown. If unsuccessfully treated, it may develop into *eclampsia*, a far more serious disorder that may lead to convulsions and coma. Both infant and maternal mortality are high in eclampsia. Rapid termination of the pregnancy by cesarean section may be indicated.

An **ectopic pregnancy** is the implantation of a preembryo anywhere other than in the uterus. A common site is in a uterine tube. Treatment involves surgical removal of the embryo.

A **miscarriage** is a spontaneous abortion. Most miscarriages occur within the first 12 weeks of development as a result of gross abnormalities of the embryo or placenta. Another cause is the untimely transfer of the production of estrogens and progesterone from the corpus luteum to the placenta at approximately 12 weeks of development.

Morning sickness is characterized by nausea and vomiting upon getting up in the morning. It usually starts around the sixth week of the pregnancy and typically lasts from one to six weeks. The exact cause is unknown, though high levels of hCG and progesterone in the blood are believed to play a role. About 60% of pregnant women experience this discomfort.

Prenatal and Postnatal Disorders

Birth defects may be inherited or may be caused by a variety of *teratogens* (ter-ah'-to-jens), environmental agents that produce physical abnormalities during prenatal development. Teratogens include alcohol, illegal drugs, some therapeutic drugs, X-rays, and certain diseases such as German measles (Rubella). Generally, the earlier the embryo is exposed, the greater the defect produced. Alcohol is the most common teratogen. It produces *fetal alcohol syndrome*, which is characterized by a small head; mental retardation; facial deformities; and abnormalities of the heart, genitals, and limbs.

Physiological jaundice, a postnatal disorder, sometimes occurs in a newborn because the destruction of RBCs occurs faster than the liver can process the bilirubin. This results in excess bilirubin in the blood. Phototherapy (exposure to UV light) is a common treatment to speed up bilirubin breakdown. Jaundice may be a more serious problem in premature infants. This type of jaundice usually resolves once the newborn's liver gains full function.

Infant respiratory distress syndrome (IRDS), or *hyaline membrane disease,* is a postnatal disorder characterized by an inability to produce surfactant within the lungs of an infant. The lack of surfactant decreases the ability of the infant to successfully inflate its lungs during inspiration. IRDS is most common in premature infants, whose lungs have not yet begun or have not completed surfactant production.

Sudden infant death syndrome (SIDS), or "crib death," is a postnatal disorder characterized by the sudden death of an infant with no medical history or explanation upon autopsy. Infants are at highest risk of SIDS during sleep and, though its exact cause is unknown, risk factors such as hypoxia while sleeping, deficits in respiratory control, and nicotine exposure during development have been identified.

18.9 Genetics

Learning Objectives
13. Explain the roles of DNA, genes, and chromosomes in inheritance.
14. Describe the basic patterns of inheritance.

Genetics is the study of heredity, the passing of inherited traits from one generation to the next. The determiners of hereditary traits are located on **chromosomes,** consisting of DNA and proteins. It is the DNA that controls inheritance and directs cellular functions.

Each human body cell contains 46 chromosomes that exist as 23 unique pairs. Chromosome pairs 1 through 22 are called *autosomes* because they control most inherited traits except gender. Gender is determined by chromosome pair 23, the *sex chromosomes*. There are two types of sex chromosomes, a large X chromosome and a small Y chromosome. Males possess one X chromosome and one Y chromosome (XY). Females possess two X chromosomes (XX).

A person's chromosomes, including the sex chromosomes, may be examined by making a *karyotype*. The

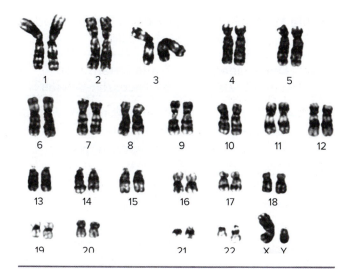

Figure 18.14 Karyotype of a Human Male.
The only difference in a female karyotype would be the presence of a pair of X chromosomes in place of the XY pair of the male.
©Biophoto Associates/Science Source

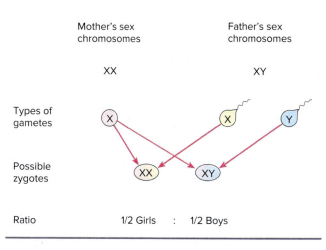

Figure 18.15 The Inheritance of Sex.

chromosomes in a dividing cell are photographed during metaphase (see chapter 3) and the photograph is enlarged. Then, the chromosomes are cut out, matched in pairs, and arranged by size and location of the centromere. Figure 18.14 is a karyotype of a healthy male. Note the X and Y chromosomes and that the chromosomes are arranged in pairs.

Sex Determination

Recall that gametes are formed by meiosis, a process that places one member of each chromosome pair in each gamete. Each human gamete contains 23 chromosomes—22 autosomes and 1 sex chromosome. We will consider only the sex chromosomes here.

Because a female has two X chromosomes in her cells, all of her gametes contain an X chromosome. A male has both an X chromosome and a Y chromosome in his cells. Therefore, half of his gametes are X-bearing, and half are Y-bearing. If a secondary oocyte is fertilized by an X-bearing sperm, the child will be a girl. If a secondary oocyte is fertilized by a Y-bearing sperm, the child will be a boy. Obviously, the probability of any zygote becoming a girl (or a boy) is one-half, or 50%. Figure 18.15 illustrates the inheritance of sex.

Genes

Recall from chapters 2 and 3 that DNA consists of a double strand of nucleotides that are joined by complementary pairing of their nitrogenous bases: adenine (A), thymine (T), cytosine (C), and guanine (G). The sequence of these bases forms the genetic code, which contains the information for producing proteins that regulate cellular functions and determines the inheritance of genetic traits.

A **gene** is a unit of inheritance. It consists of a specific sequence of DNA that codes for a unique molecule of RNA. This RNA molecule will be either directly involved in the synthesis of a polypeptide or indirectly involved in regulating the production of a polypeptide. Genes occur in a linear sequence along a chromosome, and a single chromosome may contain hundreds of genes.

Because chromosomes occur in pairs, genes also occur in pairs. An inherited trait is determined by at least one pair of genes. There may be two or more alternate forms of a gene controlling the expression of a particular trait. These alternate forms are called **alleles** (ah-lēls′), and each allele affects the expression of a trait differently. So, in the simplest case, a trait is determined by one pair of alleles present in a person's cells. If the two alleles for a trait are identical, the person is **homozygous** for that trait; if they are different, the person is **heterozygous** for that trait.

Gene Expression

Each person's chromosomes are composed of a unique assortment of genes, the **genotype** for that person. The expression of those genes yields observable traits known as the **phenotype.** Though the phenotype is what is seen, the genotype is responsible for the inheritance and expression of those traits.

Dominant and Recessive Inheritance

Some alleles are dominant, and some are recessive. A *dominant allele* is always expressed, whereas a *recessive allele* is expressed only when both alleles are recessive.

Consider the example of earlobe attachment. Unattached earlobes are controlled by a dominant allele *(A).* Attached earlobes are controlled by a recessive allele *(a).* Note that it requires only one dominant allele to express

Clinical Insight

From 1990 to 2003, the *Human Genome Project* was an international research project that determined the base sequence of more than 92% of the human genome and mapped the base sequences of known genes. This information has greatly accelerated genetic discoveries, such as identifying genes that cause human genetic disorders. Treatment regimens for genetic disorders have also advanced through development of targeted drug therapies and genetic engineering. For example, great strides have been made in preventing or modulating the effects of autoimmune disorders, such as type I diabetes mellitus, by transferring therapeutic genes into affected cells in mice. These findings have definite future human applications. Human trials involving selective destruction of the autoimmune T cells that destroy beta cells within the pancreas have also shown promising results.

the dominant trait but that both recessive alleles must be present for the recessive trait to be expressed.

Genotype	Phenotype
AA	Unattached earlobes
Aa	Unattached earlobes
aa	Attached earlobes

Table 18.4 indicates a few human traits that are determined by dominant and recessive alleles.

Incomplete Dominance

Incomplete dominance is a type of inheritance where the two alleles for a gene can create three different phenotypes. Each genotype–homozygous dominant, heterozygous, and homozygous recessive–has a different phenotype. An example is sickle-cell disease, a condition characterized by defective hemoglobin that cannot carry adequate oxygen. Erythrocytes with the defective hemoglobin assume a characteristic sickled or crescent shape. Sickle-cell disease occurs among people whose ancestors lived in central Africa. About 8.3% of black Americans possess the allele for sickle-cell disease.

A person who inherits both recessive alleles for sickle-cell disease ($H^S H^S$) produces unhealthy hemoglobin, leading to the formation of sickled cells that cannot carry sufficient oxygen. Because of their shape, the sickled cells tend to plug capillaries. Symptoms include pain in joints and the abdomen and chronic kidney disease.

In the heterozygous state (HH^S), some hemoglobin molecules are healthy but others are unhealthy. Fortunately, few RBCs become sickled when oxygen is at a healthy level and clinical symptoms are absent at such times. However, more RBCs become sickled during

Table 18.4 Examples of Traits Determined by Dominant and Recessive Alleles

Traits Determined by Dominant Alleles	Traits Determined by Recessive Alleles
Freckles	Absence of freckles
Dimples in cheeks	Absence of dimples
Dark hair	Light hair or red hair
Full lips	Thin lips
Unattached earlobes	Attached earlobes
No thalassemia	Thalassemia
Feet with arches	Flat feet
Huntington disease	No Huntington disease
Astigmatism	No astigmatism
Farsightedness	No farsightedness
Panic attacks	No tendency to panic attacks
Extra fingers or toes	Ten fingers and ten toes
No cystic fibrosis	Cystic fibrosis
No hemophilia	Hemophilia*
Type A, B, or AB blood	Type O blood
Type Rh+ blood	Type Rh− blood
Full color vision	Red-green color blindness*
No gout tendency	Gout*

*X-linked trait.

times of decreased blood oxygen level, a characteristic that allows detection of carriers of the sickle-cell allele. The heterozygote state affords some protective advantage against the pathogen causing malaria. The homozygous dominant genotype produces the phenotype of all healthy hemoglobin.

Codominance

In some traits, both alleles are expressed and affect the phenotype. This type of inheritance is referred to as **codominance.** An example of codominance can be seen with the ABO blood group. There are three alleles involved: a dominant I^A that causes the production of the A antigen; a dominant I^B that causes the production of the B antigen; a recessive i that has no function. If both I^A and I^B are present, both alleles are expressed. Because the recessive i has no function, genotype ii produces neither A nor B antigens. This is called type O blood, which simply means there are no A or B antigens. The possible genotypes and phenotypes for the ABO blood group are

Genotypes	Phenotypes
$I^A I^A$, $I^A i$	Type A blood
$I^B I^B$, $I^B i$	Type B blood
$I^A I^B$	Type AB blood
ii	Type O blood

Polygenetic Inheritance

Many traits are controlled by **polygenes,** a number of different genes that may be located on the same or different chromosomes. Each gene contributes to the phenotype, though some genes may have more influence on the trait than others. To add to the complexity of polygenic inheritance, each gene involved may possess a number of different alleles. Environmental factors may also exert influence over the expression of a phenotype. For these reasons, it is difficult to predict the inheritance of polygenic traits. Examples of traits controlled by polygenes are height, skin pigmentation, and intelligence.

The ABO blood group is also governed by polygenes. The gene for the H antigen is found on chromosome 19. The H gene possesses two alleles: a dominant *H* that causes the production of H antigen and a recessive *h* that is nonfunctional. Individuals who are homozygous dominant (*HH*) or heterozygous (*Hh*) possess the H antigen. The I^A and I^B alleles, which are located on chromosome 9, produce enzymes that add to the H antigen and produce either A or B antigens. Many people mistakenly conclude that type O blood has no antigens because the *i* alleles have no function. However, most people with blood type O actually have H antigens. Individuals with genotype *hh* do not produce the H antigen and have what is called the *Bombay phenotype*. These individuals will be type O even if their genotype contains the I^A, I^B, or both I^A and I^B alleles because, without the H antigen, A and B antigens cannot be formed.

X-Linked Traits

A few traits are determined by genes on the X chromosome. These are **X-linked,** or *sex-linked,* **traits.** Recessive X-linked traits affect males more frequently than females. Males possess only one X chromosome. If a recessive trait is carried by the X chromosome in a male, the trait will be seen. Females possess two X chromosomes. To see the recessive trait, a female must possess two recessive alleles. If the female possesses one dominant "healthy" allele, the recessive trait will not be seen.

Red-green color blindness is a common X-linked recessive trait. A color-blind male inherits the allele for color blindness from his mother, who provides his X chromosome. The mother may either have full color vision or be red-green color-blind (table 18.5). It is important to note that if the mother has full color vision, she still possesses the allele for color blindness and is considered a carrier for the color-blindness trait.

Predicting Inheritance

Parents often wonder about the chances of their child developing certain inherited traits. This can be predicted for some traits for which the inheritance pattern has been determined and if the genotypes of the parents are known. Such predictions indicate the *probability*, rather than absolute certainty, that a trait will be inherited.

Table 18.5 Possible Genotypes and Phenotypes for Red-Green Color Blindness, an X-Linked Trait

	Genotype	Phenotype
Females	$X^C X^C$	Full color vision
	$X^C X^c$	Full color vision carrier
	$X^c X^c$	Color-blind
Males	$X^C Y$	Full color vision
	$X^c Y$	Color-blind

C = Allele for full color vision; c = Allele for color blindness.

Let's consider freckles. Freckles are determined by a dominant allele *(F)*, and a nonfreckled phenotype is determined by a recessive allele *(f)*. The possible genotypes and phenotypes are

Genotypes	Phenotypes
FF	Freckled
Ff	Freckled
ff	Nonfreckled

Figure 18.16 shows how to determine the probability of the freckled or nonfreckled trait in the next generation if the genotypes of the parents are known. In this example, the

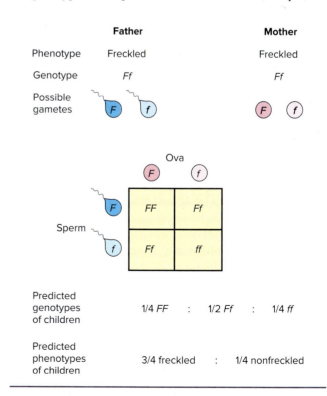

Figure 18.16 Predicting Inheritance.
This diagram shows the probability of freckles being inherited by children of parents who are heterozygous for freckles.

parents are known to be heterozygous for freckles. What is the probability that their children will be freckled?

Because each parent is heterozygous, meiotic division during gamete formation causes half of the gametes of each parent to contain an allele for freckles (*F*), and half to carry an allele for normal pigmentation (*f*). The union of sperm and secondary oocyte occurs at random (by chance), so we must allow for all possible combinations of gametes. This is accomplished by using a *Punnett square* (a chart named after Reginald Punnett, a geneticist).

The alleles in ova are placed along the horizontal axis, while the alleles in sperm are placed along the vertical axis. Next, the allele of each ovum is written in the squares below each ovum, while the allele of each sperm is written in the squares to the right of each sperm. The Punnett square now shows all possible genotypes that may occur in the next generation.

From this information, the predicted genotype ratio may be determined. Then, knowing that the trait for freckles is dominant and that the presence of a single dominant allele (*F*) produces freckles, the predicted phenotype ratio may be determined. Note in figure 18.16 that it is possible for two heterozygous freckled parents to have a child with normal pigmentation. However, if one parent is homozygous dominant for freckles and the other is heterozygous for freckles, all children would be freckled.

The inheritance of any dominant/recessive trait may be determined in a similar manner.

Check My Understanding

13. What are the relationships among chromosomes, DNA, genes, and alleles?
14. What distinguishes the dominant/recessive pattern of inheritance?
15. Why are recessive X-linked traits expressed more often in males than in females?

18.10 Inherited Diseases

Learning Objective

15. Explain the inheritance of the more common inherited disorders.

Inherited, or genetic, diseases are caused by either chromosomal abnormalities or specific alleles. The development of advanced techniques and new knowledge makes an understanding of genetic disease increasingly important.

Chromosomal Abnormalities

Some genetic diseases are related to the presence of an additional chromosome or to the absence of a chromosome. These disorders result from errors that occur during meiosis, causing some gametes to receive both members

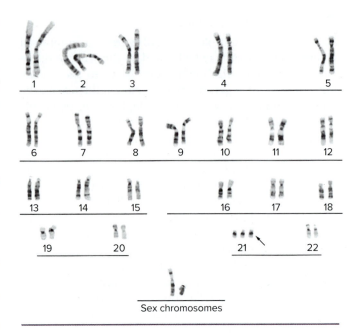

Figure 18.17 Down Syndrome Karyotype.
A karyotype of a male with Down syndrome caused by trisomy 21.
©McGraw-Hill Education/Courtesy of The National Human Genome Research Institute

of a chromosome pair while other gametes receive neither member. If such gametes are involved in zygote formation, a genetic disorder occurs. The genetic damage usually is so severe that it causes a spontaneous abortion. In some cases, the effect is not lethal but disabling.

An example of a disabling genetic disorder is Down syndrome, one of the more common genetic disorders. It is caused by the presence of an extra chromosome 21, as shown in figure 18.17. Down syndrome is characterized by mental retardation, short stature, short digits, slanted eyes, and a protruding tongue. A girl with Down syndrome is shown in figure 18.18. Infants with Down syndrome are born more often to mothers and fathers over 40 years of age.

Single-Gene Disorders

These disorders usually affect the infant's metabolism after birth, when it must depend on its own life processes, or they may appear later in life. Consider a few examples of single-gene disorders.

Cystic fibrosis, an autosomal recessive disorder, is the most common genetic disorder among Caucasians. It is caused by a missing chloride channel on mucus-secreting cells. This causes production of thick mucus that blocks respiratory airways and leads to an early death from respiratory infections.

Phenylketonuria (PKU), an autosomal recessive disorder, is due to a missing enzyme needed to metabolize

Figure 18.18 A Girl with Down Syndrome.
©George Doyle/Getty Images RF

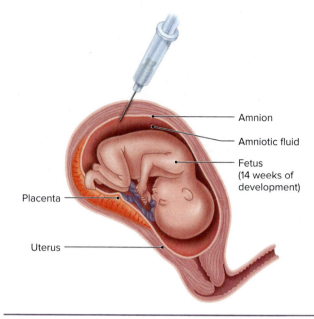

Figure 18.19 Amniocentesis.
In amniocentesis, a sample of amniotic fluid is withdrawn and the suspended fetal cells are examined for genetic abnormalities.

phenylalanine (an amino acid). Without treatment, mental and physical retardation result. A special diet that limits phenylalanine can prevent these effects if it is started at birth and continued to adulthood.

Tay-Sachs disease, an autosomal recessive disorder, primarily affects Jewish people of central European ancestry. An enzyme needed to metabolize a fatty substance associated with neurons is missing. The results are mental retardation, muscle weakness, seizures, and finally death, usually by two years of age.

Huntington disease, an autosomal dominant disorder, results from one or more missing enzymes needed in aerobic respiration. This causes a buildup of lactate in neurons in the brain. Uncontrollable muscle contractions, memory loss, and personality changes begin between 30 and 50 years of age. Death occurs within 15 years after the appearance of symptoms.

Hemophilia A and *hemophilia B,* X-linked recessive disorders, result from missing clotting factors. Prolonged bleeding can be life-threatening, and joints may be painfully disabled. Patients are dependent upon frequent transfusions of healthy plasma or intravenous injections of the missing clotting factor.

Genetic Counseling

Prospective parents who have genetic disorders in one or both of their families may benefit from genetic counseling. By collecting genetic information, a genetic counselor can inform prospective parents of the probability of a genetic disorder's appearance in their children. Genetic information may be collected from family histories and blood tests of the prospective parents and family members. If the woman is pregnant, ultrasound may be used to detect gross fetal abnormalities, and fetal cells may be obtained for examination. Fetal cells are obtained in two ways: amniocentesis and chorionic villi sampling.

In *amniocentesis,* a hollow needle is inserted through the abdominal and uterine walls of the mother and into the amnion to draw out a sample of amniotic fluid (figure 18.19). Fetal cells in the sample are grown by tissue culture and are analyzed to see if there are chromosomal abnormalities. Also, the amniotic fluid is analyzed for the presence of specific proteins that indicate serious neural defects. Amniocentesis is usually done around the 14th week of development, when ample amniotic fluid is available for sampling without injury to the fetus.

In *chorionic villi sampling,* a narrow tube is inserted through the cervix and fetal tissue from chorionic villi is suctioned out. Chromosomes in the fetal cells are examined for abnormalities. Chorionic villi sampling may be performed at 10 weeks and chromosome examination can be done immediately.

Both amniocentesis and chorionic villi sampling have inherent risks for mother and fetus. Fetal risks seem to be greater in chorionic villi sampling.

Fetal cells can also be collected for analysis through a procedure called *fetal cell sorting.* A small amount of fetal cells enter the mother's blood supply during pregnancy.

A fluorescent cell sorter can be used to identify and separate out the rare fetal cells from a maternal blood sample for analysis. This method of fetal cell collection circumvents the health risks associated with amniocentesis and chorionic villi sampling. Though these conventional methods are still more commonly used, fetal cell sorting is being used for purposes of sex determination and determination of fetal Rh status, in addition to the detection of major chromosomal abnormalities and some single-gene disorders.

Chapter Summary

18.1 Fertilization and Early Development

- Sperm usually encounter the secondary oocyte in a uterine tube. Only one sperm can enter an oocyte.
- When a secondary oocyte is penetrated by a sperm, it undergoes the second meiotic division, forming an ovum and a second polar body. Fertilization occurs with the fusion of the sperm nucleus and ovum nucleus, forming a zygote. The zygote is the single cell that leads to the development of a human being.
- The developing offspring is called a preembryo between fertilization and the end of the second week of development.
- The zygote undergoes cleavage divisions, forming a morula. Continued mitosis forms a blastocyst containing an embryoblast.
- On about the seventh day of development, the blastocyst becomes implanted in the endometrium.

18.2 Embryonic Development

- The developing offspring is called an embryo from the beginning of the third week through the end of the eighth week of development.
- The embryonic disc separates into the three germ layers: ectoderm, mesoderm, and endoderm.
- The trophoblast becomes the chorion, the first and outermost extraembryonic membrane. The chorionic villi will become the embryonic part of the placenta.
- The amnion develops behind the embryo, envelops the embryo, and becomes filled with amniotic fluid that serves as a shock absorber.
- The yolk sac develops on the front of the embryo. It subsequently branches to form the allantois. Both of these extraembryonic membranes form the first formed elements for the embryo. The allantois also brings umbilical blood vessels to the placenta.
- The placenta is formed of both embryonic and maternal tissues and is functional by the end of the eighth week. It allows the exchange of materials between embryonic blood and maternal blood.
- By the seventh week, the embryo exhibits a head, a body, limbs with digits, eyes, and ears.

18.3 Fetal Development

- The developing offspring is called a fetus at the beginning of the ninth week, and it clearly resembles a human. Organ systems are in rudimentary form and the head is as large as the body.
- Fetal development includes development of the organ systems to functional levels, an increase in size and weight, ossification of bones, and development of distinguishable gender.

18.4 Hormonal Control of Pregnancy

- Through 10 to 12 weeks of development, the trophoblast and chorion produce hCG, which maintains the corpus luteum in the ovary. The corpus luteum produces progesterone and estrogens, which in turn maintain the endometrium.
- By approximately 12 weeks, the placenta takes over the production of progesterone and estrogens. The corpus luteum degenerates and the ovaries remain inactive for the remainder of the pregnancy.

18.5 Birth

- Relaxin from the corpus luteum and placenta promotes the growth of placental blood vessels, in addition to other cardiovascular changes in the mother.
- The high blood levels of estrogens in the late stage of pregnancy counteract the inhibitory action of progesterone against uterine contractions.
- The onset of labor involves action potentials and oxytocin. Pressure of the fetus on the cervix leads to the formation of action potentials that are sent to the hypothalamus, which stimulates oxytocin release by the posterior lobe of the pituitary. Oxytocin stimulates uterine contractions, dilating the cervix, which triggers the formation of more action potentials. These interactions set up a positive-feedback mechanism that strengthens contractions until birth.
- Labor involves three stages: (a) dilation of the cervix, (b) delivery of the infant, and (c) delivery of the placenta.
- After birth, an accumulation of carbon dioxide stimulates the respiratory rhythmicity center to trigger the first breath. Surfactant in pulmonary alveoli makes breathing easier.

18.6 Cardiovascular Adaptations

- The pattern of fetal circulation is adapted to a life in which the lungs and digestive system are nonfunctional and nutrients and oxygen are derived from the mother's blood via the placenta. Embryonic blood is carried to the placenta by umbilical arteries and is returned by an umbilical vein.

- Cardiovascular adaptations pass oxygenated and nutrient-rich blood as quickly as possible from the umbilical vein to the aorta to meet the needs of body cells. These adaptations include the ductus venosus, foramen ovale, and ductus arteriosus.
- When the infant starts breathing, fetal cardiovascular adaptations are eliminated to enable an efficient separation of the pulmonary and systemic circuits.

18.7 Lactation

- The high blood levels of estrogens and progesterone during pregnancy prepare the mammary glands for lactation.
- Prolactin-releasing factors increase blood prolactin level during pregnancy. High levels of estrogens and progesterone block its action during pregnancy.
- After birth, the low blood levels of progesterone and estrogens allow prolactin to stimulate the mammary glands to secrete milk.
- Colostrum is secreted first, followed by true milk after two to three days.
- Suckling stimulates the formation of action potentials that are sent to the hypothalamus, which (a) secretes prolactin-releasing factors, ensuring the continued release of prolactin and milk secretion; and (b) stimulates the posterior lobe of the pituitary to secrete oxytocin, which results in milk ejection.

18.8 Disorders of Pregnancy, Prenatal Development, and Postnatal Development

- Pregnancy disorders include eclampsia, ectopic pregnancy, miscarriage, and morning sickness.
- Prenatal and postnatal disorders include birth defects, physiological jaundice, infant respiratory distress syndrome, and sudden infant death syndrome.

18.9 Genetics

- The determiners (genes) of hereditary traits are located on chromosomes. There are 46 chromosomes in human body cells: 22 pairs of autosomes and one pair of sex chromosomes.
- Sex chromosomes for females are XX; for males, XY.
- A gene is the unit of heredity. It is a portion of a DNA molecule that codes for a specific molecule of RNA. The RNA molecule is either directly or indirectly involved in the production of a polypeptide.
- Alternate forms of a gene are called alleles. A person with identical alleles for a trait is homozygous for that trait. If the alleles are different, the person is heterozygous.
- The genotype is the alleles present for a gene determining a trait. The phenotype is the observable characteristics determined by the genotype.
- Many genes have only two alleles whose expression is either by dominant/recessive inheritance or by incomplete dominance. Some traits are determined by codominance and others are determined by polygenetic inheritance.
- X-linked traits are determined by genes located on the X chromosome.
- The probability of transmitting traits to the next generation may be predicted using a Punnett square.

18.10 Inherited Diseases

- Genetic diseases may result from either chromosomal abnormalities or defective alleles.
- Prospective parents who have genetic diseases in their family histories may benefit from genetic counseling.

Improve Your Grade

Connect Interactive Questions Reinforce your knowledge using multiple types of questions: interactive, animation, classification, labeling, sequencing, composition, and traditional multiple choice and true/false.

SmartBook Proven to help students improve grades and study more efficiently, SmartBook contains the same content within the print book but actively tailors that content to the needs of the individual.

Anatomy & Physiology REVEALED® Dive into the human body by peeling back layers of cadaver imaging. Utilize this world-class cadaver dissection tool for a closer look at the body anytime, from anywhere.

APPENDIX A

Keys to Medical Terminology

MEDICAL TERMS MAY CONSIST OF three basic parts: a prefix, a root word, and a suffix. All terms have a root word, but some terms may lack either a prefix or a suffix. A **prefix** is the first portion of a term and comes before the root word. A **suffix** comes after the root word and is the last portion of a term. Both the prefix and the suffix modify the meaning of a root word, and they may be used with many different root words. The **root word** is the main portion of the term. Root words often occur at the beginning of a term, but they also may end a term or may be sandwiched between a prefix and a suffix. When determining the meaning of a term, you start with the suffix, then move to the prefix, and finally consider the root word. Consider these examples.

1. **Laryngitis** becomes **laryng/itis** when broken into its component parts.

    laryng- = the root word meaning *larynx*
    -itis = the suffix meaning *inflammation*

    Thus, the meaning of laryngitis is *inflammation of the larynx*.

2. **Endogastric** becomes **endo/gastr/ic** when broken into its component parts.

    endo- = the prefix meaning *within*
    gastr- = the root word meaning *stomach*
    -ic = the suffix meaning *pertaining to*

    Thus, the meaning of endogastric is *pertaining to within the stomach*.

The parts of a term are linked together in a way that aids pronunciation. This often requires the use of *combining vowels*. For example, when linking *gastr-* and *-pathy* to form a term meaning disease of the stomach, the vowel *o* is inserted to form *gastropathy*.

Some terms consist of more than one root word. For example, *gastr/o/enter/o/col/itis* means inflammation of the stomach, intestine, and colon.

You can see that once you know the meaning of common prefixes, root words, and suffixes, understanding medical terminology becomes much easier.

In the sections that follow, some common prefixes, suffixes, and root words are listed along with examples to help you understand medical terminology.

Singular and Plural Endings

Most medical terms are derived from Greek and Latin words. Therefore, changing from singular to plural is done by changing the ending of the term rather than by adding an *s* or *es* or changing a *y* to *ies* as in English terms. Examples of singular and plural endings are

Singular Ending	Plural Ending	Example
-a	-ae	pleura; pleurae
-en	-ena	lumen; lumena
-is	-es	testis; testes
-ma	-mata	carcinoma; carcinomata
-um	-a	epicardium; epicardia
-ur	-ora	femur; femora
-us	-i	glomerulus; glomeruli
-x	-ces	appendix; appendices

Common Prefixes

Prefix	Meaning	Example
a-	without, not	a/sepsis: sterile; without germs
ab-	away from, from	ab/duct: carry away from
ad-	to, toward	ad/duct: carry toward
an-	without, not	an/ergia: without energy
ante-	before	ante/cibum: before meals
anti-	against	anti/histamine: against histamine
bi-	two	bi/lateral: on two sides
bio-	life	bio/logy: study of life
brachy-	short	brachy/gnathia: shortness of the lower jaw
brady-	slow	brady/cardia: slow heart rate
cent-	hundred	centi/meter: 1/100 of a meter
circum-	around	circum/oral: around the mouth
co-, com-, con-	with, together	com/press: squeeze together
de-	from, down	de/congest: reduce congestion
dia-	through	dia/rrhea: flow through
dis-	apart	dis/infect: free from infection
dys-	bad, difficult	dys/pepsia: difficult digestion
ect-	external, outer	ecto/derm: outer skin
en-	in, on	en/cranial: in the cranium
end-	within	endo/crine: secrete within
epi-	upon	epi/dermis: upon the skin
ex-	out, away from	ex/halation: to breathe out
extra-	outside of, in addition to	extra/ocular: outside the eye
hemi-	half	hemi/plegia: paralysis of one-half of the body
hyper-	above, over	hyper/trophy: excessive growth
hypo-	below, under	hypo/dermic: under the skin
infra-	below, beneath	infra/orbital: below the orbit
inter-	between	inter/cellular: between cells
intra-	within	intra/cellular: within cells
kil-	thousand	kilo/gram: 1,000 grams
macr-	large	macro/cyst: large cyst
mal-	bad, ill, poor	mal/ady: disease, disorder
mes-	middle	meso/nasal: middle of the nose
meta-	after, beyond	meta/tarsals: beyond the tarsals
micr-	small	micro/colon: abnormally small colon
milli-	one-thousandth	milli/gram: 1/1,000 of a gram
multi-	many	multi/cellular: having many cells
neo-	new	neo/plasm: new growth
ob-	against, in the way of	ob/scure: indistinct, hidden
olig-	few	oligo/spermia: few sperm
onc-	tumor	onco/genic: tumor-causing
per-	through	per/forate: to make holes
peri-	around	peri/osteum: around a bone
poly-	many	poly/morphous: many forms
post-	after, in back of	post/ocular: behind the eye
pre-	before, in front of	pre/mature: before maturation
presby-	old	presby/cardia: old heart
pro-	before, in front of	pro/chondrial: before cartilage
re-	again, back	re/flex: bend back
retr-	backward, behind	retro/nasal: back part of nose
semi-	half	semi/lunar: half moon
sub-	under	sub/cutaneous: under the skin
super-	above, superior	super/acute: strongly acute
supra-	above, superior	supra/nasal: above the nose
sym-	together, with	sym/physis: growing together
syn-	together, with	syn/dactyly: fusion of fingers
tachy-	fast	tachy/cardia: rapid heart rate

Common Suffixes

Suffix	Meaning	Example
-algia	pain	neur/algia: pain in a nerve
-centesis	puncture to aspirate fluid	amnio/centesis: puncture amnion to obtain a sample of amniotic fluid
-cide	kill	bacterio/cide: substance killing bacteria
-cis	cut	in/cision: a cut into
-cyte	cell	erythro/cyte: red blood cell
-dynia	pain	entero/dynia: intestinal pain
-ectomy	cut out	append/ectomy: procedure to cut out the appendix
-emesis	vomiting	poly/emesis: much vomiting
-emia	blood	an/emia: without blood
-gnosis	knowledge	pro/gnosis: foreknowledge
-gram	record	myo/gram: muscle record
-graphy	making a record	myo/graphy: making a record of muscle action
-iasis	abnormal condition	candid/iasis: *Candida* infection
-itis	inflammation	sinus/itis: inflammation of sinuses
-lepsy	seizures	narco/lepsy: seizures of numbness
-logy	study of	bio/logy: study of life
-lysis, -lytic	breakdown, dissolve	myo/lysis: breakdown of muscles
-megaly	enlargement	nephro/megaly: kidney enlargement
-oid	resembling	ov/oid: resembling an egg
-oma	tumor	carcin/oma: cancerous tumor
-osis	abnormal condition	nephr/osis: abnormal kidney condition
-ostomy	make an opening	ile/ostomy: opening into the small intestine
-pathy	disease	neuro/pathy: disease of nerves
-penia	deficiency, poor	leuko/penia: deficiency of white blood cells
-pepsia	digestion	dys/pepsia: poor digestion
-philia	attraction, love	acido/philic: attracted to acid
-phobia	abnormal fear	acro/phobia: fear of heights
-plasia	formation	hypo/plasia: deficient formation
-plasty	make, shape	angio/plasty: shaping a blood vessel
-plegia	paralysis	para/plegia: paralysis of lower body and both legs
-pnea	breath	brady/pnea: slow breathing
-rrhea	discharge, flow	pyo/rrhea: pus discharge
-soma, -some	body	chromo/some: colored body
-stasis	control, stop	hemo/stasis: stop bleeding
-therapy	treatment	thermo/therapy: heat therapy
-tomy	to cut	laparo/tomy: to cut into the abdomen
-trophy	development	hyper/trophy: excessive development
-uria	urine	glucos/uria: glucose in the urine

Common Root Words

Root Word	Meaning	Example
acr-	extremity, peak	acro/phobia: fear of heights
aden-	gland	aden/oma: tumor of a gland
angi-	blood vessel	angio/pathy: diseased vessel
arthr-	joint	arthr/itis: inflammation of joints
brachi-	arm	brachi/al: pertaining to the arm
carcin-	cancer	carcin/oma: cancerous tumor
card-	heart	cardio/logy: study of the heart
carp-	wrist	carp/al: pertaining to the wrist
cephal-	head	cephal/ic: pertaining to the head
cervic-	neck	cervic/al: pertaining to the neck

Common Root Words continued

Root Word	Meaning	Example
chole-	bile	chole/cyst/itis: inflammation of the gallbladder
chondr-	cartilage	chondro/cyte: cartilage cell
colp-	vagina	colpo/dynia: vaginal pain
cost-	rib	cost/algia: rib pain
crani-	skull	cranio/malacia: softening of the skull
cutan-	skin	sub/cutan/eous: under the skin
cyan-	blue	cyan/osis: bluish skin color
cyst-	bladder	cyst/algia: pain in bladder
cyt-	cell	cyto/logy: study of cells
dactyl-	fingers, toes	dactylo/megaly: abnormally large fingers or toes
dent-	tooth	denti/form: toothlike
derm-, dermato-	skin	hypo/derm/al: under the skin
dors-	back	dors/al: pertaining to the back
duct-	carry	ad/duct: carry toward
edema	swelling	edema/tous: swollen
encephal-	brain	encephal/itis: inflammation of the brain
enter-	intestine	enter/dynia: intestinal pain
erythr-	red	erythro/cyte: red blood cell
esthe-	sensation, feeling	an/esthe/tic: substance causing an absence of sensation
esthen-	weakness	my/esthenia: muscle weakness
febr-	fever	febr/ile: feverish
gastr-	stomach	gastro/spasm: stomach spasm
gen-	to produce	patho/gen: disease-causing agent
gingiv-	gum	gingiv/itis: inflammation of the gums
glu-, glyc-	sugar	hyper/glyc/emia: excessive blood sugar
gynec-, gyno-	female	gyneco/logy: study of female disorders
hem-, hemato-	blood	hemo/genesis: blood formation
hepat-	liver	hepat/ectomy: removal of the liver
hist-	tissue	histo/logy: study of tissues

Root Word	Meaning	Example
hom-, home-	same	homo/sexual: attracted to the same sex
hydr-, hydra-	water	hydra/tion: gaining water
kerat-	horny, cornea	kerat/osis: condition of abnormal horny growths
lacrim-	tear	lacrim/al: pertaining to tears
lact-	milk	lacta/tion: producing milk
lapar-	abdomen	laparo/tomy: to cut into the abdomen
laryng-	larynx	laryngo/pathy: disease of the larynx
later-	side	uni/lateral: one-sided
leuc-, leuk-	white	leuko/cyte: white blood cell
lingu-	tongue	lingu/iform: tongue-shaped
lip-	lipids, fat	lip/oid: similar to fat
lith-	stone	oto/lith: ear stone
mamm-	breast	mammo/gram: X-ray of breast
melan-	black	melan/in: black skin pigment
men-	monthly, mensis	meno/pause: cessation of menses
metr-	uterus	myo/metr/ium: muscle layer of uterus
morph-	shape, form	morpho/logy: study of shape
my-	muscle	myo/card/itis: inflammation of the heart muscle
myel-	marrow, spinal cord	myel/algia: pain of the spinal cord or its membranes
nas-	nose	nas/al: pertaining to the nose
nephr-	kidney	nephr/itis: inflammation of a kidney
neur-	nerve	neur/ectomy: excision of a nerve
odont-	tooth	odonto/pathy: disease of the teeth
oo-	egg, ovum	oo/genesis: formation of ova
orchid-	testis	orchid/itis: inflammation of a testis
oss-, oste-	bone	osteo/malacia: softening of bones
ot-, aur-	ear	oto/lith: ear stone; aur/icular: pertaining to the ear

Common Root Words continued

Root Word	Meaning	Example
path-	disease	patho/logy: study of disease
pect-	chest	pecto/ral: pertaining to the chest
ped-	child	ped/iatrics: medical specialty dealing with children's disorders
pep-, peps-	digest	pep/tic: pertaining to digestion
phag-	eat	phago/cyt/osis: engulfment of particles by cells
pharyng-	throat, pharynx	pharnygo/rrhea: discharge from the pharynx
phleb-	vein	phleb/itis: inflammation of a vein
pneum-	air	pneumo/thorax: air in the chest
pneumon-	lung	pneumono/pathy: disease of a lung
proct-	rectum	procto/col/itis: inflammation of the rectum and colon
pseud-	false	pseudo/hernia: false rupture
psych-	mind	psycho/genic: originating in the mind
pulmo-, pulmon-	lung	pulmon/ary: pertaining to a lung
py-	pus	pyo/cele: cavity containing pus
pyel-	kidney pelvis	pyelo/gram: X-ray of kidney pelvis
quadr-	four	quadri/plegia: paralysis of both upper and lower limbs
rhin-	nose	rhin/itis: inflammation of the nose
salping-	uterine tube	salping/ectomy: removal of a uterine tube
scler-	hard	scler/osis: hardening
sect-	cut	sect/ion: process of cutting
sept-	presence of microbes	septic/emia: infection of blood
sten-	narrow	sten/osis: narrowed condition
strict-	draw tight	con/strict/ion: draw tightly together
therm-	heat	hypo/thermia: low body temperature
thorac-	chest, thorax	thoraco/dynia: chest pain
thromb-	clot	thromb/us: a blood clot
tox-	poison	tox/in: poisonous substance
vas-	vessel	vaso/dilation: expansion of a vessel
viscer-	internal organ	viscero/genic: originating in the internal organs
vita-	life	vita/l: essential for life

Common Medical Abbreviations *continued*

RLQ	right lower quadrant	TMJ	temporomandibular joint
RUQ	right upper quadrant	tPA	tissue plasminogen activator
Rx	prescription	TSS	toxic shock syndrome
SA	sinoatrial	Tx	treatment
SIDS	sudden infant death syndrome	URI	upper respiratory infection
SLE	systemic lupus erythematosus	UTI	urinary tract infection
STD	sexually transmitted disease	VD	venereal disease
TIA	transient ischemic attack	WBC	white blood cell

APPENDIX B

Common Medical Abbreviations

ABG	arterial blood gas	HAV	hepatitis A virus
ACh	acetylcholine	Hb	hemoglobin
AD	Alzheimer disease	HBV	hepatitis B virus
AIDS	acquired immunodeficiency syndrome	HCV	hepatitis C virus
ALS	amyotrophic lateral sclerosis	HDL	high-density lipoprotein
ANS	autonomic nervous system	HLA	human leukocyte antigen
ARF	acute renal failure	HDN	hemolytic disease of the newborn
AV	atrioventricular	HIV	human immunodeficiency virus
BMR	basal metabolic rate	HR	heart rate
BP	blood pressure	HSV	herpes simplex virus
BUN	blood urea nitrogen	IBD	inflammatory bowel disease
CAD	coronary artery disease	IBS	irritable bowel syndrome
CBC	complete blood count	ICF	intracellular fluid
CHD	coronary heart disease	ICP	intracranial pressure
CHF	congestive heart failure	ID	intradermal
CNS	central nervous system	IDDM	insulin-dependent diabetes mellitus
CO	cardiac output	IM	intramuscular
COPD	chronic obstructive pulmonary disease	IRDS	infant respiratory distress syndrome
CPR	cardiopulmonary resuscitation	IV	intravenous
CRF	chronic renal failure	LDL	low-density lipoprotein
C-section	cesarean section	LLQ	left lower quadrant
CSF	cerebrospinal fluid	LUQ	left upper quadrant
CT (CAT)	computed (axial) tomography	LV	left ventricle
CVA	cerebrovascular accident	MI	myocardial infarction
CVS	chorionic villi sampling	MRI	magnetic resonance imaging
D & C	dilation and curettage	MS	multiple sclerosis
DMD	Duchenne muscular dystrophy	MVP	mitral valve prolapse
DNR	do not resuscitate	NE	norepinephrine
DRI	dietary reference intake	NGU	non-gonococcal urethritis
Dx	diagnosis	NIDDM	non-insulin-dependent diabetes mellitus
EBV	Epstein-Barr virus	NPN	nonprotein nitrogen
ECF	extracellular fluid	OC	oral contraceptive
ECG (EKG)	electrocardiogram	PE	pulmonary embolism
EEG	electroencephalogram	PET	position emission tomography
EP	ectopic pregnancy	PID	pelvic inflammatory disease
ESR	erythrocyte sedimentation rate	PKU	phenylketonuria
ESRD	end stage renal disease	PMN	polymorphonuclear leukocyte
FAS	fetal alcohol syndrome	PMS	premenstrual syndrome
FUO	fever of unknown origin	PSA	prostate-specific antigen
GER	gastroesophageal reflux	PNS	peripheral nervous system
GERD	gastroesophageal reflux disease	RBC	red blood cell
GFR	glomerular filtration rate	RDA	recommended daily allowance
GI	gastrointestinal	REM	rapid eye movement

APPENDIX C

Healthy Values for Common Blood Tests

Test	Healthy Value*	Clinical Significance of Changes
Albumin	3.2–5.5 g/100 ml	Decrease in severe burns and kidney disease.
Amylase	4–25 units/ml	Increase in intestinal obstruction, mumps, and acute pancreatitis.
		Decrease in chronic pancreatitis, cirrhosis of the liver, and eclampsia.
Bilirubin, total	0–1.0 mg/100 ml	Increase in excessive RBC destruction, liver disease, and blocking of bile duct.
Calcium	8.5–10.5 mg/100 ml	Increase in hyperparathyroidism.
		Decrease in hypoparathyroidism, severe diarrhea, and malnutrition.
Chloride	100–106 mEq/liter	Increase in dehydration and nephritis.
		Decrease in severe diarrhea, severe burns, and ketoacidosis.
Cholesterol, total	120–220 mg/100 ml	Increase in diabetes mellitus and hypothyroidism.
Cholesterol, HDL	30–70 mg/100 ml	Increase reduces risk of atherosclerosis.
Cholesterol, LDL	62–185 mg/100 ml	Increase increases risk of atherosclerosis.
Clotting time	3–6 min	Increase in hemophilia.
Creatinine	0.6–1.5 mg/100 ml	Increase in kidney disease.
Globulins	2.3–3.5 g/100 ml	Increase in chronic infections.
Glucose (fasting)	70–110 mg/100 ml	Increase in diabetes mellitus, liver disease, hyperthyroidism, and pregnancy.
		Decrease in hypoglycemia and hypothyroidism.
Hemoglobin	Women: 12–16 g/100 ml	Increase in polycythemia, congestive heart failure, and chronic obstructive pulmonary disease.
	Men: 13–18 g/100 ml	Decrease in anemia, cirrhosis of the liver, hemorrhage, and hyperthyroidism.
Hematocrit	Women: 38%–47%	Increase in polycythemia and severe dehydration.
	Men: 40%–54%	Decrease in anemia, leukemia, cirrhosis of the liver, hemorrhage, and hyperthyroidism.
pH	7.35–7.45	Increase in hyperventilation and alkalosis.
		Decrease in acidosis and severe diarrhea.
Platelet count	150,000–400,000/µl	Increase in heart disease, cirrhosis of the liver, and cancer.
		Decrease in chemotherapy and leukemia.
Potassium	3.5–5.0 mEq/liter	Increase in acute kidney failure and extensive cellular destruction.
		Decrease in severe diarrhea or vomiting and kidney disease.
Prothrombin time	11–15 seconds	Increase in liver disease and vitamin K deficiency.
Red blood cell count	Women: 4.2–5.4 million/µl	Increase in polycythemia and severe dehydration.
	Men: 4.7–6.1 million/µl	Decrease in anemia, leukemia, and systemic lupus erythematosis.
Reticulocyte count	0.5%–1.5% of RBCs	Increase in leukemia and hemolytic anemia.
Sodium	136–145 mEq/liter	Increase in severe dehydration.
		Decrease in kidney disease, severe burns, and excessive diarrhea and vomiting.
Urea (BUN)	6–20 mg/100 ml	Increase in kidney disease and high protein diet.
		Decrease in kidney failure.
Uric acid	Women: 2.5–7.5 mg/100 ml	Increase in eclampsia and gout.
	Men: 3–9 mg/100 ml	
White blood cell count (total)	4,500–10,000/µl	Increase in acute infections, leukemia, and some cancers.
		Decrease in chemotherapy and radiation sickness.
Neutrophils	40%–60% of total	Increase in acute infection.
Eosinophils	1%–4% of total	Increase in allergies.
Basophils	0.5%–1% of total	Increase in diabetes mellitus, chicken pox, and myxedema.
Lymphocytes	20%–40% of total	Increase in immune reactions.
Monocytes	2%–8% of total	Increase in chronic infections.

*Healthy values may vary with type of equipment used for analysis.

APPENDIX D

Healthy Values for Common Urine Tests

Test	Healthy Value*	Clinical Significance of Changes
Acetone and ketones	0	Increase in starvation and ketoacidosis.
Albumin	0–trace	Increase in kidney disease and hypertension.
Ammonia	20–70 mEq/liter	Increase in liver disease and diabetes mellitus.
Bilirubin	0	Increase in obstruction of bile ducts.
Calcium	< 300 mg/day	Increase in hyperparathyroidism; decrease in hypoparathyroidism.
Color and clarity	Pale yellow to light amber; transparent	Smoky color indicates blood; cloudiness may be due to bacteria or food pigments.
Creatinine	0.5–2 g/day	Increase in infection; decrease in kidney disease.
Glucose	0	Increase in diabetes mellitus.
pH	4.8–7.5	Increase in urinary infections and alkalosis. Decrease in acidosis, emphysema, dehydration, starvation, and high-protein diet.
Specific gravity	1.002–1.030	Increases in dehydration. Decreases in excessive fluid intake, alcohol intake, and diabetes insipidus.
Urea	10–30 g/day	Increase in hemolytic anemia and high-protein diet; decrease in kidney disease.
Uric acid	250–750 mg/day	Increase in gout and some liver diseases; decrease in kidney disease.
Urobilinogen	0–4 mg/day	Increase in liver disease and hemolytic anemia. Decrease in obstruction of bile ducts and severe diarrhea.

*Healthy values may vary with the type of equipment used for analysis.

GLOSSARY

A

abdomen The front of the trunk located between the diaphragm and pelvis.

abdominal cavity The portion of the abdominopelvic cavity between the diaphragm and the pelvis.

abdominopelvic cavity The body cavity below the diaphragm.

abdominopelvic quadrants The four divisions of the abdominopelvic cavity formed by a median plane and a transverse plane through the umbilicus.

abdominopelvic regions The nine divisions of the abdominopelvic cavity formed by the intersection of two sagittal and two transverse planes.

abduction The movement of a body part away from the midline.

abortion The removal of an embryo or fetus from the uterus prior to birth.

absorption The uptake of substances by cells; the process by which nutrients pass from the alimentary canal into the blood.

accessory organs Organs that assist the functions of primary organs.

accommodation The focusing of light rays on the retina by the lens.

acetylcholine (ACh) A neurotransmitter secreted from the terminal bouton of many neurons.

acetylcholinesterase An enzyme promoting the breakdown of acetylcholine in synaptic clefts.

acid A substance that ionizes in water, releasing hydrogen ions.

acidosis Condition of arterial blood below pH 7.35.

acne Plugged hair follicles that form pimples due to infection by certain bacteria.

acquired immunodeficiency syndrome (AIDS) A progressive decrease in immune capability caused by infection of T cells and macrophages with HIV.

acromegaly A disorder caused by hypersecretion of growth hormone after bone growth is complete.

action potential An electrochemical signal created by and conducted along the axon of a neuron.

active immunity Immunity derived from activation of B cells and T cells by an invasion of a pathogen.

active site The location on an enzyme where the chemical reaction occurs.

active transport Movement of substances across a plasma membrane, requiring the expenditure of energy by the cell.

Addison disease An endocrine disorder caused by a hyposecretion of hormones by the adrenal cortex.

adduction The movement of a body part toward the midline.

adenine A nitrogenous base of nucleic acids that pairs with thymine in DNA and uracil in RNA.

adenosine diphosphate (ADP) A molecule used to form adenosine triphosphate.

adenosine triphosphate (ATP) Chemical energy storage molecule in the body.

adipose tissue Loose connective tissue containing large numbers of fat-storing adipocytes.

adrenal cortex The outer portion of an adrenal gland.

adrenal gland An endocrine gland located on top of each kidney.

adrenal medulla The inner portion of an adrenal gland.

adrenocorticotropic hormone (ACTH) A hormone secreted by the anterior lobe of the pituitary gland that stimulates the adrenal cortex to secrete hormones.

adult stem cells Partially specialized cells capable of producing several types of specialized cells.

aerobic respiration The part of cellular respiration that requires oxygen and mitochondria.

afferent glomerular arteriole The arteriole carrying blood to the glomerulus of a nephron.

age-related macular degeneration (AMD) The destruction of the macula resulting in loss of vision in the center of the visual field.

ageusia A loss of taste function.

agglutination The clumping of red blood cells in an antigen-antibody reaction.

agonist A muscle whose contraction moves a body part.

agranulocyte A type of white blood cell that lacks visible cytoplasmic granules.

albumin An abundant plasma protein that helps transport substances in blood, and maintain the pH and osmotic pressure of blood.

aldosterone A mineralocorticoid hormone produced by the adrenal cortex that regulates potassium and sodium concentrations in the blood.

alimentary canal The tube through which food passes from the esophagus to the anus.

alkaline Pertaining to a base.

alkalosis Condition of arterial blood above pH 7.45.

allantois An extraembryonic membrane that forms as an outpocketing from the yolk sac.

allele An alternate form of a gene.

allergen A foreign substance capable of stimulating an allergic reaction.

allergy An abnormally intense immune reaction.

all-or-none response The type of response by muscle cells and neurons to stimulation; total response or no response.

alopecia Excessive hair loss.

alveolar ducts Tiny air passages that open into pulmonary alveoli.

alveolar gas exchange The exchange of oxygen and carbon dioxide between the air in pulmonary alveoli and the blood in alveolar capillaries.

Alzheimer disease A disorder caused by a loss of cholinergic neurons in the brain and characterized by loss of memory.
amenorrhea The absence of menstruation.
amino acid The building unit of proteins.
amnion The extraembryonic membrane that envelops the embryo and fetus.
ampulla The expanded portion of a semicircular canal.
amylase An enzyme that catalyzes the digestion of polysaccharides into simple sugars.
anaerobic respiration The part of cellular respiration that does not require oxygen and occurs within the cytosol.
anaphase The stage of mitosis in which chromatids of replicated chromosomes separate and move to opposite poles of the cell.
anatomy The study of the structure of living organisms.
androgens Generic term for hormones related to testosterone.
anemia The decreased ability of the blood to carry oxygen.
aneurysm A bulging, weakened portion of a blood vessel.
anosmia An inability to detect odor.
antagonist A muscle whose contraction opposes that of the agonist.
antebrachial Pertaining to the forearm.
anterior Pertaining to the front of the body.
anterior root The anterior attachment of a spinal nerve to the spinal cord that contains axons of motor neurons.
antibody A protein produced by plasma cells that binds to a specific antigen.
antibody-mediated immunity Immunity resulting from the production of antibodies.
anticodon A group of three nucleotides of a transfer RNA molecule that pairs with a codon of a messenger RNA molecule.
antidiuretic hormone (ADH) The hormone secreted by the posterior lobe of the pituitary gland that promotes the reabsorption of water by the kidneys.
antigen A substance capable of causing an immune response.
antigen-presenting cell (APC) A phagocytic cell that engulfs a foreign antigen and displays the antigen on its plasma membrane.
anus The terminal opening of the alimentary canal.
aorta The systemic artery receiving blood directly from the left ventricle.
aortic valve A valve at the base of the aorta preventing the backflow of blood into the left ventricle.
apocrine sweat gland A sweat gland that deposits secretions into a hair follicle.
aponeurosis A broad sheet of dense regular connective tissue that attaches a muscle to another muscle, bones, the dermis, or a ligament.
appendicitis Inflammation of the appendix.
appendicular Pertaining to the upper and lower limbs.
appendicular skeleton The bones of the pectoral girdles, pelvic girdle, and upper and lower limbs.
appendix Wormlike structure extending from the cecum that contributes to the immune defense of the body.
aqueous humor A watery fluid filling the anterior and posterior chambers of the eye.
aqueous solution A solution with water as the solvent.
arachnoid mater Middle membrane of the meninges enveloping the brain and spinal cord.
areolar connective tissue Most abundant connective tissue; loose connective tissue with widely spaced protein fibers.
arrhythmia An abnormal heartbeat such as bradycardia or fibrillation.
arteriole A small artery that leads to a capillary network.
arteriosclerosis Hardening of an arterial wall due to the deposition of calcium salts.
artery A blood vessel that carries blood away from the heart.
arthritis An inflammation of joints.
articular cartilage The cartilage tissue that covers the ends of bones in freely movable joints.
association area A region of the cerebrum involved in interrelating sensory inputs and motor outputs.
asthma Chronic inflammation and constriction of the respiratory passages.
astigmatism A visual disorder caused by an unequal curvature of the cornea or lens.
astrocyte A type of neuroglia assisting in neuronal growth and formation of the blood-brain barrier.
atherosclerosis A decrease in the lumen of an artery caused by fatty deposits in the vessel wall.
athlete's foot A fungal infection of the skin.
atmospheric pressure The pressure of the air that surrounds the earth.
atom The smallest unit of an element.
atomic number The number of protons in an atom of an element.
atomic mass The sum of the neutrons and protons in an atom of an element.
atrioventricular bundle (AV bundle) Specialized muscle tissue carrying action potentials from the AV node to the right and left bundle branches.
atrioventricular node (AV node) A node of specialized muscle tissue that receives action potentials from the SA node and transmits them to the AV bundle.
atrophy The reduction in muscle size and strength due to loss of myofibrils.
atrium A heart chamber that receives blood returned to the heart by veins.
auditory Pertaining to the ear.
auditory ossicles Three tiny bones in the middle ear that transmit vibrations from the tympanic membrane to the oval window of the internal ear.
auditory tube A tube connecting the middle ear to the pharynx; the eustachian tube.
autoimmune disease A variety of diseases caused by the body's immune response attacking the body's own tissues.
autonomic division The division of the PNS involved in the involuntary control of cardiac muscle, smooth muscle, adipose tissue, and glands.
autonomic ganglion A cluster of postganglionic cell bodies located outside of the central nervous system.
axial Pertaining to the longitudinal axis of the body.
axial skeleton The portion of the skeleton that supports and protects the head, neck, and trunk.
axon A neuronal process that carries action potentials away from the cell body.

B

B lymphocyte A lymphocyte that develops into an antibody-producing plasma cell; a B cell.
base A substance that ionizes in water, releasing hydroxide ions (OH^-), or binds to hydrogen ions (H^+).
basement membrane A thin layer of noncellular material attaching epithelial tissue to underlying connective tissue.

basilar membrane The membrane supporting the spiral organ.
basophil A granulocytic WBC with large blue cytoplasmic granules.
bed bugs Microscopic parasitic insects that feed almost exclusively off human blood.
bedsores Skin ulcers caused by a deficient blood supply to localized areas.
benign prostatic hyperplasia (BPH) The enlargement of the prostate without inflammation.
benign tumor A tumor that does not spread to another site.
bile Fluid secreted by the liver and stored in the gallbladder.
bile duct A small tube that carries bile from the cystic and common hepatic ducts into the duodenum.
bilirubin A bile pigment formed by the breakdown of hemoglobin.
bipolar neuron A neuron with only one dendrite and one axon.
birth defects Physical abnormalities that are inherited or caused by teratogens during prenatal development.
blastocyst A preembryonic stage of development consisting of a hollow ball of cells.
blindness A partial or total loss of vision.
blister A fluid-filled pocket that forms when the epidermis separates from the dermis.
blood The specialized fluid connective tissue that transports substances to and from cells.
boil A bacterial infection of a hair follicle and its sebaceous gland.
bone tissue A hard connective tissue with a rigid matrix containing calcium salts; osseous tissue.
botulism Food poisoning caused by the bacterium *Clostridium botulinum*.
brachial Pertaining to the arm.
brainstem The portion of the brain including the midbrain, pons, and medulla oblongata.
bronchial tree The branching bronchi.
bronchiole A branch of the respiratory tract between bronchi and alveolar ducts.
bronchitis Inflammation of the bronchi.
bronchus A branch of the respiratory tract between trachea and bronchioles.
buffer A substance that stabilizes the pH of body fluids by combining with or releasing hydrogen ions.
bulbo-urethral glands Male accessory glands that secrete an alkaline secretion into the urethra.
burn Damage to the skin caused by heat, chemicals, or radiation.

C

calcitonin A hormone secreted by the thyroid gland that decreases the Ca^{2+} level of the blood.
callus Thickened layers of epidermis caused by chronic pressure.
canaliculi Microscopic canals between lacunae in bone tissue.
cancer A malignant tumor; a tumor that can spread to other sites.
capillary A tiny blood vessel in tissues where exchange of materials between the blood and the interstitial fluid occurs.
carbaminohemoglobin ($HbCO_2$) A molecule formed by the combination of carbon dioxide and hemoglobin.
carbohydrate An organic molecule composed of carbon, hydrogen, and oxygen in a 1:2:1 ratio.
carbonic anhydrase An enzyme in RBCs that catalyzes the combination of carbon dioxide and water to form carbonic acid.
cardiac Pertaining to the heart.
cardiac control center The center in the medulla oblongata controlling heart rate and force of contraction.
cardiac cycle The sequence of events that occur during one heartbeat.
cardiac muscle tissue The muscle in the wall of the heart.
cardiac output The volume of blood pumped from each ventricle in one minute.
carotene A yellowish pigment that occurs in the skin; a precursor of vitamin A in plant foods.
carpal bones The bones of the wrist.
carrier-mediated active transport Carrier proteins use ATP to move substances across the plasma membrane against (opposite to) their concentration gradient.
carrier-mediated diffusion Carrier proteins bind and move water-soluble substances across the plasma membrane along their concentration gradients.
carrier-mediated transport Movement of substances across the plasma membrane by carrier proteins.
carrier protein A membrane protein that physically binds to and transports a specific type of substance across a plasma membrane.
cartilage tissue A specialized connective tissue characterized by a semisolid matrix and chondrocytes located within lacunae.
cartilaginous joint Joint with two bones joined by a thin band or pad of cartilage.
cataract A vision disorder in which the lens of the eye becomes clouded.
cecum The pouchlike first segment of the large intestine.
cell The simplest structural and functional unit of an organism.
cell body The portion of a neuron that contains the nucleus.
cell cycle Time period from the separation of daughter cells of one division to the separation of daughter cells of the next division.
cell division The division of a parent cell to form new daughter cells.
cell-mediated immunity Immunity that is characterized by a direct attack on pathogens by T cells.
cellular respiration Breakdown of organic nutrients in cells to release energy and form ATP.
cement The hard substance that attaches the root of a tooth, through periodontal ligaments, to a dental alveolus in the mandible or maxilla.
central nervous system (CNS) The portion of the nervous system composed of the brain and spinal cord.
central canal A canal in the center of an osteon that contains blood vessels and nerves; osteonic canal.
centriole Paired cylindrical organelles that form the mitotic spindle during cell division.
centromere The portion of a chromosome that is attached to a spindle fiber during cell division.
cephalic Pertaining to the head.
cerebellum The part of the brain that coordinates body movements.
cerebral cortex The outer layer of the cerebrum that is composed of grey matter.
cerebral hemispheres The two major parts of the cerebrum that are separated by the longitudinal cerebral fissure.
cerebral palsy A motor disorder caused by brain damage, resulting in uncontrolled muscular contractions.
cerebrospinal fluid (CSF) The fluid filling the ventricles of the brain, the subarachnoid space of the meninges, and the central canal of the spinal cord.
cerebrovascular accident (CVA) A stroke; brain impairment caused by aneurysm, hemorrhage, embolism, or thrombus.
cerebrum The portion of the brain involved in conscious action, sensory perception, memory, and intelligence.

ceruminous gland A gland that produces cerumen (earwax).
cervical Pertaining to the neck or to the cervix of the uterus.
cervix The narrow end of the uterus that opens into the vagina.
channel protein A tunnel-shaped membrane protein that creates a pore or opening, which allows for a specific substance to pass across a plasma membrane.
chemical bond A bond that joins two atoms together.
chemical formula A shorthand designation of the kinds and numbers of atoms in a molecule.
chemical reaction A process that makes or breaks chemical bonds.
chemoreceptor Sensory receptor stimulated by certain chemicals.
chief cell A cell of a gastric gland that secretes digestive enzymes.
chlamydia A sexually transmitted disease caused by the bacterium *Chlamydia trachomatis*.
cholecystokinin (CCK) A hormone secreted by the mucosa of the small intestine that stimulates secretion of pancreatic juice and contraction of the gallbladder.
cholinergic axon An axon that releases the acetylcholine at its terminal bouton.
chondrocyte A cartilage cell.
chorion The outermost extraembryonic membrane forming from the trophoblast of the blastocyst and contributing to the development of the placenta.
chorionic villi Fingerlike projections of the chorion that penetrate into the endometrium.
choroid The middle, pigmented vascular layer of the eyeball.
choroid plexus A specialized mass of capillaries and ependymal cells in the ventricles of the brain that secrete cerebrospinal fluid.
chromatid One-half of a replicated chromosome.
chromatin Portions of extended, uncoiled chromosomes appearing as dark granules in a cell nucleus.
chromosome A threadlike or rodlike structure in the nucleus that is composed of DNA and protein.
chronic obstructive pulmonary disease (COPD) A group of disorders in which the airflow to the lungs is obstructed.
chylomicron A microscopic droplet of lipids and lipid-soluble vitamins coated with protein.

chyme The acidic, semiliquid substance that exits the stomach into the small intestine.
cilia Microscopic, hairlike projections from the free surface of ciliated epithelial cells.
ciliary body An enlarged ring, in front of the choroid, containing the ciliary muscles of the eye.
circumduction The movement of a body part in a circular path.
cirrhosis The destruction of liver tissue and its replacement with dense irregular connective tissue.
cleavage The early divisions of pre-embryotic development leading to the formation of a morula.
clitoris Two columns of corpus cavernosum that extend bilaterally near the pubis of a female.
clone A population of identical cells derived from a common ancestral cell.
coagulation The formation of a blood clot.
cochlea The coiled portion of the internal ear containing the sensory receptors for hearing.
cochlear hair cells Sensory receptors for hearing.
codon Three nucleotides of messenger RNA that code for a specific amino acid and are complementary to both the three nucleotides of DNA and an anticodon of transfer RNA.
colitis Inflammation of the colon mucosa.
collagen fibers Fibers composed of collagen protein that provide strength and flexibility to connective tissues.
collecting ducts Microscopic tubules composing renal pyramids.
colon The major segment of the large intestine.
color blindness The inability to see certain colors or all colors.
coma A state of unconsciousness.
common cold A virus-caused inflammation of air passages and associated structures.
common mole A pigmented epidermal growth produced by growing clusters of melanocytes.
compact bone Bone tissue formed of tightly packed osteons.
complement A group of plasma proteins that works with antibodies to destroy cells of pathogens.
compound A substance composed of two or more elements combined by ionic or covalent bonds.

concentration gradient The difference in the concentration of a substance at two different locations.
concussion A jarring of the brain caused by a blow to the head.
conducting system of the heart Specialized muscle tissue that transmits action potentials from the SA node to the myocardium.
conductivity The ability of neurons to conduct action potentials.
condyle A rounded process of a bone.
cones Receptors for color vision that are located in the retina.
congestive heart failure (CHF) A disorder in which the heart is unable to pump out all of the blood it receives.
conjunctiva The mucous membrane lining the eyelids and covering the front of the eye.
conjunctivitis Inflammation of the conjunctiva.
connective tissue A tissue that binds other tissues together.
constipation Difficult defecation of hard, dry feces.
contraception Devices or procedures that prevent contact of sperm and secondary oocyte in sexual intercourse.
contraction The shortening of a muscle in response to stimulation.
corn A thickened area of skin on the top of a toe created by chronic pressure.
cornea The anterior clear window of the eye.
coronary Pertaining to the heart.
coronary arteries The blood vessels that supply the myocardium.
corpus callosum A mass of myelinated and unmyelinated axons connecting the two cerebral hemispheres.
corpus luteum Gland formed from a ruptured ovarian follicle after ovulation.
cortex The outer layer of an organ.
cortisol A glucocorticoid hormone secreted by the adrenal cortex.
costal Pertaining to the ribs.
costal cartilages Cartilages attaching the ribs to the sternum or to other costal cartilages.
covalent bond A chemical bond between two atoms that is formed by the sharing of valence electrons.
coxal Pertaining to the hip.
cramp Involuntary and painful tetany.
cranial Pertaining to the cranium.
cranial cavity The cavity in which the brain resides.
cranial nerve A nerve that originates from the brain.

cranium The skull bones forming the cranial cavity and enveloping the brain.
creatine phosphate A molecule in muscle tissue that temporarily holds additional energy for forming ATP.
creatinine A nitrogenous waste produced by muscle metabolism.
cretinism A congenital disorder due to a lack of thyroid hormones.
crista ampullaris The structure in the ampulla of a semicircular canal containing sensory receptors for dynamic equilibrium.
crural Pertaining to the front of the leg.
cubital Pertaining to the elbow.
Cushing syndrome A disorder due to the hypersecretion of glucocorticoids by the adrenal cortex.
cutaneous Pertaining to the skin.
cutaneous membrane The skin.
cyclic adenosine monophosphate (cAMP) The second messenger involved with most nonsteroid hormones.
cystic duct The duct leading from the gallbladder to the bile duct.
cystitis Inflammation of the urinary bladder.
cytokine A chemical secreted by helper T cells that stimulates the division of B cells to form a clone.
cytokinesis The division of the cytoplasm during telophase of cell division.
cytoplasm The semifluid material located between the nucleus and plasma membrane of a cell.
cytosine A nitrogenous base of nucleic acids that pairs with guanine.
cytosol A gellike fluid of the cytoplasm.
cytotoxic T cell (Tc) A T cell that binds to cells with foreign antigens and injects chemicals to kill the cells.

D

dandruff Excessive shedding of dead epidermal cells from the scalp.
deafness A partial or total loss of hearing.
deciduous teeth The first set of teeth, which are lost and replaced by permanent teeth.
decomposition (catabolic) reaction The breakdown of complex molecules into simpler molecules.
defecation The expulsion of feces through the anus.
dehydration synthesis The combining of two molecules by the removal of a water molecule.

dendrite A neuronal process that carries impulses toward the cell body or axon.
dense connective tissues Supportive connective tissues with dense, thick protein fibers and few cells.
dense irregular connective tissue A dense connective tissue with tightly packed, irregularly arranged collagen fibers.
dense regular connective tissue A dense connective tissue with tightly packed, regularly arranged collagen fibers.
dental caries Tooth decay.
dentin The hard, bonelike substance that forms most of a tooth.
deoxyhemoglobin The molecule remaining after oxyhemoglobin has given up some of its oxygen.
deoxyribonucleic acid (DNA) The double-stranded nucleic acid forming the hereditary component of chromosomes.
depolarization A change that causes the membrane potential to become less negative.
dermal papillae Nipplelike projections of the dermis at the dermis-epidermis boundary.
dermis The inner layer of the skin.
detrusor The muscle in the wall of the urinary bladder.
diabetes insipidus A disorder caused by a deficiency of antidiuretic hormone and by nonfunctional ADH receptors.
diabetes mellitus A disorder caused by a deficiency of insulin and by nonfunctional insulin receptors.
diaphragm The sheetlike skeletal muscle separating the thoracic and abdominopelvic cavities.
diaphysis The long shaft of a long bone.
diarrhea Production of watery feces.
diastole The relaxation phase of the cardiac cycle.
diastolic blood pressure The blood pressure in systemic arteries during heart diastole.
diencephalon The portion of the brain containing the thalamus, hypothalamus, and epithalamus.
differentiation The process by which cells become specialized for various functions.
diffusion The passive movement of substances from an area of higher concentration to an area of lower concentration.

digestion The mechanical and chemical process that breaks down food into absorbable nutrient molecules.
dipeptide A molecule composed of two amino acids chemically combined.
disaccharide A molecule composed of two monosaccharides chemically combined.
dislocation Displacement of a bone forming a joint.
distal Farther from the origin.
diuresis The excessive production of urine.
diverticulitis Inflammation of diverticula of the colon mucosa.
dopamine A neurotransmitter secreted by certain neurons in the brain.
dorsum Pertaining to the back; the posterior surface of the trunk.
ductus arteriosus A short artery that carries blood from the pulmonary trunk to the aorta during fetal development.
ductus venosus A short vein that carries blood from the umbilical vein to the inferior vena cava during fetal development.
duodenum The first segment of the small intestine.
dura mater The outermost membrane of the meninges.
dynamic equilibrium The maintenance of balance during linear acceleration and rotational movement of the head.
dysgeusia A distorted or impaired sense of taste.
dyslexia A reading disorder in which letters and words are reversed.
dysosmia A distorted sense of smell.
dysmenorrhea Painful menstruation.

E

eating disorders Disorders resulting from obsessive concerns about weight control.
eccrine sweat gland A sweat gland depositing secretions onto the skin surface.
eclampsia A disorder of pregnancy characterized by high blood pressure, edema, and possibly convulsions and coma.
ectoderm The germ layer forming the posterior surface of the embryonic disc.
ectopic pregnancy A condition in which the embryo is implanted at a site other than the uterus.

eczema A skin disorder characterized by red, dry, scaling skin.

edema The swelling of tissues due to the excessive accumulation of interstitial fluid.

effector A structure that functions by performing an action that is directed by an integrating center.

efferent glomerular arteriole An arteriole carrying blood away from a glomerulus.

eicosanoids A group of chemicals secreted by nonendocrine cells that act as paracrine signals affecting the activity of nearby cells.

ejaculation The discharge of semen from the male urethra.

ejaculatory duct A short duct formed by the merger of the duct from a seminal vesicle and a vas deferens.

elastic cartilage Cartilage tissue containing an abundance of elastic fibers.

elastic connective tissue A dense connective tissue with tightly packed elastic fibers.

elastic fibers Fibers composed of elastin protein that provide elasticity in certain connective tissues.

electrocardiogram (ECG or EKG) A record of the electrical activity of the heart during a cardiac cycle.

electrolyte A substance that ionizes when dissolved in water.

electron A negatively charged particle that revolves around an atomic nucleus.

element A substance that cannot be broken down into a simpler substance by chemical means.

elephantiasis A disorder in which lymphatic vessels are plugged by an infestation of microscopic worms.

embolism The blockage of a blood vessel by an embolus.

embolus A moving blood clot or foreign body in the blood.

embryo Name of the developing offspring from the beginning of the third week through the end of the eighth week of development.

embryonic stem cells Unspecialized cells capable of producing all types of specialized human cells.

emphysema A lung disorder in which alveolar walls rupture, reducing the respiratory surface.

enamel The hard outer layer of a tooth crown.

endocardium The inner lining of the heart chambers.

endochondral bone Bone that is first formed of hyaline cartilage, which is replaced by bone tissue.

endochondral ossification The formation of bone tissue within a cartilage.

endocrine gland A gland that secretes hormones.

endocytosis The process by which the plasma membrane engulfs, or internalizes, solid particles and droplets of liquid.

endoderm The germ layer forming the anterior surface of the embryonic disc.

endolymph The fluid in the membranous labyrinth of the internal ear.

endometriosis The growth of endometrial tissue at sites other than in the uterus.

endometrium The inner lining of the uterus.

endoplasmic reticulum (ER) A network of membranous channels used to transport substances within a cell.

endosteum The connective tissue membrane lining of a medullary cavity and the trabeculae of spongy bone.

endothelium The inner lining of the heart, blood vessels, and lymphatic vessels.

enzyme A protein that aids and speeds up (catalyzes) a specific chemical reaction.

eosinophil A granulocytic WBC with red cytoplasmic granules.

ependymal cell A type of neuroglia that lines the ventricles of the brain and the central canal of the spinal cord.

epicardium Inner layer of the serous pericardium that adheres to the outer surface of the heart.

epicondyle A bony projection located just above a condyle.

epidermis The stratified squamous epithelium covering the dermis; the outer layer of the skin.

epididymis The highly coiled tube that carries sperm from the seminiferous tubules to the vas deferens.

epigastric region The upper middle portion of the abdominopelvic cavity.

epiglottis A cartilaginous flap that closes over the larynx during swallowing.

epilepsy A neural disorder characterized by sudden lapses of consciousness and possible convulsions.

epinephrine A hormone secreted by the adrenal medulla in response to stress.

epiphysial line The line of fusion between an epiphysis and a diaphysis when bone growth in length is complete.

epiphysial plate The hyaline cartilage between the epiphysis and diaphysis of immature long bones; growth plate.

epithalamus The part of the diencephalon located above and behind the thalamus forming the roof of the third ventricle.

epiphysis The enlarged end of a long bone.

epithelial tissue A thin tissue that covers body and organ surfaces and lines body cavities, and forms secretory portions of glands; epithelium.

erectile dysfunction The inability to attain and maintain an erection long enough for sexual intercourse.

erection The process by which erectile tissue engorges with blood.

erythropoietin (EPO) The hormone that stimulates RBC production.

esophagus The portion of the alimentary canal carrying food from the pharynx to the stomach.

essential amino acids Amino acids that must be obtained in food because they cannot be synthesized by the body.

estrogens The female sex hormones produced primarily by the ovaries.

eversion The movement that turns the sole of the foot laterally.

exchange (rearrangement) reaction A reaction in which two reactants exchange components to form two different products.

excitation-contraction coupling The pairing of an action potential and the physical contraction of the muscle fiber.

excretion The removal of metabolic wastes and excessive substances from the body.

exocrine gland A gland whose secretion is carried to a specific site by a duct.

exocytosis The process by which a cell releases substances by fusion of a secretory vesicle with the plasma membrane.

exophthalmic goiter An endocrine disorder caused by hypersecretion of thyroid hormones.

expiration The movement of air out of the lungs; exhalation.

extension A movement that increases the angle between two body parts forming the joint.

external Pertaining to the surface of the body.

external acoustic meatus A canal in the temporal bone that leads from the external environment to the tympanic membrane.

external respiration Breathing and alveolar gas exchange.
extracellular Pertaining to regions outside of cells.
extrinsic muscles of the eyeball Muscles that move the eyeball.

F

facet A small, flat surface of a bone that is for forming a joint with another bone.
facilitated transport Movement of substances across a plasma membrane with the help of carrier proteins.
fainting A brief loss of consciousness due to a sudden reduction in blood supply to the brain.
farsightedness A visual disorder in which the image is focused behind the retina.
fascia Dense irregular connective tissue supporting, covering, and separating muscles.
fat A triglyceride; adipose tissue.
fatigue The lack of response to stimulation by muscle cells or neurons.
fatty acid An organic molecule that forms part of a triglyceride.
feces Material discharged from the anus in defecation.
fertilization Union of sperm nucleus and ovum nucleus.
fetus Name of the developing offspring from the beginning of the ninth week of development to birth.
fever An elevated body temperature.
fever blister A cold sore; small vesicles on the lips caused by a *Herpes simplex* viral infection.
fibrin The insoluble protein filaments that form a blood clot.
fibrinogen A soluble plasma protein converted into insoluble fibrin during blood clot formation.
fibroblast A connective tissue cell that forms intercellular protein fibers and ground substance.
fibrocartilage Cartilage tissue containing tightly packed collagen fibers.
fibromyalgia A painful condition of the muscles and joints with no known cause.
fibrosis The replacement of muscle tissue with dense irregular connective tissue.
fibrous joint Joint with skeletal structures joined by a thin band of dense irregular connective tissue.

filtration The forcing of water and solutes across a membrane by hydrostatic pressure.
fissure A long, narrow cleft separating two parts.
flagellum A long, hairlike extension from a cell.
flexion Movement that decreases the angle between two body parts forming the joint.
follicle A cavity or saclike depression.
follicle-stimulating hormone (FSH) A hormone secreted by the anterior lobe of the pituitary gland that stimulates oogenesis and spermatogenesis.
follicular epithelial cells Cells that surround a developing oocyte.
fontanelles Membranous regions between cranial bones of an infant.
foramen A small canal or passageway in a bone or membrane.
foramen ovale The opening between the atria in a fetal heart.
formed elements The solid components of blood: red blood cells, white blood cells, and platelets.
fossa A depression in a bone or organ.
fovea centralis A small depression in the retina that contains only densely packed cones.
fracture A broken bone.
free nerve endings Sensory neuron dendrites functioning as sensory receptors.
frontal plane A plane dividing the body or organ into front and back portions; a coronal plane.

G

gallbladder A small sac under the liver that temporarily stores bile.
gallstones Crystallization of cholesterol in the bile within the gallbladder.
gamete A sex cell; a sperm or secondary oocyte.
ganglion A cluster of cell bodies located in the PNS.
gastric gland A gland of the stomach mucosa that secretes gastric juice.
gastric juice The secretion of gastric glands containing HCl and digestive enzymes.
gastrin A hormone secreted by the stomach mucosa that stimulates the secretion of gastric juice.
gene A unit of heredity; part of a DNA molecule in a chromosome.

genital herpes A sexually transmitted disease caused by the herpes simplex virus type 2.
genital warts A sexually transmitted disease caused by the human papilloma viruses.
genotype The genetic composition of an individual.
germ layers The three tissue layers of the embryonic disc that form all body tissues during embryonic development.
gigantism An endocrine disorder caused by hypersecretion of growth hormone before bone growth is complete.
gland A cell or group of cells that produces a secretion.
globulin A type of plasma protein.
glomerular capsule A double-walled membrane enclosing a glomerulus.
glomerular filtrate The fluid that enters the glomerular capsule during glomerular filtration.
glomerular filtration The forcing of water and small solutes from the blood in a glomerulus into a glomerular capsule.
glomerulonephritis Inflammation of the kidney involving the glomeruli.
glomerulus The cluster of capillaries enveloped by the glomerular capsule.
glottis The vocal folds and the opening between them within the larynx.
glucagon The pancreatic hormone that promotes the formation of glucose from glycogen.
glucocorticoids A group of hormones secreted by the adrenal medulla that influences glucose metabolism.
glucose The monosaccharide that is the primary energy source for cells.
glycerol An organic molecule that is the backbone of triglyceride and phospholipid molecules.
glycogen The polysaccharide that is the storage form for carbohydrates in the body.
goblet cell A mucus-producing epithelial cell.
goiter An enlarged thyroid gland.
Golgi complex A cellular organelle that packages substances for secretion from the cell.
gonad A primary sex gland; an ovary or testis.
gonadotropin A hormone from the anterior lobe of the pituitary gland that stimulates activity of the gonads.
gonadotropin-releasing hormone (GnRH) Hormone secreted by the hypothalamus that activates the release of gonadotropins from the anterior lobe of the pituitary gland.

gonorrhea A sexually transmitted disease caused by the bacterium *Neisseria gonorrhoeae*.
granulocyte A WBC containing visible cytoplasmic granules.
grey matter The portion of the central nervous system lacking myelin.
groin The body region at the junction of a thigh and pelvis.
growth hormone (GH) The hormone of the anterior lobe of the pituitary gland that promotes cell division and cell enlargement.
guanine A nitrogenous base of nucleic acids that pairs with cytosine.

H

hair follicle An inward, tubular extension of the epidermis containing the hair root.
head An enlarged rounded end of a bone.
heart murmur An abnormal heart sound usually caused by a leaking valve.
heart rate The number of heart contractions (beats) per minute.
helper T cell (T_H) A T cell that initiates immune responses after binding to a foreign antigen on an APC.
hematopoiesis The formation of formed elements in the blood.
hemocytoblasts Stem cells in red bone marrow from which formed elements develop.
hemoglobin The pigmented protein in red blood cells, involved in transporting oxygen and carbon dioxide.
hemolytic disease of the newborn (HDN) A fetal disorder caused by maternal antibodies attacking fetal Rh+ red blood cells.
hemophilia A disorder in which clot formation is impaired.
hemorrhage Bleeding; excessive blood loss.
hemorrhoids Swollen, inflamed veins of the anal canal.
hemostasis A positive-feedback mechanism initiated after vascular injury to stop or limit blood loss.
hepatic Pertaining to the liver.
hepatic lobules The structural and functional units of the liver.
hepatitis Inflammation of the liver.
herniated disc A bulging intervertebral disc.
heterozygous A condition in which the alleles of a gene pair are different.

hives An itching rash resulting from an allergic reaction.
homeostasis The maintenance of a relatively stable internal environment.
homozygous A condition in which the alleles of a gene pair are identical.
hormone A chemical secreted by an endocrine gland that affects the functions of target cells.
human chorionic gonadotropin (hCG) An LH-like hormone secreted by the trophoblast and chorion that maintains the corpus luteum through 10 to 12 weeks of development.
hyaline cartilage Most abundant cartilage tissue: protein fibers are not easily visible.
hydrogen bond A weak bond between a positively charged hydrogen atom and a negatively charged atom in the same or a different molecule.
hydrogen ion H^+; a single proton.
hydrolysis The breakdown of complex molecules into simpler molecules by the addition of water molecules.
hydroxide ion OH^-.
hymen A membrane that partially covers the vaginal orifice.
hyperglycemia A high concentration of glucose in the blood.
hyperparathyroidism The hypersecretion of PTH, which can lead to weakened bones and kidney disease.
hypersecretion Production of an excessive amount of a secretion.
hypertension Chronic high blood pressure.
hypertonic solution A solution with a higher concentration of impermeable solutes (lower concentration of water) than the cell.
hypertrophy An increase in muscle fiber size and strength.
hypochondriac region Either the left or right upper portion of the abdominopelvic cavity lying lateral to the epigastric region.
hypogastric The lower middle region of the abdominopelvic cavity; pubic region.
hypogeusia A reduced ability to taste.
hypoglycemia A low concentration of glucose in the blood.
hypoparathyroidism The hyposecretion of PTH, which can lead to neuromuscular insufficiency and cardiac arrest.
hyposecretion Production of an insufficient amount of secretion.
hyposmia A decreased ability to detect odor.

hypothalamus The part of the diencephalon below the thalamus to which the pituitary gland is attached.
hypotonic solution A solution with a lower concentration of impermeable solutes (higher concentration of water) than the cell.

I

immunity Resistance to specific antigens.
immunocompetent Capable of responding to a foreign antigen.
immunoglobulin Plasma proteins consisting of antibodies.
impetigo A highly contagious bacterial infection of the skin characterized by pustules that rupture and become crusted.
implantation The attachment of a blastocyst to the endometrium.
impotence The inability of a male to obtain and maintain an erection for sexual intercourse.
infant respiratory distress syndrome (IRDS) Disorder caused by the inability to produce surfactant in the lungs of an infant.
infectious mononucleosis A viral infection of lymphocytes that causes them to resemble monocytes.
inferior vena cava Large vein that returns blood to the right atrium from most structures below the diaphragm.
infertility In females, the inability to conceive; in males, the inability to produce sperm to achieve fertilization of the secondary oocyte.
inflammation A localized response to damaged or infected tissues that is characterized by swelling, redness, pain, and heat.
inflammatory bowel disease (IBD) A group of chronic disorders characterized by inflamed intestines and intestinal cramps resulting from an overactive immune response.
influenza A virus-caused disorder with coldlike symptoms.
inguinal region Either the left or right lower portion of the abdominopelvic cavity lying lateral to the hypogastric region.
inhibin A hormone produced by ovaries and testes that inhibits secretion of FSH.
inorganic substance A chemical that does not contain both carbon and hydrogen together.
insertion The movable attachment of a skeletal muscle.

inspiration Movement of air into the lungs; inhalation.
insulin The pancreatic hormone that helps glucose to enter body cells.
integrating center A structure that functions to interpret information and coordinate a response.
integument The skin.
interatrial septum Septum separating the left and right atria of the heart.
intercalated disc A dark-staining membrane at the junction of adjoining cardiac muscle cells.
intercellular Pertaining to the spaces between cells.
internal Pertaining to the inside of the body or an organ.
internal respiration Systemic gas exchange and aerobic respiration.
interneuron A neuron responsible for processing and interpreting action potentials within the central nervous system.
interphase The nondividing phase between two cell divisions.
interstitial cell A cell located between seminiferous tubules that secretes testosterone.
interstitial fluid Intercellular fluid; tissue fluid.
interventricular septum Septum separating the left and right ventricles of the heart.
intervertebral disc The fibrocartilaginous pad separating two adjacent vertebrae.
intestinal gland A gland of the mucosa of the small intestine that secretes intestinal juice.
intestinal juice The secretion of intestinal glands.
intestinal villus A small, fingerlike projection of the intestinal mucosa.
intracellular Pertaining to within cells.
intramembranous bone Bone that develops within layers of membranes.
intramembranous ossification The formation of bone tissue within embryonic connective tissue.
intrinsic factor A substance secreted by gastric glands that is essential for adequate absorption of vitamin B12.
inversion The movement that turns the sole of the foot medially.
involuntary Without conscious control.
ion An atom or group of atoms with an electrical charge.
ionic bond A chemical bond formed between two ions with opposing electrical charges.
ionization The dissociation of ionic compounds forming ions.
iris The colored circular muscle that controls the amount of light entering the lens of the eye.
irritability The ability of a neuron to respond to a stimulus by forming an action potential.
irritable bowel syndrome Intestinal disorder causing abdominal bloating, cramping pain, constipation, and diarrhea.
isotonic solution A solution that has the same concentration of impermeable solutes (same concentration of water) as the cell.
isotope An atom with a different number of neutrons than atoms of the same element.

J

joint The junction between two bones or between a bone and a tooth; an articulation.
juxtaglomerular complex Specialized cells of the afferent and efferent glomerular arterioles and ascending limb of the nephron loop that are involved in controlling glomerular blood pressure.

K

keratin Tough, fibrous protein that provides waterproofing and abrasion protection for the epidermis.
keratinization The process by which cells form large amounts of keratin.

L

labia majora The lateral folds of the female external genitalia.
labia minora The medial folds of the female external genitalia.
labor The series of events associated with childbirth.
labyrinth The interconnecting tubes and chambers of the internal ear.
labyrinthine disease A group of internal ear disorders that produce symptoms of dizziness and nausea.
lacrimal apparatus Structures involved in the production and removal of tears.
lacrimal gland A gland that secretes tears.
lactase The brush border enzyme that catalyzes the digestion of lactose into glucose and galactose.
lactate An organic by-product formed from pyruvic acid during anaerobic respiration.
lactation Milk secretion.
lacteal A lymphatic capillary within a villus of the small intestine.
lacuna A fluid-filled cavity in bone or cartilage tissue that contains an osteocyte or chondrocyte.
lamellae Layers of solid matrix in bone.
laryngitis Inflammation of the mucosae of the larynx.
larynx Cartilaginous box providing a passageway for air between the pharynx and the trachea and containing the vocal folds.
lateral Pertaining to the side.
lens The structure that focuses images on the retina of the eye.
leukemia A cancerous blood disorder characterized by an excess production of certain WBCs.
leukotrienes Secretions by cells that promote inflammation in nearby cells.
ligament A band or cord of dense regular connective tissue that joins bones together at joints.
limbic system The portions of the cerebrum and diencephalon involved in emotions and moods.
lingual Pertaining to the tongue.
lipase The enzyme that catalyzes the digestion of triglycerides into monoglycerides and fatty acids.
lipid A class of organic macromolecules that includes steroids, triglycerides, and phospholipids.
liver The digestive gland that secretes bile and processes absorbed nutrients prior to their entrance into the general circulation.
longitudinal cerebral fissure The deep fissure that separates the left and right cerebral hemispheres.
loose connective tissues Supportive connective tissues with widely spaced protein fibers intertwined between cells.
lumbar Pertaining to the lower back between the ribs and hips.
lumen An opening within a hollow internal organ.
lungs The respiratory organs for gas exchange.
luteinizing hormone (LH) The hormone secreted from the anterior lobe of the pituitary gland that causes ovulation and controls the functions of the corpus luteum in females and testosterone secretion in males.
lymph A fluid connective tissue in lymphatic vessels.
lymphadenitis Inflammation of the lymph nodes.
lymph node A lymph-filtering secondary lymphoid organ.

lymphatic vessel A vessel transporting lymph.
lymphocyte A type of white blood cell that is involved in immune reactions.
lymphoma A cancer of lymphoid tissue.
lysis Rupture of a cell due to a rapid uptake of water.
lysosome A cellular organelle consisting of a sac of digestive enzymes.

M

macrophage A modified monocyte that has entered tissue spaces and is involved in phagocytosis and immune reactions.
macula Sense organ of the internal ear containing sensory receptors for static and dynamic equilibrium; the yellow spot on the retina containing the fovea centralis.
major calyx Urine-collecting passageway formed by the convergence of two or three minor calyces.
malignant tumor A tumor with the capability of spreading (metastasizing) to other sites; a cancer.
maltase The brush border enzyme that catalyzes the digestion of maltose into glucose and fructose.
mammary gland A milk-producing gland located within a female breast.
mast cell A modified basophil in tissue spaces that releases histamine and serotonin in allergic reactions.
mastication The process of chewing food.
matrix Extracellular substance in connective tissues.
matter Anything that has weight (mass) and occupies space.
maximal stimulus A stimulus producing a maximum response by a muscle.
mechanoreceptor Sensory receptor stimulated by mechanical forces such as pressure and touch.
medial Toward the midline.
median plane A plane that divides the body or an organ into equal left and right halves.
mediastinum The medial space that separates the thoracic cavity into left and right portions.
medulla The inner or central portion of an organ.
medulla oblongata The part of the brainstem that is continuous with the spinal cord.

medullary cavity The central cavity in the shaft of a long bone.
meiosis A form of cell division in which the daughter cells contain one-half the number of chromosomes as the parent cell.
melanin The brown-black pigment found in the epidermis.
melanocyte A cell of the epidermis that produces melanin.
melatonin A hormone secreted by the pineal gland that influences biorhythms.
membrane potential Voltage created by electrical charge differences across the plasma membrane.
memory cell A dormant B or T cell produced in an initial immune response that can respond quickly if the same antigen reappears.
meninges A group of three membranes that envelops the brain and spinal cord.
meningitis Inflammation of the meninges.
menopause The cessation of menstruation.
menstrual cycle The repetitive monthly changes in the endometrium.
menstruation The breakdown and discharge of the endometrium.
mental illness A group of disorders characterized by abnormal behaviors.
mesentery A fold of the peritoneum that supports abdominal organs.
mesoderm The middle germ layer of the embryonic disc.
messenger RNA (mRNA) The type of RNA that carries information for protein synthesis from DNA to the ribosomes.
metabolism The sum of the chemical reactions in the body.
metacarpals The bones of the palm.
metaphase The phase of cell division characterized by the chromosomes arranged along the equator of the cell.
metatarsals The bones in the foot between the tarsal bones and the phalanges.
microfilament A microscopic protein strand within cells that is part of the cytoskeleton.
microglial cell A type of neuroglia involved in phagocytosis.
microtubule A microscopic tubule of protein within cells that is part of the cytoskeleton.
microvilli Microscopic projections of the plasma membrane on the free surfaces of certain epithelial cells.
micturition Urination.
midbrain The part of the brainstem between the pons and the thalamus.
mineralocorticoids A group of hormones secreted by the adrenal cortex that influence the concentration of electrolytes in body fluids.

minor calyx A funnellike receptacle receiving urine from a renal papilla.
miscarriage A spontaneous abortion.
mitochondrion A cellular organelle that is the site of aerobic respiration.
mitosis Process by which a cell divides to form two new daughter cells with the same number and composition of chromosomes as the parent cell.
mitotic spindle A spool-shaped arrangement of spindle fibers formed during prophase of cell division.
mitral valve The valve between the left atrium and the left ventricle; the left atrioventricular valve.
mixed nerve A nerve consisting of axons of both sensory and motor neurons.
molecule Two or more similar or dissimilar atoms bonded together by covalent bonds.
monocyte A large agranulocytic WBC that functions in phagocytosis.
monosaccharide A simple sugar; a structural unit of carbohydrates.
morning sickness A temporary disorder of pregnancy characterized by nausea upon arising.
motion sickness Nausea triggered by repetitive equilibrium receptor stimulation.
motor area The region of the cerebrum that initiates actions of skeletal muscles.
motor division Portion of the nervous system carrying action potentials from the CNS to muscles, glands, and adipose tissue.
motor nerve A nerve consisting mostly of axons of motor neurons.
motor neuron A neuron that activates a muscle, gland, or adipose tissue.
motor speech area The portion of the frontal lobe of the cerebrum that coordinates muscles involved in speech.
motor unit A somatic motor neuron and the muscle fibers that it innervates.
mucous membrane An epithelial membrane that lines tubes or cavities of organ systems with openings to the external environment; mucosa.
mucus Thick fluid produced by goblet cells.
multiple sclerosis (MS) A disorder characterized by the degeneration of the myelin sheath surrounding neurons in the central nervous system.
multipolar neuron A neuron with several dendrites and one axon.
muscle fiber A skeletal muscle cell.
muscle tissue The type of body tissue that is specialized for contraction.
muscle tone The state of slight contraction in a skeletal muscle.
muscular dystrophy A disorder characterized by the atrophy of muscles.

mutation A spontaneous change in a gene.
myasthenia gravis A disorder characterized by severe muscular weakness.
myelin sheath An insulating layer, formed by neuroglia, that surrounds an axon.
myelin sheath gaps Tiny spaces between adjacent myelin-forming cells, where the axon is exposed; nodes of Ranvier.
myocardial infarction Death of a portion of the myocardium due to blockage of a coronary artery; a heart attack.
myocardium The muscle layer of the heart wall.
myofibril A thin contractile element within a muscle cell.
myofilament Contractile filament within a myofibril; composed of either actin or myosin.
myoglobin A protein in muscles that temporarily stores a small amount of oxygen.
myositis Inflammation of a muscle.
myxedema An adult disorder caused by a hyposecretion of thyroid hormones.

N

nasal cavity The cavity of the nose.
nasal septum The partition of bone and nasal cartilage separating the nasal cavity into left and right portions.
natriuretic peptides Hormones secreted by the heart that promote excretion of water to reduce blood volume.
nearsightedness A vision disorder in which the image is focused in front of the retina.
negative-feedback mechanism A mechanism used to keep a variable within its normal range, thereby maintaining homeostasis.
nephron The structural and functional unit of the kidney.
nerve A bundle of myelinated and unmyelinated axons in the peripheral nervous system.
nerve tissue Tissue specialized for the formation and conduction of action potentials.
nerve tract A bundle of myelinated and unmyelinated axons in the central nervous system.
neuralgia Pain in a nerve.
neurilemma The outermost layer, formed by a Schwann cell, that wraps around a myelinated axon in the peripheral nervous system.
neuritis Inflammation of a nerve.

neuroglia Cells that support and protect neurons.
neuromuscular junction The junction of the terminal bouton of a motor neuron with a muscle fiber.
neuron A cell capable of producing an action potential; a nerve cell.
neurotransmitter A chemical secreted by a terminal bouton that can excite or inhibit the postsynaptic cell.
neutron A noncharged particle in an atomic nucleus.
neutrophil A phagocytic WBC with pale lavender cytoplasmic granules.
nociceptor Sensory receptor stimulated by tissue damage; pain receptor.
nonspecific resistance Resistance mechanisms that act against all types of pathogens.
norepinephrine A neurotransmitter released by terminal boutons of sympathetic postganglionic neurons; a hormone secreted by the adrenal medulla.
nuclear envelope The double membrane surrounding the nucleus of a cell.
nucleic acid A macromolecule composed of a series of nucleotides; either DNA or RNA.
nucleolus A dark-staining spherical structure within a cell nucleus that is composed of protein and rRNA.
nucleotide The building unit of nucleic acids; consists of a simple sugar, a phosphate group, and a nitrogenous base.
nucleus The spherical cellular organelle containing the chromosomes; the core of an atom; or a mass of cell bodies in the central nervous system.
nutrient A substance required for body cells to function.

O

olfactory Pertaining to the sense of smell.
oligodendrocyte A type of neuroglia that forms a myelin sheath around axons within the central nervous system.
oocyte An immature ovum (egg cell).
oogenesis The process of ovum formation.
oogonia The stem cells in an ovary that divide to form primary oocytes.
ophthalmic Pertaining to the eye.
optic Pertaining to the eye.
optic chiasma The site where axons from the medial half of the retina cross over to the opposite side of the brain.

optic disc The site where retinal ganglion cell axons exit the eye and form the optic nerve; the blind spot.
oral Pertaining to the mouth.
organ A structure formed of two or more tissues that performs specific functions.
organ system A group of organs that work in a coordinated fashion to carry out specialized functions.
organelle A complex of macromolecules acting like a "mini-organ" within a cell that performs specific functions.
organic substance Large, complex substance that contains both carbon and hydrogen in the same molecule, usually with oxygen too.
orgasm The culmination of sexual stimulation.
origin The immovable attachment of a skeletal muscle.
osmosis The passive movement of water across a selectively permeable membrane.
ossification The process of bone tissue formation.
osteoblast A cell that deposits bone matrix.
osteoclast A cell that removes bone matrix.
osteocyte A mature bone cell.
osteomyelitis An inflammation of bone and bone marrow caused by bacterial infection.
osteon The structural unit of compact bone consisting of lamellae and osteocytes around a central canal.
osteoporosis A disorder in which bone matrix is reabsorbed, producing weakened bones.
otitis media A middle ear infection.
otolith A granule of calcium carbonate associated with receptors for equilibrium.
oval window The membrane-covered opening in the vestibule of the internal ear into which the stapes is inserted.
ovarian Pertaining to an ovary.
ovarian cycle The monthly formation and release of a secondary oocyte and the ovarian events that take place in preparation for pregnancy.
ovarian follicle A spherical structure composed of an oocyte enveloped by supporting cells.
ovary The female gonad producing oocytes and female sex hormones.
ovulation The rupture of a mature ovarian follicle, releasing a secondary oocyte.

ovum A female reproductive cell; a female gamete.
oxyhemoglobin (HbO$_2$) The molecule carrying oxygen to body cells.
oxytocin (OT) The hormone secreted from the posterior lobe of the pituitary gland that stimulates uterine contractions and milk ejection.

P

palate The roof of the mouth formed of the hard and soft palates.
palmar Pertaining to the palm of the hand.
pancreas An abdominal organ secreting both digestive secretions and hormones.
pancreatic duct The tube carrying pancreatic juice to the duodenum.
pancreatic islets Clumps of pancreatic cells that secrete insulin and glucagon.
papilla A small, nipplelike projection.
papillary muscle A projection of the myocardium to which tendinous cords are attached.
paralysis A disorder in which a muscle is unable to contract.
paramedian plane A vertical plane that divides the body or a body part into unequal right and left portions.
paranasal sinus An air-filled cavity in a bone located near the nasal cavity.
parasympathetic part The part of the autonomic division arising from the brainstem and sacral region of the spinal cord.
parathyroid glands Small endocrine glands embedded in the back of the thyroid gland.
parathyroid hormone (PTH) The hormone secreted by the parathyroid glands that causes an increase in blood Ca^{2+} level.
parietal Pertaining to the wall of a cavity.
parietal cell A type of cell in gastric glands that secretes hydrochloric acid and intrinsic factor.
parietal pericardium The outer membrane of the serous pericardium.
parietal peritoneum The membrane lining the wall of the abdominal cavity.
parietal pleura The layer of pleura lining the inner surface of the thoracic cavity.
Parkinson disease A neural disorder characterized by muscular weakness, tremor, and rigidity.

parotid glands The largest salivary glands located just in front of and below the ears.
parturition The process of giving birth.
passive transport Movement of substances across plasma membranes without the expenditure of energy.
pathogen A disease-causing substance, virus, or organism.
pectoral Pertaining to the chest.
pectoral girdle The bones that attach the upper limbs to the trunk at the shoulder.
pelvic Pertaining to the pelvis.
pelvic cavity The lower portion of the abdominopelvic cavity.
pelvic girdle The bones that attach the lower limbs to the trunk at the hips.
pelvic inflammatory disease (PID) Inflammation of the female reproductive organs and pelvic tissues.
pelvis The ring of bones formed by the coxal bones and sacrum.
penis The male copulatory organ containing the urethra.
pepsin An enzyme secreted by gastric glands that catalyzes the digestion of proteins into peptides.
peptic ulcers Digestion of part of the stomach mucosa.
peptidase The enzyme that catalyzes the digestion of peptides into amino acids.
peptide A molecule composed of two or more amino acids.
peptide bond A bond that joins two amino acids.
perception The conscious awareness of a sensation.
pericardium The membrane surrounding the heart.
pericardial cavity The space between the epicardium and parietal pericardium.
pericarditis Inflammation of the serous pericardium.
perichondrium A connective tissue membrane covering the outer surface of cartilage tissue.
perilymph The fluid in the bony labyrinth of the internal ear.
perineum In males, the region between the anus and the scrotum; in females, the region between the anus and the mons pubis.
periodontal disease A disorder in which gums are inflamed and associated periodontal ligaments and bone degenerate.
periosteum The membrane formed of dense irregular connective tissue that covers the outer surfaces of bones.
peripheral nervous system (PNS) Portion of the nervous system composed of cranial and spinal nerves, ganglia, and sensory receptors.

peripheral resistance Resistance slowing the flow of blood; friction between the blood and the walls of blood vessels.
peristalsis The wavelike contractions that move materials through tubular organs.
peritoneal cavity The space between the visceral and parietal peritonea.
peritoneum The membrane lining the abdominal cavity and covering abdominal organs.
peritonitis Inflammation of the peritoneum.
peritubular capillaries The capillary network that surrounds a renal tubule.
pH A measure of the hydrogen ion concentration of a solution.
pH scale A scale that establishes the values of pH from 0 to 14.
phagocytosis The process by which cells engulf particles.
pharynx Passageway behind the nasal and oral cavities that extends downward to the larynx and esophagus; the throat.
phenotype The observable traits determined by genes.
phlebitis Inflammation of a vein.
phospholipid A molecule containing two fatty acids and a phosphate group attached to glycerol.
photoreceptor Sensory receptor stimulated by light energy; rod or cone.
physiological jaundice A disorder of newborn infants caused by the rapid destruction of fetal RBCs.
physiology The study of the function of living organisms.
pia mater The delicate, innermost membrane of the meninges.
pineal gland A small endocrine gland within the brain that is involved in biorhythms.
pinocytosis The process by which cells engulf liquids.
pituitary dwarfism Abnormal growth caused by hyposecretion of growth hormone from the anterior lobe of the pituitary gland.
pituitary gland The endocrine gland attached to the hypothalamus; the hypophysis.
placenta Structure formed from embryonic and maternal tissues that allows for the exchange of substances between maternal and embryonic/fetal blood.
plantar Pertaining to the sole of the foot.
plasma The liquid portion of the blood.
plasma cell An activated B cell that produces antibodies.
plasma membrane Outer boundary of a cell.
plasma protein One of several proteins dissolved in the plasma.

platelet A formed element involved in blood clot formation.
pleura The membrane lining the thoracic cavity and covering the lungs.
pleural cavity The space between the visceral and parietal pleurae.
pleurisy Inflammation of the pleurae.
plexus A network of nerves or blood vessels.
pneumonia An acute inflammation of the pulmonary alveoli caused by bacterial or viral infection.
polar body A small, nonfunctional cell formed during oogenesis.
polar molecule A molecule with slightly positive or negative charges on its surface.
polarization The formation of an electrical charge on a plasma membrane due to unequal concentrations of ions on either side of the membrane.
poliomyelitis A viral disease in which motor neurons are destroyed, causing paralysis.
polycythemia A disorder in which there is an excessive number of RBCs.
polypeptide An organic macromolecule formed of many amino acids.
polysaccharide An organic macromolecule formed of many monosaccharide units.
pons The part of the brainstem located between the midbrain and the medulla oblongata.
popliteal The body region composed of the back of the knee.
positive-feedback mechanism A mechanism used when the originating stimulus needs to be amplified and continued in order for the desired result to occur.
posterior Toward the back; dorsal.
posterior root The posterior attachment of a spinal nerve to the spinal cord that contains axons of sensory neurons.
postganglionic axon An autonomic axon extending from an autonomic ganglion to an effector.
postnatal The period after birth; opposite, *prenatal*.
postsynaptic Pertaining to the cell that is activated by a signal at a synapse.
preembryo Name of the developing offspring from fertilization through the end of the second week of development.
preganglionic axon An autonomic axon extending from the central nervous system to an autonomic ganglion.
pregnancy The female condition in which a developing offspring is in the uterus.
premenstrual syndrome (PMS) A female disorder occurring just prior to menstruation characterized by pain and emotional stress.

prenatal The period from conception to birth; opposite, *postnatal*.
presbyopia The visual condition in which the near-point distance becomes greater with age.
presynaptic Pertaining to the neuron that releases a signal at a synapse.
primary lymphoid organ An organ where lymphocytes become immunocompetent.
prime mover A muscle whose contraction is primarily responsible for a particular movement.
process A projection on a bone.
progesterone A female sex hormone secreted by the corpus luteum and the placenta that maintains the endometrium.
projection The brain mechanism that makes a sensation seem to come from the body part being stimulated.
prolactin (PRL) The hormone secreted from the anterior lobe of the pituitary gland that stimulates the production of milk by mammary glands.
pronation Turning the forearm so that the palm faces downward.
prophase The phase of mitosis in which the chromosomes coil, appearing as rodlike structures.
proprioceptor Sensory receptor stimulated by changes in body position or movements of the body or its parts.
prostaglandin A type of paracrine signal.
prostate An accessory gland that surrounds the proximal portion of the male urethra.
prostatitis The inflammation of the prostate.
protein A complex, nitrogen-containing organic macromolecule that usually consists of more than 50 amino acids.
prothrombin An inactive plasma protein that is converted to thrombin in blood clot formation.
prothrombin activator Substance released by platelets that converts prothrombin into thrombin.
proton A positively charged particle in an atomic nucleus.
protraction The movement of a body part forward.
proximal Pertaining to nearer the origin.
pseudostratified An arrangement of cells that appears layered but is not.
pseudounipolar neuron A neuron with a single process that divides into two branches extending in opposite directions.
psoriasis A chronic skin disorder characterized by redness, itching, and excessive scaling.
puberty The developmental stage in which the reproductive organs mature and become functional.
pulmonary Pertaining to the lungs.

pulmonary alveolus A microscopic air sac in a lung.
pulmonary circuit The blood pathway that transports blood to and from the lungs.
pulmonary edema The accumulation of fluid in the lungs.
pulmonary embolism A blood clot or a foreign body that blocks an artery in the lung.
pulmonary trunk Large artery that carries blood from the right ventricle and divides to form the pulmonary arteries.
pulmonary valve The valve preventing a backflow of blood from the pulmonary trunk into the right ventricle.
pulmonary ventilation The movement of air into and out of the lungs; breathing.
pulse The expansion of an artery resulting from a surge of blood generated by ventricular contraction.
pupil The opening in the iris that allows light to pass through the lens to the retina.
pus A thick fluid composed of WBCs and bacteria.
pyelonephritis Inflammation of the kidney involving the renal pelvis.
pyruvic acid An organic molecule formed by the breakdown of glucose in anaerobic respiration.

R

receptor A structure that functions to collect information.
rectum The straight portion of the large intestine that follows the sigmoid colon.
red blood cell A hemoglobin-containing formed element that transports respiratory gases; erythrocyte.
red bone marrow Primary lymphoid organ responsible for the production of all formed elements.
referred pain Pain that seems to originate from a site that is different from the site being stimulated.
reflex An involuntary, rapid, and predictable response to a stimulus.
reflex arc The neural pathway of a reflex; usually involves sensory neuron, interneuron, and motor neuron.
refraction The bending of light rays.
relaxin Hormone secreted by the corpus luteum and placenta that aids in the development of placental blood vessels and other cardiovascular changes in the mother.
renal Pertaining to a kidney.
renal calculi Kidney stones.
renal corpuscle The portion of a nephron composed of a glomerulus and its enveloping glomerular capsule.

renal cortex The outer layer of a kidney.
renal failure Reduction or cessation of urine formation by the kidneys.
renal medulla The inner layer of a kidney.
renal pelvis A cavity within a kidney that is continuous with a ureter.
renal tubule The portion of a nephron composed of a proximal convoluted tubule, a nephron loop, a distal convoluted tubule, and the collecting duct.
renin The enzyme released by the juxtaglomerular complex that converts a plasma protein into angiotensin I.
renin-angiotensin mechanism A chemical control mechanism that regulates the blood pressure.
rennin A gastric enzyme that curdles milk proteins in infants.
respiratory membrane Membrane formed of alveolar and capillary cells through which gas exchange occurs between air in pulmonary alveoli and blood in capillaries.
respiratory rhythmicity center The area in the medulla oblongata that triggers each cycle of inhale and exhale.
resting membrane potential Membrane potential in cells with irritability that are inactive.
reticular fibers Fibers composed of collagen protein that form supporting frameworks for other tissues and organs.
reticular formation A neural network in the diencephalon, brainstem, and spinal cord that is connected to higher brain centers and arouses the cerebrum to wakefulness.
reticular tissue A loose connective tissue containing reticular cells and reticular fibers.
retina The inner layer of the eyeball that contains the photoreceptors.
retinoblastoma A cancer of immature retinal cells.
retraction The movement of a body part backward.
rhinitis Inflammation of the nasal cavity.
ribonucleic acid (RNA) A single-stranded nucleic acid whose nucleotides contain ribose sugar.
ribosomal RNA (rRNA) The RNA composing ribosomes.
ribosome A tiny cellular organelle composed of protein and rRNA and serving as the site of protein synthesis.
rickets A disorder caused by a deficiency of vitamin D resulting in weakened bones due to insufficient deposition of calcium salts.
rod A photoreceptor associated with black and white vision.
rotation The turning of a body part on its longitudinal axis.
round window A membrane-covered opening in the vestibule of the internal ear.

S

saccule Enlargement within the vestibule of the internal ear containing sensory receptors for static and dynamic equilibrium.
sacral Pertaining to the sacrum.
sagittal plane A plane dividing the body or an organ into left and right portions.
saliva The secretion of salivary glands.
salivary glands Glands secreting saliva into the mouth.
sarcolemma The plasma membrane of a muscle cell.
sarcomere The contractional unit of a myofibril.
sarcoplasm The cytoplasm of a muscle cell.
sarcoplasmic reticulum The smooth endoplasmic reticulum of a muscle cell.
saturated fats Triglycerides whose fatty acids do not have double carbon-carbon bonds.
Schwann cell A type of neuroglia that forms the myelin sheath and neurilemma around an axon in the peripheral nervous system.
sciatica Neuralgia of a sciatic nerve.
sclera The fibrous outer layer of the eyeball; the white of the eye.
scrotum The pouch containing the testes in males.
sebaceous gland An epithelial gland in the dermis that secretes sebum into a hair follicle.
sebum The oily secretion of a sebaceous gland.
secondary lymphoid organ An organ where immunocompetent lymphocytes proliferate and immune responses occur.
second messenger An intracellular messenger that activates or inactivates enzymes to produce the characteristic effect for the hormone.
secretin A hormone secreted by the intestinal mucosa that stimulates the pancreas to secrete pancreatic juice.
segmentation Localized contractions used to mix luminal contents with digestive secretions.
selectively permeable membrane A membrane that allows only certain substances to enter or exit the cell.
semen Fluid composed of sperm and secretions of male accessory glands.
semicircular canal A loop of the membranous labyrinth within the internal ear that contains sensory receptors detecting rotational head movement.
semilunar valve A valve located at the base of the aorta or pulmonary trunk.
seminal vesicles Male accessory glands whose secretions contribute to semen.
seminiferous tubule A tubule within which sperm are formed in a testis.
sensation Awareness as a result of the brain interpreting sensory action potentials.
sensory adaptation A decrease in action potential formation due to repetitive stimulation of a sensory receptor.
sensory area A region of the cerebrum that interprets sensory action potentials as sensations.
sensory division Portion of the nervous system carrying action potentials from sensory receptors to the central nervous system.
sensory nerve A nerve composed of axons of sensory neurons.
sensory neuron A neuron that carries sensory action potentials from a sensory receptor to the central nervous system.
septum A partition such as the interventricular septum.
serotonin A substance released by platelets that constricts blood vessels.
serous fluid Fluid secreted by serous membranes.
serous membrane Epithelial membrane that lines the thoracic and abdominopelvic cavities and covers most of the internal organs within these cavities.
sesamoid bone A small bone embedded in a tendon.
severe combined immunodeficiency (SCID) A group of disorders that result from a deficit or absence of both B and T cells.
sex hormones Estrogens, progesterone, and testosterone.
sexually transmitted diseases Various infections caused by bacteria or viruses transmitted via sexual intercourse.
shingles Inflammation of a peripheral nerve due to reactivation of the chicken pox virus.
simple diffusion Movement of small, lipid-soluble molecules across a plasma membrane down their concentration gradient, without the use of energy and membrane proteins.

simple epithelium Epithelium composed of a single layer of cells.
simple goiter Enlargement of the thyroid gland caused by a deficiency of iodine in the diet.
simple sugar A monosaccharide or disaccharide.
sinoatrial node (SA node) Specialized muscle tissue in the right atrium that initiates heart contractions; the pacemaker of the heart.
sinus An air-filled cavity in a bone.
sinusitis Inflammation of the mucosae of a sinus.
skeletal muscle tissue The type of muscle tissue that is attached to bones and skin.
skull Bones of the cranium and face.
smooth muscle tissue The type of muscle tissue in the walls of hollow organs except the heart.
solute A substance dissolved in a solvent.
solution A fluid composed of solutes dissolved in a solvent.
solvent A fluid that can dissolve solutes.
somatic division The division of the PNS involved in the voluntary and involuntary control of skeletal muscles.
spasm A sudden, involuntary contraction of a muscle.
sperm A male reproductive cell; a male gamete.
spermatogenesis The process of sperm production.
spermatogonium A stem cell in a testis that divides to form primary spermatocytes.
sphincter A ring of smooth or skeletal muscle tissue that contracts to close an opening and relaxes to create an opening.
spinal Pertaining to the spinal cord.
spinal cord The portion of the central nervous system that occupies the vertebral canal.
spinal nerve A nerve that branches from the spinal cord.
spinal plexus A network of nerves in which axons of the anterior rami of the spinal nerves are sorted and recombined before continuing on.
spiral organ The sense organ in the cochlea of the internal ear that contains sensory receptors for hearing.
spleen Secondary lymphoid organ located behind the stomach that stores formed elements and clears pathogens from the blood.
spongy bone Bone tissue composed of interconnected bony plates surrounded by red or yellow bone marrow; trabecular bone.
sprain An injury caused by the tearing of a ligament.

starch A common polysaccharide in foods derived from plants.
static equilibrium Maintenance of balance during movement of the head with respect to gravitational force.
steroid A group of lipids that includes sex hormones and cholesterol.
stimulus A change in the environment.
stomach The expanded portion of the alimentary canal that receives food from the esophagus.
strabismus The inability to focus both eyes simultaneously on the same object.
strain An injury caused by overuse or overstretching of a muscle.
stratified Layered.
stratified epithelium Epithelium composed of more than one layer of cells.
stratum basale The innermost layer of the epidermis whose cells are involved in cell division.
stratum corneum The outermost layer of the epidermis composed of dead keratinized cells.
stroke volume The volume of blood pumped from each ventricle per heartbeat.
subarachnoid space The space under the arachnoid mater that is filled with cerebrospinal fluid.
subcutaneous tissue A tissue composed of areolar connective tissue and adipose tissue and located just beneath the skin; hypodermis.
subendocardiac conducting network Specialized cardiac muscle tissue that transmits action potentials from the bundle branches to the myocardium.
submucosa Connective tissue beneath a mucous membrane.
substrate A substance acted upon by an enzyme.
sucrase The brush border enzyme that catalyzes the digestion of sucrose into glucose and fructose.
sudden infant death syndrome (SIDS) The sudden death of an infant with no medical history or explanation upon autopsy; crib death.
superficial Near the surface of the body or organ.
superior Toward the head.
superior vena cava Large vein that returns blood to the right atrium from most structures above the diaphragm.
supination Turning the forearm so that the palm faces upward.
surfactant A chemical in pulmonary alveoli that reduces surface tension and prevents pulmonary alveolar collapse.
suture An immovable joint joining skull bones.
sympathetic part The part of the autonomic division arising from the thoracic and lumbar regions of the spinal cord.

symphysis A cartilaginous joint, formed by fibrocartilage, allowing only a slight degree of movement.
synapse The junction between an axon and another neuron or effector cell.
synaptic cleft The space between a terminal bouton and the postsynaptic cell within a synapse.
synaptic transmission The passage of an action potential from one neuron to another cell across a synapse.
synergist A muscle whose contraction helps the prime mover perform its action.
synovial fluid Fluid secreted by synovial membranes within a synovial joint.
synovial joint Joint with two bones separated by a small fluid-filled cavity and surrounded by a connective tissue capsule.
synovial membrane The inner layer of the joint capsule in a synovial joint.
synthesis (anabolic) reaction The combining of smaller molecules to form more complex molecules.
syphilis A sexually transmitted disease caused by the bacterium *Treponema pallidum*.
systemic circuit The blood pathway that transports blood to and from all parts of the body except the lungs.
systemic gas exchange The exchange of oxygen and carbon dioxide between the blood in systemic capillaries and tissue cells.
systole The contraction phase of the cardiac cycle.
systolic blood pressure Arterial blood pressure during ventricular systole.

T

T lymphocyte A type of lymphocyte involved in cell-mediated immunity; a T cell.
tactile corpuscle A sensory receptor for light touch located in the skin.
target cell A cell that possesses receptors specific for that hormone.
tarsal bones The bones of the ankle.
taste bud An organ on the tongue containing receptors for taste.
telophase The final phase of cell division in which daughter nuclei and cells are formed.
tendinous cords Cords of dense regular connective tissue that anchor AV valve cusps to papillary muscles.
tendon A band or cord of dense regular connective tissue that attaches a

muscle to a bone, other muscles, dermis, and ligaments.
terminal boutons Enlargements at the tips of an axon's terminal arborization.
testis The male gonad that produces sperm and testosterone.
testosterone The primary male sex hormone produced by the testes.
tetanic contraction A sustained muscle contraction.
tetanus A potentially fatal disorder caused by a bacterial toxin; lockjaw.
tetany A continuous, maximal muscle contraction.
thalamus The part of the diencephalon located above the midbrain and forming the lateral walls of the third ventricle.
thermoreceptor A sensory receptor stimulated by temperature changes.
thick myofilament A thick protein filament composed of myosin that interacts with actin to produce contraction of a myofibril.
thin myofilament A thin protein filament composed of actin that interacts with myosin to produce contraction of a myofilament.
thoracic Pertaining to the trunk.
thoracic cavity The body cavity located above the diaphragm.
threshold stimulus The minimal stimulus that produces formation of an action potential.
thrombin The enzyme that converts fibrinogen into fibrin.
thrombocytopenia A disorder in which there is an abnormally low number of platelets.
thrombosis A condition in which a thrombus blocks a blood vessel.
thrombus A stationary blood clot formed in an unbroken blood vessel.
thymine A nitrogenous base found only in DNA that pairs with adenine.
thymosins The hormones secreted by the thymus.
thymus Primary lymphoid organ responsible for T cell maturation.
thyroid gland A bilobed endocrine gland located just below the larynx attached to the front of the trachea.
thyroid hormones Hormones secreted by the thyroid that increase the rate of cellular metabolism.
thyroid-stimulating hormone (TSH) The hormone secreted from the anterior lobe of the pituitary gland that stimulates the secretion of thyroid hormones.
thyroxine (T_4) The primary thyroid hormone.
tissue A group of similar cells performing a similar function.

tonsillitis Inflammation of the tonsils.
tonsils Masses of lymphoid tissue located around the opening to the pharynx.
toxic shock syndrome A female disorder caused by toxins released by bacteria growing in a tampon.
trachea Airway extending between the larynx and the bronchi.
transcription The rewriting of the DNA code into the codon sequence of messenger RNA.
transfer RNA (tRNA) A type of RNA that carries amino acids to a ribosome during protein synthesis.
translation The process in which the codons of mRNA place amino acids of a forming protein in a specific sequence.
transverse plane A plane that divides the body or organ into upper and lower portions.
transverse (T) tubules Invaginations of the sarcolemma into a muscle cell.
tremor Involuntary, repetitive weak contractions.
tricuspid valve The valve between the right atrium and the right ventricle; the right atrioventricular valve.
triglyceride A lipid molecule composed of three fatty acids attached to glycerol; a fat molecule.
triiodothyronine (T_3) The secondary thyroid hormone.
trochanter A large, broad process of a bone.
trypsin A pancreatic enzyme that catalyzes the digestion of proteins and peptides into amino acids.
tubal ligation A sterilization procedure in which the uterine tubes are cut and blocked.
tubercle A small, rounded process of a bone.
tuberculosis Inflammation of the lungs caused by the bacterium *Mycobacterium tuberculosis*.
tuberosity A large, roughened process.
tubular fluid Fluid within the renal tubule of the nephron.
tubular reabsorption The movement of substances from the tubular fluid into the blood plasma.
tubular secretion The movement of substances from the blood plasma into the tubular fluid.
tympanic membrane The thin membrane between the external ear and middle ear.

U

umbilical cord The connecting link between the placenta and the developing embryo or fetus.
umbilical region The central portion of the abdominopelvic cavity surrounding the umbilicus.
unsaturated fats Triglycerides whose fatty acids have one or more double carbon-carbon bonds.
uracil A nitrogenous base found only in RNA that pairs with adenine.
urea The nitrogenous waste formed by the liver as a result of protein metabolism.
ureter A narrow tube that carries urine from a kidney to the urinary bladder.
urethra A tube carrying urine from the urinary bladder to the external environment.
urethritis Inflammation of the urethra.
uric acid A nitrogenous waste produced from nucleic acid metabolism.
urinary bladder The hollow, muscular-walled organ that temporarily stores urine.
urine Waste and excessive substances removed from the blood by the kidneys and excreted from the body.
uterine Pertaining to the uterus.
uterine tube A tube extending from the uterus to an ovary; it carries secondary oocytes and preembryos toward the uterus.
uterus The female organ that contains the embryo and fetus until birth; the womb.
utricle Enlargement within the vestibule of the internal ear containing sensory receptors for static and dynamic equilibrium.
uvula The cone-shaped structure extending downward at the back of the soft palate.

V

vagina A tubular organ that extends from the vestibule of the vulva to the uterus.
valence shell An atom's outermost shell of electrons.
varicose veins Dilated, swollen veins whose valves are nonfunctional.
vas deferens A slender tube leading from an epididymis to the ejaculatory duct in males.

vascular Pertaining to blood vessels.

vasectomy A sterilization procedure in which the vasa deferentia are cut and blocked.

vasoconstriction The narrowing of the lumen of a blood vessel.

vasodilation Expansion of the lumen of a blood vessel.

vasomotor center The portion of the medulla oblongata that controls the diameter of blood vessels.

vein A blood vessel that carries blood toward the heart.

ventricle A cavity, such as a pumping chamber of the heart or a fluid-filled space in the brain.

venule A small blood vessel that carries blood from a capillary to a vein.

vertebral canal The cavity that contains the spinal cord.

vesicle A fluid-filled, membranous sac in the cytoplasm of a cell.

vestibular hair cells Sensory receptors for equilibrium.

vestibule The space between the labia minora.

visceral Pertaining to organs in a body cavity.

visceral peritoneum The portion of the peritoneum that covers the surfaces of organs in the abdominal cavity.

visceral pleura The portion of a pleura that covers the outer surface of a lung.

viscosity The resistance of a fluid to flow.

vitamin An organic nutrient other than a protein, lipid, or carbohydrate that is needed in small amounts from food for healthy functioning of the body.

vitreous body The gellike substance filling the vitreous chamber of the eye.

vocal folds Folds of tissue within the larynx that vibrate to produce sounds as air passes over them.

voluntary Consciously controlled.

vulva The female external genitalia.

W

water balance The state of equilibrium between water intake and water output.

white blood cells Formed elements that have defensive and immune functions; leukocytes.

X

X-linked traits Traits determined by genes on the X chromosome.

Y

yellow bone marrow Adipose tissue in the medullary cavity of a long bone and the spaces of spongy bone of some bones.

yolk sac An extraembryonic membrane that forms the early formed elements and germ cells for the embryo.

Z

zygote Cell formed by the union of the sperm nucleus and ovum nucleus that leads to the development of a human being.

INDEX

A

A band, 137
abdominal aorta, 276
abdominal cavity, 9
abdominal viscera, veins in, 282-284
abdominal wall
 muscles, 149-150
 veins, 282-284
abdominopelvic cavity, 9
 subdivisions, 13-14
ABO blood group, 250-254, 419
abortion, 398
absorption, 70, 88, 327
 dietary lipids, 344
 in mouth, 333-334
 small intestine, 342-344
 stomach, 335-336
accessory ducts, 380
accessory glands, 378, 381-382
accessory organs, 327
accommodation, 193, 208
ACE inhibitors, 363
acetabulum, 121-123
acetaminophen, 219
acetylcholine, 139-140, 187
acetylcholinesterase, 140, 142
acid-base balance, 371-373
acidosis, 35, 357, 371-373
acids, 34-35, 371-373
acquired immunodeficiency syndrome (AIDS), 303-304, 399
acromegaly, 223
acromion, 119-120
actin, 137
action potentials, 137, 162
 breathing and, 319
 conduction, 168-170
 formation, 168
 taste, 198
active immunity, 302-303
active site, enzymes, 43
active transport, 49, 58-60
adaptation, 194
Addison disease, 232

adductor longus, 153-155
adductor magnus, 153-155
adenine, 44-45, 61
adenosine diphosphate (ADP), 44, 46
adenosine triphosphate (ATP), 44, 46
 in muscle fibers, 142
adipocytes, 74
adipose tissue, 69, 74-75
adrenal cortex, 230-232
adrenal glands, 230-232
adrenal medulla, 230
adrenergic axons, 187
adrenocorticotropic hormone (ACTH), 225
adult stem cells, 69
aerobic respiration, 54, 60, 327, 347
afferent glomerular arterioles, 360-361
age-related macular degeneration (AMD), 213
ageusia, 212
agglutination, 240, 251-254
aging, of skin, 98
agonists, 135, 146
agranulocytes, 244-247
albumins, 247, 249
aldosterone, 230, 370-371
alimentary canal, 327-330
alkalosis, 35, 357, 371-373
allantois, 407
alleles, 403, 417-420
allergens, 290, 304
allergy, 304
all-or-none response, 144
alopecia, 99
alveolar arch, 114
alveolar ducts, 311-312
alveolar gas exchange, 308, 320
Alzheimer disease (AD), 189
amenorrhea, 398
amino acids, 41-44
 nutrients, 349
amniocentesis, 421-422
amnion, 407
amniotic cavity, 407
amniotic fluid, 407

amphiphilic phospholipids, 37, 39-41
ampulla, 204
ampullary cupula, 204
anabolic steroids, 139
anabolism, 13
 chemical reactions, 31-32
anaerobic respiration, 60, 142-143, 327, 347
anal canal, 345-346
anal columns, 345-346
analgesia, 196
anal sphincters, 345-346
anaphase, mitosis, 63-64
anatomical position, 6
anatomy, 2
androgens, 232, 384
anemia, 254
aneurysm, 285
angiotensin-converting enzyme (ACE), 363-364
angiotensin II, 363-364
anions, 29
anorexia nervosa, 352
anosmia, 212
antagonists, 135, 146
anterior chamber (eyeball), 209
anterior horns, 179
anterior lobe, pituitary gland, 223-226
anterior median fissure, 179
anterior root, 180-183
anterior sacral foramina, 118
anterior tibial artery, 278
antibodies, 290, 300-302
antibody-mediated immunity, 298, 300-302
anticodons, 61-62
antidiuretic hormone (ADH), 226, 370-371
antigen-antibody complex, 301-302
antigen-presenting cell, 298-299
antigens, 290, 298
anti-implantation devices, 396, 398
antiserum injections, 302
anuria, 373

anus, 345-346
aorta, 263, 275-276
aortic valve, 262
aplastic anemia, 254
apocrine sweat glands, 88, 95
aponeurosis, 135-136
apoptosis, 91
appendicitis, 351
appendicular portion, 2, 6-7
appendicular skeleton, 109, 119-124
appendix, 345-346
aqueous chamber (eyeball), 209
aqueous humor, 209
aqueous solution, 33-34
arachnoid mater, 172
areola, 395
areolar connective tissue, 74-75
arm, muscles of, 151-153
arrhythmias, 285
arteries, 258
 structure, 269-270
 systemic circuit, 275-279
arterioles, 270
arteriosclerosis, 285
arthritis, 128
articular cartilage, 104
articular processes, 116-117
artificially acquired active immunity, 302
artificially acquired passive immunity, 302
ascending aorta, 275-276
ascending colon, 345-346
ascending lumbar vein, 282
aspirin, 219
association area, 174
asthma, 322
astigmatism, 212
astrocytes, 166
atherosclerosis, 266, 285
athlete's foot, 99
atlas, 116-117
atmospheric pressure, 13, 314-315
atomic mass, 27
atomic number, 27

450

atoms, 3, 25
  structure, 25-27
atria, 260
atrioventricular (AV) bundle, 266
atrioventricular (AV) node, 266
atrioventricular valves, 260-261
atrium, 258
atrophy, 143
auditory ossicles, 114, 199-200
auditory tubes, 199-200, 309-310
auricular surface, 121-123
autoimmune diseases, 304
autonomic division, 162-163, 184-188
autonomic ganglion, 185
autonomic heart regulation, 268-269
autonomic neurotransmitters, 187
autonomic reflexes, 182-184
autoregulation, 275
autosomes, 416-417
axial portion, 2, 6-7
axial skeleton, 108-119
axillary artery, 278
axillary vein, 280, 282
axis, 116-117
axons, 82, 162, 164
  preganglionic, 187
azygos vein, 282

B

bacteria, 298
ball-and-socket joints, 125-126
baroreceptors, 194-195
  breathing and, 319
Barrett esophagus, 334
barrier contraceptives, 397-398
basement membrane, 70
bases, 34-35, 371-373
basilar membrane, 201
basilic vein, 280, 282
basophils, 244-247
bed bugs, 99
bedsores, 99
behavioral birth control methods, 397
benign prostatic hyperplasia (BPH), 398

benign tumors, 64
bicarbonate buffer system, 371-373
bicarbonate ions, 322
  pancreatic secretions, 338
biceps femoris, 153-155, 157-158
bile, 340-341
bile duct, 339-340
bilirubin, 242, 340, 416
bipolar neurons, 164-165
birth control, 395-398
  patch, 397
birth defects, 416
birth process, 411-413
  cardiovascular adaptations, 413-415
blastocyst, 404-405
blindness, 212
blisters, 99
blood, 79, 81, 239-255
  chemicals in, 358
  disorders, 254-255
  flow, 272-273
  human blood types, 250-254
  plasma composition, 369-373
  renal supply, 360-361
  supply to heart, 263
  umbilical cord, 413
blood clots, 250
blood (serum) creatinine level, 368
blood doping, 254
blood glucose, homeostasis, 15, 234-235
blood groups, 250-254, 419
blood pressure, 273-275, 278, 358
  glomerular, 362
blood urea nitrogen (BUN) values, 368
blood vessels, 269-272
  disorders of, 285-286
  renal system, 360-361
blood viscosity, 274
blood volume, 273-274
B lymphocytes, 245, 291, 299, 300-302
body cavities, 9-10
body communication, 220
body fluid composition, 357
  hydrogen ion concentration, 372
body mass index (BMI), 352
body membranes, 83-85
body regions, 2, 6, 8

body substances, 33-46
body temperature, 13, 96-97
  breathing and, 319
boils, 99
bone
  compact bone, 79-80, 103, 105
  formation, 106-108
  fractures, 127-128, 130-131
  gross structure, 103-105
  homeostasis, 108
  long bone, 103-106
  microscopic structure, 105
  spongy (trabecular) bone, 79-80, 103-104
  structure, 103-106
  surface features, 110
  tissue, 69, 79-80
  types, 104
bony labyrinth, 200-201
botulism, 157
brachial artery, 278
brachial plexus, 181-183
brachial vein, 280, 282
brachiocephalic trunk, 276
brachiocephalic vein, 281
brain
  anatomy and function, 172-178
  functions, 176
brain death, 174
brainstem, 175-176
Braxton Hicks contractions, 411
breast cancer, 399
breastfeeding, 415-416
breasts, 395
breathing, 308
  control, 318
  cycle, 316
  factors influencing, 319
  mechanisms of, 314-316
  muscles of, 149-151
breech births, 412
bronchi, 311-312
bronchial tree, 308, 311-312
bronchioles, 311-312
bronchitis, 322
bronchoconstriction, 311-312
bronchodilation, 311-312
brush border enzymes, 343
bubos, 304
buffers, 34, 371-373
building units, 37

bulbo-urethral glands, 378, 381-382
bulbs of the vestibule, 390-391
bulimia, 352
burns, 99

C

calcaneal tendon, 157-158
calcaneus, 158
calcitonin, 228-229, 371
calcium homeostasis, 228-229
calluses, 99
canaliculi, 79, 105
cancer
  cell division and, 64
  esophageal, 334
  female reproductive system, 399
  immunotherapy, 300
  lung cancer, 323
  lymphatic system and, 295
  nuclear medicine and, 28
  prostate, 398
  skin cancer, 98
  testicular, 398
canine teeth, 331-332
capillaries, 258, 270-272
capitulum, 120
carbaminohemoglobin ($HbCO_2$), 322
carbohydrates, 25, 36-38
  digestion, 343
  liver and metabolism of, 339
  nutrients, 347-348
carbon, atomic structure, 27
carbon dioxide, transport, 322
carbonic acid (H_2CO_3), 322
carbonic acid-bicarbonate buffer system, 35
carbonic anhydrase, 322
carbon monoxide poisoning, 322
cardia (stomach), 334-335
cardiac control center, 176, 268-269
cardiac cycle, 258, 265
cardiac muscle tissue, 81-82
cardiac output, 258, 267-269, 273-274
cardiac sphincter, 334
cardiac troponin complex, 142
cardiac veins, 263

cardiovascular system, 4, 258–286
  anatomy, 258–259
  postnatal adaptation of, 413–415
carotenes, 92–93
carpal bones, 120–121
carrier-mediated active transport, 58
carrier-mediated diffusion, 56–57
carrier proteins, 49, 51
cartilage tissue, 69, 77–79
  articular cartilage, 104
  costal cartilage, 118
cartilaginous joint, 103, 125–126
catabolism, 13
  decomposition (catabolic) reaction, 32
cataracts, 214
cations, 29
  electrolyte balance, 370–371
cauda equina, 180–183
cecum, 345–346
celiac trunk, 276
cell body, neurons, 82, 163–164
cell cycle, 63
cell identity markers, 51
cell-mediated immunity, 298–300
cells, 3, 49–65
  division, 62–65
  epithelial, 70
  in humans, 49
  parts of, 51
  reticular cells, 74–75
  structure, 49–55
cellular dehydration, 371
cellular level, 3
cellular respiration, 49, 59–60, 142–143, 347–348
cellulose, 347
central (osteonic) canals, 79, 105, 179
central nervous system, 162
  protection, 171–172
centrioles, 49, 54
centromeres, 63, 378
cephalic vein, 280, 282
cerebellum, 173, 176–177
cerebral cortex, 173–174
cerebral hemispheres, 173–174
cerebral palsy, 189
cerebrospinal fluid (CSF), 172, 177–178
cerebrovascular accidents, 189
cerebrum, 172–175
cerumen, 95
ceruminous glands, 95
Cervarix, 399
cervical cancer, 399
cervical cap, 396–397
cervical nerves, 180–183
cervical plexus, 181–183
cervical spinal region, 179
cervical vertebrae, 116–117
cervix, 389–390
  labor and delivery, 411–413
cesarean section, 123
chancre, 399
channel-mediated diffusion, 56–57
channel proteins, 49
cheeks, 330
chemical action, nonspecific resistance, 296
chemical bonds, 25, 29–31
chemical digestion, 327
chemical formulae, 25, 28–29
chemical level, 3
chemical reactions, 25, 31–33
chemicals
  in blood, glomerular filtrate and urine, 358
  breathing and, 319
  of life, 24–46
chemical symbol, 27
chemoreceptors, 193, 195
  breathing and, 319
chief cells, 335
chlamydia, 399
cholecystokinin (CCK), 335, 338, 341
cholesterol, 349
cholinergic axons, 187
chondrocytes, 77
chorion, 406–407
chorionic villi, 406–407
  sampling, 421–422
choroid, eye, 207
choroid plexus, 177–178
chromatids, 63, 378
chromatin, 52
chromosomes, 49, 52, 416–417
  abnormalities, 420–422
  cell division, 63
  spermatogenesis, 378–379
chronic obstructive pulmonary disease (COPD), 322
chylomicrons, 343
chyme, 327, 336, 338, 341, 342, 346
cilia, 54–55
ciliary body, 208
circulation pathways, 275
circumcision, 384
cirrhosis, 352
cisterna chyli, 291
citric acid cycle, 347
clavicle, 119–120
cleft palate, 113
clitoris, 390–391
cloning, of T cells, 299
coagulation, 240, 249–250
cocaine, parasympathetic nervous system and, 188
coccygeal nerves, 180–183
coccyx, 118
cochlea, 201
cochlear hair cells, 193, 202
codominance, 418–419
coitus interruptus, 397
collagen, fibers, 74
collateral ganglion, 187
collecting ducts, 359–361, 365–366
colon, 345–346
color blindness, 214, 419
comas, 189
comminuted fracture, 127, 130
common cold, 322
common hepatic artery, 276
common iliac arteries, 277
common integrative, 174
common mole, 99
communication, modes of, 220
compact bone, 79–80, 103, 105
complementary base pairing, 61
complement fixation, 296
complement system, 290, 296
complete blood count (CBC), 245
complete fracture, 127
complex carbohydrates, 37
compound fracture, 127
compounds, 25, 28–33
computerized tomography (CT), 11, 28
concentration gradient, 56
conchae (superior and nasal), 110, 114, 309
concussion, 189
condoms, 396–397
conductivity, of neurons, 167
condylar joints, 125–126
cones, 193, 208
congestive heart failure, 285
conjunctiva, 206
conjunctivitis, 214
connective tissues, 69, 74–79
  dense, 75–76
  loose, 74–75
  membranes, 84–85
constipation, 352
continuous ambulatory peritoneal dialysis (CAPD), 372
contraception, 396–398
contraction mechanism, 140–145
  energy, 142
contraction phase, 144
coracoid process, 119–120
cornea, 207
corns, 99
coronal suture, 108
coronary angioplasty, 266
coronary arteries, 263
coronary bypass surgery, 266, 282
coronary sinus, 263, 275
coronoid fossa, 120
coronoid process, 114, 120
corpus callosum, 173–174
corpus luteum, 391
cortical nephrons, 359–361
cortisol, 232
cortisone, 232
costal cartilages, 118
costal facets, 116–117
covalent bonds, 29–31
coxal bones, 121–123
cranial cavity, 9
cranial nerves, 180–181
  eyeball, 206–207
  smell, 198–199
  taste, 198
cranial reflexes, 182–184
cranium, 108–110
creatine phosphate, 135, 142
creatinine, 358, 368
cremaster, 382–383
cretinism, 228
cribriform plates, 110
cricoid cartilage, 310
crista ampullaris, 204–205
Crohn disease, 351
crossover, 378
crown, 332
cryptorchidism, 383
Cushing syndrome, 232
cutaneous membrane, 84, 88
cuticle, 95–96
cyclic adenosine monophosphate (cAMP), 221
cyclosporine, 303
cystic fibrosis, 420

cystitis, 373
cytokinesis, 64
cytoplasm, 49, 51
cytosine, 44–45, 61
cytoskeleton, 54–55
cytosol, 49, 51
cytotoxic T cell, 300

D

dandruff, 99
dartos muscle, 382–383
deafness, 212
deciduous teeth, 331
decomposition (catabolic) reaction, 32–33
decubitus ulcers, 99
defecation reflex, 346
dehydration synthesis, 36, 39
delayed allergic reactions, 304
deltoid tuberosity, 119–121
dendrites, 82, 162, 164–165
dendritic cells, 91
dens, 116–117
dense connective tissue, 75–76
  irregular tissue, 76–77
  regular tissue, 75–76
dental alveolus, 332
dental caries, 332, 352
dental pulp, 332
dentin, 332
deoxyhemoglobin, 241
deoxyribonucleic acid (DNA)
  characteristics, 61
  protein synthesis, 61
  replication, 64
  structure, 44–45
depolarization, 168–169
Depo-Provera, 397
dermal papillae, 88, 91
dermal ridges, 90–92
dermis, 88, 90–92
descending colon, 345–346
desmopressin, 226
diabetes insipidus, 226
diabetes mellitus, 234–235
diagnostic imaging, 11
diaphragm, 9, 314–315
diaphragm (contraceptive), 396–397
diaphysis, 103, 104
diarrhea, 352
diastole, 258, 265
diastolic blood pressure, 273–275
diencephalon, 175

Dietary Reference Intake (DRI), 327, 346–347
diffusion, 49, 56–57
digestion, 327
  in mouth, 333–334
  organs of, 328
  overview, 327–328
  pancreatic enzymes, 338–339
  small intestine, 342–344
  stomach, 335–336
digestive system, 5, 327–352
  disorders, 351–352
dipeptide, 42–43
directional terms, 2, 6–7
disaccharides, 37–38
dislocation, of joint, 130
distal convoluted tubule (DCT), 359–361, 364–366
diuresis, 373
diuretics, 370
diverticulitis, 351
dizygotic twins, 404
dominant inheritance, 417
dopamine, 189
dorsalis pedis, 278
dorsal respiratory group (DRG), 318
Down syndrome, 420–422
ductus arteriosus, 413
ductus venosus, 413
duodenum, 341–344
dura mater, 172
dynamic equilibrium, 13, 193, 203–205
dysgeusia, 212
dyslexia, 189
dysmenorrhea, 398
dysosmia, 212
dyspnea, 322

E

ear
  anatomy and function, 199–203
  disorders of, 212
eating disorders, 352
eccrine sweat glands, 88, 95
eclampsia, 416
ectoderm, 406
ectopic pregnancy, 416
eczema, 99
edema, 291
effector, 2
  negative-feedback mechanism, 14–15

efferent glomerular arteriole, 360–361
eicosanoids, 219
ejaculation, 384
ejaculatory duct, 378, 381–382
elastic cartilage, 78
elastic connective tissue, 76–77
elastic fibers, 74
electrocardiogram (ECG/EKG), 267
electroencephalogram (EEG), 174
electrolytes, 34, 249
  water and balance of, 369–370
electron, 25
electron shells, 25
electron transport chain, 347
elements, 25–27
  formed elements, 79, 81, 240
  table of, 26
elephantiasis, 304
embolism, 255
embolus, 240, 255
embryo, 403, 405
embryoblast, 404
embryonic development, 405–410
  disorders, 416
embryonic disk, 405–406
embryonic stem cells, 69
emmetropia, 211
empendymal cells, 166–167
emphysema, 322
emulsification, 340
enamel, 332
encephalitis, 188
endochondral ossification, 103, 106–107
endocrine glands, 70, 218, 236
endocrine system, 5, 217–236
  structure and function, 218–219
endocytosis, 49, 59
endoderm, 406
endolymph, 200–201
endometriosis, 399
endometrium, 389–390
endoplasmic reticulum (ER), 52–53
endosteum, 104
endothelium, 71
  arteries and veins, 269
energy
  foods, 347–348
  for muscle contraction, 142

enzymes, 25, 42–44
  digestive, 343–344
eosinophils, 244–247
epicardium, 259
epicranial aponeurosis, 148
epidermal ridges, 92
epidermis, 88–90
epididymis, 379–380
epidural anesthetic, 184
epidural space, 172
epiglottis, 310
epilepsy, 189
epinephrine, 230
  heart function and, 269
epiphysial (growth) plate, 103–104
epiphysis, 103, 104
epithalamus, 175
epithelial cells, 70
epithelial membranes, 83–84
epithelial tissues, 69–73
equilibrium, 203–205
erectile dysfunction (ED), 398
erection, 384
erythropoietin (EPO), 242, 357
esophagus, 334
essential amino acids, 349
essential nutrients, 346
estrogens, 236, 377, 391
ethmoidal cells, 110
ethmoid bone, 110
excess post-exercise oxygen consumption, 143
exchange (rearrangement) reactions, 32
excitation-contraction coupling, 140
excitatory neurotransmitters, 171
excretion, skin, 88
excretory ducts, 206
exercise, muscles and, 143
exocrine glands, 70, 218
exocytosis, 49, 59
exophthalmic goiter, 228
expiration, 308, 315–316
expiratory reserve volume, 316
extensor carpi radialis longus, 153–154
extensor carpi ulnaris, 153–154
extensor digitorum, 153–154
extensor digitorum longus, 155–157
external acoustic meatus, 109, 199–200
external carotid artery, 278
external ear, 199–200

external genitalia
  female, 389–391
  male, 383
external iliac artery, 278
external iliac vein, 282
external intercostal muscles, 314–315
external jugular vein, 281
external respiration, 308
external urethral sphincter, 369
extracellular fluid (ECF), 33–34
extraembryonic membranes, 406–407
extrinsic factors, heart function, 268
extrinsic muscles, eye, 206–207
eye
  anatomy, 207–208
  disorders of, 212–213
  functions, 210
eyeball
  chambers, 209
  extrinsic muscles, 206–207
eyebrows, 206
eyelashes, 206
eyelids, 206

F

facial bones, 110, 113
facial expression muscles, 147–148
fad diets, 349
fainting, 189
falciform ligament, 339
false ribs, 118–119
farsightedness, 211, 214
fats, 37, 39–41
fatty acids, 37, 39–41, 335, 339
feces, 346
feet, muscles of, 154–156
female pelvis, 123
female reproductive cycles, 391–394
female reproductive system, 5, 376–399, 386–396
  disorders, 398–399
  sex hormones, 235–236, 391–392
female sexual response, 391
femoral artery, 278
femoral vein, 282
femur, 123–124

fertilization, 403–406
fetal cell sorting, 421–422
fetal development, 408–410, 411
  disorders, 416
fetal skeleton, 106
fetal skull, 114–115
fetal transmission, 407
fetus, 403
  cardiovascular adaptations after birth, 413–415
fever, 297
fever blisters, 99
fibrin, 250
fibrinogen, 249
fibroblasts, 69, 74
fibrocartilage, 78–79
fibromyalgia, 157
fibrous joint, 103, 125
fibrous layer, eye, 207
fibula, 123–124
fibular artery, 278
fibularis longus, 155–158
fibular vein, 282
fight-or-flight response, 188
fillings (teeth), 332
filtration, 70, 271
fimbriae, 389
fingers, muscles, 153–154
first-degree burn, 99
fissured fracture, 128, 130
flagella, 54–55
flat bones, 103–104
flexor carpi radialis, 153–154
flexor carpi ulnaris, 153–154
floating ribs, 118–119
follicle-stimulating hormone (FSH), 225–226, 384–386, 392–394, 410
fontanelles, 114–115
food, survival and, 13
foramen magnum, 109
foramen ovale, 413
forearm, muscles of, 151–153
formed elements, 79, 81, 240
fossa ovalis, 415
fovea centralis, 208
fractures, 127–128, 130–131
fraternal twins, 404
freckles, 93
free edge (nail), 95–96
free nerve endings, thermoreceptors, 94
frenulum, 330
friction reduction, 70
frontal belly, 148
frontal bone, 108
frontal lobe, 173–174

frontal plane, 8
fundus (stomach), 335
fundus (uterine), 389–390

G

gallbladder, 339–340
gallstones, 352
gamete, 377
ganglia, 162, 165
Gardasil, 399
gas exchange, 320
gastric glands, 335–336
gastric juice, 335
gastric lipase, 335
gastric pits, 335
gastric ulcers, 336
gastric veins, 282
gastrin, 335, 338
gastrocnemius, 154–158
gastroesophageal reflux (GER), 334
gene expression, 218, 220–221, 417–418
genes, 44–45, 403, 417
genetic counseling, 421–422
genetics, development and, 416–420
genital herpes, 399
genotypes, 403, 417–419
germ layer, 403, 405–406
GH-inhibiting hormone, 223
GH-releasing hormone, 223
gigantism, 223
gingiva, 332
glands. *See also* specific glands
  integumentary system, 88, 94–95
glandular epithelium, 70
glans of clitoris, 390–391
glaucoma, 210
glenoid cavity, 119–120
globulins, 249
glomerular capillaries, 362
glomerular capsule, 357, 359–361
glomerular filtrate, 357–358, 361–367
glomerular filtration, 357, 361–367
glomerulonephritis, 373
glomerulus, 357, 359–361
glottis, 308, 311
glucagon, 232
glucocorticoids, 232
glucose, 36–37, 347
gluteus maximus, 153–155
gluteus minimus, 153–155

glycerol, 37, 39–41
glycogen, 37, 339, 347
  muscle contraction, 142–143
glycolysis, 347
goblet cells, 71
goiter, 228
Golgi complex, 53–54
gonadotropin-releasing hormone (GnRH), 225–226, 384–386, 394, 410
gonadotropins, 225–226
gonads, 235–236, 377
gonorrhea, 399
gout, 358
graded response, 15–16, 144–145
grand mal epilepsy, 189
granular cells, 361–362
granulocytes, 244–247
granulosa cells, 394
Graves disease, 228, 304
gravity, blood flow, 272
greater sciatic notch, 121–123
greater trochanter, 123–124
greater tubercle, 119–121
great saphenous vein, 282
greenstick fracture, 128, 130
grey matter, spinal cord, 179
gross anatomy, 2
ground substance, 74
growth hormone (GH), 223
guanine, 44–45, 61
gustatory epithelial cells, 197

H

hair, 93–94
hair follicle, 88, 93–94
hamstrings, 153–155
hard palate, 309
Hashimoto disease, 304
haustra, 346
H band, 137
head
  arteries to, 278
  digestive structures, 331
  muscles of, 148–149
  veins, 281
headaches, 189
hearing, 199–203
  physiology of, 202–203
heart
  anatomy, 258–265
  autonomic regulation, 268–269
  blood flow through, 263–265
  chambers, 260–261
  conducting system, 266–267

disorders of, 285–286
protective coverings, 259–260
regulation of function, 267–269
sounds, 265
valves, 260–263
wall, 260
heart attack, cardiac troponin complex, 142
heartburn, 334
heart murmurs, 285
heart rate, 267–269
heat production, muscular activity, 143
Helicobacter pylori, 336
helper T cell, 299
hematocytoblasts, 242
hematopoiesis, 240, 242–243
hemisphere specialization, 175
hemodialysis, 372
hemoglobin, 92, 240
  anaerobic respiration, 142–143
  structure, 241
hemolytic anemia, 254
hemolytic disease of the newborn (HDN), 252–254
hemophilia, 254, 421
hemorrhagic anemia, 254
hemorrhoids, 286, 351
hemostasis, 240, 248–250
  disorders of, 254–255
hepatic artery proper, 339
hepatic ducts, 339
hepatic lobules, 339
hepatic portal system, 282
hepatic portal vein, 282, 339
hepatic sinusoids, 339
hepatic triads, 339
hepatic veins, 282, 339
hepatitis, 351–352
hepatocytes, 339
herniated disc, 130
Herpes simplex virus, 99
Herpes simplex virus type 2, 399
heterozygous, 403, 417
high-density lipoproteins (HDLs), 349
higher brain centers, 319
hinge joints, 125–126
hip bones, 121–123
histology, 2
hives, 100
Hodgkin lymphoma, 304
homeostasis, 2, 13–16
  blood glucose, 15, 234–235
  bone, 108

calcium, 228–229
hormone production, 222
sweat glands, 95
homozygous, 403, 417
hormonal control, 222
  bile secretion, 341
  contraception, 396–398
  femal reproductive cycle, 391–394
  gastric secretions, 336
  kidneys, 371
  male reproduction, 384–386
  pancreatic secretions, 338
  pregnancy, 409–411
  water and electrolyte balance, 369–371
hormone replacement therapy (HRT), 395
hormones, 218–219
  adrenal glands, 230–232
  chemistry, 219–220
  mechanism of action, 220–221
  pancreatic, 232–235
  parathyroid, 228–230
  pituitary gland, 223–226
  production, 222
  sex hormones, 235–236, 384, 391–392
  thyroid gland, 226–228
human chorionic gonadotropin (hCG), 410
Human Genome Project, 418
human immunodeficiency syndrome, 303–304
human leukocyte antigens (HLAs), 303
human papilloma viruses, 399
humerus, 119–120
Huntington disease, 421
hyaline cartilage, 77–78
hyaline membrane disease, 323
hydrocephalus, 180
hydrochloric acid, 335
hydrogen
  atomic structure, 27
  bonds, 31–32
  ions, 34, 372
  isotope, 27
hydrogenation, 40
hydrolysis, 36, 327
hydrophilic molecules, 31
hydrophobic molecules, 31
  fats, 37, 39–41
hydroxyl group, 36
hyoid bone, 114
hyperglycemia, 235
hyperparathyroidism, 229–230

hypersecretion, 218, 222
hypersensitivity, 304
hypertension, 274, 286, 363
hyperthermia, 96–97
hypertonic solution, 57–58
hypertrophy, 143
hypodermis, 92
hypogeusia, 212
hypoglycemia, 235
hypoparathyroidism, 229
hyposecretion, 218, 222
hyposmia, 212
hypothalamus, 175, 223
  stress and, 231
hypothermia, 96–97
hypotonic solution, 57–58

I

identical twins, 404
ileal orifice, 341–344
ileum, 341–344
iliac crest, 121–123
iliacus, 153–155
iliac vein, 282
ilium, 121–123
immediate allergic reactions, 304
immune response, 299, 302–303
immunity, 290, 298–302
  types of, 302–303
immunocompetence, 290–291
immunoglobulins, 301–302
immunotherapy, 300
impetigo, 99
implantation, 403, 405
incisor teeth, 331–332
incomplete dominance, 418
incomplete fracture, 128, 130
incus, 199–200
induced abortion, 398
infant respiratory distress syndrome (IRDS), 323, 416
infant skull, 114–115
infectious disease
  lymphoid system disorders, 303–304
  pelvic tract, 389
  skin, 99
infectious mononucleosis, 254
inferior articular process, 116–117
inferior mesenteric arteries, 277
inferior mesenteric vein, 282

inferior vena cava, 263, 280, 282–283
infertility, 398–399
inflammation, 290, 296–297
inflammatory bowel disease (IBD), 351
inflammatory disorders
  digestive system, 351
  nervous system, 188
  respiratory system, 322–323
  urinary system, 373
inflation reflex, 319
influenza, 323
infraspinatus muscle, 151–152
infundibulum, 389
inheritance, 417–420
inherited diseases, 420–422
inhibin, 386
inhibiting hormones, 223
inhibitory neurotransmitters, 171
inner cell mass, 404
inorganic substances, 25, 33–36
insertion, 135, 146
inspiration, 308, 314–315
inspiratory reserve volume, 316
insula, 173–174
insulin, 232, 347
insulin-dependent diabetes, 234–235
insulin-independent diabetes, 235
integrating center, 2, 14
integument, 88
integumentary system, 4, 88–100
  accessory structures, 93–96
  skin, 88–93
interarterial septum, 260
intercalated discs, 81–82
interferons, 296
interlobar arteries, 360–361
interlobar veins, 360–361
interlobular bile ductule, 339
internal carotid artery, 278
internal ear, 200–201
internal iliac artery, 278
internal iliac vein, 282
internal intercostal muscles, 315–316
internal jugular vein, 281
internal respiration, 308

internal urethral sphincter, 369
interneurons, 165
interphase, cell cycle, 63
interstitial cells, 377-378
interstitial cell stimulating hormone (ICSH), 226
interstitial fluid, 33, 270
interthalamic adhesion, 175
intertubercular sulcus, 119-121
interventricular septum, 260
intervertebral discs, 115-117
intervertebral foramina, 116-117, 180
intestinal glands, 342-344
intestinal juice, 342
intestinal villi, 342-344
intra-alveolar (intrapulmonary) pressure, 314-315
intracellular fluid (ICF), 33-34
intramembranous ossification, 103, 107
intrapleural pressure, 314
intrauterine device (IUD), 396, 398
intrinsic factors
 heart function, 268
 stomach, 335-336
ionic bonds, 29
ionization
 inorganic substances, 33
 water, 33-34
ions, 29
 inorganic, 36
 membrane potential, 167-168
iris, 207-208
irregular bones, 105
irritability, of neurons, 167
irritable bowel syndrome (IBS), 351
irritant reflexes, 319
ischium, 122-123
isomers, 37
isotonic contractions, 146
isotonic solution, 57-58
isotopes, 27

J

jaundice, 340, 416
jejunum, 341-344
joints, 103, 124-127
 disorders, 128

juxtaglomerular complex, 357, 361-363
juxtamedullary nephrons, 359-361

K

Kaposi sarcoma, 303
karyotyping, chromosomal, 417-420
keratin, 72, 88, 90
keratinocytes, 89-90
kidneys, 359-361
kidney stones, 373
kyphosis, 131

L

labia majora, 390-391
labia minora, 390-391
labor and delivery, 411-413
labyrinthine disease, 212
lacaunae, 77
lacrimal apparatus, 206
lacrimal bones, 113
lacrimal gland, 206
lacrimal sac, 206
lactase, 343
lactation, 415-416
lacteal capillary, 342-344
lactiferous sinuses, 395
lactose, 37
lactose intolerance, 343
lambdoid suture, 109
lamellae, 79, 105
lamellar corpuscles, 92, 194
Langerhans cells, 91
lanugo, 409
large intestine, 344-346
laryngitis, 323
laryngopharynx, 309-310
larynx, 308
 anatomy, 310-311
Lasik surgery, 211
latent phase, muscle contraction, 144
lateral condyle, 123-124
lateral epicondyle, 120
lateral horns, 179
latissimus dorsi, 151-152
left common carotid artery, 276
left gastric artery, 276
left subclavian artery, 276
leg, muscles, 153-158
lens, 208
lesser trochanter, 123-124

lesser tubercle, 119-121
leukocytes, 243-247
leukotrienes, 219
levator scapulae, 150-152
levels of organization, 3-4
life, maintenance of, 13-16
ligaments, 103, 125-126
 patellar, 123-124, 157
ligamentum arteriosum, 415
ligamentum venosum, 414
limbic system, 175
lingual glands, 331
lingual lipase, 331
lingual papillae, 197, 331
lingual tonsils, 294, 310
lipids, 25, 37, 39-41
 absorption, 344
 liver metabolism of, 339
 nutrients, 347-349
lipid-soluble vitamins, 347-348, 350
lips, 330
liver
 nutrient absorption, 347
 structure and function, 339-341
lobar bronchi, 311-312
long bones, 103-106
longitudinal cerebral fissure, 173-174
longitudinal sections, 8
loose connective tissue, 74-75
lordosis, 131
loudness, 202-203
low-density lipoproteins (LDLs), 349
lower esophageal sphincter, 334
lower limb, 123-124
 arteries to, 278
lower respiratory tract, 308-309, 312
lumbar arteries, 277
lumbar nerves, 180-183
lumbar plexus, 181-183
lumbar spinal region, 179
lumbar vertebrae, 116-117
lumen, 70
lung cancer, 323
lungs
 anatomy, 313
 microscopic organization, 313
lunula, 95-96
luteal phase, 394
luteinizing hormone (LH), 225-226, 384-386, 392-394, 410
lymph, 33, 290-291
lymphadenitis, 304

lymphatic capillaries, 290-291
lymphatic system, 290
lymphatic trunks, 291
lymphatic vessels, 290-291
lymph nodes, 290, 292-293
lymph nodule, 293
lymphocytes, 244-247
 immunity and specialization of, 298
lymphoid organs, 290-294
lymphoid system, 5, 290-292
 components, 295
 disorders, 303-304
lymphoid tissues, 294-295
lymphoma, 304
lysosomes, 54
lysozyme, 333

M

macromolecules, 3
macrophages, 242, 247
macula, 203, 208
macula densa, 361-362
macular degeneration, 213
magnetic resonance imaging (MRI), 11
maintenance of life, 13-16
major calyx, 359-361
male infertility, 381
male pelvis, 123
male reproductive system, 5, 376-386
 disorders, 398
 sex hormones, 236, 384-386
male reproductive tract, 380-382
male sexual response, 384
malignant melanoma, 98
malignant tumors, 64
malleus, 199-200
maltase, 343
maltose, 37, 339
mammary glands, 395
mandible, 114
mandibular condyle, 114
mandibular fossa, 109
masseter, 148
mass movements, 346
mastication, 327, 331
 muscles, 147-149
mastoid process, 110
maternal infection, 407
matrix, 69, 74
matter, 5
maxillae, 113
maximal stimulus, 144-145

mechanical barriers, nonspecific resistance, 296
mechanical digestion, 327
mechanoreceptors, 193–194
medial condyle, 123–124
medial epicondyle, 120
medial malleolus, 123–124
medial umbilical ligaments, 414
median cubital vein, 280, 282
median plane, 8
median sacral crest, 118
medulla oblongata, 176
medullary cavity, 103, 105
meiosis, 62–65, 377–378, 387–389
meiosis II, 387–389
Meissner corpuscles, 91, 194
melanin, 88, 92–93
melanocytes, 90–91, 92–93
melatonin, 175, 236
membrane potential, 167–168
membranes
 body cavities, 10–11
 body membranes, 83–85
membranous labyrinth, 200–201
memory B cell, 301
memory T cell, 299
meninges, 10, 84, 171–172
meningitis, 188
menopause, 377, 394–395
menstrual cycle, 377, 392–394
mental illness, 189
Merkel cells, 91
mesenteries, 11, 342–344
mesoderm, 406
mesothelium, 70–71, 83–84
messenger RNA (mRNA), 61
metabolism, 2, 13
 hydrogen ion concentration, 372
metacarpals, 120–121
metaphase, mitosis, 63, 65
metastasis, 64
 lymphatic system and, 295
metatarsals, 123–124
micelles, 343
microanatomy, 2
microfilaments, 54
microglial cells, 166
microtubules, 54
microvilli, 55
micturition, 357, 369
micturition reflex, 369
midbrain, 175
middle ear, 199–200
mifepristone (RU 486), 398
mineralocorticoids, 230

minerals, 349–350
 bone storage of, 103
minor calyx, 359–361
miscarriage, 410, 416
mitochondria, 53–54
mitosis, 49, 62–65
mitotic spindle, 63, 65
mitral valve, 260–261
 prolapse, 265
mixed nerves, 180–181
M line, 137
molar teeth, 331–332
molecular formula, 28
molecules, 3, 25, 28–33
 nonpolar, 31
 polar, 31
 solute molecule, 58
 water, 32
monocytes, 244–247
monoglycerides, 335, 339
monosaccharides, 36–38, 37–38
monounsaturated fats, 40
monozygotic twins, 404
morning sickness, 416
morula, 404
motion sickness, 212
motor division, 163
 cerebrum, 174
motor end plates, 139–140
motor nerves, 180–181
motor neurons, 137, 140, 165
motor unit, 135, 137, 139
 recruitment, 144–145
mouth
 anatomy, 330–332
 digestion and absorption in, 333–334
movements, 327, 346
mucosa
 alimentary canal, 329–330
 lymphoid tissue, 294–295
 urinary bladder, 368–369
mucosa associated lymphoid tissue (MALT), 295
mucous membrane, 69, 84
mucous neck cells, 335
mucus, 84
multiple sclerosis, 189, 304
multipolar neurons, 164–165
muscle fibers, 80, 135
 contraction, 144
 skeletal muscle, 136–137
muscles, 136–155
 abdominal wall, 149–150
 actions, 146
 alimentary canal, 329
 anterior view, 147
 breathing, 149–150

contraction physiology, 140–145
facial expression and mastication, 147–148
head, 148–149
major muscles, 146–155
naming criteria, 146
pectoral girdle, 149–153
posterior view, 147
respiratory, 315–316
skeletal, 136–155
structure, 136–140
tissue, 80–81
muscle spindles, 195
muscle tissues, 69, 79–82
 cardiac, 81–82
 skeletal muscle, 80–81
 smooth muscle, 81–82
 types, 135
muscle tone, 135, 144–145
muscular dystrophy, 157
muscular pump, blood flow, 273
muscular system, 4, 134–158
 disorders, 156–158
myasthenia gravis, 157, 304
myelin sheath, 162, 166
myocardial infarction, 285
myocardium, 260
myofibrils, 136–137
myofilaments, 137–138
myoglobin, 135, 142–143
myogram, 144
myometrium, 389–390
myosin, 137
myosin binding sites, 137–138
MyPlate Guide, 349, 351
myxedema, 228

N

nails, 95–96
nasal bones, 113
nasal cavity, 308–309
nasal conchae, 110, 114, 309
nasal septum, 309
nasolacrimal duct, 206
nasopharynx, 309–310
natriuretic peptides, 364, 371
natural killer cells, 297
naturally acquired active immunity, 302
naturally acquired passive immunity, 302
nearsightedness, 211, 214

neck
 arteries to, 278
 digestive structures, 331
 veins, 281
negative-feedback mechanism, 13–16
 blood glucose homeostasis, 234–235
 body temperature regulation, 96–97
 hormone production, 218, 222
 male sex hormone secretion, 385–386
nephron loop, 359–361, 364–366
nephrons, 357, 359–361, 367
nerves, 162, 180–181
 vision physiology, 210–211
nerve tissue, 69, 82–83, 163–167
nerve tracts, 179–180
nervous system, 5, 161–189
 anatomical divisions, 162
 components, 162–163
 disorders, 188–189
neural control
 gastric secretions, 336
 pancreatic secretions, 338
neuralgia, 189
neurilemma, 166
neuritis, 188
neuroglia, 82, 162, 166–167
neurological disorders, 183
 muscles, 157–158
neuromuscular interaction, 137
neuromuscular junction, 139–140
neuronal processes, 82
neurons, 82–83, 162
 depolarization/ repolarization, 168–169
 physiology, 167–171
 postganglionic, 185
 preganglionic, 185
 structure, 163–164
 types, 164–166
neuroses, 189
neurotransmitters, 135, 139, 169–171
 autonomic, 187
neutron, 25
neutrophils, 244–247
newborn, first breath of, 412–413
nipple, 395
nitrogenous wastes, 249, 357–358

nociceptors, 193, 196-197
nonelectrolytes, 34
noninfectious disorders, 304
noninflammatory disorders
  digestive system, 352
  nervous system, 189
  respiratory system, 323
  urinary system, 373
nonpolar covalent bonds, 30-31
nonpolar molecules, 31
nonspecific resistance, 296-297
nonsteroid hormones, 219, 221
norepinephrine, 230
  cardiac control, 269
normal range, negative-feedback mechanism, 15
norplant, 397
nose, anatomy, 308-309
nostrils (nares), 308-309
nuclear envelope, 52
nuclear imaging, 28
nuclear medicine, 28
nuclei, cerebral, 173-174
nucleic acids, 25, 44-45
nucleolus, 52
nucleotides, 44-45
nucleus
  atomic, 25
  cellular, 49, 52
nutrient foramen, 105
nutrients, 327, 346-351
nutritional anemia, 254

O

obesity, 352
oblique fracture, 128, 130
obstructive pattern, 317
obturator foramen, 122-123
occipital belly, 148
occipital bone, 108-109
occipital condyles, 109
occipital lobe, 173-174
occipitofrontalis, 148
olecranon, 120
olecranon fossa, 120
olfactory mucosa, 310
olfactory receptors, 193
olfactory sensory neurons, 198-199
oligodendrocytes, 166
oocytes, 386-388
oogenesis, 377, 386-389
oogonia, 388
ophthalmoscopy, 209

opsin, 210
optic chiasma, 210-211
optic disc, 207-208
oral contraceptives, 396-397
oral vestibule, 333
orbit, anatomy, 206
organelles, 3, 49, 51
organic substances, 25, 35-45
  reactions involving, 37
organismal level, 4
organ level, 3
organ of Corti, 193, 202
organ systems, 3-5
organ transplants, rejection of, 303
orgasm, 384, 391
origin, 135, 146
oropharynx, 309-310
osmosis, 49, 57, 70
ossification, 106
osteoarthritis, 128
osteoblasts, 106
osteoclasts, 107
osteocytes, 79, 105
osteomyelitis, 128
osteon, 79, 105
osteoporosis, 126, 128
otitis media, 212
otoliths, 203
ova, 386-387
ovarian arteries, 277
ovarian cycle, 377, 392-394
ovarian follicles, 386-387, 394
ovarian mesothelium, 386-387
ovarian veins, 282
ovaries, 235, 386-389
ovulation, 377, 389
ovum, 404
oxygen
  muscle contraction, 142-143
  survival and, 13
  transport, 320-322
oxyhemoglobin, 241, 321
oxytocin, 226, 411-413

P

pacemakers, 267
Pacinian corpuscles, 92, 194
pain, 195-196
  relief, 219
palate, 309, 330
palatine bones, 113
palatine tonsils, 294, 310
palmaris longus, 153-154

pancreas, 232-235
  anatomy, 232-233
  structure and function, 337-339
pancreatic amylase, 339
pancreatic duct, 337
pancreatic islets, 232
pancreatic juice, 337
pancreatic lipase, 339
pancreatic vein, 282
papillary ducts, 359-361, 367
papillary layer, 91
paracrine signals, 218-219
paralysis, 189
paramedian plane, 8
paranasal sinuses, 108, 111, 310
parasympathetic division, 185-188
  functions, 187-188
parathyroid glands, 228-230
parathyroid hormone (PTH), 228-230, 365, 371
paricardial cavity, 11
parietal bone, 108
parietal cells, 335
parietal layers, 2, 11
parietal lobe, 173-174
parietal peritoneum, 11, 329
parietal pleura, 313-314
Parkinson disease, 189
parotid glands, 333
parturition, 403, 411
passive immunity, 302-303
passive transport, 49, 56-58, 60
patella, 123-124, 157
patellar ligament, 123-124, 157
pathogens, 290, 293, 298
pectoral girdle, 119-120
  muscles, 149-153
pectoralis major, 151-152
pectoralis minor, 150-152
pelvic cavity, 9
pelvic girdle, 121-123
pelvic inflammatory disease, 389, 399
pelvic inlet, 123
pelvis, 121-123
  arteries to, 278
  veins to, 282
penis, 378, 382-383
pepsin, 296, 335
peptidases, 343
peptide bonds, 42-43
peptides, 335, 339
perception, 193-194
pericardial cavity, 259-260

pericarditis, 285
pericardium, 2, 259
perichondrium, 84
perilymph, 200-201
perimetrium, 389-390
perineum, 389-390
Periodic Table of Elements, 26
periodontal disease, 352
periodontal ligaments, 332
periosteum, 84, 104
peripheral nervous system (PNS), 162
  anatomy, 180-181
  functional division, 162-163
peripheral resistance, 273-275
peristalsis, 327, 330
peritoneal cavity, 12
peritoneum, 2, 11, 328-329
peritonitis, 336, 352
peritubular capillaries, 357, 360-361
permanent teeth, 331
pernicious anemia, 254
perpendicular plate, 110
petit mal epilepsy, 189
phagocytosis, 59, 296
phalanges, 120-121, 123-124
pharyngeal tonsils, 294
pharynx, 308, 334
  anatomy, 309-310
phenotypes, 403, 417-419
phenylketonuria (PKU), 420-421
phlebitis, 286
phosphate buffer system, 372
phosphate group, 37, 39-41
phospholipids, 37, 39-41, 347-348
photoreceptors, 193, 197
phrenic nerves, 181-183
pH scale, 34
physiology, 2
pia mater, 172
pineal gland, 236
pinocytosis, 59
pitch, 202-203
pitocin, 226, 415
pituitary dwarfism, 224
pituitary gland, 223-226
pivot joint, 125-126
placenta, 403, 407-408
  delivery of, 412-413
plane joint, 125-126
planes, 2
planes, anatomical, 6, 8-9
plasma, 33, 240, 247-249
  composition, 369-373
  solutes, 247
plasma cells, 301

plasma membrane
    structure, 49–51
    transport, 55–60
platelets, 79, 81, 240, 247
    plug formation, 249
pleurae, 2, 11, 313–314
pleural cavity, 11, 313–314
pleural fluid, 11
pleurisy, 323
Pneumocystis carinii, 303
pneumonia, 323
pneumothorax, 315
polar body, 389, 403
polar covalent bonds, 31
polar molecules, 31
polycythemia, 254
polygenes, 419
polygenic inheritance, 419
poliomyelitis, 158
polypeptides, 42–43
polysaccharides, 37–38, 347
polyunsaturated fats, 40
pons, 176
pontine respiratory group
    (PRG), 318
popliteal artery, 278
popliteal vein, 282
positive-feedback mechanism,
    16, 218, 222
    hemostasis, 249–250
positron emission tomography
    (PET), 11, 28
postcentral gyri, 173–174
posterior chamber (eyeball),
    209
posterior horns, 179
posterior inferior iliac spine,
    121–123
posterior intercostal arteries,
    276
posterior intercostal veins,
    282
posterior language area, 174
posterior lobe hormones, 226
posterior median fissure, 179
posterior root, 180–183
posterior sacral foramina, 118
posterior tibial artery, 278
postganglionic neuron, 185
postnatal cardiovascular
    adaptation, 413–415
postsynaptic signal, 162, 169
precapillary sphincters, 271
precentral gyri, 174
preembryo, 403
    development, 404–405
prefrontal area, 174
preganglionic axons, 187
preganglionic neurons, 185

pregnancy
    disorders of, 416
    hormonal control in,
        409–411
    labor and delivery, 411–413
premature infants, 413
premenstrual phase, 394
premenstrual syndrome, 399
premolar teeth, 332
premotor area, 174
prenatal disorders, 416
presbyopia, 214
presynaptic signal, 162, 169
primary immune response,
    302–303
primary lymphoid organ,
    290–291
primary motor areas, 174
primary oocytes, 388, 403
primary ossification center,
    107
primary spermatocyte, 378
primary structure, proteins,
    42–43
probability, inheritance and,
    419–420
progesterone, 236, 377,
    391–392
programmed cell death, 91
projection, 193–194
prolactin (PRL), 226
proliferative phase, menstrual
    cycle, 394
prophase, mitosis, 63, 65
proprioceptors, 193, 195
prostaglandins, 218–219
prostate, 378, 381–382
prostate cancer, 398
prostatitis, 398
protection, 70
    skin, 88
protein buffer system,
    372–373
proteins, 25, 41–44
    carrier proteins, 49, 51
    channel proteins, 49
    digestion and absorption,
        339, 343
    nutrients, 349
    plasma, 247, 249
    receptor proteins, 51
protein synthesis, 61–62
prothrombin activator, 249
proton, 25
proximal convoluted tubule
    (PCT), 359–361, 364–365
pseudostratified ciliated
    columnar epithelium,
    71–72

pseudounipolar neurons,
    164–165
psoas major, 153–155
psoriasis, 100
psychoses, 189
puberty, 377–378, 384
pubis, 122–123
pubis symphysis, 122–123
pulmonary alveoli, 308,
    311–313
pulmonary arteries, 263
pulmonary circuit, 258,
    263–264, 275
pulmonary edema, 323
pulmonary embolism, 323
pulmonary trunk, 263
pulmonary valve, 262
pulmonary veins, 263
pulp cavity, 332
pulse points, 274
pulse pressure, 273
Punnett square, 420
pupil, 207–208
pus, 297
pyelonephritis, 373
pyloric sphincter, 335

Q

QRS complex, 267
quadriceps femoris,
    153–155
quaternary structure, proteins,
    42–43

R

radial artery, 278
radial tuberosity, 120
radial veins, 280, 282
radioisotopes, 27
    nuclear medicine, 28
radius, 120
random alignment, 378
RBC count, 241
reabsorption, 271
receptor proteins, 51
receptors, 2
    negative-feedback
        mechanism, 13–14
    sensory, 162, 193
    touch and pressure,
        194–195
recessive inheritance, 417
Recommended Daily
    Allowance (RDA),
    346–347

rectum, 345–346
red blood cells, 79, 81, 240
    disorders, 254–255
    respiratory gas transport,
        320–322
    structure and function,
        241–242
    tonicity, 58
red bone marrow, 104, 118,
    290–291
    red blood cell production,
        241–242
red pulp (spleen), 293–294
referred pain, 196
reflex arcs, 182–184
reflexes, 162, 182–184
relaxation phase, muscle
    contraction, 144
relaxin, 411
releasing hormones, 223
renal arteries, 276–277,
    360–361
renal autoregulation,
    363–367
renal calculi, 373
renal columns, 359–361
renal corpuscles, 357,
    359–361, 367
renal cortex, 359–361
renal failure, 373
renal insufficiency, 372
renal mechanism, 371, 373
renal medulla, 359–361
renal papilla, 359–361
renal pelvis, 359–361
renal pyramids, 359–361
renal tubule, 357, 359–361,
    367
renal veins, 282, 360–361
renin, 357, 363
renin-angiotensin mechanism,
    357, 363–367
rennin, 335
repolarization, 168–169
reproductive systems, 376–
    399, 402–422
    disorders, 398–399
residual volume, 316
respiratory centers, 318,
    412–413
respiratory gases, transport of,
    320–322
respiratory mechanism, 371,
    373
respiratory membrane,
    312–313
respiratory movement, blood
    flow, 272–273
respiratory mucosa, 310

respiratory rhythmicity center, 176, 412-413
respiratory system, 4, 308-323
  disorders, 322-323
  organization, 309
  structures, 308-314
respiratory volumes and capacities, 316-317
resting heart rate, 269
resting membrane potential (RMP), 167-168
restrictive pattern, 317
reticular cells, 74-75
reticular fibers, 74
reticular formation, 176
reticular layer, 92
reticular tissue, 74-75
retina, 193, 207-209
retinoblastoma, 214
retroperitoneal space, 11
Rh (D) antigen, 251-252
Rh blood group, 250-254
rheumatoid arthritis, 128, 304
rhinitis, 323
rhodopsin, 210
RhoGAM test, 254
rhomboid major and minor, 150-152
ribonucleic acid (RNA), 44
  characteristics, 61
  protein synthesis, 61
ribosomal RNA (rRNA), 61
ribosomes, 52
ribs, 118-119
rickets, 128
right bundle branches, 266
right common carotid artery, 278
right lymphatic duct, 291
right subclavian artery, 278
rods, 193, 208
root (teeth), 332
root canal, 332
rotator cuff, muscles, 150-152
rough endoplasmic reticulum (RER), 53
round ligament, 414
round window, 201

S

saccule, 203
sacral canal, 118
sacral foramina, 180
sacral hiatus, 118, 180
sacral nerves, 180-183
sacral plexus, 181-183
sacral promontory, 118
sacral spinal region, 179
sacroiliac joint, 121-123
sacrum, 118
saddle joint, 125-126
sagittal plane, 8
sagittal suture, 108
saliva, 333
salivary amylase, 333
salivary glands, 333
salts, 35
sarcolemma, 137
sarcomeres, 137-138
sarcoplasm, 137
sarcoplasmic reticulum, 137, 139
saturated fats, 40
scala tympani, 201
scala vestibuli, 201
scapulae, 119-120
Schwann cells, 166
sciatica, 188
sclera, 207
scoliosis, 131
scrotum, 378, 383
sebaceous glands, 88, 94-95
sebum, 95
secondary immune response, 302-303
secondary lymphoid organ, 290-291
secondary oocyte, 377, 403-404
secondary sex characteristics, 384-385
secondary spermatocytes, 378
secondary structures, proteins, 42-43
second-degree burn, 99
second messenger, 218, 221
secretin, 335, 338
secretion, 70
secretory phase, menstrual cycle, 394
secretory vesicles, 53
section, 2, 6
segmental arteries, 360-361
segmental fracture, 128, 130
segmentation, 327, 330
selectively permeable membrane, 49, 51
sella turcica, 110
semen, 377, 383
semicircular canals, 193, 204
semilunar valves, 262-263
semimembranosus, 153-155, 158
seminal vesicles, 378, 381-382
seminiferous tubules, 377-379
semitendinosus, 153-155, 158
sensations, 193-194
senses, 193-214
  disorders of, 212-214
  general senses, 193-196
  special senses, 193, 197-212
sensory adaptation, 193-194
sensory division, 162-163
  cerebrum, 173-174
sensory nerves, 180-181
sensory neurons, 165
sensory perception, 88
sensory receptors, 162, 193
serosa, alimentary canal, 329
serous fluid, 11, 83-84
serous membranes (serosae), 2, 10-12, 69, 83-84
serous pericardium, 11
  parietal layer, 259
serratus anterior, 150-152
serum, 251
sesamoid bones, 105
set point, negative-feedback mechanism, 15
severe combined immunodeficiency (SCID), 304
sex chromosomes, 416-417
sex determination, 417
sex hormones, 235-236
  female, 391-392
  male, 236, 384-386
sex inheritance, 417
sexually transmitted diseases (STDs), 399
sexual response
  female, 391
  male, 384
shingles, 188-189
short bones, 103-104
shoulder girdle, 119-121
  arteries to, 278
  veins to, 280, 282
sickle-cell disease, 244, 254, 418
sigmoid colon, 345-346
simple columnar epithelium, 71-72
simple cuboidal epithelium, 71
simple diffusion, 56
simple epithelium, 70-71
simple fracture, 127
simple squamous epithelium, 70-71
simple sugars, 37
single-gene disorders, 420-422
sinoatrial node (SA node), 266
sinuses, paranasal, 108, 111
sinusitis, 323
skeletal muscles, 136-155
  abdominal wall, 149-150
  actions, 146
  anterior view, 147
  breathing, 149-150
  contraction physiology, 140-145, 272
  facial expression and mastication, 147-148
  head, 148-149
  major muscles, 146-155
  pectoral girdle, 149-153
  posterior view, 147
  structure, 136-140
  tissue, 80-81
skeletal system, 4, 103-131
  appendicular skeleton, 109, 119-124
  axial skeleton, 108-119
  bones, 103-108
  disorders, 126-127 130-131
  divisions, 108
  fetal skeleton, 106
  functions, 103
  joints, 124-127
skin
  aging, 98
  color, 92-93
  disorders, 99-100
  functions, 88
  structure, 88-92
skin cancer, 98
skull, 108-115
  infant, 114-115
sliding-filament muscle contraction model, 140-142
small intestine, structure and function, 341-344
small saphenous vein, 282
smell, sense of, 198-199, 212
smoking, 309
smooth endoplasmic reticulum (SER), 53
smooth muscle tissue, 81-82
  urinary bladder, 369
sodium chloride
  in body, 35
  ionic bond, 29
sodium-potassium pump, 58
soft palate, 309
soleus, 154-158
solutes, 33-34
  molecules, 58
solvents, 33-34
somatic division, 162-163, 185

somatic reflexes, 182–184
space-filling models, water molecule, 32
spasms
    muscular, 158
    vascular, 249
sperm, 377, 380–381
    fertilization, 403
spermatids, 378
spermatogenesis, 377–378, 380
spermatogenic cells, 377–378
spermatogenic epithelium, 377–378
spermatogonia, 378
spermicides, 396–397
sphenoid, 110
sphincters, 327
    anal, 345–346
    lower esophageal, 334
    pyloric, 335
    urethral, 369
spinal cord, 172
    anatomy, 178–179
    functions, 179–180
spinal ganglion, 180–183
spinal nerves, 180–183
    autonomic and somatic motor pathways, 185–186
spinal plexuses, 180–183
spinal reflexes, 182–184
spinal tap, 184
spine
    curvatures, 131
    of scapula, 119
spiral fracture, 128, 130
spiral organ, 193, 202
spirogram, 316
spirometer, 316
spleen, 290, 293–294
splenic artery, 276
splenic vein, 282
spongy (trabecular) bone, 79–80, 103, 104
spontaneous abortion, 398
sprains, 130
squamous sutures, 109
stapes, 199–200
starch, 37, 339
static equilibrium, 193, 203–205
stem cells, 69
sterilization, 398
sternal puncture, 118
sternum, 119
steroids, 41, 219–221, 347–348
stomach, structure and function, 334–336
strabismus, 214
strains, 157

stratified columnar epithelium, 73
stratified cuboidal epithelium, 73
stratified epithelium, 72–73
stratified squamous epithelium, 72–73
stratum basale, 90
stratum corneum, 91
stratum granulosum, 91
stratum lucidum, 90–91
stratum spinosum, 90
stress, 231
striations, 137
stroke volume, 258, 267–269
structural formula, 28
structural proteins, 42–43
styloid process, 110, 120
subclavian vein, 280–282
subcutaneous tissue, 88
    structure, 92
subendocardial conducting network, 266
sublingual glands, 333
submandibular glands, 333
submucosa, 329
subscapularis muscle, 151–152
substrates, 43, 327
sucrase, 343
sucrose, 37
sudden infant death syndrome (SIDS), 416
summation, 144
superior articular process, 116–117
superior conchae, 110
superior mesenteric artery, 276
superior mesenteric vein, 282
superior vena cava, 263, 281
superior vesical arteries, 414
supraspinatus muscle, 151–152
surfactant, 308, 312
surgical neck, 119–121
survival needs, 13
sutural bones, 103–104
sutures (skull), 108–109
swallowing reflex, 334
sweat glands, 88, 95
sympathetic division, 185–188
    functions, 187–188, 363
sympathetic trunk, 187
synapses, 162, 169
synaptic cleft, 139–140, 169
synaptic transmission, 169–170
synovial joints, 103, 125–129
synovial membranes, 85

synthesis (anabolic) reactions, 31–33
syphilis, 399
systemic circuit, 258, 263–264, 275
    arteries, 275–279
    blood pressure in, 272
    veins, 280–284
systemic gas exchange, 308, 320
systole, 258, 265
systolic blood pressure, 273–275

T

tactile corpuscles, 91, 194
tactile epithelial cells, 91
taeniae coli, 346
target cells, 218–219
tarsal bones, 123–124
taste, sense of, 197–198, 212
taste buds, 193, 197
taste pore, 197
taste receptors, 197
Tay-Sachs disease, 421
teeth, 331–332
    decay, 332
telophase, mitosis, 64
temperature regulation, 88, 96–97, 194
temporal bones, 109
temporalis muscle, 148
temporal lobe, 173–174
temporomandibular fossa, 109
temporomandibular joint, 109, 114
tendon, 135–136
tendon organs, 195
teres major muscle, 151–152
teres minor muscle, 151–152
terminal boutons, 139–140
tertiary structures, proteins, 42–43
testes, 236, 377–379
    undescended, 383
testicular arteries, 277
testicular cancer, 398
testicular veins, 282
testosterone, 236, 377, 384–386
tetanus, 158
tetany, 135, 144–145
thalamus, 175
thermoreceptors, 193–194
thigh, muscles, 153
third-degree burn, 99
thoracic cage, 118–119

thoracic cavity, 9, 259
    veins, 282–283
thoracic duct, 291
thoracic nerves, 180–183
thoracic spinal region, 179
thoracic vertebrae, 116–117
threshold stimulus, 144, 168
thrombin, 249–250
thrombocytopenia, 255
thrombophlebitis, 255
thrombosis, 255
thrombus, 240, 250
thymine, 44–45, 61
thymus, 236, 290, 292
thyroid cartilage, 310
thyroid gland, 226–228
thyroid-stimulating hormone (TSH), 224–225
thyrotropin-releasing hormone (TRH), 224–225
thyroxine, 227–228
tibia, 123–124, 157
tibialis anterior, 155–157
tibial tuberosity, 123–124
tibial veins, 282
tidal volume, 316
tissue, 69
tissue level, 3
tissue macrophage system, 296
tissue plasminogen activator (tPA), 250
T lymphocytes, 245, 291, 299–300
toes, muscles of, 154–156
tongue, 197–198, 330–331
tonsillectomy, 294
tonsillitis, 294
tonsils, 294
total hip replacement, 123
total lung capacity (TLC), 316–317
total vessel length, 274
toxic shock syndrome, 399
trabeculae, 79, 104
trachea, 308, 311
tracheal cartilages, 311
transcellular fluids, 33
transcription, 61–62
transfer RNA (tRNA), 61
transfusions, blood compatibility, 253–254
transitional epithelium, 73
translation, 61–62
transport
    active transport, 49, 58–60
    passive transport, 56–58, 60
    plasma membrane, 55–60
    respiratory gases, 320–322

transverse cerebral fissure, 173, 176
transverse colon, 345–346
transverse fracture, 128, 130
transverse planes, 8
transverse tubules, 137, 139
trapezius, 150–152
tricuspid valve, 260–261
triglycerides, 37, 39, 335, 339, 343, 347–348
triiodothyronine, 227–228
trochlear notch, 120
trophoblast, 404
tropic hormones, 222
tropomyosin, 137
troponin complex, 137
true ribs, 118–119
trypsin, 339
tubal ligation, 398
tubercle, 118
tuberculosis, 323
tubular fluid, 357, 361–367
tubular reabsorption, 357, 361–367
tubular secretion, 357, 361–367
tunica externa, 269
tunica intima, 269
tunica media, 269
twin births, 404
tympanic cavity, 199–200
tympanic membrane, 199–200
type I diabetes, 234–235
type II diabetes, 235

U

ulcerative colitis, 351
ulna, 120
ulnar artery, 278
ulnar veins, 280, 282
umbilical arteries, 413
umbilical cord, 407
  blood, 413
unsaturated fats, 40
upper limb, 119–121
  arteries to, 278
  veins to, 280, 282
upper respiratory tract, 308–309
uracil, 44–45, 61
urea, 348
uremia, 373
ureter, 367–368
urethra, 367–369, 378, 381–382
urethritis, 373
uric acid, 358
urinalysis, 368
urinary bladder, 367–369
urinary system, 5, 356–373
  disorders, 373
  functions, 357–358
urine, 358
  characteristics, 364–367
  excretion, 367–369
  formation, 361–367
uterine tubes, 389
uterus, 389–390
utricle, 203
UV radiation, skin cancer, 98
uvula, 330

V

vagina, 389–390
vaginal ring, 397
vagus nerve, cardiac control, 269
valence, atomic, 25, 29
valence shell, 29
varicose veins, 286
vasa recta, 357, 360–361
vascular layer, eye, 207
vascular spasm, 249
vas deferens, 378–381
vasectomy, 398
vasoconstriction, 258, 274–275
vasodilation, 258, 274–275
vasomotor center, 176, 274–275
vastus lateralis, 155–157
veins, 258
  structure, 269, 271–272
  systemic circuit, 280–284
velocity, blood flow, 273
venous return, 268
ventral respiratory group (VRG), 318
ventricles, 177–178, 258, 260
venules, 271
vernix caseosa, 409
vertebra
  structure, 115–117
  thoracic, 118–119
vertebral arch, 115–117
vertebral arteries, 278
vertebral body, 115–117
vertebral canal, 9
vertebral column, 115–118
vertebral foramen, 115–117
vertebral veins, 281
vesicles, 53
vessel diameter, 274
vestibular folds, 311
vestibular glands, 390–391
vestibular membrane, 201
vestibule, 390–391
viruses, 296
visceral layer, 2, 11
visceral organs, autonomic division, 186–187
visceral peritoneum, 11, 329
visceral pleura, 313–314
vision, 205–212
  mechanics of, 211
  optical disorders, 211
  physiology of, 210–211
vital capacity, 316
vitamin D
  conversion, 357, 365, 371
  synthesis, 88
vitamins
  classification, 349–350
  lipid-soluble, 347–348, 350
  water-soluble, 349–350
vitreous body, 209
vitreous chamber (eyeball), 209
vocal folds, 310–311
voltage, membrane potential, 167
vomer, 113

W

water
  composition, 33–34
  conservation, 357, 361–367
  electrolyte balance and, 369–370
  hydrogen bonds in, 31–32
  hydrolysis, 36
  ionization, 33–34
  loss pathways, 370
  survival and, 13
water compartments, 33–34
water-soluble vitamins, 350
white blood cells, 79, 81, 240, 243–247
  disorders, 254
  inflammation, 297
white pulp (spleen), 293–294
wisdom teeth, 331
withdrawal, birth control, 397
wrist, muscles, 153–154

X

X-linked traits, 419

Y

yellow bone marrow, 105
yolk sac, 407

Z

Z lines, 137
zygomatic bones, 113
zygomatic process, 109–110
zygote, 389, 403–404